STANDARD PLANT OPERATORS' MANUAL Third Edition

A basic reference on the operation and maintenance of energy systems equipment, complete with license examination questions and answers

Also by Stephen Michael Elonka (and coauthors)

Electrical Systems and Equipment for Industry, Arthur H. Moore and Stephen M. Elonka, Van Nostrand Reinhold, New York, 1971. Reprint 1977 by Robert E. Krieger Publishing Company, Inc., Huntington, New York.

Standard Instrumentation Questions and Answers, Stephen M. Elonka and Alonzo R. Parsons, McGraw-Hill Book Company, New York, 1962. Reprint 1976 (in Portuguese) Editora McGraw-Hill do Brasil, Ltda, São Paulo, Brasil. Reprint 1979, Robert E. Krieger Publishing Company, Inc., Huntington, New York.

Standard Electronics Questions and Answers, Stephen M. Elonka and Julian L. Bernstein. McGraw-Hill Book Company, New York, 1964. Reprint 1968 (in Polish) Wydawnictwa Naukowo-Techniczne, Warszawa, Poland. Reprint, 1968, Mei Ya Publishing Company, Inc., Taipei, Taiwan.

Standard Plant Operators' Questions and Answers, Stephen M. Elonka and Joseph F. Robinson, McGraw-Hill Book Company, New York, 1959.

Standard Refrigeration and Air Conditioning Questions and Answers, Stephen M. Elonka and Quaid W. Minich, McGraw-Hill Book Company, New York, 1961, 1973. Reprint 1973, Rainbow Bridge Book Company, Taipei, Taiwan. Reprint 1978 (in Portuguese) Editora McGraw-Hill do Brasil, Ltda, São Paulo, Brasil. Reprint, 1975, 1977, 1978, Tata McGraw-Hill Publishing Company, Ltd, New Delhi, India. Reprint, 1977, National Book Store, Manila, Philippines.

Standard Boiler Operators' Questions and Answers, Stephen M. Elonka and Anthony L. Kohan, McGraw-Hill Book Company, New York, 1969. Reprint 1976, Tata McGraw-Hill Publishing Company Ltd, New Delhi, India.

Standard Plant Operators' Manual, Stephen M. Elonka, McGraw-Hill Book Company, New York, 1956, 1965, 1975.

Standard Basic Math and Applied Plant Calculations, Stephen M. Elonka, McGraw-Hill Book Company, New York, 1978. Reprint 1979 (in Portuguese, Editoria McGraw-Hill do Brasil, Ltda, São Paulo, Brasil.

Standard Industrial Hydraulics Questions and Answers, Stephen M. Elonka and Orville H. Johnson, McGraw-Hill Book Company, New York, 1967.

Boiler Room Questions and Answers, Alex Higgins and Stephen M. Elonka, McGraw-Hill Book Company, New York, 1945, 1976.

Marmaduke Surfaceblow's Salty Technical Romances, (122 stories), Stephen M. Elonka, Krieger Publishing Company, Inc., Huntington, New York, 1979.

STANDARD PLANT OPERATORS' MANUAL Third Edition

Stephen Michael Elonka

Contributing Editor, Power magazine. Licensed Chief Marine Steam Engineer, Oceans, Unlimited Horsepower.
Licensed as Regular Instructor of Vocational High School, New York State.
Member: Instrument Society of America, National Association of Power Engineers (life, honorary).
Author: The Marmaduke Surfaceblow Story. *Co-author:* Standard Plant Operator's Questions and Answers, Volumes I and II; Standard Refrigeration and Air Conditioning Questions and Answers; Standard Instrumentation Questions and Answers, Volumes I and II; Standard Electronics Questions and Answers, Volumes I and II; Handbook of Mechanical Packings; Standard Industrial Hydraulics Questions and Answers; Standard Boiler Operators' Questions and Answers; Electrical Systems and Equipment for Industry

McGRAW-HILL BOOK COMPANY

New York   St. Louis   San Francisco   Auckland   Bogotá
Hamburg   Johannesburg   London   Madrid   Mexico
Montreal   New Delhi   Panama   Paris   São Paulo
Singapore   Sydney   Tokyo   Toronto

Library of Congress Cataloging in Publication Data
Elonka, Stephen Michael.
  Standard plant operators' manual.

  Includes index.
  1. Power-plants—Handbooks, manuals, etc.
2. Plant maintenance—Handbooks, manuals, etc.
I. Title.
TJ164.E4   1980   621.4   79-22089
ISBN   0-07-019298-7

1234567890   KPKP   89876543210

The editors for this book were Tyler G. Hicks and Joan Matthews, and the
production supervisor was Paul A. Malchow.

It was printed and bound by The Kingsport Press.

CONTENTS

PREFACE

Our energy crisis makes this book a *must* for everyone responsible for squeezing more Btu out of each ounce of fuel in his power plant. This third edition, greatly expanded and revised, provides more complete information on safe and efficient operation and maintenance of boilers, air conditioning systems, electrical and many other major pieces of energy systems equipment than did the previous edition. New material has been added to help operators zero in on malfunctioning machinery, thus save costly outage. New pages (155) of questions and answers, including *ten* trouble-shooting charts, and a new chapter on Well Water have been added.

Never before has so much practical information on plant operation and maintenance been crammed into one book; the cream of my most popular *basic* articles, and also some I generated and edited for authorities in my many years with *Power* magazine.

The 2,092 illustrations given here will help more than volumes of words to make clear the many practical operations needed to keep power services shipshape. This book will help the reader to pass examinations for steam operating engineer's licenses, both stationary and marine, and for passing refrigeration and air conditioning examinations, as well as for operating diesel engine plants. Thus, it will round out any library used for training, operating, and maintaining energy systems equipment.

Plant and power engineers write me frequently, asking for study material for their operators and maintenance men. Now, here is a volume that fits well into training programs. Firms will profit by placing a copy of *Standard Plant Operators' Manual* in the hands of every employee who is entrusted with their expensive machinery. Mistakes cost money, especially in lost production and repair bills. Any one of these chapters should help technicians advance to higher pay.

Steve Elonka

ACKNOWLEDGMENT

In addition to the numerous manufacturers credited in the individual articles, I especially thank the following: H. Carnes, T. C. Caskey, L. J. Cormack, T. G. Hicks, I. J. Karassik, A. L. Kohan, J. M. Mahoney, C. J. McCann, J. R. McDonnell, W. E. Mooney, J. J. O'Connor, N. Peach, G. I. Reed, H. R. Schanck, S. Spence, and G. J. Timmer for giving permission to reproduce a few pages of material I handled for them in *Power* magazine.

Steve Elonka

1
Boilers, refractories, water treatment

<u>Who</u> else cares about boiler safety?

Back in April, 1889, a regular column in POWER, *This month's accidents*, listed 14 boiler explosions in which 27 people had been killed or injured. On the same page was this announcement: "A convention of boilermakers will be held in Pittsburgh this month." At that meeting, the American Boiler Manufacturers Association set out to establish standards to "secure safety to the lives and property of all communities where boilers are used."

Since then, the ASME and other organizations have become involved. Though boilers have been exploding since, the work of these engineers has saved countless lives and property damage in America. And today the good work started 91 years ago is being carried on at an accelerated rate by various organizations. So let's see who they are and how they can help assure long and safe economical life for your boilers and pressure vessels.

1   What is the ASME Code?

A   The American Society of Mechanical Engineers Code is the recognized standard for the construction of boilers and pressure vessels. Its use ensures the owner or operator that his boiler and pressure vessel have been built to the best engineering practices and checked throughout its construction by highly qualified independent inspectors.

2   What is the role of the ASME Boiler and Pressure Vessel Committee?

A   The ASME Boiler and Pressure Vessel Committee is responsible for formulating ASME Boiler and Pressure Vessel Code standards for construction, revising them as dictated by technical progress and interpreting them to achieve uniform understanding and application. The recent Code Section X, Nuclear Vessels, which covers fiberglass reinforced plastic pressure vessels, shows how the Code Committee keeps up with technical advancement.

The Main Committee, its Subcommittees and Task Groups (over 300 competent engineers) meet regularly six times each year. There are many special meetings, in addition, to discuss important problems.

3   How can you support the Code Committee?

A   The Main Committee and all of its Subcommittees welcome the attendance and participation of persons having an interest in the matters under consideration and discussion. Or, if unable to attend in person, you may send in comments and suggestions; they are always welcome and will receive thorough consideration.

4   Who are the inspectors?

A   There are *three* types of inspectors. One type is employed by the state, province, or city. His responsibility is to see that all provisions of the boiler and pressure vessel law, and all the rules and regulations of his jurisdiction are observed. Any order of this inspector must be complied with unless the owner or operator petitions for relief or exception. A state inspector qualified under Code rules may also make shop or field erection ASME Code inspections.

The second type is the insurance company inspector, who has been qualified to make these ASME Code inspections. If he holds a commission under the law of the jurisdiction where the unit is located, he can also make the required periodic reinspection. As a commissioned inspector, he requires compliance with all the provisions of the law and rules and regulations of the authorities. In addition, he may recommend changes designed to prolong the life of the boiler or pressure vessel.

The owner-user inspector—the third type—is continuously employed by a company to inspect *unfired pressure vessels* for direct use and not for resale by such a company. He must also be qualified under the rules of any state or municipality which has adopted the Code.

Because some states do not permit this group of inspectors to serve in lieu of state or insurance company inspectors, it's important to check with the authorities before using their services.

5   How do the inspectors serve you?

A   The inspectors carefully check the construction of every ASME Code stamped boiler or pressure vessel. Then they certify that the object, parts and assembly meet all ASME Code standards. Also, at frequencies established by years of experience, inspectors examine the boiler and pressure vessel equipment and accessories, advising the owner and operator of their condition. He also recommends ways to correct unsafe or improper physical conditions and operating practices.

The inspector can also advise the owner or operator about how to apply for acceptance of specially designed objects not covered by the Code or laws. This is why inspectors are classed separately as a basic source of protection for your boiler.

6   Why a boiler and pressure vessel law?

A   No matter how good the Code, or how high the quality of inspection, they are essentially ineffective without means to enforce their use. Thus, laws have been enacted by the legislative bodies in most of the states, many municipalities, and every one of the provinces of Canada, which give the Code legal standing and establish inspection procedures and enforce Code use.

At first these laws varied in many essential details. However—through the efforts of the Uniform Boiler and Pressure Vessel Laws Society and the National Board of Boiler and Pressure Vessel Inspectors—today there is a reasonable degree of uniformity. These laws permit economical construction and installation of such Code-approved equipment in the U.S. and Canada.

Because some differences still remain, an owner or operator should be familiar with any special requirements of the jurisdiction in which he operates or plans to operate boiler or pressure vessel equipment. For example, in 6 of the states having boiler laws, nuclear vessels are not covered. In 2 states, these laws do not cover low pressure boilers. Nebraska's law, under the low-pressure boiler section, covers *only* low-pressure boilers in *schools*. The difference in coverage of various laws in the pressure-vessel field is even wider.

7   What is required for nuclear installations?

A   At present all new installations of nuclear equipment must have design and construction approved by the Nuclear Regulatory Commission (NRC) and be licensed by them prior to operation. In addition, any state or municipal provisions must be complied with, if required by local laws, rules and regulations.

If the equipment is to be insured, the insurance pools which provide third party liability and property damage coverage may also ask that certain safety standards be met.

Obviously, the ordinary periodic reinspection requirements of the existing laws are not applicable to nuclear vessels, due to radiation exposures. While much thought is being given to this problem, as yet there is no satisfactory solution. Each installation has to be considered individually, and a special reinspection program agreed upon by the owner, operator and authorities.

8   Are nonstandard or foreign-built boilers and pressure vessels acceptable?

A   Most states having boiler and pressure vessel laws based on the ASME Code make provision for *special* acceptance of vessels that do not carry the Code stamp. Such vessels might fall into any one of the three categories:

(1) vessels of unusual design for which Code provisions do not apply; (2) vessels built abroad to codes having design criteria and material specifications that differ from the ASME Code; (3) vessels built to ASME Code requirements, but where an approved Code inspector can not, without special arrangements, inspect the unit under construction and certify it.

Therefore, before any owner or operator buys a nonstandard or foreign-built unit, he should make certain of the conditions under which local authorities will permit installation and operation of the boiler.

9 Are Canadian requirements the same as U.S.?

A No, *all* Canadian provinces, unlike the United States, today have boiler and pressure vessel laws, rules and regulations administered by a most competent group of inspectors. These regulations closely follow the pattern of those in the United States and accept the ASME Boiler and Pressure Vessel Code as a standard.

As in the United States, provincial regulations include some deviations; so, when in doubt, refer to the Canadian Standards Association B51 Code for the construction of boilers and pressure vessels. Or request further information from the Chief Inspector of the province in which the object is to be installed.

The one important difference that requires special mention is that *manufacturers must register and submit designs and specifications to the Chief Inspector of the province for approval before fabricating boilers or pressure vessels for installation in that province.*

Your own state's enforcement official can usually tell you where to address inquiries about other states, cities or Canadian provinces.

10 What are the basic goals of the Uniform Boiler and Pressure Vessel Laws Society?

A The Society is a nonpolitical, noncommercial, nonprofit, technical body supported by the voluntary contributions of its membership consisting of manufacturers, insurers and users. The Society believes that any laws, rules or regulations should follow nationally accepted codes and standards.

The UBPVLS keeps owner and operator members abreast of all boiler and pressure vessel law developments. On special request, members are provided with specific information or assistance regarding any matter pertaining to boiler and pressure vessel laws, rules or regulations.

The Society's most important function is working with the authorities to secure and maintain uniform laws, rules and regulations affecting the use and operation of boilers and pressure vessels. One important advantage of uniformity is the assurance it gives the owner or operator that equipment will be acceptable in other jurisdictions, should it be relocated. However, much work remains to be done. While the Code is widely accepted, all sections are not legal standards. Five states have done little or nothing in the way of legislation to recognize the Code as the standard for construction. Not only that, but in some states and cities, regulations and procedures have not even been updated.

11 Who can join in the work of UBPVLS?

A Any owner or operator who wishes to promote the objectives of this Society may become a member by making a small annual contribution to its support. In return, he will receive bulletins to keep him abreast of changes and of proposals to change laws, rules and regulations that affect boiler and pressure vessel construction, installation, operation and maintenance. He will also receive an annual Data Sheet showing the status of ASME Code acceptance, officials in charge of boiler and pressure vessel laws, and other information of interest.

12 How does the National Board of Boiler and Pressure Vessel Inspectors benefit you?

A The stated objectives of the National Board are as follows: (a) to promote uniform enforcement of boiler and pressure vessel laws and rules; (b) to secure uniform approval of specific designs and structural details of boilers and pressure vessels, as well as of appurtenances and devices instrumental in the safe operation of such vessels; (c) to promote one uniform code of rules, including one standard stamp for all boilers and pressure vessels constructed in accordance with the requirements of that code, and one standard of qualifications and examinations for inspectors who are to enforce the requirements of said code; (d) to gather, compile and distribute data and statistics useful to the members in their enforcement work on such boiler laws and rules.

Three services deserve special mention: (1) By developing requirements for potential inspectors, the National Board assures the owner and operator of the most highly qualified and competent inspection service.

(2) By serving as a clearing house with particular reference to the capacity certification of safety valves and relief valves, the National Board *supplements* the Code and is invaluable to owners, operators and inspection agencies. Keep the word *supplements* in mind.

(3) By promulgating inspection, maintenance and repair standards, the National Board serves the owner and operator in securing uniform practices and administration throughout the United States and Canada.

Membership is voluntary, and is restricted to the chief inspector, or other official charged with enforcing boiler and pressure vessel inspection regulations. In case of municipal inspectors, membership is restricted to cities of at least 1 million population. Members may be of any political subdivision of the United States or Canada that has adopted one or more sections of the Boiler and Pressure Vessel Code of ASME, including Section 1.

13 How can you support the National Board?

A By insisting that all Code vessels be built by manufacturers registered with the National Board and that they be stamped with the official National Board symbol. Also by attending and participating in the Board's annual meeting.

If you, as an owner or operator, are not now making full use of these available means to secure safe, economical use of your boiler and pressure vessel equipment, don't you now think it is good business to do so? ●

1 Pipes out in the open and easily seen can be traced without trouble, but don't overlook those hidden in trenches just because they're hard to get at Lift the cover plates so you can follow the pipes. Traps are often hidden in trenches

Training Yourself to Be a Good Operating Engineer

It takes time and effort to get thoroughly acquainted with a plant, but such know-how pays off when things go haywire. Learn what makes things tick by following a plan like this

► At a National Power Exposition held recently, Jenkins Bros exhibited the model of a power plant to show where and how their valves can serve the power engineer. Let's see if we can't make the model serve equally well to demonstrate for young engineers how major items connect together in a modern power plant. Here, in the photographs, is a complete hookup without obstructing floor beams, building columns, or floor trench plates. Everything is out in the open where pipe runs can be traced without confusion, Fig. 1.

Sketch Each Line. Now you might ask, why spend time tracing pipe runs?

The answer is simple: To be a good operator you must know every valve and switch in the plant, be able to locate each one instantly, and understand how its manipulation affects individual pieces of equipment as well as the plant as a whole. Piping diagrams and blueprints, easily understood by an experienced engineer, only confuse a beginner.

The easiest, quickest and soundest way to get to know a pipe run is to arm yourself with a writing board, paper and pencil, and sketch each line as you trace it out physically. Follow each branch under floor plates and behind

beams so every tap-off and valve can be put on your sketch.

One-line diagrams are sufficient: use an X to represent valves. Completeness is more important than a fancy drawing. Get the details down on paper and save finishing touches until later.

Drain and Vent Lines. Don't overlook drain and vent lines. In one plant, it was common practice to drain an outside fire line before cold weather set in. After being asked to drain this line, the workman reported the job completed. When tested a month later the pipe was found frozen, and every section and branch valve broken. The man

2 Learn the purpose of every valve connected in a pipe line. There is a reason for its being there. When opportunity arises, study its internal construction

Condensate Line. Going back to the fuel-oil heaters *A* and *B*, Fig. 2, trace condensate line to traps *L* and *M* and main discharge line *N*. Below *M* we find another trap *O*, which drains the heating coil fed by valve *H*. The circuit inside vault *I* should be investigated at the first opportunity. Now surprise yourself by seeing how easy it is to sit down and draw the layout without referring to your original sketch of the actual piping.

Having in mind a clear picture of the system, what shall we look for if the furnace flame begins to show sparks and smoke streamers? A quick look at the oil thermometer shows a temperature drop. What has happened? If you can't eliminate the trouble instantly, best move is to cut in the spare oil heater first thing to save time. Then start looking for the trouble.

Fig. 4, next page, shows a close-up view of Fig. 2. Starting with main stop valve *A*, we find it open. Next check pressure gage *B*. If reading is correct we know reducing valve *C* is working and the associated manually-operated valve properly set.

Suppose the oil temperature now shows signs of returning to normal. This points an accusing finger at the condensate trap on the main oil heater or tank coil unless someone has accidentally closed one or more valves at these units. Regardless of which trap is giving trouble, cutting in the spare heater provides sufficient capacity to carry the load.

But suppose the oil temperature still remains low. Next place to look for the trouble is temperature-regulating valve *D*, Fig. 4. Check it by cracking open its bypass valve and listening for steam hissing through. Another place to look for trouble is the condensate line. Has the discharge been obstructed?

Trouble Shooting. To become a competent trouble shooter, you must know your equipment. Having gained this knowledge, you can spot trouble without wasting valuable time by starting methodically at one end of the equipment and working to the other end.

For instance, suppose you know definitely that no one tampered with any valves and no cold oil was added to the fuel tank in the vault when oil temperature began to drop. This confines possible trouble spots to the pressure-reducing valve, temperature-regulating valve, the two traps then in service, and the condensate discharge line. A glance at the pressure gage either indicates or eliminates the pressure-reducing valve, and opening the bypass checks the temperature-regulating

had shut the main supply valve, but had forgotten to open the drain and vent valves. The line did not empty. Fortunately, no fires occurred in the meantime.

Depending on plant location, floor drain lines can introduce operating hazards if the operator is not acquainted with this built-in protection. Where there is danger of flood water backing up into the plant, valves are, or should be, installed where possible or plates provided so floor openings can be closed.

Such protection is valueless if the person responsible for applying it has not concerned himself with learning and practicing the job of preparing for high water. Overlooking one opening can prove extremely embarrassing to say the least.

Small Troublemakers. As is usually true, small things cause trouble oftener than big ones. Seldom is a turbine shut down by accidental closing of the main steam-header stop valve. Connection from header to turbine is simple, with-

out confusing branch lines, therefore easy to remember.

On the other hand, closing the stop valve in a cooling-water line that feeds several pieces of equipment can prove disastrous if such machines as running air compressors or turbine-bearing oil coolers are overlooked. Damaged bearings and scored compressor cylinders are hard to explain away if they result from carelessness.

Tracing a Pipe Run. Tracing out a pipe run is easy. Consider the steam supply for fuel-oil heaters *A* and *B* in Fig. 2. We know this steam comes from the boiler, so we follow both steam lines *C* and *D* back and find they come together at *E*. Just beyond junction *E* is the temperature-regulating valve *F* with its shutoff and bypass valve, and a branch line *G* with its temperature-regulating valve *H*, which feeds a set of oil-tank heating coils inside vault *I*. Entire system is fed by pressure-reducing valve *J* supplied by stop valve *K*.

3 Become familiar with boiler valves. Main steam valve usually operates automatically to stop flow into a burst header or backflow because of a ruptured boiler tube. Feedwater line always contains a check valve near the boiler

4 The study of automatic regulating valves is a subject in itself. Learn their purpose in a system and what care they need to keep them working smoothly

comes from, Fig 3. Feedwater from heater A comes through line B and interconnected valves, and enters the boiler drum at C. Saturated steam leaves the drum and travels through line D to the far side of the boiler where it enters the superheater. Leaving this unit through valve E, steam enters the ring header connecting the two boilers. Oil-heater supply comes off at F and turbine takeoff is at G.

A ring header such as this can get you into trouble if the various valves are manipulated improperly. Best precaution is to check and double-check your actions if, upon closing a certain valve, steam is heard hissing through as the valve begins to throttle near closed position. This can mean that steam to some piece of running equipment is being shut off because you have left a valve closed that should be open.

Another Caution: Never get into the habit of spinning a valve open or closed. Take it easy. On large manually operated valves fitted with a bypass, open the bypass and let downstream pipe get up to pressure before opening the main valve. On the reverse operation, have the bypass open before closing main valve; then close bypass.

Remember that tracing pipe runs is only one phase of learning power-plant operation, but it is just as important as any other. Another important step is to study every piece of equipment in the piping system. Learn what makes a pressure-reducing or temperature-regulating valve work, and what trouble signals to look for.

valve. Thus within a minute everything up to the trap and condensate line can be checked.

Effect of cutting in the auxiliary heater should soon appear on the oil temperature to indicate whether condensate is getting away from the main heater. Accurate thinking and quick action in a case of this kind mean the difference between a plant shutdown and a minor incident.

Quick action, without taking time to think, or without knowing what to think about, invites disaster. Suppose upon noticing the flame action you immediately attempt to cure the trouble by throttling oil flow to the burner. This appears to help, but you have failed to check all the angles, and meanwhile the oil is getting cooler. Eventually the flame cannot be maintained. Result is that steam pressure falls, load is dropped, and the plant has to be shut down completely.

More delay occurs before the trouble is finally located and the oil brought back to atomizing temperature so the fire can be relighted. Such avoidable interruptions make a black mark on an engineer's record.

Tracing Steam Travel. While on the subject of steam, let's see where it

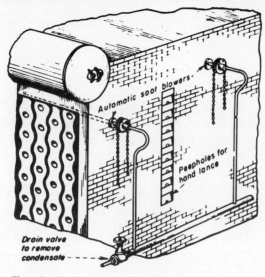

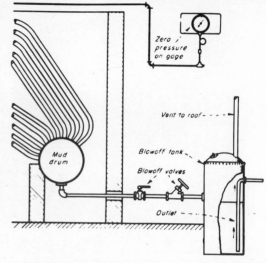

1 The tubes should be cleaned by blowing the soot with hand lance or automatic soot blowers. For thorough job, use both automatic soot blowers and lance

2 The boiler should be emptied by opening the bottom blowoff valves only after the pressure reaches zero on the gage. Then your boiler is safe to dump

Steps In Preparing Boilers For Inspection

Water-tube boilers need a lot of attention before inspection time. Tubes must be blown, boiler drained and washed. Clean the fireside thoroughly also, so the inspector can find flaws and defects easily. A good engineer also cleans his boiler room before the inspector arrives. Things to look for in boiler are oil, scale, and corrosion. Follow these steps carefully for best results

▶ MOST STATES AND CITIES require that all steam boilers of a certain size and operated above 15 psi be inspected internally at least once a year. This inspection is made either by a state or city boiler inspector, or by an insurance company inspector authorized for that territory. In some cities, the municipality requires their own inspection, entirely independent of any insurance firm's inspection.

These "double inspections" don't happen often. When they do, both inspectors usually arrange to be present on the same day.

When the inspection date has been set it is the engineer's duty to have the boiler opened and ready at the appointed time. Preparation is important, so follow these steps to save time.

Blowing Tubes. Before taking the boiler out of service, clean the soot from the fireside of the boiler. This is usually done by blowing air or steam against the tubes, either with a hand lance or automatic soot blowers, Fig. 1. Most old boilers are cleaned with a hand lance while newer ones have automatic soot blowers.

Blowing tubes with wet steam is bad for the tubes. The moisture cakes the soot, plugging the spaces between tubes. This not only stops up the gas passages and cuts down the tube heat absorption surface, but forms sulphuric acid, which corrodes the tubes.

To avoid this, use superheated steam if possible. In any case, make sure all condensate is drained before using the soot blowers by opening the drain valve on the header.

Draining Boiler. After taking the boiler off the line and blocking the fuel supply, close all fire and ashpit doors, dampers, and other air inlets. Allow the unit to cool slowly.

When the pressure drops to zero on the gage, and a reasonable time has passed to allow the boiler and furnace walls to cool, open the blowoff valves so the water drains to the blowoff tank, Fig. 2.

Caution: Never empty a boiler through the blowoff connections while it shows steam pressure. That's because the water protects the shell and tubes against heat stored in the furnace brickwork. Otherwise, unequal strains might set in, causing riveted joints and tube ends to leak.

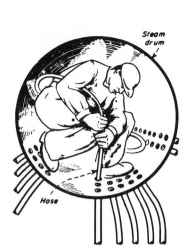

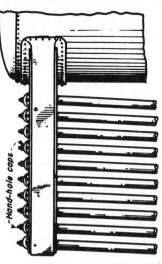

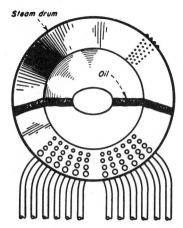

3 Wash drums and tubes with a hose before sediment has chance to harden. That makes job easy

4 Be sure to remove all handhole plates on header boilers for tube inspection for the inspector

5 The steam-drum water-level line is the best place to check for traces of oil. Rub your hand along shell

Washing Out Boiler. After opening the water-side, wash out the shells and tubes with a spray of high-pressure water. Be sure all tubes are washed thoroughly, Fig. 3.

If it is a header-type boiler, either sectional or solid, remove all handhole caps for washing and examining the tubes for scale, Fig. 4.

Caution: Wash immediately after draining the water from the boiler. This prevents any heat contained in the brick setting or furnace walls baking the soft muddy scale to the shell and tubes.

Look for Oil. Check along the steam-drum water level for signs of oil. It's usually hard to detect, especially if only a very small amount is present. One sure way to tell is to run the back of your finger along the water line, Fig. 5. If your fingers stain and the stain can't be washed off with soap and water, oil is getting into the boiler and that's a dangerous condition in any boiler.

Oil enters with feedwater. Then it's distributed throughout the boiler by circulation. Because it's lighter than water, it gradually rises and forms a scum along the water level. The real danger comes from tiny solid particles sticking to the oil before it adheres to the drum sides. Gradually this weighted oil settles to the heating surfaces. That's when tubes blister, or fail completely.

Removing Scale. Examine the inner shell and tubes for scale. If it's heavier

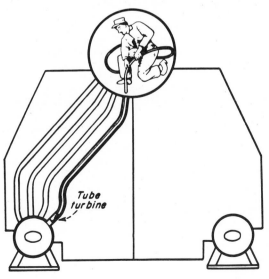

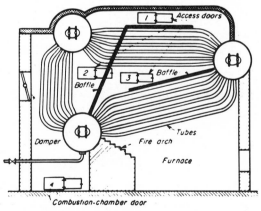

6 Boilers of this type have access doors for fireside inspection, cleaning, and maintenance. Check baffles and dampers

7 Turbine the boiler tubes for inspection. If tubes show no trace of scale, turbine them at least once in five years

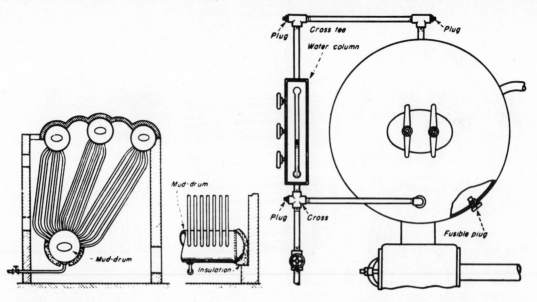

8 If the mud drum is insulated, as in this boiler, be sure to remove the insulation from drum ends so metal is visible. Then inspector can do a better job for you

9 Open gage-glass connections and renew fusible plu before the inspector arrives. He wants them in to notch shape so you do not get into trouble

than "eggshell' thickness, remove it. This is done by hand or mechanical means, or both.

If you use the right feedwater treatment, the thin scale should peel off easily by running a hand scraper lightly over the metal. The tubes should be turbined, Fig. 7. Even if they are free of scale, it's good practice to turbine tubes at least once every five years.

If the boiler has been neglected and is badly scaled, the drums will have to be hand-chipped. Chipping is a long and tough job. It requires patience. If you're pressed for time, several concerns are equipped to remove all scale completely by the acidizing process.

Cleaning Fireside. Enter the combustion chamber through the access doors and clean the soot from tubes, mud drum, blowoff lines and combustion-chamber floor, Fig. 6.

Next enter the furnace and wire-brush the lower bank of tubes. Be sure to examine the fire arch and furnace side walls for needed repairs and leaks. Then open all access doors in the side-wall settings, and remove the soot from tubes and steam drum. While you're in there, examine the baffles for breaks. Make sure they haven't shifted.

In bent-tube boilers, remove the insulation from the outside surfaces of both mud-drum heads and between the brick setting and the heads, Fig. 8. The ends are now ready for a complete inspection.

That's important because these drum are subject to severe corrosion. Of course their condition depends on the upkee they get and the boiler's age.

Fusible Plugs. Remove the cross te plugs from the water-column connectin piping so the lines can be examined Fig. 9. They often get plugged and caus a false gage-glass reading. Running brush through them every time the boile is cleaned internally is good practice.

If the steam drums have fusible-plug: renew them. This should be done once year, regardless of their condition.

Your boiler is now in good shape an it's ready for the boiler inspector. It your duty to give the inspector any infor mation that will help him do the job righ

Don't Block Safety Devices

▶ CERTAIN TYPES of automatic oil-fired boilers have a push-button bypass switch connected in the burner flame-failure circuit. Making a habit of blocking the switch closed when lighting off, instead of holding it by hand, can lead to serious trouble if the attendant walks away and forgets to release the switch.

The danger lies in the fact that rising steam pressure may reduce the oil feed until the flame goes out. But the bypass switch is closed and flame failure does not shut the burner down. What happens? Steam pressure falls and oil flow is restored to a hot firebox without a lighting-off torch being inserted. Result—a bad furnace explosion and serious injury.

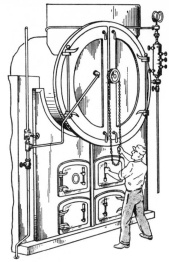

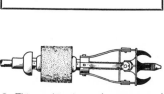

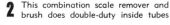

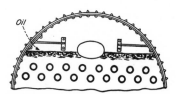

1 Automatic soot blowers remove soot from inside tubes while boiler's hot

2 This combination scale remover and brush does double-duty inside tubes

3 Brush tubes in cold boiler by hand to help inspector find any trouble

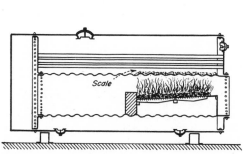

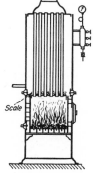

9 Scotch-marine-type boiler's corrugated furnace has tough scale above the fire

10 Look for scale in upright boiler on tubesheet and tube ends above fire

11 Rub back of your clean fingers along water line to test for oily deposit

Firetube Boilers..

. . . need plenty of preparing before the boiler inspector shows up.

▶ REMOVE SOOT FIRST. Whatever type fire-tube boiler you have, blow tubes before removing boiler from service.

Use steam or air, but don't blow moisture on tubes. Hand lance, automatic soot blower, Fig. 1, or both are needed. If boiler is "dead" Fig. 3, blow tubes, brush, scrape, or do all three. Brush and scrape if soot is caked. Combination tube scraper and brush, Fig. 2, cuts hardened soot loose and brushes tube at same time.

Always drain condensate from steam before blowing tubes. Soot caked on tubes causes corrosion.

Draining Boiler. *Caution! Don't drain water from boiler before boiler setting has cooled enough to prevent damage, Fig. 4.*

Fire doors and dampers open while cooling cause unequal expansion, mak-ing riveted seams leak. Keep them closed.

Remove Plates. Remove all manhole and handhole plates to wash waterside.

ASME Code specifies all boilers must have manholes, handholes, or other openings for examining and cleaning boiler's inside. Insert high-pressure spray hose if it's impossible to enter waterside.

Start washing as soon as plates are off. Then soft scale is mushy and easy to wash out. If allowed to dry, it hardens and makes removal tough.

Removing Scale. Run scraper lightly over softer deposits in hrt boiler. Remove hard scale with hand chipping hammer or mechanically. In either case, don't injure boiler material.

Look for scale on crown sheet in firebox boiler, Fig. 5. In hrt boiler,

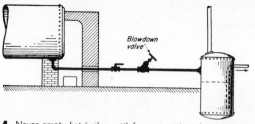

4 Never empty hot boiler until furnace settings have cooled. Empty boiler heats up unevenly and riveted seams will leak

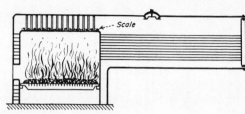

5 Scale forms in this firebox boiler on crown sheet above fire. Remove scale thoroughly with scraper or wirebrush

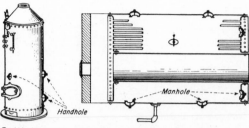

6 Wash small upright boiler through these handholes

7 Remove all manhole plates and clean waterside well

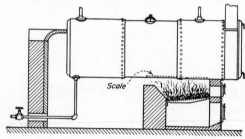

8 Scale in the boiler is heaviest directly above the fire. Remove all scale so boiler inspector knows metal's condition

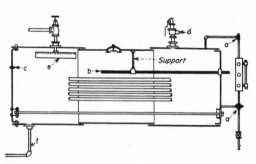

12 Check water column **a**, internal feed line **b**, fusible plug **c**, safety valve **d**, dry pipe **e**, and blow-off pipe, **f**

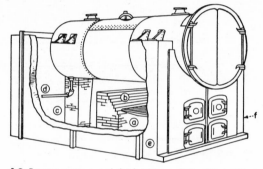

13 Remove grates **a**, and check bridge wall **b**, setting **e**, blow-off piping **d**, and clean combustion chamber **c** thoroughly

Fig. 8, look on lower shell. In scotch marine, Fig. 9, look above corrugated furnace. In vertical boiler, Fig. 10, look on crown sheets. Trouble-spot is directly above fire in these boilers. Remove *all* scale.

Removing scale from fire-tube boilers is a tedious job. Tubes may be rattled with pneumatic "rattler" from fireside. Or flat metal pushed back and forth between tube banks. If scale is heavy, it may be worth while to cut out and remove all tubes, then scale shell by hand.

Oil In Boiler. Oil is a non-conductor of heat so remove it from waterside. In boiler, oil eventually causes overheating and plate bulges when it gets in with feed water. Exception is new boilers not cleaned properly. Usually oil is in tiny particles over

heating surface, or an emulsion along water level, Fig. 11.

Check water gage glass for oil while boiler's steaming. If boiler "foams" open and check for oil. Then remedy the cause at once.

Boiler Accessories. (1) Remove plug *a*, on water column across tees, Fig. 12. Check for plugged connecting lines by poking rod through. (2) Check internal feed line, *b*, for holes. See that support is okay and line's end is not plugged with scale. (3) Examine fusible plug, *c*, and renew annually. (4) Test safety valve, *d*, for working condition. See that blow-off opening isn't plugged. (5) Examine dry pipe, *e*, for holes or slots being plugged.

Furnace and Fire Side. (1) If coal-fired boiler, Fig. 13, remove ashes from grates, *a*, and from combustion cham-

ber. If internally-fired boiler, such as vertical-tubular, scotch-marine, and locomotive-firebox, remove grates. (2) Check furnace walls and bridge-wall condition, *b*, boiler setting, *e*, brick pier of hrt boiler for protecting blow-off line *d*. (3) Remove blowoff-pipe insulation, *d*, if exposed to combustion, whether rope asbestos, cast-iron sleeve, or fire brick. (4) Brush soot from both tube sheets and any portion of shell exposed to combustion. (5) Remove all soot in combustion chamber, *c*. (6) If small "package-type" scotch-marine boiler, where furnace cannot be entered from front, remove rear plate for access to furnace and rear tube sheet.

By doing these things before the boiler inspector arrives, you will get the inspection you should have.

Are you <u>sure</u> your pressure gages show the correct reading?

Even if you check your pressure gages with dead-weight testers, a water leg can throw the reading way off. I've run into the problem often enough to offer this advice: You'd better make sure your readings are correct

One of the first problems that came up when I took charge of a 100,000-lb-per-hr 250-psi saturated-steam generating station was the deaerating heater. "It hasn't worked right since it was installed six years ago," the plant operator informed me.

They could never reach saturated temperature of water in the storage portion of the heater. They blamed the trouble on improper venting of the noncondensable gases.

I was told that the ¾-in. vent valve was wide open at all times and a plume of steam was always visible. The operator added that the thermometer in the storage area of the heater and the pressure gage on the operating level had been checked and rechecked by each operator, time and time again during those six years.

They used dead-weight testers for the pressure gage, and freezing- and boiling-point references for the thermometer to insure accuracy. They had even calibrated the thermometer with a modern iron-constantan thermocouple for accuracy.

I listened while the operator explained how they had always maintained a pressure of at least 5 psi on the gage, and yet water temperature had never reached saturation conditions of 227 F. Best temperature they had recorded was only around 213-218 F. Once in a while they had increased plant backpressure to 7 or 8

psi by adjusting the spring-loaded multiport vent valve. But this was always done with extreme care and fear, because the manufacturer's rep had cautioned them never to go above 5 psi or flashing would occur at the feed-pump suction.

Just then one of the young graduate engineers came in. He went through an explanation of Dalton's Law of partial pressure, the error correction for Keenan & Keyes steam tables, and then waded into coefficients of rolling friction and sliding friction to be considered when using the dead-weight tester. For good measure he topped all this off with a highly sophisticated dissertation on mercury thermometers. I was most impressed.

While listening, I remembered one of the common questions in marine engineers' license examinations. It had to do with the relationship between steam pressure indicated below decks in the engine room, and pressure reading on the steam gage in the chief engineer's office located as much as 60 ft above deck.

The pressure gage for this central-station heater was located about 10 ft *below* the heater. No allowance had been made for the water leg.

Figuring that the pressure gage had been calibrated with the dead-weight tester often enough to be accurate, I asked the operator to take

the gage off the panel and install it on the heater's steam-inlet piping, about 10 ft above its existing location.

I then adjusted the multiport back-pressure valve so the gage read 5 psi. As I walked out, I told the operator to check the thermometer on the heater in about 20 minutes and then let me know the reading. "It should be very close to 227 F," I told him.

Twenty minutes later my phone rang. It was the operator. "We can't change that gage," he told me, "we need it on the operating platform."

"OK," I said, "then get the instrument department to install a predetermined error in the gage to compensate for the 10-ft water leg, and replace the gage on the operating panel."

During the past few years I have run across this water-leg situation in many plants. It's caused by the weight of condensed steam collecting in the dead-end line—adding to actual steam pressure in the vessel. Sometimes boiler gages are the problem, where the boiler drum is 20 to 25 ft above the firing level. Another problem comes up with diaphragm-operated control valves where the signal line is tied into the steam header some 10 to 20 ft above the valve.

Plant engineers often lose sight of the fact that weight changes when steam condenses back into water in steam lines.

Watch boiler inspector

1 How do you go inside the steam drum? That depends on many things. Here the inspector goes in head first. Take only low-voltage electrical equipment inside

2 Check shell carefully for corrosion, such as pits, by tapping with hammer

6 Small holes in dry-pipe top and tiny drain-holes at bottom must be free

7 Head seams in drum often groove or corrode. Check all suspected spots

8 Openings in drum for water column must be free. Watch lower hole

▶ REGISTERED POWER BOILERS must be inspected at regular intervals, says the National Board Inspection Code. Owner or user has to prepare boiler for inspection when notified by inspector.

Beforehand, boiler must be opened and prepared for internal and external inspection. Caution! In cooling down a boiler, don't dump water until unit has cooled slowly, or you will damage settings and the tubes. Let it cool down naturally, if possible. Remove all manhole plates and other openings.

If deterioration of inaccessible parts is suspected, you may have to remove in-sulation material, masonry or fixed parts of boiler. Where moisture or vapor shows through covering, remove covering for inspection.

Inspector must get as close to boiler parts as possible with strong light. If attachments or apparatus need testing, plant operator does it in inspector's presence if he orders it.

How to prepare boiler for inspector. 1: Dump water and wash waterside with hose. 2: Remove manhole and handhole plates, wash out plugs and water-column connection. 3: Clean furnace and combustion chambers, blow

soot from tubes and all surfaces to be inspected. 4: Remove fuel grates. 5: Remove section of brickwork if ordered by inspector. 6: Remove steam gage for testing. 7: Disconnect pipe or valve if steam or water leaks into boiler.

Certificate of Inspection can be withheld by inspector until boiler is properly prepared. You might have to remove enough jacketing or setting along longitudinal seams of shells, drums or domes so he can inspect rivets, pitch of rivets for data needed in determining safety of boiler. Or you might have to drill suspected thin spots at his request so he

check your boiler's waterside: I

3 Look into all downcomers at drum's rear for scale, also condition of tubes

4 Upcomers under baffles may corrode or have scale deposits. Hammer **test**

5 Inspect longitudinal seams inside the shell carefully for corrosion as here

9 Check manhole plate for corrosion on seat, also cracks from bellows effect

10 Tubes' inside headers may be heavily scaled. Check lower tubes well

11 Mud-drum header may be badly scaled. Check condition of scale

can gage thickness for possibly reducing steam pressure in older boilers.

Here are inspector's internal checks: Scale, oil, etc. Inspector examines all surfaces of exposed metal inside to check action caused by water treatment, scale solvents, oil or other substances that might have entered with feedwater. Oil inside a boiler is dangerous; you must prevent more entering at once. Oil or scale in tubes of watertube boilers or on plates over fire of any boiler weakens metal, causing bagging or rupture.

Corrosion along or next to a seam is more serious than in solid plate away from seams. Strength of longitudinal joint is less than solid sheet, so thinning here is dangerous. Severe corrosion is likely where water circulation is poor. Inspector checks these places.

Grooving and cracks along longitudinal seams are bad because they usually occur when material is highly stressed. Internal grooving, in fillets of unstayed heads, and external grooving, in outer surfaces of heads concave to pressure, are common because of slight movement in heads.

Stays are inspected for even tension, fastened ends are examined for cracks

where stays are punched or drilled for rivets or bolts.

Manholes and other openings are examined internally and externally for deformity or cracks. Opening to water-column connections, dry pipes and safety valves must be free of obstructions, such as mud, scale, etc.

Ligaments between tube holes in heads of all boiler types often crack, then leak and weaken boiler. Also check beading on tube ends for erosion and corrosion, especially in type of boiler here. Remember, the cleaner the boiler, the better job the inspector does.

Inspect the fire side

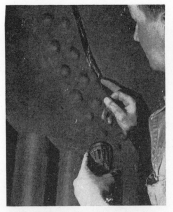

1 Check the cross box and down-comers in rear pass for leakage—corrosion

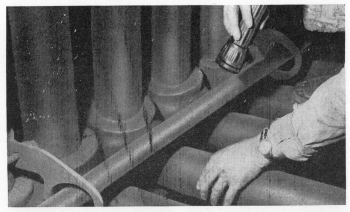

2 Trouble often starts on generating tubes near soot blowers. Look for signs of thinning and erosion from steam impinging on tubes. This is an important check

6 Look along the tubes for signs of overheating, bulging and also sagging

7 Easiest way to find bowed tubes in many boilers is to use straightedge

9 Leakage between tube and wall here can cause a blasting effect on tube

8 Inspect overhead baffles and bridge-wall for leakage, tubes for corrosion

10 Check brickwork for cracks, spalling, and for its general condition

# of your boiler carefully: 2

3 Check position of soot-blower holes for their even spacing between tubes

4 Look into superheater header to see condition of tubes, also the headers

5 Check refractory around burner throat for spalling—for general condition

Here are important trouble spots for you to check

▶ WE INSPECTED water side of boiler on pages 18 and 19. The article points out that the National Board Inspection Code requires registered power boilers to be inspected at regular intervals. It's up to you to prepare boilers.

After boiler has been emptied, cooled and dumped, open for internal and external inspection. You can't find all defects if boiler is not clean.

Note: Insurance men tell me some of the largest firms won't cooperate in cleaning boiler thoroughly for inspector. No marine engineer thinks of having a marine inspector come aboard until each pass of fire side is blown with hand lance, tubes inside furnace wire-brushed, water side of tubes turbined and boiler thoroughly washed out. Unless boiler is prepared properly, inspection is not made. Why don't insurance firms stick together and refuse to insure a boiler unless it is properly prepared for inspection?

After inspecting water side, inspect fire side thoroughly as shown here.

Bulging, blistering, leaks. Be sure to inspect carefully plate and tube surfaces that are exposed to fire. Look for places that might become deformed by bulging or blistering during operation. Solids in water side of lower generating tubes cause blisters when sludge settles in tubes and water cannot carry away heat. Inspect water side of such de-formed areas, then remove whatever solids caused the bulges or blisters.

Bulges and blisters can seriously weaken the plates or tubes. Boiler must be taken out of service until defective part or parts have been properly repaired. Blistered tubes must be cut out and replaced with new, and not hammered into shape.

In vicinity of seams and tube ends, observe boiler structure carefully for leakage. Firetubes sometimes blister, but rarely collapse.

Lap joints, fire cracks. Lap-joint boilers are apt to crack where plates lap in longitudinal or straight seam. If there is evidence of leakage or trouble at this point, you might have to remove rivets or slot the plate to determine whether cracks exist in the seam. Cracks in shell plates are usually dangerous, except fire cracks that run from edge of plate into the rivet holes of girth seams. Usually, a limited number of such fire cracks are not very serious.

Test staybolts by tapping one end of each bolt with a hammer. You get better results by holding hammer or heavy tool at opposite end while tapping.

Tube defects. Tubes in horizontal firetube boilers deteriorate faster at ends toward the fire. Tapping outer surface with light hammer shows if there's serious thinness. Tubes of vertical tubular boilers usually thin at upper ends when exposed to products of combustion. Lack of water cooling is the cause here.

Tubes subject to strong draft often thin from erosion caused by impingement of fuel and ash particles. Soot blowers, improperly used will also thin the tubes. Leaky tube often causes tubes in its vicinity to corrode seriously from water striking soot, causing acid condition. You can lose many tubes in this way.

Short tubes or nipples joining drums or headers lodge fuel and ash, then cause corrosion if moisture is present. Clean, then examine all such places thoroughly.

Baffling in watertube boiler. Check to see if baffling is in place. Combustion gas, short circuiting through baffles, raises temperatures on portions of boiler, causing trouble. Check location of combustion arches for right position with relation to tube surfaces.

Heat localization from improper or defective burner or stoker, or operation causing a blowpipe effect, must be corrected to prevent overheating.

When repairs are made, inspect workmanship. Excessive tube rolling is common. Then tube ends are thinned at tube sheets, tube-sheet openings enlarged—both harmful to boiler. Under-rolling is also common, especially if tubes are hard to get to. Hydrostatic test will show such leaks and leaky tubes can be rerolled (see how to roll and reroll tubes, pp 10, 11, 18, 19).

Now we'll inspect external

1 Safety valve is most important fitting on boiler. Examine spring to see that no foreign material is wedged between coils. Check the general outside condition

2 Valve lever must be in working order; but don't lift valve on dead boiler

6 If stop valve is in water column, install the outside screw-and-yoke type

7 Globe valve must be between drum and check valve, pressure under seat

8 Main stop must work freely. No leak around packing gland or insulation

11 Asbestos rope between the sinuous headers must be tight; no air leaks

12 Blowdown valve and fittings must be shipshape: can shut down boiler

Inspection Code is

► WE INSPECTED boiler waterside and fireside on pages 18 to 21 and explained that National Board Inspection Code requires registered power boiler to be inspected at regular intervals. Now let's wind up this job by inspecting the external fittings with Inspector Ed Zeman, of Hartford Boiler Inspection and Insurance Co.

Safety valves are the most important attachments on boiler; so they must always be in perfect condition to work. There should be no rust, scale or foreign matter in casings to hinder free operation. Best way to test setting and freedom of safety valves is by pop-

fittings of your boiler: 3

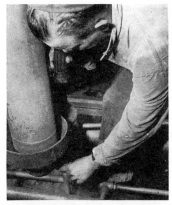

3 Escape pipe has to be well-bracketed; drain hole clear to carry condensate

4 Superheater safety must blow first to keep flow through the superheater

5 Gage-glass shut-off valves should not bind or leak. Check condition

9 U bolts through channel-iron supports help hold weight of steam drum. Examine bolt threads and everything connected with support to insure first-class order

10 Soot blowers may not work or they may be ready to fail. Inspect well

very firm on these twelve important points—you must know them

ping valves with steam. If this can't be done, test by try-levers. Inspect discharge pipe to make sure it's secure, won't blow away when valve pops. Operators have been killed as valve discharging into fire room fills space with steam in few seconds. Drain-opening in discharge line must not be plugged.

Steam gages have to be removed to test by comparing with a standard test gage. Steam gage must not be located so it's exposed to externally high temperature or to internal heat. Blow out steam pipe leading to steam gage.

Water column. Steam pipe should drain toward column, water-pipe con-nection should drain toward boiler so water cannot accumulate in steam con-nection, giving false reading.

Observe relative position of water col-umn to fire surfaces of boiler for con-forming to Code requirements for your boiler. Examine condition of attach-ments.

Suspended boilers. Examine supports where boiler structure comes near set-ting walls or floor. Make sure ash and soot won't bind boiler structure to pro-duce excessive strains from expansion under operating conditions.

Check-valve and stop-valve in feed-water inlet line must be placed so stop valve is next to drum, with pressure under seat of globe valve.

Main stop must be in working condi-tion, check externally.

Soot blowers need checking for doing their job right. They can cut tubes.

Sinuous headers must be calked or air leaks into the boiler between headers and cools it, raising excess air.

Blowdown valves must work freely, be packed, and all external piping and fittings be inspected to make sure they are not corroded or otherwise damaged.

Study National Board Inspection Code and avoid outage and accidents by keep-ing your boiler safe—and efficient.

Laying Up Boilers?

DRY METHOD

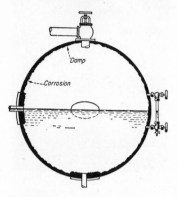

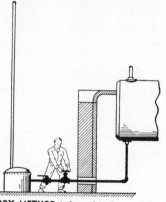

CORROSION works fast in idle boilers. Water side is attacked if no precautions are taken. Idle boilers are laid up either dry or wet to prevent this damage

DRY METHOD is best if boiler won't be needed for a month or more. First step after boiler is cool is to drain boiler completely dry by opening blowdown valve

CLEAN BOILER'S fire side thoroughly by blowing tubes, removing soot and inspecting thoroughly. Soot on idle boilers collects moisture and corrodes metal

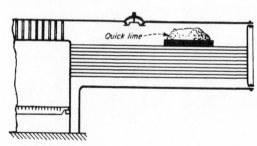

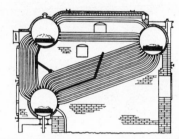

QUICK LIME in wooden tray placed on top of tubes in locomotive-type boiler is next step. Use about 30 pounds of lime for each 100-hp boiler capacity. Place lime inside as soon as boiler has dried out. Then close up boiler tightly

MULTI-DRUM BOILER like this must have wooden tray of quick lime in each drum. Reason is most moisture collects on large drum areas. Lime is very hygroscopic, and absorbs moisture rapidly, before it collects on internal surfaces

WET METHOD

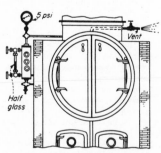

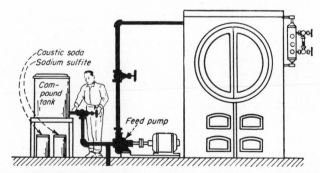

USE WET METHOD if standby boiler is idle only a week. Drain, clean, inspect. Fill to normal level, raise two to five psi steam to vent dissolved gases

MAKE WATER ALKALINE in excess of 400 ppm by three pounds of caustic soda for each 1000 gallons of water in boiler. Then add about 1.5 pounds of sodium sulfite per 1000 gallons of water. Test to see if water has minimum residual sulfite content of 100 ppm. Refer to page 54, Apr. 1, 1948, **OE** for test method

Wet or Dry?

INSPECT WATERSIDE carefully. Enter the steam drum or shell with low-voltage electric light. Remove any corrosion found, with a scraper. Scrape down to bare metal. Make sure valves are tight so no steam or water enters boiler

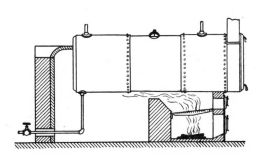

DRY BOILER next. This is important. One way is to make light wood fire in empty boiler's furnace with thin packing-box boards. Because there is no water in boiler, burn only a few boards at a time. Another way is to use a hot-air stove

HRT BOILERS like this usually need only one wooden tray placed on shell's bottom through manhole. Wooden tray is easy to build, holds enough lime and fits through manhole. Lime is inexpensive, can be bought at most building-supply firms

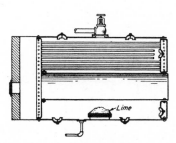

CLOSE UP BOILER after trays are in place. That means all manhole, handhole and other connections must be tight. If valves leak, blank them off. If boiler is idle for long, open it every three months and renew damp lime with new lime

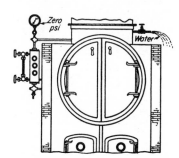

FILL BOILER with water until it comes out of vent (after boiler cools enough so pressure returns to zero, but before vacuum forms). Close all connections

TEST WATER in idle boiler every week. Keep akalinity and sulfite concentration where it belongs. The colder water is in idle boiler the better for the boiler

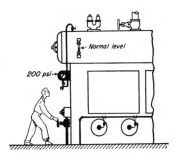

BEFORE CUTTING boiler in on line, give blowdown to reduce alkalinity. This prevents carryover. Test water and give blowdowns until water is right

This Spring, Lay Up

HOW BOILER IS KNOCKED OUT

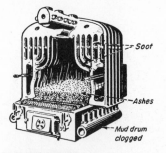

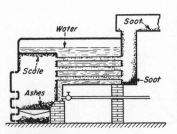

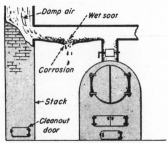

CAST-IRON BOILER corrodes very badly if left dirty in summer. Soot absorbs moisture, forms corrosive sulfuric acid

STEEL BOILER corrodes quickly in the summer if boiler is dirty and is not laid up properly for the idle period

UPTAKES and breechings coated with soot corrode in summer months when there's not enough heat to keep moisture out

HOW TO ADD YEARS OF LIFE

CLEAN boiler by removing ashes and punching tubes with wire brush until all heating surfaces are perfectly clean

REMOVE GRATES if coal burner. Clean grates with wire brush, renew burned or warped grates and swab with lube oil

WIRE BRUSH all heating surfaces of every type boiler. Remove thoroughly all soot and scale from between sections

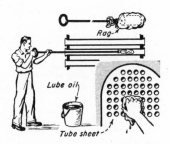

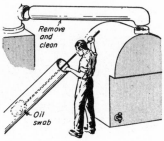

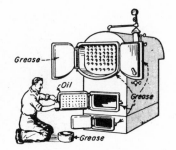

SWAB all tubes with used lube oil by placing a rag over brush. Then swab the tubesheet thoroughly with used lube oil

REMOVE SMOKE PIPES on smaller boilers. Clean soot, then oil swab inside. Stack in dry place until heating season

GREASE all machined surfaces on both steel and cast-iron boilers. This prevents air leakage past corroded parts

Heating Boilers Right

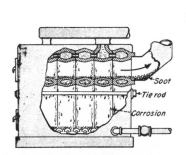

DAMP BOILER ROOM causes sweating in idle boiler that's filled with water. Then corrosion knocks out dirty boiler fast

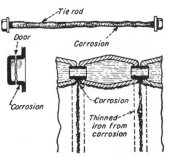

CAST-IRON BOILER is attacked by corrosion between sooty sections, at sooty tie-rods, and at machined surfaces of doors

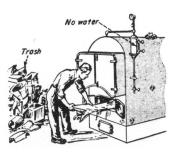

BURNING TRASH in idle boiler is bad. Many boilers are badly damaged from firing with low water or no water at all

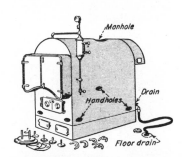

REMOVE MANHOLE and handhole plates on steel boilers (cast iron boilers don't have them) after draining off all water

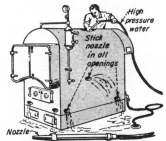

WASH INSIDE with highest pressure water you have soon as boiler is dumped. Then sediment is soft and washes out easily

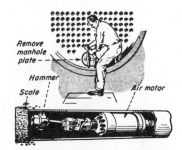

IF AFTER OPENING and washing large boiler, lime and mineral scale are on tubes outside, remove with this tube cleaner

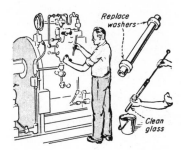

REMOVE gage glass and wash in household ammonia. Renew washers and make sure the gage glass lines are not plugged

HANG SIGN on boiler saying whether filled or dry. Operators Notebook 35 told how to lay up waterside WET or DRY

CLEANED, oiled and left open for air circulation, boiler rates long life. Cast-iron boiler full; steel boiler usually dry

Removing, Rolling, Beading

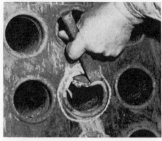

1 Chisel bead from smoke box end so tube can be removed from tube sheet

2 Collapse end of the tube with "oyster knife" until the expanded part is free

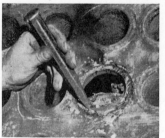

6 Remove the welded tube end and then drive tube out so it can be removed

7 Rolled end may still not be free to clear tube sheet hole. Collapse it

8 Hammer lightly along tube while withdrawing if scale gives trouble

11 With tube marked to extend ⅜ in. from both ends, pull out and wedge

12 With tube wedged to keep it from turning, cut at mark with tube cutter

13 Position the tube in tube sheets and hold it in place with a thin wedge

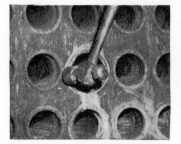

17 When tube is rolled just enough to make it tight, peen end over lightly

18 Ductile tube is peened over quickly with hammer but finish with this tool

19 Move beading tool about half a tool width after each light hammer blow

And Welding Your Boiler Tubes

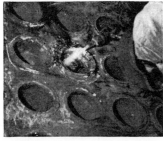

3 Burn bead off welded end of same tube with torch. Be sure it's same tube

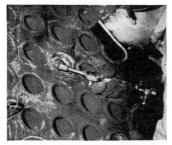

4 Cut tube all around with torch about two inches inside of the tube sheet

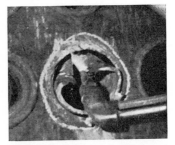

5 Split the collapsed end of tube with acetylene torch if it bulges inside

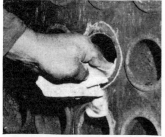

9 Sand tube-sheet bearing-surface so new tube bears against smooth surface

PUTTING IN NEW TUBE ▶

10 Slide new tube in. You might have to guide other end with a long rod

14 This end extends 3/8 in., is ready for rolling. Wedged tube won't turn

15 Set tube roller so the small flaring rollers are at right place from stop

16 Oil rollers, insert the tool and keep steady push against tool while rolling

20 After rolling tube's other end, this power impact hammer beads it fast

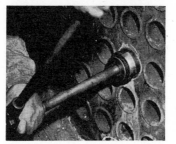

21 Finish rolling after beading to make sure tube didn't loosen from beading

22 Here's finished job. Furnace end of tube is welded to prevent leakage

Declare War Now On

WHY TUBES ARE ROLLED

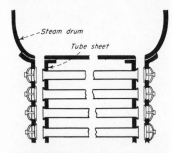

BOILER TUBES are expanded into tube sheets to seal opening around tube and to stay tube firmly in its position

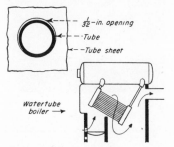

NEW WATERTUBE BOILER, has about 1/32-in. clearance between tube and tube-sheet hole at both ends of the tube

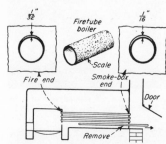

FIRETUBE BOILER tubes become scaled on outside. To help remove tube, clearance is 1/16 in. at end shown

HOW TO REMOVE AND ROLL

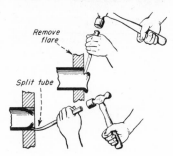

TO REMOVE TUBE, chisel off flare, flush with tube sheet. Then split tube with diamond point chisel like this

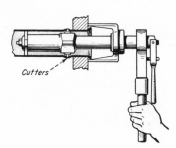

CUT BOTTOM-ROW TUBES inside tube sheet with cutter so they drop into furnace. Collapse and remove ends

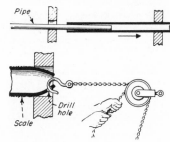

SUPPORT LONG TUBE with pipe. If tube is badly scaled, drill hole in collapsed end and pull with chainfall

DON'T OVERROLL

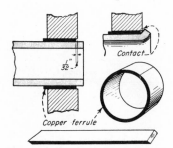

OUTSIDE FERRULE fills loose space around enlarged holes. Buy, or make from thin copper sheet; use as shown

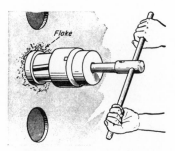

UNDER-ROLL, never over-roll tubes. Roll slowly and STOP when tiny flakes start leaving tube sheet around the tube

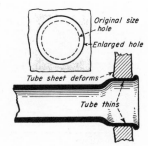

OVER-ROLLING enlarges and deforms tube sheet. Tube is thinned and weakened inside tube sheet to a dangerous point

Sloppy Tube Rolling!

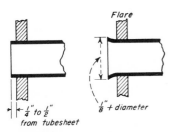

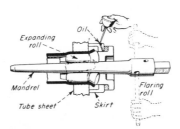

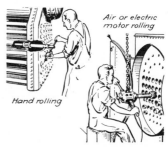

ASME CODE requires ends of watertube boiler tubes extend ¼ to ½ in. and flared ⅛-in. dia larger than tube

FLARING is usually done with tube expander that has belling or flaring rolls, with skirt to control flare depth

EXPANDERS are hand or motor operated. Hand rolling is harder work but there's less chance of damaging costly tubes

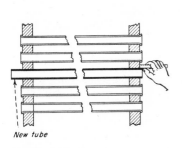

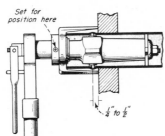

BEFORE INSTALLING new tube, check tube-sheet hole for damage. Stick in new tube and extend ¼ to ½ in. on ends

CUT TUBE at other end with inside tube cutter. If no cutter, mark tube in place (remove before wedging), and saw

TO HOLD TUBE in place while rolling one end, drive thin tapered brass wedge between tube and tube-sheet hole

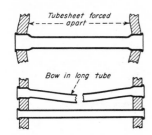

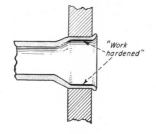

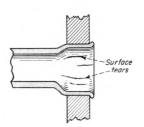

OVER-ROLLING forces tube sheets apart. Long tubes of small diameter bow, then collect sediment and bag-fail

OVER-ROLLING 'work hardens' inside of tube at tube sheet. Hardness makes metal brittle and causes early failure

OVER-ROLLING causes surface tears inside rolled part. Tears weaken tube in time and again cause early failure

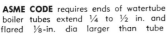

Replace Your Tubes and Nipples Right

When you roll tubes or nipples, know just what you're doing. Here's the right way to replace both your fire- and watertubes

1 Removing a firetube can be done in the four steps shown above: Use a sharp cold chisel, shear off the bead at each end. Slot the tube end with cape chisel. Use hammer and punch to collapse it partly. Tube is then ready to pull out

▶ YOU NEED a special skill developed through first-hand experience to roll tube ends in right. Without this experience it's easy to over-roll tube ends easy to get at, and easy to under-roll the tough ones. A good experienced boiler mechanic can spot an over-rolled tube at a glance. If the tell-tale signs are so easy to read with a practiced eye let's follow through the steps of tube removal and replacement for both firetube and watertube boilers and see what these signs are.

Firetube Removal. There is a right and a wrong way to pull out a firetube. Use the right tools the right way and tube removal is no problem. But with a badly scaled tube you may have to change the procedure.

Four steps in Fig. 1 show how to pull out a firetube ordinarily. (a) Begin with a sharp cold chisel and cut off the bead at each end. (b) Next use a cape chisel to slot the tube end across the rolled joint. A skilled gas cutter could do both these steps with an acetylene torch but you run the danger of scoring tube sheet. (c) After slotting tube ends, partially collapse them with hammer and punch. (d) Now pull tube out. A threaded rod and washer as a clamp for the tube or a block and fall may be needed to work the tube out. To complete the job, it may help to ream the tube hole to get a good surface.

In the case of a badly scaled tube, too tough to pull out with the usual procedure, cut it up with an acetylene torch and remove the pieces through a manhole or handhole.

For watertubes collapse the flare with a hammer, Fig 3. If you can get at the tube, you might choose to cut the flare flush with the tube sheet instead. Then follow procedure in Fig. 1. for firetubes.

Installing Firetubes. Cut new tube to length so it protrudes at each end $\frac{1}{4}$ to $\frac{3}{8}$ in. Roll tube with an expander just enough to give a tight joint under hydrostatic pressure of at least safety-valve setting. Experience is the best teacher. There are some sensitive indicators that determine the correct amount of rolling, especially for high-pressure tubes. And there are tell-tale signs that the eye can read.

One valuable guide is a measurement of the inside tube diameter. As the tube-rolling progresses, the metal in the rolled wall tends to flow away from the inside of the tube sheet or header, Fig. 2. An extruded lip forms. So check the inside diameter and look for the lip. With these guides you soon develop the "feel of the roller." Under no condition let the extruded lip stand out more than $\frac{3}{32}$ in. Inside diameter after rolling should be no more than original diameter plus clearance plus one half tube thickness.

People used to think the Prosser joint, Fig. 5, gave a stronger joint. But because there is a chance of cutting into the tube too much, the Prosser expander is no longer in common use. The ASME code limits this method to

Rolling In Firetubes .. Removing Watertubes .. Rolling In Watertubes

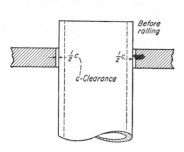

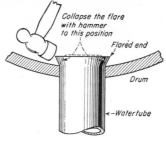

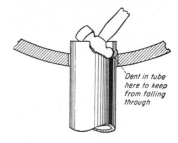

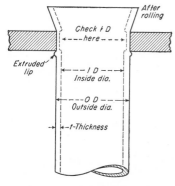

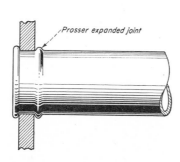

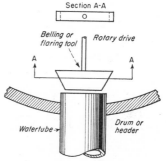

2 Rolling tube ends in right requires experience. Check inside diameter and extruded lip to insure good job. Max ID at the tube sheet = ID + C + 0.5t

3 Remove water tubes by collapsing the flare with a hammer. If you can get at tube you may cut off the flare flush with tube sheet; proceed as in Fig. 1

4 A ball-peen hammer can dent in tube side just enough to keep tube from slipping. Then complete rolling the tube into position with an expander

5 The so-called Prosser expanded joint, now limited to tubes of 1½-in. dia and smaller, was once held to give a stronger joint but overcutting may occur

6 Belling or flaring tools, when they are used, should flare the tube to not less than ⅛ in. larger than dia. If joint "weeps" under hydrostatic test, reroll

tubes of 1½-in. diameter and smaller. Once you have expanded firetube ends, bead them over into tight beads so they won't burn off or waste away.

Suppose you get an occasional tube just a little short of the desired length. Insert a steam hose to heat it up and expand it. While tube is still hot, roll it into place. This method is not recommended for a large number of tubes because it will put too much pull on the tube sheet.

Installing Watertubes. Cut watertubes so they stick out from ¼ to ½ in. beyond the tube sheet or header. Set the tube (you may have to use a ball-peen hammer, Fig. 4, to keep it from slipping out of position) and then roll in the usual manner. A belling or flaring tool, Fig. 6, should be used to flare the tube to not less than ⅛ in. larger than the diameter.

How About Nipples? Mud-drum nipples are replaced in the same way you handle watertubes. Usually they are much harder to get at. So you may

need special tools. For instance, to get around a corner at a 45-deg angle or more, such as you face when working through a handhole or a tube-cap hole, you need a universal joint in the tube expander and flaring tool.

Interdeck nipples require the same methods. But the close clearances call for special tricks. You can't shift an entire header in position very easily, so you may have to cut out the interdeck nipple and install the new one in two or more pieces. Before fusion welding became widely used, interdeck nipples were made of extra-heavy tubing with one end recessed and threaded to receive the threaded end of the second piece. If the new nipple proves too long to get around a corner through handhole, you can now use two or three pieces and butt-weld the joints. Bevel the ends and insert chill rings or backing rings before welding.

Final Checks. Subject all tube or nipple replacements to hydrostatic tests. Pressure can be about working pressure

of boiler and water temperature from 80 to 120 F. Very slight "weeps" will disappear at operating temperatures. Mark any tube or nipple where the expanded end shows greater leakage, so you can tighten with light re-rolling.

If you have to enlarge tube holes to overcome poor wall surfaces, use copper ferrules. With firetube boilers under severe service, electric-weld the bead with a light seal weld. This gives good results particularly with locomotives and at the upper ends of vertical firetubes and the rear ends of hrt boilers where flame scrubbing occurs.

Do you know . . .

that carbon on atomizer tips is the surest sign of improper location of tips? If atomizer is too far in, eddy is formed in entering air currents. A fine fog of oil is drawn back behind the tip's face; is deposited back on atomizer pipe, forming carbon.

1 Before superheater tubes were washed with hot water, they looked like this

2 After washing, same tubes were factory clean and ready to absorb heat

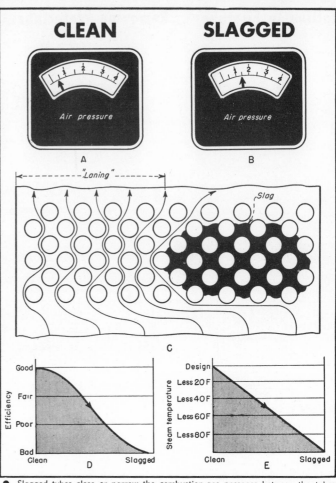

CLEAN   **SLAGGED**

Air pressure   Air pressure

A   B

"Laning"   Slag

C

Efficiency: Good / Fair / Poor / Bad — Clean ... Slagged

D

Steam temperature: Design / Less 20 F / Less 40 F / Less 60 F / Less 80 F — Clean ... Slagged

E

3 Slagged tubes close or narrow the combustion-gas passages between the tubes. To keep steam steady, more air is needed and steam superheat temperature drops

Water Washing Boiler Fire Side

▶ WATER WASHING boiler fire sides removes slag quickly and easily. Both stationary and marine boilers are washed today. There's no damage to brickwork if you use the following method. When a boiler is washed the first time, the "factory-clean" tubes always surprise the engineers.

Keep Clean. For maximum efficiency, a boiler's heat-exchange surfaces must be kept clean. Remember, manufacturers count on boilers being kept fairly clean when designing them for a given job. So thorough cleaning pays big dividends.

Why should boilers cleaned with soot blowers be washed? Experience shows that soot collects in places blowers can't reach. Most operators supplement mechanical soot blowing with hand lancing.

While soot blowing and hand lancing removes most soot and ash in tube banks, slag still forms near furnace-row tubes, on top of mud drums and on superheater tubes. Fig. 1 shows slagged tube bank. This accumulation is hard to remove by either soot blowers or hand lance while boiler is steaming. It's just as hard to remove mechanically with boiler cold.

Troubles from Slag. What effect do slagged boiler and superheater tubes have on boiler operation? (1) Higher air pressure is needed for same steam load, Fig. 3B. (2) "Laning" caused by high gas velocities through remaining open lanes can overheat tube metal, Fig. 3C. (3) Efficiency and outlet steam temperature both drop off sharply from design points. Fig. 3D, 3E. Lower steam temperature reduces turbine and other steam-consuming equipment's efficiency greatly. (4) Gas forced to bypass around slagged areas can overheat and burn out baffles, protection plates, casings, soot-blower bearings and elements, Fig. 6.

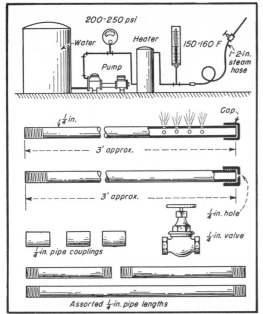

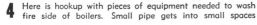

4 Here is hookup with pieces of equipment needed to wash fire side of boilers. Small pipe gets into small spaces

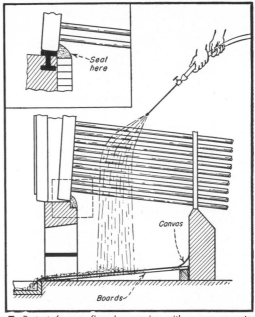

5 Protect furnace floor by covering with canvas so water runs out. Protect back of walls by waterproofing as shown

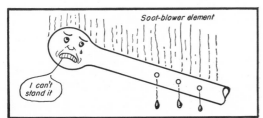

6 Soot-blower tubes, boiler baffles, plates, etc. overheat, burn from hot gas when slagged tubes can't absorb the heat

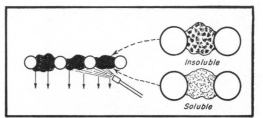

7 Slag is fused soluble and insoluble material. Hot water dissolves soluble. Water pressure knocks loose insoluble

Here's how to wash the fire side of boilers to remove slag that can't be cleaned mechanically

Water Washing. How is the fire side washed with water? This job is simple. Spray water at 150 to 160 F into the tube banks at 200 to 250 psi. Wash from the top down to avoid direct impingement against brickwork. Use hookup and equipment shown in Fig. 4.

Slag is an ash composed of soluble and insoluble material. It fuses into a solid rock-like substance between the tubes when heated above a certain temperature

When high-pressure hot water hits the slag, soluble material is dissolved from the bond and flushed away, Fig. 7. Insoluble is knocked loose by water

pressure. Then tubes are really clean.

Protecting Brickwork. Two questions always come up when washing boilers: (1) How are water and sludge removed from the furnace? (2) How should brickwork be protected while washing? Because no two plants are the same, there's no set answer for all boilers.

A tarpaulin rigged in furnace, Fig. 5, collects water and sludge so it flows through an access door. This drains into strainer-covered sump or to sewer.

Fasten the tarpaulin at four corners. Seal and cap brick wall's top with waterproofing paint or other material to prevent water getting behind walls.

This material will burn out after fire is lighted.

The average boiler can be completely water washed in 4 to 10 hours, depending on the slag accumulation.

Drying Out. To eliminate corrosion when job is finished, dry out furnace immediately. On oil-fired boilers, light a small burner. On coal-burning boilers, use a wood fire or a very light coal fire. Caution: Make sure there's water in boiler.

The drying out time is usually from 8 to 12 hours for a thorough job.

Once your boiler is water washed, Fig. 2, it'll shine "factory clean."

Fifteen Q&A

ANCHORS spaced properly will add to plastic wall's life. Rough finish helps wall dry; cut lines prevent face cracks

Q   What is spalling?

A   Spalling occurs when the "hot end" of refractory lining gets hot, expands from rapid heating, breaking off at ends. This condition can be especially severe when the other end of brick or plastic is relatively cold. Thus, rapid temperature change in only one end of the refractory has same effect as pouring hot water into a cold glass. Spalling is more common with dense brick or dense refractory of any kind, while loosely bonded brick has tendency to resist spalling.

Spalling is probably more common in oil-fired furnaces because of rapid *on* and *off* firing, causing thermal shock. You can also have trouble in furnaces when burning wood refuse, corn cobs or when incinerating paper.

If brick spalling is bad enough so refractory must be renewed, say every six months, change to next higher grade of brick. If that is not complete solution, go another grade higher. Because plastic refractory is one of best spall-resistant linings when properly installed, let's learn more about it here.

Q   What are the four general classifications of furnace lining in common use today?

A   Low duty, high duty, super duty and high alumina.

Q   How many kinds of plastic refractory (specialty refractories) are there?

A   Two major types. One is known as plastic refractory, which is a mixture of grog, consisting of ground firebrick, clay and water. This has consistency of stiff mud. It's installed by pounding into the wall with a 2½-pound hammer and then trimming off to required thickness.

Second kind is known as a castable. This is dry material of ground firebrick, clay and a hydraulic setting cement, usually of the high-alumina type. This material is mixed with water and poured into a wood form, like any ordinary cement mixture.

Q   What is the most common cause of plastic refractory failure experienced in furnace linings?

A   Improper anchorage. Never pound this material onto a thin, glazed wall. First remove old lining, whether 4½ or 9 in. deep. Be sure to enlarge back opening in old wall so new material is keyed into place.

Another cause of failure is improper spacing of anchors. Best practice is to place one anchor per 3 sq ft of wall area. Another reason is heating new wall too rapidly. Plastic is more sensitive to rapid heating than firebrick.

Q   How do you avoid irregular cracks in plastic walls?

A   Plastic shrinks when it dries, then it cracks. You can control cracking by cutting lines in wall with trowel, to depth of a third of plastic thickness. Make these cuttings

on plastic refractory

Sure, you **buy** plastic refractory—but do you **apply** it right?
We've seen so much of this excellent material wasted by
improper handling that we asked Jack Harris, who supervises its
installation, to answer a few important questions. Be
sure all your refractory boys get these answers down pat

in 4-ft squares. Such a wall also cracks, but cracks all take place in back of cuts and will not run in all directions over wall area.

Q Where should plastic and castable walls be used?

A Plastic walls stand a much higher temperature than castable walls. They are also more spall resistant. Use plastic in combustion chamber, use castable in same furnace but in areas that are much cooler and away from flame.

Q Are there castables with different properties?

A Yes. There are high-temperature refractory castables, as well as insulating castables. Remember, the higher the recommended temperature, the lower the insulating properties. A great deal of heat lost through the side walls can usually be saved with the proper selection of castable lining.

Q Should wood form used for pouring be left in place to burn out while drying out a castable wall?

A Always try to remove the wooden form before lighting off the furnace. If left in, moisture is prevented from evaporating from surface as heat is applied. This defeats the purpose of drying out in the first place. When the wooden form starts burning, heat is concentrated on wall's surface. That has tendency to glaze the surface, sealing off escape paths from green plastic inside the wall. Then as steam forms, the green wall's surface breaks off from tiny explosions beneath this glazed face. Always dry out your wall with a small wood fire of packing cases if possible, in center of furnace. This is more satisfactory than using one small burner in center of furnace as heat is not as intense.

Q Should a plastic wall be smoothed with a trowel after it's hammered and trimmed to required thickness?

A No. Leave surface as left by trimmer—which has a rough texture. This rough texture greatly increases the surface area available for evaporation when drying out the wall. That allows moisture to evaporate from plastic rapidly and evenly.

Slick surfaces, on the other hand, trap moisture and slow down evaporation, causing surface cracks. Also, if high temperatures contact wall, there will be explosions or popping effect under the surface.

Q Some engineers perforate a green plastic wall with a spike. Is this good practice?

A No. Usually this is not necessary if wall has been left with proper texture. But if you must light off the furnace, perforate wall about 4 in. deep every 5 in. with a $\frac{1}{4}$-in.-diameter spike. To see how effective this is, look into furnace while it's heating up. You will see a tiny jet of steam coming from each perforation as the wall heats.

Q How much time is needed to cure a specialty setting wall for best service life?

A Good practice is to keep wall temperature below boiling point of water (212 F). Remember, all moisture must come out through furnace face of wall. You must help establish flow of moisture from back to front. Once the moisture flowing to surface establishes path through refractory, it will continue to flow so long as there is heat. This critical time in curing takes about four hours with small wood fire.

Be sure to have wood fire in center of floor or grates, and not against walls. After four hours, increase your fire of coal, oil, gas, etc, gradually, so you reach normal operating temperature in about 24 hours.

Q What is meant by "burning in" a wall?

A A Wall is burned in when it has been heated to a point where a ceramic bond has been formed, which begins at about 1800 F. For best results, drying-out and burning-in should be one continuous operation, with temperatures climbing steadily higher.

Q Is air drying of plastic refractory harmful?

A For castables, no—for plastics, yes. Air drying means leaving surface set for a length of time before drying with fire. Trouble is that air evaporates moisture from surface only—and so flow of moisture from back is not started. Then surface cracks as it shrinks. Constant danger here is from moisture flow weeks later when wall looks dry. Then if boiler is fired without proper precautions, wall will fail. Note that this is one of the most common causes of failure. It is a constant headache to refractory contractors who supply the best materials and do high-class work. Their material or workmanship is usually blamed.

Q If a wall cannot be dried out with fire as soon as it's completed, what precautions must be taken?

A Cover walls completely by nailing burlap to wall, then keep burlap damp. But don't throw water on burlap so it washes off plastic. After fastening the wet burlap, close all doors and dampers so there's no air circulation through furnace. That helps burlap stay moist longest possible time.

Q How long can refractory specialties be stored?

A Castables can be stored in dry places indefinitely. Plastic shipped in usual cardboard of moistureproof packing is good for about six months, in a cool dry place.

If plastic dries out from long storage, you might save it by slicing in slabs a few inches thick and placing wet burlap between all layers. But this will not always save the plastic. Best practice is to order it as needed. Then you won't have to store it. Be sure to follow these hints carefully.

WHY BOILERS CRACK

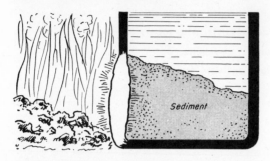

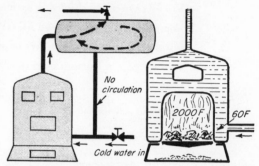

SEDIMENT PILES UP month after month, until hottest part of casting next to fire is cut off from circulation. Casting is distorted and finally cracks, causing costly repair and shutdown

RUNNING COLD WATER directly into boiler is another reason why cast-iron boilers crack. With 60 F water striking hot cast iron, expansion cracks boiler even without sediment in water

HOW TO PREVENT CRACKS

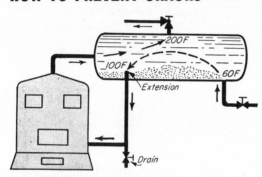

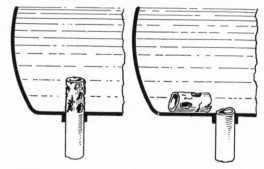

EXTENSION keeps most sediment from reaching boiler. Run cold water into tank as shown so 200-F water near top takes chill off 60-F water before it reaches boiler near hottest part

BEWARE—just because your hot-water tank has extension, don't neglect tank or boiler. Extension might be badly corroded or broken off. Examine once a year to check condition

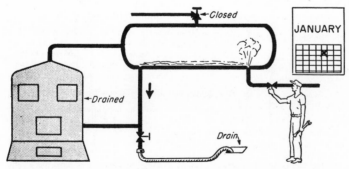

FLUSH TANK (if it has no manhole) before installing extension. First close discharge to building supply. Then drain boiler completely dry. Open cold water to tank and flush tank until discharge from hose shows that water is completely clear

FLUSH BOILER by hooking high-pressure hose to one side after draining boiler. Run water until its discharge clears up

Water-Boiler Cracks

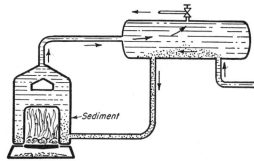

SEDIMENT AGAINST HEATING SURFACE of cast-iron, sectional hot-water-heating boilers cracks many boilers yearly. Owners and operators suddenly find ashpit flooded like this

ALL WATER HAS SEDIMENT, whether from well, lake or other sources. Sediment is often minute particles. It travels through tank and settles in lowest part of the boiler to do much harm

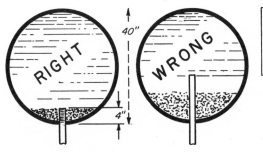

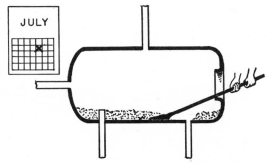

ONE TENTH OF TANK DIAMETER is right extension length. Longer extension is bad because space under end is dead space. it cuts down circulation, cools hot water too much besides

CLEAN TANK AT SET PERIODS after installing extension. Good practice is once a year. With manhole on tank, it's easy to remove sediment and also examine tank's inside thoroughly

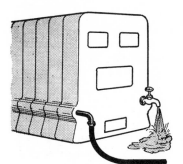

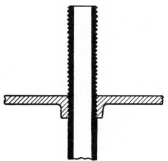

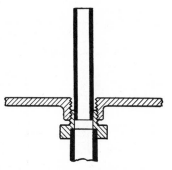

FLUSH FROM OTHER SIDE like this until water clears. Cast-iron boilers have no manhole or handhole plates to take off

SIMPLE EXTENSION is made by running pipe die over pipe until extension is right length. Then screw into the tank

SWEATED-FITTING EXTENSION is made by sweating pipe into bushing screwed into tank. Then sweat line into bushing

Handle Blowoff Valves

ASME CODE REQUIREMENTS

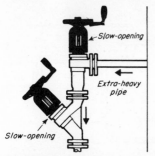

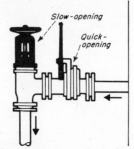

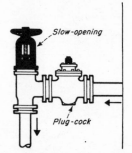

Slow-opening
Extra-heavy pipe
Slow-opening

Slow-opening
Quick-opening

Slow-opening
Plug-cock

PARAGRAPH 308, ASME Code for Power Boilers, requires all boilers carrying over 100-psi working pressure, except traction or portable boilers, to have two blowoff valves on each blowoff pipe. These may be two slow-opening valves, or one slow-open- ing and one quick-opening valve, or one slow-opening valve and one plug cock. Traction and portable boilers must have one slow- or one quick-opening blowoff valve. On all types boilers, connect the blowoff valves with only extra-heavy pipe

DETAILS AND OPERATION

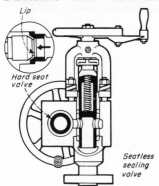

Lip
Hard seat valve
Seatless sealing valve

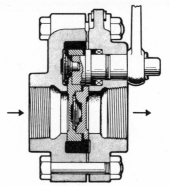

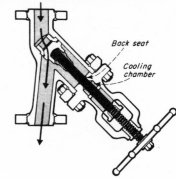

Back seat
Cooling chamber

COMBINATION seatless sealing valve in same steel block with hard-seat valve, 400-2500 psi. Lip protects seat from wire-drawing. Ported valve needs no seat

QUICK-OPENING valve, up to 600 psi. Spring holds disk to sealing surface of port. That keeps grit off sealing faces. Disk rotates in use, keeps joints tight

STRAIGHTWAY flow, good to 1800 psi. For back-seating to pack under pressure with valve open. Straight passage of sedi- ment prevents scale clogging or eroding

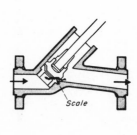

Scale

BE CAREFUL when blowing boiler. First open quick-opening valve slowly. Then open slow-opening valve slowly enough to prevent shock, but fast enough so seat won't wiredraw. To stop blowing, close slow-opening quickly, then fast-opening

NEVER jam valve if it won't close. Open few turns fast to clear, close again slow. Jamming on scale wire-draws

BLOW when boiler has banked fire or low loads — sediment settles more then. Watch glass, don't leave open valve

With Kid Gloves

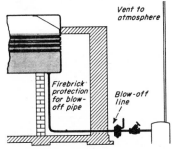

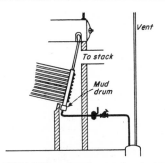

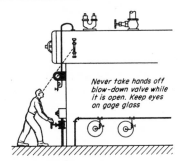

EMPTY cold boiler through blowdown line for cleaning, inspection or for repairs. Don't empty until water is fairly cool, or boiler seams and joints warp, leak

BLOW OUT mud, scale or sediment while boiler is steaming. *Caution:* Make sure blowdown valves are closed on idle units or scalding water will blow into them

BEFORE CUTTING boiler in on line, give blowdown to reduce the alkalinity. This prevents carryover. Test the water and give blowdowns until water is just right

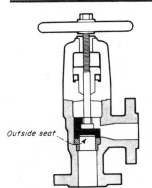

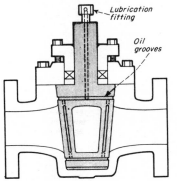

ANGLE blowoff valve, 200 psi. Seating surface is on *outside* of seat ring. That protects seat from cutting effect of the water, scale and sediment passing through

PLUGCOCK, 175 psi. Plug is pressure-lubricated through stem so tapered cock turns easily. Since valve opens, closes by turning, seating surfaces stay clean

DON'T use globe valve on blowdown line. Pocket in valve clogs with mud, sediment and scale. Then boiler cannot be blown, valve can't be cleared under pressure

HERE ARE BLOWDOWN TANK DETAILS

Tank prevents damage. Blowing directly to sewer instead of to tank blows hot water and steam in other sewer connections. This tank is always full of water. When hot water blows in, cooler water in bottom overflows to sewer as it's displaced. Large open vent prevents backpressure. Small siphon breaker keeps tank from emptying. Manhole is for internal inspections

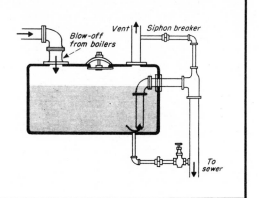

Know your basic controls

What you must know

► Your HOT-WATER BOILER must have a safety relief valve to keep the pressure at or below the maximum allowable working pressure. But, until recently, ways of doing this were not clearly understood. Today, the ASME Boiler Code defines the correct procedure in detail. But, in spite of this information, many hot-water jobs are still fouled up, Fig. 1. Is yours fouled? Insurance magazines are filled with reports of life and property loss from needless explosions.

What causes a relief valve to open? The average hot-water boiler is built for maximum working pressure of 30 psi. Relief valve is set for same pressure. So anything that ups boiler pressure to 30 psi causes valve to open, Fig. 3.

Critical demand is placed on the safety relief valve during emergency when it must discharge both high-temperature water and steam, Fig. 4. When relief valve discharges 212 F, sudden drop causes higher temperature water to flash into steam—that's why ASME relief valve is tested and rated on *steam*. Usually valve lifts when heating system can't dissipate heat as fast as it's developed by boiler.

How about low-water conditions? Many hot-water boilers are knocked out by this condition, Fig. 5. Erroneous belief that the pressure-reducing valve used to fill the hot-water system initially will keep boiler and system full causes trouble. Pressure-reducing valve is usually set at 12 to 18 psi while safety relief opens at 30 psi and closes at 26 psi —so pressure reducing valve is ineffective during time relief valve is working. If a hand-fill valve is used, any leak in system can quickly cause a low-water condition.

Minimum basic control you should install on a hot-water boiler is a float-operated low-water fuel cutoff, Fig. 6. Because of higher working pressures of hot-water boiler, low-water cutoff is for working pressures over 30 psi.

Fig. 7 shows low-water cutoff action in emergency. Lowering boiler water and water in float chamber at same time causes float to drop. That opens electrical circuit and stops automatic burner.

Low-water control, Fig. 8, is about foolproof. But you can't rely 100% on low-water cutoff to stop automatic burner each time there's low water. Reason is that under some conditions low-water cutoff cannot work. Then you must use combination boiler water-feeder and low-water cutoff shown in Figs. 8, 9 and 10.

Several kinds of overfiring can occur to make low-water cutoff useless. In such emergencies only way your boiler remains safe is to feed water to boiler as fast as water discharges through the relief valve, Fig. 9.

Large leak or drained boiler causes low-water cutoff switch to stop burner, Fig. 10. How about firing a dry boiler? Fired continuously by a runaway burner and carrying no heating load, this boiler may not explode because it has a modern ASME approved relief valve to protect against over pressure. But the boiler may be ruined because it goes dry, or because it tries to operate as a steam boiler though it isn't hooked up for this service. Check your boiler today.

—Courtesy, McDonnell & Miller, Inc, Chicago, Ill.

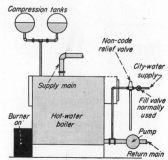

1 Non-code relief valve—capacity isn't marked, valve's in wrong place, scale in feed line isolates valve from boiler. Don't use relief valve as a reducing valve

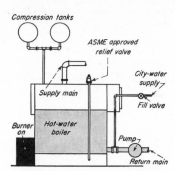

5 Causes of low water: Carelessness in firing dry boiler, drawing hot water; piping, boiler, pump, etc leaks; relief-valve discharge caused by overfiring

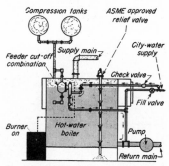

9 Low-water cutoff's useless if boiler's hand fired, fuel feed sticks open, burner's put in 'on' position, zone controls isolate boiler from hot-fuel bed

for hot-water heating boilers

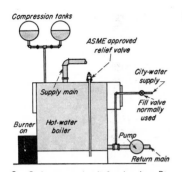

2 Code-approved relief valve has Btu rated capacity, also try-lever for testing. Valve is atop boiler, rated capacity is matched to gross output of boiler

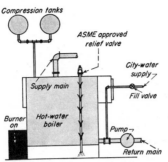

3 Relief valve opens from city-water pressure, hydrostatic test, water-logged or small compression tanks, high-static head or from high pump pressure

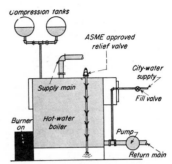

4 Causes of overfiring. Limit control doesn't stop burner, fuel valve fails, burner on manual, residual heat from coal firing, burner is greatly oversized

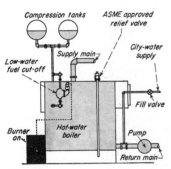

6 Low-water cutoff. Locate control above lowest permissible water level as there's no normal water line in hot-water boiler. Sketch shows safe hookup

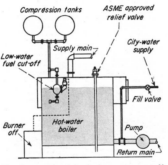

7 Remember: attach low-water cutoff that cuts off fuel before water level hits danger point, or attach feeding device with cutoff to steam-water boilers

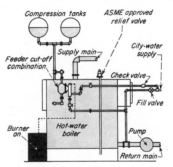

8 Here, water's fed mechanically as fast as relief valve discharges it, burner stops electrically for low water. Setup is best for steam-water boilers

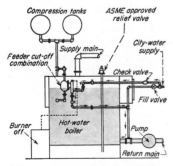

10 Safe hot-water boiler has ASME relief valve to prevent over-pressure, low water cutoff or preferably feeder cutoff combination for maximum safety

Right way... and common piping errors

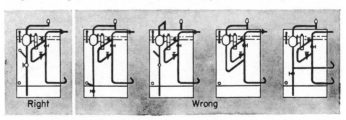

Right   Wrong

FIRST SKETCH is correct, all others wrong. Check your water feeders and low-water cutouts for these common mistakes. Piping errors shown cause everything from high water in gage glass to low water—regardless of where water is in your boiler. Unless corrected, your boiler will eventually fail. Don't be surprised to find one of these errors in your piping. Install an ASME approved safety valve.

Install Gage Glasses

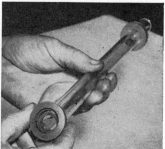

1 Glass thins mostly above water line; also breaks with fittings out of line

2 First shut off top and bottom cocks with chains when the glass breaks

3 Then open drain valve on column that is used for blowing down the glass

4 Remove broken glass, then open cocks from floor to blow out broken pieces

5 Cut glass to right length by revolving between cutters until it breaks

6 Place top washer, and then nuts (not shown); bottom washer; then install

7 Add graphite to washer so nut does not twist glass; hand tighten only

8 Heat slowly by blowing steam at top. Open cocks from floor; close drain

Round Glass

LIKE OTHER CERAMICS, glass is brittle, does not deform before failure. It fractures only from tensile stresses, never from shear or compression. Stress-strain curve is straight line to the breaking point.

Glass has high degree of "notch-sensitivity." Tiny surface scratches, bruises or imperfections cause failure. So be careful when handling gage glasses. Keep them in wrapper until ready to install. If you must put them down, place on soft, clean surfaces.

Glass can be heat-treated to raise its endurance limit. This is known as "tempering." First, glass temperature is raised high, then it's chilled. That sets up residual stresses that are tensile in interior and compressive at surface. And it's at surface where tension usually starts. Tempering also tends to reduce harmful effects of surface bruises and scratches. But tubular gage glasses are not as "adaptable" to tempering as flat glasses.

Corrosion is tough on gage glasses—caused by alkalinity and temperature. Most corrosion (thinning) on tubular glasses is above water line at upper half. Here steam condenses and continually flows down glass, eroding it. Boiler water with pH of 11.5, for example, attacks glass 30 times faster than water with pH of 8.5. That shows importance of low pH if you have glass-thinning problem. Rate of attack from temperature increases 100 times from 265 to 500 F. Solution then is to use flat glass, protected by mica so water does not contact glass.

Of course, water in gage is cooler than water in boiler. Some condensate, formed in boiler connection above gage, flows into gage and lowers alkalinity. This condensate dissolves some silica in glass because of decomposition.

Silica content increases as condensate flows down over glass. Rate of corrosion below water level depends on silica remaining in gage water. Rate of water attack above water level is less as distance increases from top end of glass to bottom. Boiler water, not treated for silica removal, corrodes glass.

For over 300 psi, it's best to use flat glass, protected with mica. Pressure limits for tubular glass decrease with glass length. Buckling stress is the result of eccentric column loading.

Right for Long Service

1 Mica, protecting glass from steam and water, turns opaque before failure

2 Remove clamps, glass gaskets, mica. Graphite the threads, run them down

3 Remove old gasket but don't scrape low spots in joint surface of gage

Polish surface perfectly smooth or the tiny high spots will cause failure

5 Clean both ends of gage so scrapings won't plug cocks when back on drum

6 Clean surface thoroughly, oil with molybdenum disulfide so glass slips

7 Place new gasket, mica, new clean glass, another gasket; place clamps

8 Glass mustn't touch metal. Use only molybdenum disulfide on high temp

Flat Glass

ON HIGH-PRESSURE boilers above 1000 psi, there have been many costly gage-glass failures running into thousands of dollars. But fault has been from gage design, as well as from method of installing glass, rough handling and glass imperfections.

At Corning Glass Works, I saw a nail driven into wood with a flat glass. This is possible because high residual stresses push out against the outside high compressive-stress, which was set by tempering. Here are failure causes, ways to prevent:

(1) Never reuse gage glasses. (2) Use special glasses only for high-temperature service. (3) Under extreme pressure, gage metal tends to bulge out, so gage bends if designed poorly—starts a crack because glass won't bend. (4) Scraping gasket off metal raises burr —causing high unit loading. Always polish gage surfaces perfectly smooth. (5) Numerous scrapings to remove gaskets create high and low spots. Level old surfaces. (6) Pulling up unevenly on bolts loads glass unevenly. (7) Asbestos is "lumpy" in gasket, causing high spots. Use only best gasket material.

(8) Many failures occur on high pressures when starting up because glass was not heated gradually. Don't open gage cocks until new glass has chance to heat by conduction. Then crack top-cock only. Also, cool down slowly. (9) Don't blow down oftener than needed. (10) Use two valves for blowing down. Keep one throttled for slow flow through glass, blow with other valve only.

(11) Lubricate contact surfaces of glass with molybdenum disulfide only. This does not crystallize, allows travel between glass and fitting. (12) Don't use cover plate if it has permanent deflection, if it's of poor design, or if clamping-device design is poor. (13) Mica discoloration means that mica has failed and water contacts glass. Don't wait for glass to fail—change it.

High pressures and temperatures in newer boilers (around 2000 psi) bring new problems. Flat glasses are strengthened by tempering, but residual stresses or strain set up may release at high temperatures. Over long periods, glass loses much strength that was added by tempering. So when ordering glass, consider pressure and temperature.

To POP...
or not
to POP?

Once a day? Once a month?
During the full moon?
Just on ground-hog day?
Here is the straight dope

▶ *Power* magazine printed a reader's question asking HOW and HOW OFTEN he should test his boiler safety valves.

We got a flock of answers. All agreed that safety valves must be tested. All agreed pretty much on the HOW. Big difference of opinion was the HOW OFTEN.

We think the matter is important. Since it turned out to be so controversial, we did some research. We talked with engineers: consultants, and chief engineers in large plants and small. With valve manufacturers and with boiler inspectors. We got a cross section of opinion. Here's what they say:

Boiler inspector. Years ago, when weighted-lever safety valves were used, firemen moved weights, added weights, and even wedged timbers between the levers and the boiler-room roof. Held the valves shut. Easier to keep an even head of steam that way. But selfish chief engineers, thinking more of preserving their boilers than the feelings of the firemen, ordered daily testing of safety valves. This became established practice, more or less. With the advent of springloaded valves, daily testing continued from habit rather than from reason.

ASME Code recommends all boilers operating above 15 psi receive at least one annual internal and external inspection. Since it isn't practical to pop safety valves on high-pressure boilers, as in large utility and industrial plants, annual inspection should include removal of safety valves for complete overhaul and inspection.

Below 50 psi, pop valves once a month if plant is in poor condition. If in good shape, every six months is OK up to 600 psi. If you use the hand testing lever, have steam pressure at least 75% of safety-valve setting.

Chief engineer. Safety valves should be dismantled annually for cleaning and inspection. If plant operation leaves much dirt on valves, 6-month intervals are better. If plant is clean and valves in good condition, popping is not necessary oftener than every two months or so, up to about 650 psi. At lower pressures, increase the time between tests in proportion, but inspect often for clogged drains and dirt build-up in bodies.

Testing is a necessary evil, but should not be done oftener than absolutely necessary. Always a chance a piece of scale will be caught between disk and seat when the valve slams shut. Play it smart and have plenty of steam, at least 75% of set pressure, if you lift valves by hand. Best bet is to run pressure up to popping point. You know if the valves are working right—and you clear the seating surfaces. Another good angle: valves should be popped before a boiler outage. This checks need for repair or adjustment.

Valve manufacturer. Popping safety valves every day may keep the log right on the ball, but it's unnecessary —and costly. If valves are clean, and this depends on good housekeeping, including feedwater treatment, don't induce valve leakage by frequent testing.

Tendency to leak involves two factors: (1) possibility of dirt or scale being trapped when valve closes (2) more important, the differential between setting and operating pressures. The nearer the operating pressure to setting pressure, the greater the possibility of potential leakage.

Setting up a testing schedule to cover all types of plants wouldn't work any better than a schedule for rainfall. Tailor testing to suit your own conditions. If you're in a chemical plant, chances are your valves take a beating from "gunk" in the atmosphere, and should be tested every two months or so. If you're in a spick-and-span plant, testing every six months should be enough. Best guide to a test schedule is experience. Try a set interval. Then a longer interval. So long as popping pressure remains constant, and valves show no inclination to stick, you're doing all right. Somewhere along the line you can establish a pattern. Since safety is most important, satisfy yourself that selected frequency meets this need.

Consulting engineer. Your testing schedule is dictated by your individual operating conditions. ASME *Suggested Rules for Care of Power Boilers* means just this by saying, "Test the valves in accordance with the instructions given for the particular plant."

It's no secret that frequent testing is an invitation to leakage or damage. Particularly if the valve setting is little higher than operating pressure. During normal operation, loading of the valve disk, being proportional to differential in operating pressure, is relatively small. The less valves open, the less chance of difficulty.

Safety valves, particularly those in dirty atmospheres, should be tested at intervals founded upon experience. In clean plants, at pressures up to 650 psi, testing at 6-month intervals is practical. Once a year, valves should be removed for inspection and testing. Plants in the high-pressure range, using complex controls, can't permit haphazard opening of safety valves. Such plants have rigid outage and inspection schedules. Prior to an outage, pressure is usually run up to popping point. During outage, valves are removed, inspected and overhauled.

In speaking of valve damage, we usually think of wiredrawing, or entrapment of foreign matter. However, we can't ignore the effect of thrust. Where discharge is perpendicular to stem axis, developed thrust can be high, depending on design and capacity. Mountings, particularly the flange bolts, are subjected to severe loading.

Temperature cycling, plus shear stress could add up to bolt failure. That's another reason for less popping—and it points up the importance of inspection.

Testing tells we have protection against overpressure. But the act of testing is a form of valve maintenance and should be thought of as such. Proper function within design limits is worthy of more attention.

If slight leakage occurs, tendency too often is to increase spring compression just enough to stop leakage. Fact that this increases the setting is dismissed as negligible. Cause of leakage isn't eliminated. Though increase doesn't conflict with designed working pressure of boiler, or exceed spring-adjustment range, the end does not justify the means. Smart upkeep calls for valve overhaul, not fiddling with the spring.

More common disregard for designed function: continued use of higher-pressure springs following a permanent reduction of boiler operating pressure. ASME Code permits spring-adjustment range of 10% up to 250 psi; 5% on higher pressures. Example: reduction from 160 to 140 psi requires a change to a 140-psi spring, not merely reducing compression of the 160-psi spring to a 140-psi setting.

Popping safety valves

Here's the story . . . in a nutshell

- **Frequent popping is harmful, costly**

- **Weekly testing rarely necessary; daily—never**

- **Dirty or clean plant, high or low pressure determine how often**

- **With hand lever, have steam pressure not less than 75% of valve setting**

- **Best way: Run pressure up to popping point**

- **Remove, test and inspect valves yearly**

Have you examined your boiler lately? Regular inspection is best way to prove effectiveness of feedwater treatment. Check your water program with help of these . . .

22 Q & A's on today's feedwater treatment for packaged boilers

1 How do packaged boilers differ from designs?

A Most packaged boilers are merely modifications of conventional designs. But, since they are completely fabricated before shipment, installation is quick and easy. Large amounts of steam are produced with equipment of comparatively small physical size. This calls for relatively high heat-transfer rates and high steam-generating area per unit of packaged-boiler volume. Water holding capacity is much smaller than conventional boilers.

2 Do packaged boilers need high-quality feedwater?

A Packaged boilers don't require higher quality feedwater than conventional boilers run at equivalent ratings. But clean boiler surfaces are more critical when operating at high evaporation rates. Scale builds up at a more rapid pace. Smaller steam space promotes priming and foaming. Therefore treatment must be carefully selected and controlled to produce high-quality feedwater. Don't compare packaged boilers to older steam-generators run at much lower ratings. These units sometimes do get by with minimum attention to feedwater treatment.

3 What is the first step in any water-treatment problem?

A Analyze the water. Chemically pure water is rare. Few water supplies are suitable for domestic or industrial use without treatment. And chemical composition of different water supplies vary greatly. So it is impractical to prescribe any one ideal treatment. The water must be completely analyzed before treatment can be sensibly chosen.

4 What are the basic aims of feedwater treatment?

A Basic aims are continuity of boiler service, maximum boiler and thermal efficiency and minimum maintenance charges. Chemical composition of makeup, boiler water and condensate returns must be carefully controlled.

5 How many treatment methods are available?

A There are many methods available. All come under three broad classifications—mechanical, heat or chemical treatment. Mechanical treatment includes filtration and boiler blowdown. Heat treatment includes makeup distillation and steam purification. Distillation is limited to boilers with small amounts of makeup. Steam purifiers are used where the process demands very dry steam. Chemical treatment, both internal and external, is most widely used. External treatment adjusts raw-water analysis before entering the boiler. Internal treatment adjusts boiler-water analysis by feeding chemicals directly to the boiler.

6 What is major treatment problem with heating boilers?

A Corrosion or pitting is the main problem. Scale is not a problem because the same feedwater is used continuously. Initial treatment is usually good for the entire year.

7 Why does boiler feedwater with high percent of makeup magnify the treatment problem?

A Often live steam cannot be returned to the boiler as condensate. Makeup water is added in its place. When steam is formed, any solids present in the feedwater remain behind in the boiler. Every gallon of makeup water gradually increases the concentration of the boiler water.

8 Is a small amount of scale formation harmful?

A Scale causes localized overheating that can seriously damage your boiler. It definitely impairs heat transfer. Here's what scale does to heating efficiency.

Thickness of scale inches	Percent loss of heat		
	Soft carbonate	Hard carbonate	Hard sulfate
1/50	3.5	5.2	3.0
1/32	7.0	8.3	6.0
1/25	8.0	9.9	9.0
1/20	10.0	11.2	11.0
1/16	12.5	12.6	12.6
1/11	15.0	14.3	14.3
1/9	—	16.0	16.0

9 How is scale prevented?

A Scale-forming ingredients are calcium and magnesium which form carbonate and sulfate scale when heated in the boiler. It can be prevented by external zeolite softening of makeup plus supplementary phosphate treatment in the boiler. Internal treatment can carry the full load if external softener is too costly. Phosphate precipitates calcium and magnesium phosphate, a nonadherent sludge, removed by blowing down the boiler. Soda ash, sodium aluminate, sodium silicate and various organic chemicals are also used to prevent scale.

10 Why are dissolved gases important?

A Dissolved oxygen is the chief cause of piping and boiler corrosion. Carbon dioxide, in combination with oxygen, greatly increases rate of corrosion. Oxygen and carbon dioxide carried over with the steam are dissolved in the condensate to cause return-line corrosion.

11 What is the difference between fireside and waterside corrosion?

A Fireside corrosion usually occurs in summer when the boiler is not in use. Look for concentrated pitting along bottom of firetubes or a general thinning of metal at the entrance of direct tubes. Damage is caused by sulfuric acid, formed by reaction of moisture and sulfur on carbon or soot-covered surfaces.

Waterside corrosion oxidizes the boiler metal forming pit marks or holes in the tubes. Examine for small mounds of powdered black iron oxide along the top and sides of tubes. Concentrated pitting takes place there.

12 How do you combat corrosion?

A Prevent fireside corrosion by keeping fire surfaces clean and dry during any boiler shutdown. Waterside corrosion is generally treated chemically. Sodium sulfite is very effective. It combines readily with oxygen to form sodium sulfate. Hydrazine and organic chemicals are also used. Filming amines and ammonia combat return-line corrosion.

Oxygen and carbon dioxide can also be removed by deaeration. But most small installations cannot justify the cost of equipment. A small packaged deaerator for this service is not yet available. Larger installations, or those where percent makeup is high, can often justify deaeration.

13 What is priming and foaming?

A Priming is the discharge of water slugs with steam caused by violent bursting of steam bubbles. Causes include high water level, boiler overload, highly concentrated boiler water and dirt or oil in water.

Foaming takes place on the surface of the boiler water. Causes include finely divided suspended matter and high boiler-water concentrations.

14 Foaming and priming—How is it prevented?

A Assuming that boiler is properly designed, check for possible overload. Next see if steam separator is working. Then analyze boiler salines for alkalinity and total dissolved solids. Use blowdown if solids are too high. If alkalinity is too high, reduced feed of caustic may be the answer. Antifoam chemicals will help prevent foaming.

15 Is embrittlement a problem?

A Embrittlement or intercrystalline cracking of boiler metal causes damage that's expensive to repair. It may mean replacing whole sections of the boiler. Many feel that excess boiler-water caustic is the cause. To prevent this, maintain recommended ratios of sodium sulfate to alkalinity in the boiler. Compounds containing nitrate, lignin or tannin are often helpful.

16 What are boiler compounds?

A Compounds are chemicals used for internal treatment. Organic and/or inorganic materials are compounded into liquid, paste, solid or powder form. Common ingredients include: bark and wood extracts, starch, agar, flaxseed, gelatine, sodium silicate, sodium hydroxide, sodium phosphate,

soda ash and sodium chromate. Plant operators find them a real asset in feedwater treatment. Standard compounds are available in easy to handle form for almost every feedwater problem. Operators' job then boils down to feeding recommended amounts according to routine control tests.

17 How are compounds fed to the boiler?

A Method used depends on feedwater cycle, amount of makeup, and boiler pressure. Cost of feeding equipment is also a factor. Chemical solution is often pumped into the boiler drum distribution line. When makeup is low, pump can be operated manually at regular intervals or automatically on a timed cycle. Where the makeup is high and control more critical, use a pump arranged to automatically feed chemicals in proportion to boiler makeup flow.

Pressure solution feeders or pot type feeds are often used where accuracy of feed is not critical. Chemical can also be added in batches to heater or hot well if care is taken to avoid deposits in feed lines to the boiler.

18 What external feedwater treatment is commonly used?

A External treatment usually has a place where the percent makeup is high. Some installations use a zeolite softener supplemented by internal treatment. Other boilers require more elaborate treatment. Newest method is salt regenerated anion dealkalizer following zeolite softener. Alkalinity is reduced by exchanging bicarbonates and sulfates for chlorides. Effluent has about 15 to 30 ppm alkalinity.

If water supply is turbid and contains bacteria, a packaged type coagulator is the answer. Follow this unit with filters and zeolite softeners.

Demineralizing process may be called for if the boiler pressure is unusually high. High makeup may also call for its use. But this equipment is expensive compared to other treatments and regeneration requires acid and caustic handling. So, although its use for the packaged-boiler field is limited, keep it in mind when you *do* need high-quality solids-free water.

19 What maximum total dissolved solids is allowed in the concentrated boiler water?

A Most packaged boilers allow from 1000 to 3500 ppm. Actual figure for a given job depends on the type of boiler, operating pressure, percent makeup and makeup analysis.

20 How do you estimate amount of blowdown?

A Blowdown is about equal to total-dissolved feedwater solids divided by total-dissolved boiler solids, usually expressed as a percentage of the total feedwater flow.

21 Blowdown—How often?

A Blowdown low-pressure heating boilers very intermittently to remove sediment from bottom of the boiler and lower boiler solids during startup. Since there is no makeup, the blowdown valve should seldom be used.

Regular blowdown schedule is needed for high-pressure boilers and those with high makeup. Frequency depends on rate of boiler operation, amount and analysis of makeup.

22 Who supervises the feedwater treatment program?

A Boiler-water treatment involves more than just scale prevention. Kind and control of chemical treatment, regulating preheating and pretreating of makeup, regulation and control of boiler blowdown to keep right boiler-water solids are all just as important. Supervision and control of this program calls for a qualified operator or a firm specializing in boiler-chemical service. Prevention of feedwater trouble usually costs far less than the repair of neglected equipment. Final proof of any feedwater program is to observe its effect on the boiler waterside. Make internal boiler inspections a regular part of your routine maintenance program.

Let's hook up your

Here's how to install indirect water heaters

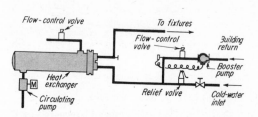

INSTANTANEOUS HOOKUP for indirect water-heater systems uses either pumped boiler water or steam as the heating medium

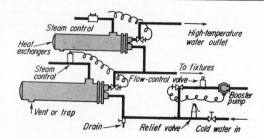

TWO TEMPERATURES of hot water—large volume in 140-F range. small volume in 180 to 200-F range—often are needed in buildings

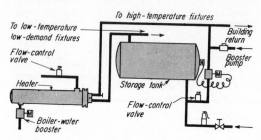

SEMITANKLESS installation provides large volumes of extra hot water that periodically are in excess of boiler and heater capacity

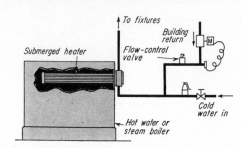

ECONOMICAL instantaneous hookup has a heater in the boiler's water chamber. Coils are installed below water line in boiler

And here's the direct and indirect water-heater picture

Water heaters come in two classes: direct and indirect. Direct-fired units, such as gas heaters and small boilers, heat by direct application of flame. Indirect heaters heat by transfer of heat through copper tubing from hot water or live saturated steam.

Indirect heating has some important advantages:
1. It furnishes hot water the year around from the same boiler that provides for space heating or process work.
2. Initial cost of an indirect heater is usually less than the cost of a direct-fired unit of equal capacity. Too, because of high-efficiency rating of most heating boilers, operating costs are less.
3. Indirect heaters need no separate fuel, electrical or chimney connections.
4. All water heaters must be cleaned periodically to maintain proper heating efficiency. But, because the heating surfaces of indirect units are not subjected to high surface temperatures produced by a direct flame, there is less tendency for deposits to precipitate out of the water being heated. For this reason indirect water heaters don't have to be cleaned as often as direct-fired units.
5. Indirect water heating is best suited to large build-

ings such as apartments, hotels, industrial buildings and institutions of all types. Usually the hot-water demand in these buildings is heavy. Therefore, the most practical and economical heat source is the main space-heating boiler for the building. Most space-heating boilers used in large residential and commercial buildings have enough heating capacity to provide the entire hot-faucet water needs on an instantaneous basis.

Direct versus storage operation

Indirect water heaters are available for use with or without storage tanks. Storage-tank installations require less boiler capacity, therefore have smaller indirect heater than tankless- or instantaneous-heater installations. But **there are many places where direct heaters can be used** more effectively and economically than a storage type unit.

Lower cost. Biggest savings is because tankless system doesn't need a hot-water storage tank. This can save hundreds, or thousands of dollars initially. Too, because of small size of tankless job, amount of standby-heat loss is reduced, cutting operating costs. When built-

indirect water heaters right

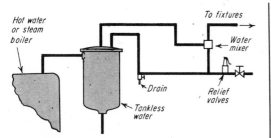

SMALL VOLUMES of hot water are gotten instantaneously with this tankless hookup. It is usually best for very hard water

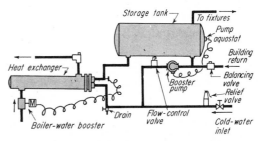

HEAVY RATES of draw hookup are in excess of the boiler's capacity. On draw, cold water enters heater instead of going to the tank

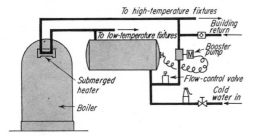

ANOTHER economical hookup gives fast temperature pickup and circulated building returns, allows you to floor mount tanks

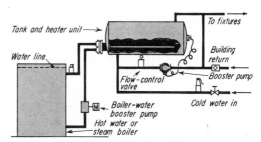

MORE HEATING and operating economy is obtained with booster pump hookup, but tank and heater installation isn't instantaneous

---

in or submerged direct heaters are used, costs are slashed to an even greater extent.

Efficiency. Instantaneous heaters can be installed so the water is heated only as it's needed. They can also be installed in conjunction with smaller storage tanks on a semitankless hookup. Instead of heating the tank by gravity circulation, a booster pump is used between the heater and the tank. Booster pump helps to reduce size of heater, tank and piping. By increasing heating efficiency it reduces pickup time. Booster pump increases heating efficiency and thereby reduces costs.

Versatility. Commercial and industrial processing applications require higher water temperatures than are normally needed for general use. But with a storage system the higher water temperatures often make it necessary to use special tank linings. Direct heaters can be installed so hot water at different temperatures will be supplied as needed without the need of storage space.

Less maintenance. Instantaneous type heaters are a blessing to both the building engineer and the owner. For the building engineer, periodic back-flushing by means of a few valves is all that's needed. There's no

lengthy shutdown period, no messy cleaning and repairing jobs and no waiting for reheating.

But, despite all the apparent advantages of direct heaters over the storage tank type, there are two very important limitations to the instantaneous method: (1) Use them only when adequate boiler-heating capacity is available and (2) don't use instantaneous water heaters in extremely hard-water areas except where water filtration or softening is used.

Where do you use which heater?

When adequate boiler capacity is available and unless water hardness is a restricting factor, the indirect heater (in either internal or external models on an instantaneous hookup) is usually a practical and economical method of heating. Even more effective for many buildings is the semi-tankless installation. This uses both an instantaneous heater and a small storage tank interconnected by an all-bronze booster pump.

For more information on this subject, request 8-page bulletin *Instantaneous Water Heaters* from the Bell & Gossett Company, Morton Grove, Illinois.

Hot-Water Service Tips

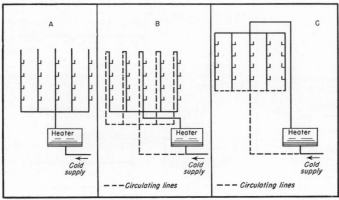

1 **A** is wasteful because upper-floor guests run off stagnant water to get it hot. **B** keeps hot water at all faucets, but needs too much expensive piping. **C** is best—gives good service with moderate piping cost

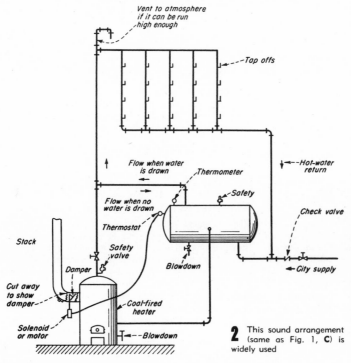

2 This sound arrangement (same as Fig. 1, **C**) is widely used

▶ THE DESIGN OF A HOT-WATER SYSTEM varies with the size and height of a hotel or apartment house. If the incoming cold-water pressure from the city main will push water to the top floor and operate flushometers, you won't need a pump.

Fig. 1 shows the three most common systems. System *A* is wasteful, but I have some fairly large examples of it. The upper floors always have to run the water off until it becomes hot.

Suppose (in winter) you heat water from 45 F to 125 F and the upper floors draw off water at 80 F, or thereabouts. If that water could be returned to the heater, instead of wasted, you could save half the heat—also a lot of water.

In *B* small circulating lines take off the tops of the risers. This is better than *A* but takes a lot of expensive piping. You will need larger pipe risers to get enough pressure and volume for the top floor.

System *C* is best of all, in my opinion. It uses less piping and smaller piping, because the risers form part of the circulating system.

Direct-Heater Used. Fig. 2 shows a complete layout of system *C* with heater and storage tank. Here city pressure is enough to reach the top floor. If city pressure is much higher than needed, install a reducing valve on the line to the storage tank.

This system is used quite a lot. Being separate from the steam heating system is an advantage. One disadvantage is that the storage tanks are small and have no manholes for cleaning. Another is that often no temperature control is provided. Usually the damper is set by hand, and not closely watched because the man in charge does other work around the building.

The control I suggest in the sketch is an aquastat in the storage tank running a motor- or solenoid-controlled damper. This will make the system more efficient and hold a steady water temperature.

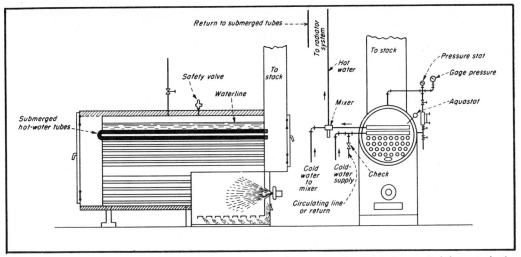

3 A row of service water-heating tubes is submerged above the fire tubes in a steam heating boiler. In winter, when the boiler water is over 212 F, the excessively hot water leaving the heating coil is tempered with cold water before delivery

Combination Heater. Fig. 3 shows an improved layout particularly for medium or large apartment houses. It might also work well in hotels that don't need steam pressure for kitchen use.

Water heating tubes are installed in place of the two top rows of the tubes in an "economic" or locomotive-type steam boiler used for the steam heating system. The sketch shows an oilfield boiler. Of course, the water tubes are submerged.

This system uses no storage tank. After mixing with the cold city water the return circulating water goes back through the lower row of water-heating tubes and returns hot through the upper row.

On the line to the house is inserted an automatically controlled mixer which lets in enough cold city water to produce the desired temperature regardless of the initial temperature.

How it Works. In winter the boiler pressure needed to satisfy the house heating controls the starting and stop-of the oil burner.

If, say, 5 psi pressure is needed for heating, the water in the boiler is at a temperature of about 227 F. The service water will be heated to between 160 F and 180 F. Then the mixer will let in enough cold water to lower it to, say, 140 F.

In summer, when no steam is needed for heating, an aquastat takes the place of the steam pressure controller for the starting and stopping of the oil burner. The aquastat keeps the boiler water at

180 F. This heats the service water in the water tubes to about 140 F to 150 F, depending on the hot-water demand. The cold-water mixer then tones this down and adjusts the amount of cold water to the desired delivery temperature.

This system has the advantage of no storage tank to maintain, but it also has its limitations: In case of trouble with either the boiler or the hot-water tubes, you have no steam or hot water unless you have another boiler standing by. Also you have to shut the boiler down at least once a year for cleaning and inspection.

For Tall Hotels. Fig. 4 shows the complete system now in general use for tall modern hotels. Here again piping system *C* is used.

The cold city water enters the suction tank through a float valve. The pump lifts the water to a roof tank, which is also used for cold-water supply and standpipe. The cold water flows down to mix with the return circulating water as it comes to the storage heater.

Either boiler or central supply steam supplies steam a 1 to 5 psi gage to an air operated steam valve of the bellows type, controlled by a thermostat inserted in the tank. After the steam condenses in the tank coils to heat the water, the condensate is trapped to the boiler—or to a preheater to warm the incoming cold water if central steam is used.

A standard air valve vents air accumulated in the coil. The vacuum breaker

is used to break the vacuum (with accompanying "crackling" in the coils) when the steam valve closes at the right temperature.

The float switch in the roof tank starts and stops the pump. Another float switch indicates high or low water by lighting lamps at the pump control board. The tank has an indicating thermometer. The bulb of a recording thermometer is inserted in the outgoing hot-water line.

Practical Tips. Recent years have seen the insertion of copper linings in the storage heaters. Without such liners you can protect against tank corrosion by draining, scraping and brushing the inside surface and coating it with a good preservative.

I mix lumnite with just enough water to make a thick paste, brush it on, let it harden a bit, then close the tank and fill. I have found very good results after two years of service. Several concerns specialize in tank coatings that will largely eliminate troubles from corrosion.

When tanks are in service and not much water is being drawn it is good practice to blow out the sludge.

A high-temperature alarm is a good protection where air operated temperature controllers are used. If the air fails, your control is lost and the water temperature may shoot up to a dangerous point.

Modern practice uses instantaneous heaters instead of storage heaters. The tanks are smaller and have much more steam coiling.

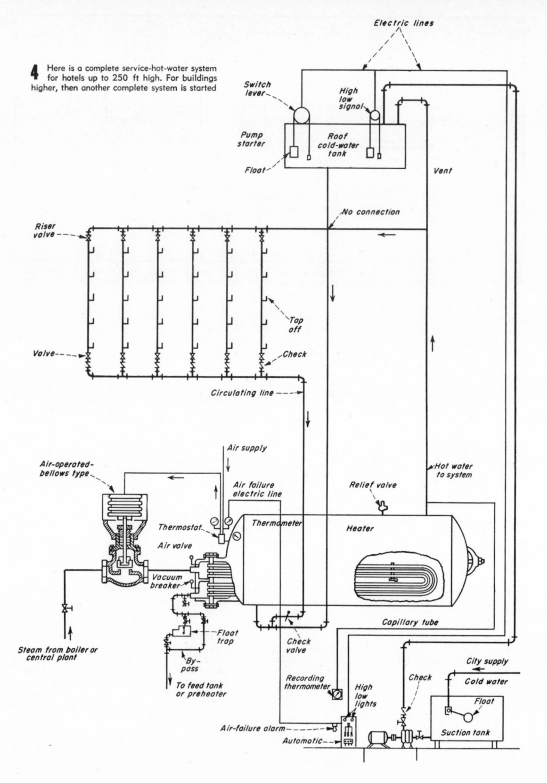

4 Here is a complete service-hot-water system for hotels up to 250 ft high. For buildings higher, then another complete system is started

Electric lines

Switch lever

High low signal

Pump starter

Roof cold-water tank

Float

Vent

No connection

Riser valve

Tap off

Valve

Check

Circulating line

Air supply

Air-operated-bellows type

Air failure electric line

Relief valve

Thermostat

Thermometer

Heater

Air valve

Hot water to system

Vacuum breaker

Capillary tube

Float trap

Steam from boiler or central plant

By-pass

Check valve

To feed tank or preheater

Recording thermometer

High low lights

City supply

Cold water

Check

Float

Air-failure alarm

Automatic

Suction tank

HOW TO CLEAN BOILER GAGE GLASS

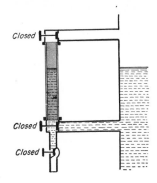

1 Close top and bottom gage valves of working boiler when gage glass gets dirty and you want to clean in a hurry

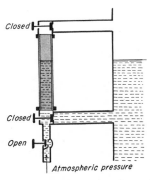

2 Open drain valve on drain line under glass. Water may not drain because atmospheric press. is working against it

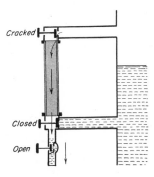

3 Crack top gage valve so steam pressure from boiler blows water out drain line. Idea is to have only steam in glass

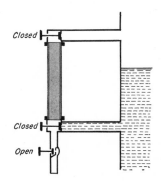

4 Close top gage valve soon as all water is out of glass and drain line. Fill cup with ordinary household ammonia

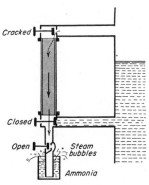

5 Hold cup under drain line and crack top gage-glass valve slightly until steam bubbles up through the ammonia

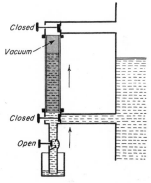

6 Close top gage valve. Steam condenses in glass quickly and ammonia is forced up into glass right up to the top

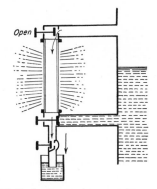

7 If glass is not clean enough, repeat these operations a few times until glass is perfectly clean and sparkling

'TRICK' CLEANING IN PLACE

► Gage glasses on small boilers are often coated with rust or some other opaque matter. And the glass is usually left that way year after year. Finally it gets so bad that telling the right water level is a tough job. Glasses on larger boilers usually get more attention.

There's no reason for dirty glasses because they can be cleaned in a jiffy *without removing glass from column*. How? First fill a cup or tumbler with ordinary household ammonia. With steam on boiler, close both top and bottom gage valves and open drain cock. Then crack upper gage valve until water leaves glass, then close. Immerse drain line in ammonia, then crack upper gage valve until steam bubbles out of ammonia, then close valve.

Steam in glass will condense and create a slight vacuum. That causes ammonia to be forced up into glass by atmospheric pressure. Repeat until glass sparkles. This operation takes only a few minutes—leaves no excuse for dirty glasses.

MAKE MINOR WALL REPAIRS

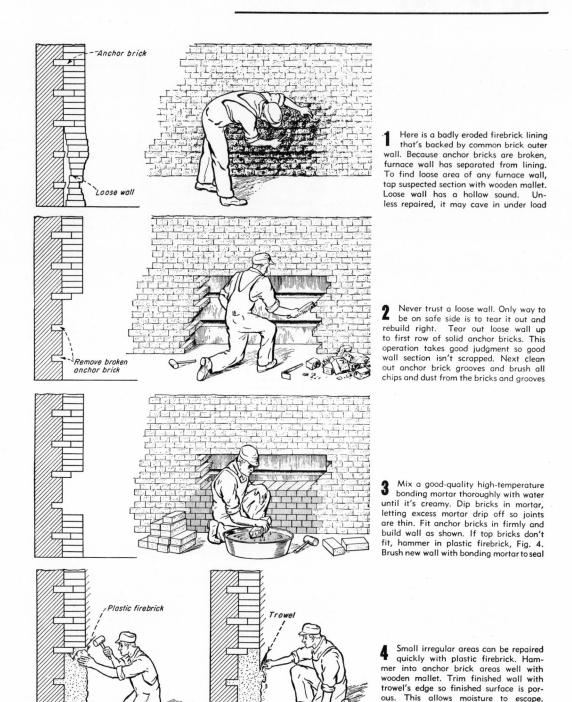

1 Here is a badly eroded firebrick lining that's backed by common brick outer wall. Because anchor bricks are broken, furnace wall has separated from lining. To find loose area of any furnace wall, tap suspected section with wooden mallet. Loose wall has a hollow sound. Unless repaired, it may cave in under load

2 Never trust a loose wall. Only way to be on safe side is to tear it out and rebuild right. Tear out loose wall up to first row of solid anchor bricks. This operation takes good judgment so good wall section isn't scrapped. Next clean out anchor brick grooves and brush all chips and dust from the bricks and grooves

3 Mix a good-quality high-temperature bonding mortar thoroughly with water until it's creamy. Dip bricks in mortar, letting excess mortar drip off so joints are thin. Fit anchor bricks in firmly and build wall as shown. If top bricks don't fit, hammer in plastic firebrick, Fig. 4. Brush new wall with bonding mortar to seal

4 Small irregular areas can be repaired quickly with plastic firebrick. Hammer into anchor brick areas well with wooden mallet. Trim finished wall with trowel's edge so finished surface is porous. This allows moisture to escape. Low- and high-temperature plastic firebrick is wrapped in 100-lb cartons

Courtesy—A P Green Fire Brick Co, Mexico, Mo.

2
Fuel: oil, gas, measuring combustion

Save fuel and your product

 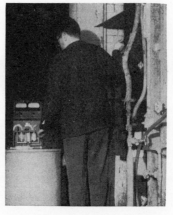 

STEAM BOILERS. Boilers that generate power or process steam usually have CO_2 recorders for controlling combustion. But power engineers know that a high CO_2 reading alone isn't proof of efficient combustion. To get the full story, you must also check for presence of unburned gas. Instrument shown takes readings in last boiler-pass to check accuracy of your CO_2 recorder. It also shows completeness of combustion. Charts and tables supplied with instrument find exact combustion efficiency. These charts and tables also show exact CO_2 and excess air in combustion products. They are easy to use.

STEEL FURNACES. Open hearth furnace, steel industries' largest fuel user, has many operating variables. Furnace deterioration causes air to leak in. Variations in fuels and other combustion components also give more trouble. Waste-gas analysis measures over-all effect of these variables. This instrument gives information for adjusting furnace controls. It analyzes gases, disregards fuel variations and isn't affected by gases from the charge. Continuous indication of oxygen and combustible content allows operator to make correction at all ranges of fuel flow needed for top efficiency.

CERAMIC, REFRACTORY KILNS. Steady flow of high-quality product through kilns is important in making brick, refractory and abrasive products, pottery and porcelain. Kiln's atmosphere has a direct bearing on product's quality. So for more accurate control, nature of atmosphere must be correct in all portions of kiln that work passes through. Here instrument serves as accurate, time-saving combustion tool to check atmospheres in vicinity of work, or in all parts of kiln, for that matter. For this service, instrument helps bring each burner into right combustion alinement. Readings can be taken by furnace operators.

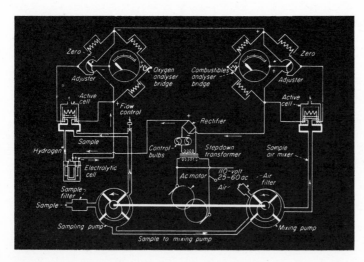

Wheatstone-bridge

▶ GONE ARE THE DAYS of guessing at combustion efficiency by glancing at character and color of flame in furnace, measuring temperature and then looking at smoke coming from stack. Today, we depend on instruments to control combustion components: fuel, air, temperature and pressure. But you must make chemical gas-analysis to check efficiency of control process.

Heat prover shown is portable, weighs 25 lb. It has sampling tip and hose, with thermocouple for measuring temperature. Instrument operates from an ac power source, 110 v, 25-60 cycles. It analyzes gas sample for oxygen and

dollars with better combustion

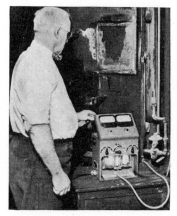

GLASS TANKS. Total fuel cost to operate modern glass tanks is very large. And, of course, economical generation and application of heat calls for efficient combustion practice. Here, again, production rates and product quality benefit from precise control of glass-tank atmosphere. Problems before combustion engineers in glass industry are like those of open-hearth fuel engineer we tell you about in second item here. Correct control settings are found from analysis of tank gases. So here, again, this instrument accurately analyzes combustion products in the presence of contaminating gases of the bath.

INDUSTRIAL FURNACES. Heating and heat-treating furnaces in metals and metals-processing industries need right combustion. Reason: Work is in direct contact with combustion products in many industrial furnaces. So burner controls must be set to produce oxidizing or reducing atmosphere in line with metallurgy of furnace product. These furnaces cover wide range in size, construction. Material is usually put in one side and comes out other, ready to process. Air-fuel ratio is generally regulated with temperature regulator. Quality of oxidizing or reducing atmosphere is important, making instrument invaluable.

CEMENT KILNS. Cement is highly competitive and cost per barrel depends largely on fuel used for process. As in steel and glass production, combustion products are diluted with the gases evolved from the chemical reactions within the kiln. So direct analysis for oxygen is needed to evaluate excess air in waste gases. In general, burners are adjusted to low oxygen and for zero combustible content. Oxygen analysis and check on combustion efficiency guide the kiln operator. It helps him set his controls so he saves fuel, yet operation is efficient, troublefree. Again, charts and tables show exact efficiency.

circuit measures resistance change

combustible-gas content. Percents by volume of these sample gases are shown continuously and simultaneously on instrument meters.

Steady flow of sample is drawn into instrument by rotary pump. After passing through filter, pump delivers sample simultaneously to both oxygen and combustible analyzing sections.

Sample flowing into combustible section is mixed with air and pumped through the active cell of combustible analyzer. Hot platinum-wire catalyst, suspended inside cell, forms one arm of a wheatstone-bridge circuit. Air-sample mixture burns at surface of wire, in-

creasing its temperature and electrical resistance in proportion to volume of combustible gas in sample. This change in electrical resistance is measured by wheatstone bridge. Bridge meter is calibrated to read directly in *Percent Combustibles.*

Sample flowing into oxygen section is mixed with hydrogen generated in electrolytic cell. Hydrogen-sample mixture passes into active cell of oxygen analyzer. Any oxygen in sample burns with hydrogen at surface of cell's platinum wire. Wheatstone-bridge circuit again measures change in electrical resistance of cell wire. Bridge meter here is cali-

brated to read in *Percent Oxygen.*

Oxygen and combustible meters have both 4% and 20% scale ranges. Dual-range feature affords greater instrument accuracy and sensitivity. Oxygen meter has additional scale for measuring output of instrument thermocouple in degrees F.

Operation is entirely automatic—all parts housed in unit. Scale ranges are selected by operating range switches below each meter. Instruments are checked periodically against calibrated cylinders of laboratory-analyzed gases by firm lending the instrument.
—Courtesy, Cities Service Oil Company

These piping systems for

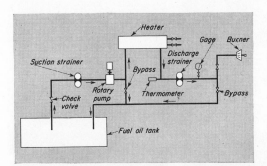

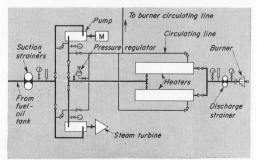

BASIC SYSTEM has only the equipment you need in moving oil from storage tank to burners, to raise oil temperature, pressure. Heart of system is a pump, usually buried below floor surface. This pump handles pressures up to 150 psi at 120 F. Use this hookup only where forced shutdowns won't affect production as there is only one fuel-oil pump in system—no spare

DUPLEX SYSTEM has a spare of each important piece of equipment needed for standby service in emergency. Adjustment diaphragm valve regulates fuel-oil pressure to burners automatically. Here pump handles pressures up to 350 psi at 120 to 130 F. One pump is driven by motor, the other by steam turbine to prevent shutdown in case of electric power failure

Rotary pump is ideal for fuel

► CHANCES ARE that you burn No. 6 (bunker C) fuel oil in your plant. Because this oil is slow flowing at normal temperatures, your piping must be right. Then the oil, to burn, has to be delivered to the burner at a high enough pressure for correct atomization.

Operating characteristic of the rotary pump shown makes this type ideal for fuel-oil service. Reason is it combines the rotary motion of a centrifugal pump with the positive-pressure features of a reciprocating pump. Because of positive displacement, the rotary pump delivers a given quantity of fluid with each revolution. This is important for fuel oil service.

While only the rotary screw-type is shown here, there are six basic types of rotary pumps. This screw-type permits higher rotative speeds because of lower fluid speeds of its axial-flow characteristic. Six basic rotary types are: screw, gear, vane, cam and piston, shuttle block, multiple piston.

In your fuel-oil system you also need strainers, stop checks, relief valves, control valves and the other devices shown. They are needed to make your system self-operative and reliable.

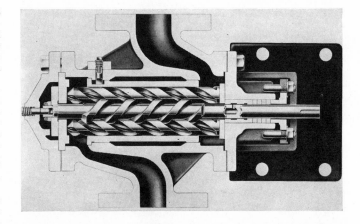

Guard against air leaks in the system. These are from improper pipe joints in suction lines, valve packing or strainers. When running pump under negative pressure, air leaks are especially troublesome.

If you have a suction lift, place this pump as close as possible to the fuel-supply tank. That prevents vapor lock from excessive suction lift. Air leaks are critical in these systems and cause cavitation. Piping systems shown for various types of installations will help you get started on the right track.
—*Courtesy, De Laval Steam Turbine Company, Trenton, N. J.*

fuel-oil service will save you $$$

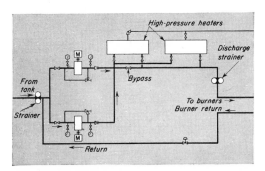

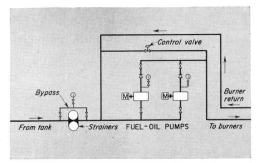

H-P SYSTEM with recirculation keeps constant pressure at burners by actuated control valve. Higher pressures are for wide-range burners. Pumps handle up to 1150 psi at 220 F. Inlet-oil temperature to pump varies with recirculation, depends on quantity of oil burned. Temperature of oil goes up as more fuel oil circulates—goes down as oil recirculation goes down

AUXILIARY START-UP SYSTEM is for No. 2 fuel oil. Because No. 2 grade oil has a low flash point, heaters are not needed in this system. Hookup is only for initial start-ups or for good-insurance start-ups after forced shutdowns. You can also use a hookup like this for an oil-burning start-up system in your pulverized-coal-burning installation. One pump may be enough

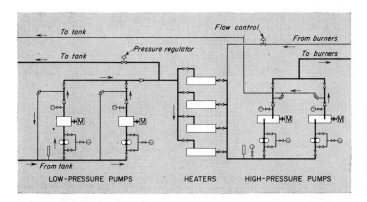

MODIFIED H-P SYSTEM for wide-range service uses l-p pumps on suction side of heaters, h-p pumps on the discharge side. Then you can use all l-p heaters. L-p pumps are larger so h-p pumps won't starve. Thus you have positive pressure on the inlet side of h-p pumps. These h-p pumps send the oil directly to the wide-range burners on the boiler fronts

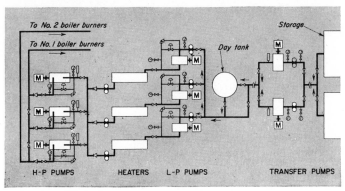

LARGE - CAPACITY SYSTEM has large storage tanks above ground (usually) some distance from burners. Transfer pump is needed to move oil to a small 24-hour-capacity day tank, usually below ground. L-p pumps in the burner service-system handle pressures up to 250 psi at 120 to 130 F. H-p pumps work 900 to 1150 psi at 200 to 220 F

You must know how

Oil must be warm_____

Fuel oils must be stored and pumped through piping while warm. Too little thought is usually given to this important fact, often causing trouble after system is installed. Today, plant operators have a bigger headache since heavy fuel oils are becoming heavier with higher pour points and viscosities. This means your present fuel system won't do its job, although it once might have worked.

Three points must be considered. (1) You must store oil warm enough so it can flow toward your suction line for removing it from tank. (2) Oil must be easily handled in piping system from fuel oil storage tank by fuel pump and in quantities demanded by burner. (3) Oil must be delivered to the burner in a state ready to burn. So you can see that these are three separate steps.

Pumpability. Experience has shown that 5000 ssu is the limit of easy pumpability. Also that a No. 6 oil of 100 ssf at 122 F will reach the 5000 ssu when its temperature is about 83 F. Today, No. 6 oils have a commercial range of between 100 and 200 ssf at 122 F. Often this reaches 300 ssf at 122 F. For heavier grades, storage temp must be close to 110 F instead of 83 F.

Some heavy fuel oils may have a maximum pour point as high as 90 F although standard ASTM pour test may call for a much lower pour point. This is because pour point of residual fuels is not certain; it depends on initial temperature before cooling and time held at this initial temperature. So anything less than 110 F suction temperature might give pour trouble even if chart viscosity temperature characteristics show pumpability at lower temperature. Then multiple-tank installations, tanks exposed or in wet soil and tanks buried below ground in cold climates must have some means of heating oil at suction inlet to 110 F. And heating must be at rate you withdraw oil. Return unburned hot oil to tank bell at suction. But this may not solve problem on cold starts. *Courtesy, Preferred Utilities Mfg Corp, New York, N. Y.*

Heating in the storage

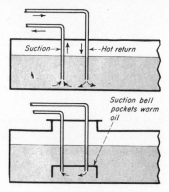

Suction → *Hot return*

Suction bell pockets warm oil

Suction bell

Most Codes require oil to be withdrawn from top of buried or vaulted tanks through suction stub reaching 6 to 8 in. from tank's bottom. Where heated oil is frequently pumped into smaller tanks, you may have no trouble if suction is near return, top sketch. But second sketch shows a better method. Here inverted bell traps the hot return oil that suction and return lines run into. This simple change of installing a bell may solve a serious pumping problem.

Ribbon heat_____

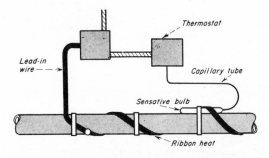

Thermostat

Lead-in wire →

Capillary tube

Sensative bulb

Ribbon heat

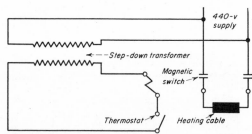

440-v supply

Step-down transformer

Magnetic switch

Thermostat   *Heating cable*

Now, electric line heaters keep your heavy oil hot on startup

Ribbon heat is a low-temperature lineal heater in form of flexible tape ½-in. wide and 1/10-in. thick. Heat is from electricity passing through four special resistance wires. All are insulated and waterproofed.

The end of this special heating tape is attached to waterproof connection of 3-ft non-metallic sheathed cable. Entire assembly is sealed at each end, not harmed by mild acids or alkalies. Use where maximum fuel temperature flowing through line is not above 170 F. Plants have been shut down where hot-storage oil flashed in the pump. All of ribbon-heat units are single phase of 110-220 or 440 volts. And they must be installed across one phase of the polyphase supply. The 440-v units in this line have a single-pole pilot thermostat operating the coil of a two-pole relay.

to heat residual fuel oil

tank is important

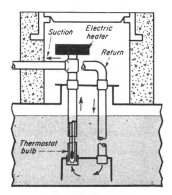

Electric heater

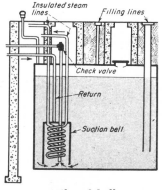

Coil and bell

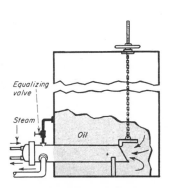

Field-storage tank heater

Suction-stub electric heater shown keeps oil at needed temperature inside suction-stub pipe section during shutdowns. This is a low-watt density, thermostatically-controlled immersion heater. Wattage available in this suction heater is low. It is not for heating oil other than to keep it pumpable during shutdowns. Thermostat is usually set for 100 to 110 F. Place a drain in concrete enclosure around manhole if water collects. This prevents flooding.

Coil heater combined with suction bell is very efficient. Coil in bell heats oil as it is withdrawn by pump. Here very little heat is needed as heated oil does not mix with large quantities in the tank. Besides, hot return from burner is also to bell. Hot water may be used instead of steam in coil, depending on local ordinance and hot water available. If high-pressure steam is used, control with temperature control valve. To play safe, send exhaust to waste.

Large field storage tanks should have a suction- or tank-outlet heater. Open end of heater shell extends into tank. Fuel suction is on closed end, outside of shell. If you have oil return line, connect outer part of shell so hot return oil does not dissipate with stored oil. To repair coils, close large clapper valve so coils can be removed without emptying tank. Equalizing valve fills the heater with oil, helps you open clapper valve when heater is empty.

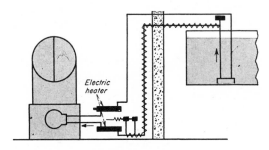

All electric heating

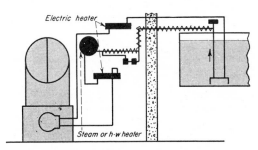

Electricity plus steam or water

Electric suction-stub heater is in vertical suction line in storage tank. Ribbon heat around long suction piping keeps oil in pumpable state. Electric heaters in tank bring oil temperature to fire point needed for burning. Data for selecting proper unit or ribbon heat sizes are based on piping covered with at least 1-in. thick sponge-felted asbestos or its equivalent. Such insulation fits firmly over flat coils.

Suction-type steam or hot-water heater is added here. Ribbon heat is for holding oil temperatures in insulated pipe lines during shutdown or out-of-service periods. Their purpose is to maintain the pipe line at a suitable temperature by making up for radiation and convection or conduction losses from piping. So don't expect ribbon heat alone to maintain oil temperature when flowing through fuel line.

Pipe Your Fuel Tank...

THE OIL SYSTEM

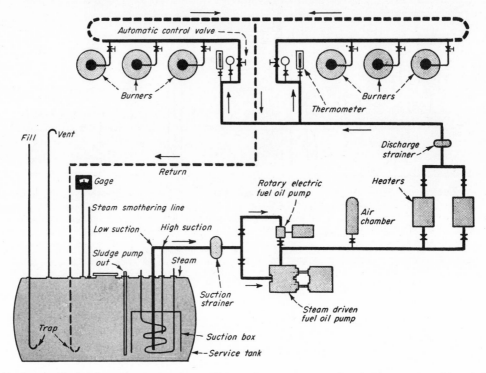

1 Here's a good hookup for industrial system burning heavy oil. Oil is preheated in storage tank for easy pumping. Suction and discharge strainers keep burner tips from plugging. Both steam and electric service pumps safeguard against most plant emergencies. Air chamber prevents joints leaking with burners closed, also dampens pulsations. Oil heaters keep oil at fire point. Return lines purge air into storage tank with unused oil. System is preferable to one with dead ends

ABOVEGROUND

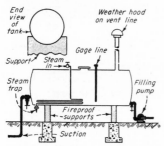

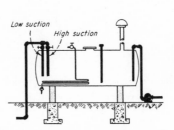

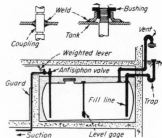

7 Horizontal aboveground tank must have a fireproof support under one third of circumference. Here oil enters, leaves tank near bottom from both ends

8 Safer way to hook up same tank is to run fill line above tank and down inside. High and low suction lines with valve at the top assure against leakage

9 Inclose indoor tank in concrete box and cover with sand. Protect suction line from damage. Connect piping to tank with welded couplings or bushings

...For Best Results

PIPING DETAILS

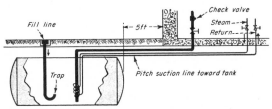

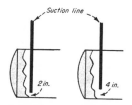

2 Buried-tank capacity is usually less than 30,000 gal. It must be at least 5 ft. from building foundation, should be below all outlets attached to tank. Fill line should be below suction-line level, or it must have a trap so it won't act as vent. Pitch suction line toward tank; install check and shutoff valves near tank

3 Have the suction 2 in. from tank bottom for light oil, 4 in. for heavy oil. That keeps sludge, water, sediment out of the system. Remove impurities yearly.

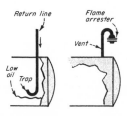

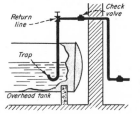

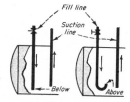

4 Trap on return line prevents vapors escaping when oil is low. Tie vent in top of tank, have flame arrester. Size vent to prevent pressure when filling

5 For aboveground tank, place check valve in return oil line as shown. That prevents siphoning oil out of the tank. Also have trap on end of the line

6 Extend fill line below suction line or place a trap on fill line's end so it cannot vent. Don't cross-connect fill line with any other lines, or to vent

BURIED

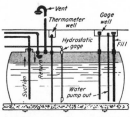

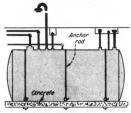

10 Safest oil storage is underground tank. Line for pumping out water and sludge is needed in large tank. Piping hookups shown are based on good practice

11 Heavy underground tanks sink, cause trouble unless buried in good spot. To prevent floating in wet ground, lay on a concrete slab, anchored with rods

Do you know...

when preparing to expand tubes— you should select expander for size and gage of tube? Thickness of seats must be taken into consideration. Expander with correct length rolls should only expand tubes about $\frac{1}{4}$-in. past the outside of the seat. This is an important dimension to remember. Tubes are often improperly rolled because this item has been overlooked. *Caution:* Don't over roll tubes. Stop rolling when scale **starts** to flake off the tube sheet.

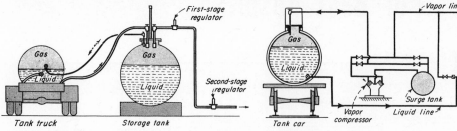

1 **TANK-TRUCK** deliveries often use truck pump or one near storage tanks. Vapor connection between tanks is needed

2 **TANK-CAR** deliveries used for large industrial plants. Vapor-compressor transfer method, widely used, is safe and fast

18 Answers to Your Questions on

Liquefied-Petroleum Fuel Gases

1 What are liquefied-petroleum fuels?

Propane and butane; they may be a single gas or a mixture of several. Lighter and easier to vaporize than gasoline, they are gases at ordinary atmospheric temperature and pressure. But when pressure is increased they liquefy readily. So it's easy to burn, store and transport them. Liquefied-petroleum gas fuels, LPG for short, have many trade names.

2 What's the source of LPG fuels?

Actively-producing oil and natural-gas wells are main source, but some is obtained in oil-refining operations.

3 How do propane and butane differ?

Propane, a little lighter, gasifies at a lower temperature, has lower heat value than butane. Table, facing page, summarizes important properties.

4 Do LPG properties affect use?

Yes. Propane boils at −44 F so it's used in cold climates. Butane boils at about 32 F, is used in warmer regions. Often mixture properties are adjusted for season of year; more propane is put in a mixture in winter than in summer.

5 Is heat needed to gasify fuels?

Yes, latent heat of vaporization. This heat must come from liquid itself or from some outside source. Study each installation to see if a controlled, external heat source is needed.

6 What heat sources are used?

External steam, hot-water, or gas-burning heaters may be used. Vaporizers, Fig. 3, are of various designs.

Suitable for a variety of industrial jobs, these fuels are rapidly finding favor in all parts of the country. Learn what they are, where they're used, and how to apply them in your plant. This summary gives data from a number of sources

Fired units, direct or indirect, used too.

7 What are industrial uses of LPG?

These fuels are used for boiler firing, heat-treating furnaces, incinerators, direct-flame sterilizing, space heating and many other jobs.

8 How is LPG transported?

High-pressure tank cars and trucks carry bulk shipments, also a few tankers. ICC approved cylinders, 20-300 lb, are often used for smaller shipments, like those between bulk plant and industrial user. Tank cars hold 8000 to 12,000 gal, tank trucks 550 to 8500 gal. Pipelines also used to transport gas from producing area to large bulk-storage farms.

9 How is LPG transferred from tank car or truck to storage tank?

Three main methods are: (1) gravity (2) pump and (3) vapor compressor. In each, LPG fuel is received in liquid form and transferred that way.

10 How do transfer methods work?

Gravity system uses difference in height between delivery and storage tanks to produce flow. Pipe connects vapor spaces of tanks so flow is pos-

sible. Method is slow and undesirable. Most tank trucks are fitted with a pump for transferring fuel to storage tank, Fig. 1. Or pump may be installed near storage tank. Equalizing vapor connection between tanks is needed.

Vapor-compressor method is often used. Instead of pumping liquid into storage tank, vapor is pumped into tank car, Fig. 2. This forces liquid from tank car to storage. Equipment arranged to pump vapor in opposite direction, too. After liquid is transferred, vapor in tank car is pumped to storage tank. In extremely cold weather compressed air may have to be supplied delivery tank. Use thoroughly dry air to prevent entrance of moisture into tank.

11 Do LPGs need special valves, fittings in transfer and other pipes?

Propane and butane are "sweet" gases having almost no sulfur content; they are relatively noncorrosive. They have solvent action on rubber and only appropriate valve-seat and gasket material should be used. Pipe compounds safe from solvent action must also be used. Experienced designers are needed for system layout and planning. Submit all designs to proper authorities for approval before starting construction.

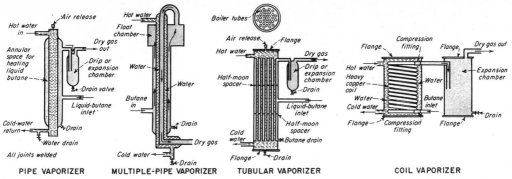

| PIPE VAPORIZER | MULTIPLE-PIPE VAPORIZER | TUBULAR VAPORIZER | COIL VAPORIZER |

3 **VAPORIZERS** turn liquefied petroleum to gas for safe transmission to various parts of plant. Vaporizers must be designed and constructed for safety, so homemade units cannot be used. Consult safety recommendations for details

PROPERTIES OF LIQUEFIED PETROLEUM FUELS

	Propane	Normal Butane
Specific gravity of liquid at 60 F; water = 1.00	0.509	0.584
Vapor pressure, psig		
At 60 F	92	12
At 100 F	172	37
Wt per gal of liquid at 60 F, lb	4.24	4.86
Boiling temp of liquid at atmospheric pressure, F	—44	32
Cu ft of gas (60 F, atmospheric pressure) per lb liquid	8.59	6.51
Cu ft of gas (60 F, atmospheric pressure) per gal liquid	36.5	31.8
Specific gravity of gas; air = 1.00	1.52	2.00
Range of flammability of gas in air, % gas in air-gas mixture	2.4 to 9.5	1.8 to 8.4
Btu per lb gas	21,600	21,300
Btu per cu ft gas at 60 F, atmospheric pressure	2520	3270

12 Is LPG always used undiluted as it evaporates from the liquid?

Gas may be used without dilution at its normal Btu value or diluted with air at lower Btu values. When diluted with air, amount used for dilution is always outside range of flammability, more gas being present than would form an explosive mixture.

13 Is any gas burner suitable?

Since a gas burner supplies a given amount of gas mixed with chosen amounts of air, any burner suitably adjustable for supply of proper air-gas mixture can be used. In addition, burner must be suited to existing furnace conditions. There are a variety of burners designed for LPG and their use is desirable. Adaptation of a burner for one gas to another, or one with a different heating value, should be done only by an experienced burner man. Whether it can be done satisfactorily depends upon burner.

14 How is LPG stored?

Tanks above and below ground are used; also portable tanks. All tanks must be designed for the working pressure to be used. National Fire Protection Assn with cooperation of insurance and other industrial agencies has developed recommended practices for design, construction and installation of storage tanks. Insurance companies have adopted these recommendations. Publications of some of these firms are listed at end of article.

15 What accessories are used on LPG storage tanks in industrial plants?

Liquid-level gages, a pressure gage, shutoff valves, safety-relief valves, excess-flow valves and backpressure check valves.

16 Do LPGs have distinct odors?

No. They are practically odorless unless an artificial odor has been put in the gas. Always check to see that gas you purchase has been artificially odorized. Also check from time to time to see that odor strength hasn't decreased markedly. Most states require warning stench in LPG.

17 What properties of LPGs must be understood for safe handling and use?

Briefly, these: *Gases* are heavier than air; gases won't diffuse rapidly into atmosphere unless air velocity is high; air-gas mixtures readily ignitable in flammable range; mix with inert gases like nitrogen, carbon dioxide, steam or air to bring below flammable limit; using fine water spray reduces probability of ignition; vapor pressure is greater than gasoline at same temperature; closed pressure vessels built to regulations and fitted with required safety devices are only safe storage place; liquids expand in storage tank with increase in temperature so tanks shouldn't be filled with liquid to top.

18 What happens to LPGs when released at normal atmospheric temperature and pressure?

You can expect that: liquids will vaporize readily though atmospheric temperature is below liquid boiling point; liquid may remain in fluid state for relatively long period if volume is sufficient even though exposed to summer heat; liquids discharged from leaks in pipes or tanks must absorb heat from surroundings to gasify, so use recommended dikes or ditches to confine or divert liquid flow; flammable air-gas mixtures exist above surface of released liquids; flammable-vapor indicators give quick check on presence of leaks; rapid removal of vapor from tank reduces liquid temperature and tank pressure, but rapid removal of liquid by normal means doesn't reduce tank pressure; liquid from storage tank may freeze hands on contact; explosive ranges are generally narrower than for other fuel gases.

Have you heard...

of the world famous Murphy's Law?
- If it can fail, it will.
- If it can be hooked up wrong, someone will do it that way.
- If it can be operated incorrectly, someone will run it that way.

Stop Gas Leaks Now...

WRECKAGE AND ITS CAUSES

SERIOUS ACCIDENTS like this mount yearly as gas leaking into basement from the high-pressure main explodes. Cost to industry runs into millions now that gas is being widely used

FURNACE EXPLODES under all types of boilers when gas leaking into gas-fired boiler is not purged. Accumulated gas from leaking burner was ignited by hot shutdown furnace here

PREVENT WITH RIGHT PIPING

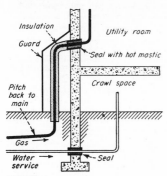

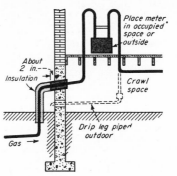

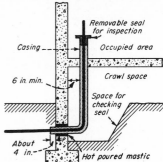

BYPASS gas around crawl space from outside so leaking gas won't follow the piping inside. Insulate against freezing

IF YOU MUST run long piping under crawl space, do as shown. Put meter in occupied space or in vented building

CASING around gas line coming up from crawl should have removable seal for inspection. Make space for checking

SAFEGUARD METER . . .

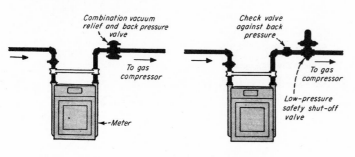

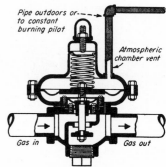

PLACE vacuum relief valve between gas compressor, or booster pump, and meter if meter is for less than 5 psig, *left*. Check-valve guards against backpressure where air or oxygen exceeds gas pressure or where the gas from the h-p system may back up

PIPE atmospheric chamber of pressure regulator to outdoors or to a constant burning pilot. *Never leave vent unpiped*

...From Wrecking Your Plant

SMALL ARC from switch wrecked this compressor house when operator flipped on light switch. Slow-leaking gas valves let in enough gas to damage many plants like this one yearly

CORRODED PIPE (above) at ground level is main hazard. Gas leaks into unused part of building. Cracked main (right) under building lets in gas, often is ignited by a pilot light

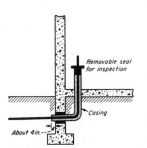

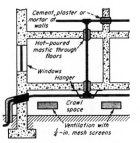

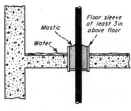

BURIED piping coming up through floor must be encased, sealed with removable seal at one end. Other end is as shown

VENTILATE crawl spaces or unused spaces if piped for gas. Seal walls, floors around piping to localize escaped gas

WET FLOORS corrode piping. Use floor sleeve high enough above floor to keep water from line. Paint all gas piping

... AND FURNACE

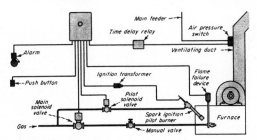

GAS-BURNING furnace needs this hookup for purging. Air-pressure switch in ventilating duct prevents ignition of accumulated gas before furnace is properly purged before relighting. Flame failure device shuts off gas if flame fails unexpectedly

Safety-first signs

Every plant should have several dozen "Men working" signs. Make them of fire-resistant fiberboard or any of the fire-resistant plastics. Ours are circular, 6- and 12-in. diameters. They have holes for wire clips or hooks to make hanging easy.

Paint both sides red, with white letters. We use the small ones for all valves, large ones for all openings. Always keep them clean and use on all connecting valves, manholes, drains etc. Make sure your men get into the habit of using them.

Here's How To Run An Orsat

▶ This is no place for a lecture on the uses of a flue-gas analysis, except to remind the reader that high carbon-dioxide (CO_2) means low excess air (hence low direct heat loss up the stack); that carbon monoxide (CO) reveals incomplete combustion, another loss; that the oxygen (O_2) measurement is often used chiefly as a check on accuracy of the CO_2 analysis.

Orsats have been used for years to determine, from a sample of flue gas, the percentages by volume of CO_2, O_2 and CO. The equipment was loaned by the Hays Corp;

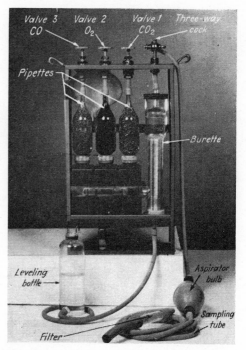

1 Front view of orsat. Gas is measured in water-jacketed burette. Gas constituents are absorbed in three pipettes in the order shown. Before starting, be sure liquid in pipettes is at mark.

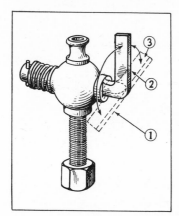

1A The 3-way cock at 1 connects header and burette to gas-sampling tube; at 2 connects header and burette to atmosphere; at 3 shuts them off from air and gas.

2 Remove breather tubes from solid plugs. That for CO_2 is left open to air. Connect other two to the bag, after bringing chemical to marks on necks of pipettes.

3 With leveling bottle hung from bottom of case and nearly full of tap water, and the cock at 1, pump bulb to force gas into burette and through bottle.

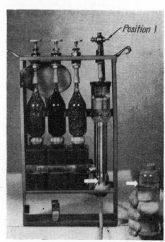

4 Raise bottle slowly until water shows well below zero mark near bottom of burette. Then hold bottle steady until both water levels are equal.

Step-by-step instructions for measuring CO_2, CO and oxygen in boiler flue gas—a sure way to improve combustion and save fuel dollars

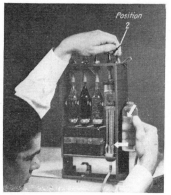

5 Clamp rubber tube tight with heel of thumb. Raise bottle about 6 in. Swing cock to 2. Release tube slightly to allow gradual rise of water in burette.

6 When level reaches exactly zero, clamp tube tightly and throw cock to 3. Measure gas by the bottom of liquid meniscus (curved surface) and have eye level

7 Raise bottle. Open CO_2 pipette needle valve 3 or 4 turns. Water will rise in burette, pushing gas into pipette. Keep eye on rising water in burette.

8 Restrict tube as water approaches neck of burette. Stop it exactly at scratch on neck. Close valve 1. Place bottle on table for few seconds to let CO_2 absorb.

9 Hold bottle low, open needle valve. As rising fluid nears pipette neck, slow down with pressure on rubber tube and stop it exactly at mark. Then close valve 1.

10 Hold bottle so liquid level in bottle is same as in burette. With eye on same level, read burette at bottom of meniscus as percentage of CO_2.

11 (NO PHOTO) to get percentage of O_2 repeat this process with pipette 2. Record burette reading. The difference between second and first reading is the percentage of O_2. Then do the same for pipette 3. The difference between third and second reading is percentage of CO.

NOTE 1—After the first absorption and measurement of CO_2, repeat to

see if you get the same reading. If the amount increase on second pass, take a third pass to make sure absorption is complete. Completely absorb each constituent of the flue gas before starting the next absorption. Unabsorbed oxygen in step 2 will show up as CO in step 3.

NOTE 2—To be comparable, all gas volumes must be measured at the same temperature and pressure. The

burette water jacket gives the constant temperature. The steps pictured measure each gas volume at atmospheric pressure.

NOTE 3—To save space it has been necessary to omit the steps for seeting up the equipment and charging the solution chambers. This is described in detail in the equipment instructions, or an experienced friend can help.

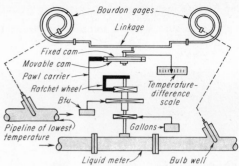

MECHANICAL instrument has two temperature-sensitive bulbs which deflect Bourdon gages to work linkage, meters

Here are 3 modern instruments to measure Btu's

1 Mechanical Btu meter . . .

Central heating and cooling in large shopping centers, airports, giant industrial, commercial or residential developments, as well as product quality in the process industries, all involve accurate measurements of heat energy used. The data is needed for accounting and quality control, too.

Today, these measurements are made with the *Btu meter,* an instrumentation system that measures the rate at which a flowing fluid gains or releases heat.

One Btu is the quantity of heat needed to raise the temperature of one lb of water through one deg F. Water is the frame of reference—6/10 Btu is enough to heat one lb of alcohol one deg F; 1/11 Btu is enough for copper. Industrial Btu meters obtain the information they need for calculating Btu by measuring the temperature difference (degree change) across a specific portion of the process. Mass flowrate of a liquid in lb per hr is proportional to square root of the pressure differential.

Specific heat, which is the thermal capacity of the fluid measured, also comes into the picture. Specific heat is simply the amount of Btu's needed to raise one lb of the particular substance one deg F. By the definition of the Btu, specific heat of water at 59 F is 1.

Multiplying specific heat by temperature differential by rate of flow, you get the rate of Btu exchange:

Specific heat (Btu per lb, F) × *temperature difference* (F) × *mass flowrate* (lb per hr) = *Btu per hr*

Thus to measure amount of heat change in a fluid, we must know characteristics of the fluid, then measure temperature change and flow rate. To obtain total Btu exchanged in the process, Btu rate is integrated with respect to time. Either mechanical or electrical integrators are used to do the job.

All the integrator does is register instantaneous Btu

measures rate of fluid flow with a positive-displacement meter which turns a meter shaft. Total flow is indicated on a flow integrator driven by a gear connected to the shaft. Temperature differential is measured by two mercury-filled bulb thermometers. They operate two spiral elements, linked to a cam hookup and a temperature-difference indicator.

A ratchet pawl connected to end of the meter shaft turns with the shaft. The pawl is also connected to a cam follower, which determines what percentage of rotation of the meter shaft it takes for the ratchet pawl to engage the ratchet wheel. Ratchet wheel is coupled to a gear which drives a Btu totalizer. That records total Btu exchanged in measured flow.

As temperature difference increases, pawl engages the ratchet wheel for a greater portion of each revolution of flowmeter shaft. A bigger Btu reading is added to the totalizer indicator for each unit of flow. Cam and linkage fix the proper relationship between flow, differential temperature and specific heat. Output: direct indication of heat exchanged in system.

rate against a time-reference base. For example, suppose in the course of one hr the Btu rate is 10,000 Btu per hr for ½ hr, 40,000 Btu per hr for ¼ hr and 20,000 Btu per hr for another ¼ hr. At the end of that hr the totalizer would have recorded a total of 20,000 Btu: 5000 for the

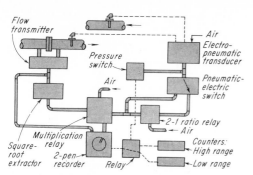

ELECTROMECHANICAL instrument has flowmeter, temperature-measuring elements, servomechanism, units shown

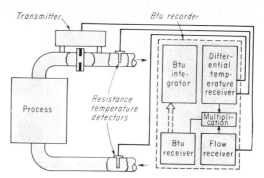

ELECTRONIC instrument is similar to electromechanical, but this one multiplies flow and temperature electronically

2 Electromechanical Btu meter...

may utilize any type of flowmetering system, although the differential flowmeter (sketch above) is most common. This flowmeter operates in the conventional way. As sketch indicates, flowmeter is coupled to the differential-temperature receiver through mechanical linkage.

Measurement of differential temperature is made either by two resistance temperature detectors or two thermocouples installed in inlet and outlet lines of the system. Output of the temperature-measuring elements operates a servo-mechanism, which positions members of a mechanical multiplying linkage.

Multiplying linkage indicates rate of heat exchange and, with suitable integration, amount of total heat exchange.

This type of Btu-metering system has one advantage over the completely mechanical meter at left. A continuous record of rate of Btu exchange is available: data often needed in the process industries.

With suitable retransmitting accessories incorporated in the system, this meter is used for control.

3 Electronic Btu meter . . .

has same basic components as the electromechanical type. But the multiplication of flow and temperature differential is done through electronic circuitry. This meter has a differential flowmeter to measure rate of fluid flow in the process.

Linear flow signal from transmitter to flow receiver is sent over electrical lines via a differential-transformer-transmission system (sketch above).

Inlet and outlet temperatures are measured with resistance temperature detectors. Flow and differential temperature are multiplied in the Btu-receiver system, using retransmitting potentiometers on the differential-temperature and flow receivers as the source of input signal to Btu receiver.

Btu receiver is a slidewire type. It rebalances output from the multiplication network, indicates and records rate of heat exchange in the metered system. This slidewire receiver also operates an integrator which continuously integrates rate of Btu exchanged to maintain a record of total Btu transfer. This electronic system can also do control jobs.

---

first $\frac{1}{2}$ hr, 10,000 for the next $\frac{1}{4}$ hr and 5000 Btu for the final 15 minutes of that given hr.

Three basic types of Btu meters are mechanical, electromechanical and electronic (above). But whichever of the three you use for measuring heat energy consumed

in any system, it must do the following: (1) measure flowrate of the fluid (2) measure temperature difference between incoming and outgoing fluid (presuming no change of state between these two points) (3) compute and integrate product of (1) and (2) to give total Btu used.

The ABC of Combustion

ELEMENTS

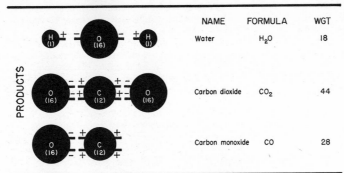

NAME	SYMBOL	VALENCE	WGT
Hydrogen	H	1	1
Carbon	C	4	12
Oxygen	O	2	16

PRODUCTS

NAME	FORMULA	WGT
Water	H_2O	18
Carbon dioxide	CO_2	44
Carbon monoxide	CO	28

REACTIONS

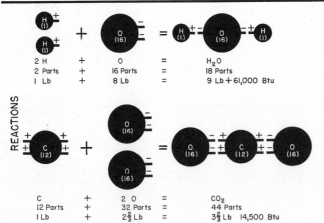

2 H	+	O	=	H_2O
2 Parts	+	16 Parts	=	18 Parts
1 Lb	+	8 Lb	=	9 Lb + 61,000 Btu

C	+	2 O	=	CO_2
12 Parts	+	32 Parts	=	44 Parts
1 Lb	+	$2\frac{2}{3}$ Lb	=	$3\frac{2}{3}$ Lb 14,500 Btu

▶ THE BURNING of coal, oil or gas is a chemical reaction. These fuels contain many different atoms in many different combinations, but the only important heat-producing atoms in commercial fuels are carbon (C), hydrogen (H).

The oxygen for combustion comes from the air supplied to the furnace. Air is 23% oxygen by weight and 21% by volume. The remainder of the air is mostly nitrogen which takes no actual chemical part in combustion.

The "Elements" table (top left) shows the atoms of hydrogen, carbon and oxygen. These pictures are highly idealized, but are all right for the present purpose. In this table "Wgt" means the atomic weight (weight of one atom), using hydrogen as the standard. Thus carbon weighs 12 and oxygen 16.

Valence explained. "Valence" takes some explaining: Try picturing the valence number as the number of hooks or bonds available to tie this atom to other atoms. The plus bonds pictured will pair with the minus bonds on other atoms.

Under "Products" the water molecule contains two atoms of hydrogen and one of oxygen, so its formula is H_2O. Note that all the bonds are paired up (plus to minus)—also that the weight of the molecule is 18, the sum of the weights of its atoms.

In the formula for carbon monoxide, two of the carbon hooks are unused, indicating that carbon monoxide is the product of imperfect combustion.

Burning H and C. Now turn to "Reactions" to see how hydrogen and carbon burn. The first line shows how two atoms of hydrogen combine with one of oxygen to make one of water (H_2O). The formula balances, meaning that the weight of products equals the weight of combining elements. Also every atom on the left shows up on the right, and every valence bond is paired.

The same is true of the combining of carbon and oxygen to form CO_2.

To fix these important relations in mind why not turn this book face down and see if you can sketch (from memory and common sense) these atoms, compounds and combustion reactions?

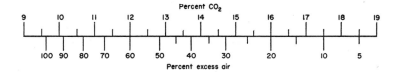

ANTHRACITE COAL

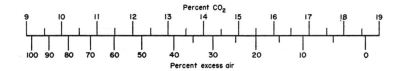

BITUMINOUS COAL

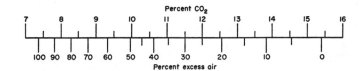

FUEL OIL

CO₂ vs Excess Air

 MANY OPERATING ENGINEERS are directly responsible for the operation of boilers fired by coal, oil or natural gas.

It is not always realized that it takes more than ten tons of air to burn one ton of 13,500-Btu coal. All this air eventually goes up the stack, although in a different chemical form.

To estimate the theoretical amount of air required to burn one pound of coal, divide the heating value of the coal by 1350. Thus, for the 13,500-Btu coal just mentioned, the theoretical air requirement is $13,500 \div 1350 = 10$ lb. Now furnaces and firing can never be perfect, so no actual furnace could burn this coal with only ten pounds of air. There must be some excess.

Twenty or thirty per cent excess might be reasonable in a small boiler furnace. With 30-percent excess air the furnace would use up 13 lb of air for every pound of coal fired. Since no matter is destroyed, and since very little material goes through the grate, nearly 14 lb of hot gas would go up the stack for every pound of coal fired.

To heat one pound of flue gas one degree it takes about one quarter Btu. So, consider a case where the boiler-room temperature is 80 F and the stack temperature is 530 F. Then the temperature rise is $530 - 80 = 450$ degrees, and the stack loss in the hot dry gas is $450 \times \frac{1}{4} \times 14 = 1575$ Btu per pound of coal fired.

If, through poor equipment, poorly operated, the stack temperature were 580 F and the excess air were 80 per cent, the air supplied per pound of coal would be 18 lb and the stack gases would weigh 19 lb per pound of coal fired. The temperature rise would be $580 - 80 = 500$ deg. Then the loss in the hot dry flue gas would be $19 \times \frac{1}{4} \times 500 = 2375$ Btu per lb coal.

The standard way to determine excess air is to analyze the flue gas for CO₂, using an "Orsat" or a CO₂ recorder The sample is taken at the point where gas leaves the boiler setting. Flue temperature should be measured at the same point. Knowing the CO₂ percentage, you can closely estimate the excess air from one of the charts above.

Then you can use a short cut to approximate the percentage of the coal's heat lost up the stack in the hot dry gases.

Here's the Rule: For the theoretical amount of air allow 1.8 per cent loss for every hundred degrees rise above boiler-room temperature.

For example, if the flue temperature is 520 F when the boiler room is 90 F, the rise will be $520 - 90 = 430$ deg. With the theoretical air supply the stack loss would be $4.3 \times 1.8 = 7.7$ per cent.

Increase this figure proportionately to allow for the excess air. For example, if the fuel is anthracite coal and the CO₂ runs 12 per cent, excess air is 60 percent from chart. Then the actual flue loss is $1.6 \times 7.7 = 12.3$ per cent.

Here's an Easy Rule For CO₂

Percent CO_2	Percent Excess Air	Percent loss for room temperature of 70 F and flue temperature of					
		400F	450F	500F	550F	600F	650F
5	314	28	32	36	40	44	48
6	245	23	27	30	34	37	41
7	196	20	23	26	29	32	35
8	159	18	20	23	25	28	31
9	130	16	18	20	23	25	27
10	107	14	16	18	20	23	25
11	88	13	15	17	19	21	23
12	72	12	14	15	17	19	21
13	59	11	13	14	16	18	19
14	48	10	12	13	15	16	18
15	38	10	11	12	14	15	17
16	29	9	10	12	13	14	16
17	22	8	9	11	12	14	15
21	0	7	8	9	10	11	12

The table above is a handy reference, but you won't really need it if you follow these easily remembered instructions

► COMBUSTION ARITHMETIC is one of those things you can make as hard or easy as you wish. Its largely a matter of how fine you want to figure. The minute you begin splitting hairs, the simplest combustion problem gets complicated.

On the other hand, if you view combustion from a practical operating angle, and are willing to accept close approximations, some of the important calculations will get about as simple as buying a sack of potatoes.

Rule for Coal and Coke.

For this I shall take only one page to explain a simple rule for estimating the stack loss from the CO_2 (carbon dioxide) percentage without referring to any table or formula. This rule is for coal and coke only. Don't use it for wood, oil, gas, bagasse, etc.

It holds because the burnable part of coal and coke is mostly carbon, with only a small percentage of the burnable hydrogen which would throw the rule off in the case of fuel oil.

The exact amount of air required to burn one pound of carbon is 150 cu ft, consisting of 31 cu ft of oxygen and 119 cu ft of nitrogen. The nitrogen doesn't burn. The oxygen burns to the same volume of CO_2, so the products of the combustion are still 150 cu ft (measured cold), consisting of 31 cu ft of CO_2 and 119 cu ft of nitrogen.

The Unavoidable Loss. Now there isn't any way to avoid the waste of heating up the necessary 150 cu ft of flue gas from room temperature to the temperature at which it is thrown out of the stack.

To heat 150 cu ft of flue gas 100 deg takes 27 Btu, which is just under 2 percent of the 14,500 Btu released by the burning of one pound of carbon.

So now we see that if the room temperature is 70 F, for example, and the stack temperature 570 F, then the temperature rise is 5.0 hundred degrees and the stack loss is $5.0 \times 2 = 10$ percent, if the coal is burned without any excess air at all.

Now suppose, with these same temperatures, that 50% excess air is used. Then the total air is 150 percent of the theoretical or 1.5 times the theoretical. So the percent loss up the stack will increase to $1.5 \times 10 = 15\%$ of all the heat in the coal fired.

Total Air from CO₂. The only thing needed to complete the rule is some way of figuring total air from the CO_2 percentage reading of an orsat or CO_2 meter.

Whatever part of the oxygen in the air actually burns with carbon produces the same volume of CO_2. The remaining volume of oxygen goes through unchanged.

The percentage of oxygen in the air by volume is 21. If this number can represent the total air supplied, the CO_2 percentage will represent the part really needed for combustion.

To make clear, take a case where the excess air is 50 percent, meaning that *two thirds* of the air *burns* and *one third is excess.* Then, of the 21 percent oxygen, 14 shows up as CO_2. Working this backward, we divide 21 by 14 to get 1.5, which means that total air is 1.5 times the theoretical requirement.

Now we can set down the rule:

A. Divide 21 by the CO_2 percentage to get the *total-air factor.*

B. Subtract room temperature from flue temperature, and divide by 100 to get the number of hundreds of degrees temperature rise.

C. For the theoretical air volume allow a flue loss of 2 percent for every 100 deg temperature rise.

D. Then actual flue loss is $A \times B \times C$

Example: Room temperature is 80F. Flue temperature is 520F. CO_2 in flue gas is 12%. What is flue loss?

A. $21 \div 12 = 1.75$, total air factor

B. $520 - 80 = 440$, or 4.40 hundred degrees rise

C. Loss $= 1.75 \times 4.40 \times 2 = 15.4\%$

Don't try to carry these computations out to a lot of decimal places. The method isn't that accurate. But check it against several values in the attached table to convince yourself that it does give a result close enough for most practical purposes.

With a little practice you can work such problems through from CO_2 percentage of flue loss in less than one minute.

How modern burners match their jobs

Characteristic new features of industrial oil and gas burners

Burner type	Fuel	Principal new features	Type of flame	Excess air	Applications	Remarks
Mechanical atomization	Oil	Higher fan and pump speeds for compacter units. Better pump cut-off. Safer flame control. Smaller capacities	Constant size	Low to medium	Steam and hot water boilers. Air heaters. Conventional industrial processes	Air heating, re-circulation of gases cut smoke, save fuel. Europe setting pace
Steam or air atomization (high-pressure)	Oil	Wide turndown	Strong spray pattern	Low	"	"
Steam or air atomization (low-pressure)	Oil	Even wider turndown	"	Low, added air may be entrained	"	"
Air atomiza-tion (medium pressure)	Oil	Widest turndown of all	"	High, with designs to maintain stable flame	"	Recently intro-duced into the U.S.
Ultrasonic	Oil	Several types using ultra-sonic or resonance princi-ple, 0.3 to 25 gal/hr	Constant	Low to medium	"	Vibration obtained electrically or by steam or air
Vertical rotary (central or wall-flame)	Oil	Improvements over old burner types. For example, preheating of wall rings	Constant, blue or luminous	Low	Mostly space heating	Many of these, or combinations, used in Europe
Vaporization	Oil	Improvements over old types. Residence time of oil in burner reduced; ex-cess oil drained out	Variable within narrow limits	Low to medium	Space heating where noise is strongly objectionable	In the U.S., API and oil companies contribute to R&D
Infrared	Gas	Protection against burn-ing of product obtained by quick-cooling radiant re-fractory, or retractable burner elements	Radiant	Low	Drying, curing, bak-ing, finishing—in textile, paper, plas-tics, veneers and ceramics industries	Must be applied judiciously to spe-cific industrial pro-cesses
Torch, premix or nozzle mix	Gas	Variable heat input. Turn-down of 40 to 1 possible. Good combustion, with lit-tle CO and NO_x	Tailored at will	Low to high	Direct heating of air for space heat-ing or industrial processes	
High intensity	Gas or Oil	Good heat penetration even at 15 ft from tip. Ad-vantageous with recupera-tive generators	Long, very stable, may develop 4×10^4 Btu/hr per cu ft	Very high, can be low with sec-ondary air	Ceramic industry. Heating of castings, bars and billets. Rotary dryers, cal-cining and roasting	Used in current revival of batch ceramics kiln
Short-flame or disc-type	Gas or Oil	Simplifies construction of utilization equipment. Strong radiation or con-vection heating	Flat flame; no forward travel	Low to medium	Melting pots, cruci-bles, forge furnaces; also batch kilns with side-burners	
Variable luminosity	Gas and Oil	Radiant or convection heating. Small amount of oil cracked	10% of oil in gas gives luminosity	High	Application to proc-esses when load or pattern changes	

These Five Steps Help Prevent Furnace Explosion

1 Get competent firemen, instruct them properly on equipment and controls. They're your first line of defense

2 Follow lighting-off procedure, below, faithfully

3 Provide good-sized inspection window so lighting torch, pilot, main burner flame, flame rod, if used, can be easily seen

4 Set up periodic inspection and maintenance program as follows (Include checking fuel-air ratio, draft adjustments, pilot flame settings):
Test safety shutoff fuel valves for tight shutoff
See how combustion safeguard responds to loss of flame contact
Test fan interlocks
Keep test records for insurance inspector's review

5 Get the insurance company's approval in advance for new installations or proposed changes. Submit plans showing burners, fuel piping, draft fans

LIGHTING-OFF TIME for oil- or gas-fired boilers represents a danger period for possible furnace-explosion build-ups

► THERE ARE REALLY only two danger periods when oil- or gas-fired furnaces can explode. These periods cover a good share of the time when a boiler should be getting the most attention. They begin with lighting off, and carry over to light-load firing. But they can happen only when there has been a build-up to dangerous proportions of unburned fuel or flammable products of incomplete combustion.

According to one insurance company's* findings 87% of all boiler furnace losses for a 13-year period (1936-1948) took place in manual-lighted boilers. These are units where fuel to the main burner can be turned on only by hand. Once lighted, though, the firing rate can be automatically controlled.

What's even more interesting, the waterwall type of boiler furnace construction is the most likely to suffer from a furnace explosion. Look at figures, facing page. Waterwall types have only relatively weak wall heads for support. Refractory furnace, though, is of much more rugged construction.

Manual-lighted boilers begin, usually, at about 10,000-lb-per-hr steaming capacity and 100 psig, and go up from there. Automatic-lighted units take in the smaller sizes.

What can you do to make sure your boiler furnace never builds up an explosive mixture? Five steps, above, give some basic precautions you can take. In addition, here are helpful operating tips.

Lighting Off. Let's establish this most important point first. A high-caliber well-trained fireman is the primary factor in safe lighting off. He instinctively goes through these steps in lighting off a boiler:

1. See that all drain valves are closed; water is in the boiler; access and observation doors are closed.

2. Check fuel valves at all burners for closure.

3. Thoroughly purge the furnace and passes for 5 minutes at ¼ maximum air-flow rate.

4. Insert a reliable lighting torch. (A dozen turns of ½-in. asbestos rope, secured by a hairpin bend, on a 5-ft length of ⅜-in.-dia steel rod is considered reliable.) Be sure torch has a substantial flame before putting it into the burner lighting part, and put it in place in such a way that main or pilot burners, if present, light off promptly.

5. Open the fuel valves at the burner, and fire the burner in keeping with the manufacturer's instructions.

6. If a main burner fails to ignite within 5 seconds after opening the fuel valve, or if the torch, pilot or main-burner flames blow out:

(a) Close the fuel valve at the main burner right away.

(b) Pull out the torch and turn off fuel to pilot, if present.

(c) Adjust dampers, burner registers and pass for no less than ¼ rated air flow through setting.

(d) Purge at this rate for 5 minutes or more before readjusting dampers, etc, and inserting the lighting torch a second time.

General Operation. A few operating conditions stand out in a study of gas- and oil-fired furnace explosions. Flame failures nearly always occur while firing with only one burner at low ratings —say below 30%—under low loads or while warming up shortly after lighting off. Under these conditions the boiler-furnace temperature drops off to a point below the prompt reignition temperature for the fuel.

Because of this, when boilers are fired below 30% rating, reduce the number of burners. Never try to cut back on burners once you get below the stable minimum firing rate. Instead, completely shut off individual burners so remaining ones have enough load to stay within stable operating range.

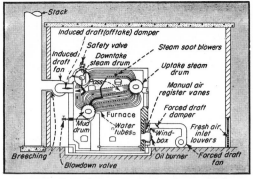

REFRACTORY-FURNACE design, *above*, will give the furnace a ruggedness that is helpful in the event of any explosion

WATERWALL TYPE of furnace suffers most from any explosion because the wall supports are relatively few and far apart →

How about flame failure or other combustion safeguards? Quick-acting electronic devices of the "flame-rod" type with gas, or "photoelectric" with oil, help if properly used. The flame-rod cannot sense unburned gas during lighting off, but it can see a pilot flame. While firing, both types respond to flame failure, but neither can do anything about dangerous build-ups of flammable products.

Since each and every boiler has its own characteristics that can bring on unforeseen operating troubles, let your combustion safeguards just sound an alarm to start off with. Then when you have reliable operation and you've made some necessary adjustments in firing and control equipment you can connect your combustion safeguards to shut off fuel.

As far as automatic combustion-control adjustments go there are some definite rules you ought to follow. For instance, make sure air-flow rate always follows steam flow and does not lead it. Then see that the fuel-flow rate is automatically adjusted to follow the air-flow rate. This cuts down the chance of over-rich combustion and gives safe response if fans fail.

Gas Hazards. Manual-lighted gas-fired boilers, particularly with large waterwall furnaces, have figured in some very severe explosions. The lighting-off explosion damage runs much higher than that during regular firing operation. In fact, there've been so many more lighting-off explosions with gas-firing compared to those with other fuels that previous practice has been closely checked. One thing stands out and that

is a failure in the past to put in reliable pilots for each main burner. These pilots prove their worth not only in lighting off but in low-load service and warm-up periods when the boiler is bringing up steam pressure.

Chief among the hazards in gas-fired operation is leaving gas-burner cocks open too long without purging, especially during unsuccessful attempts to light off. Next comes accidentally leaving burner cocks open.

High among any list of hazards is accidental flame failure while warming up after lighting off or while firing at low ratings. Biggest reason for this is unfavorable air-gas ratio adjustments.

As more or less of a follow-up of unfavorable air-gas adjustments is the danger of not enough air for complete combustion.

Oil Hazards. Similarly there are several outstanding hazards in oil-burning operation. Most important is accidental flame-failure while warming up shortly after lighting off or firing at low rating. Its causes are many. They include unfavorable oil-air ratio, poor atomization, interruption of oil flow by oil-pressure failure or a slug of water.

Next in order of hazard importance is incomplete combustion of oil. It may result from (1) too low an atomizing pressure—either from too low a steam pressure or too large a sprayer plate (2) too low a temperature, or (3) firing at too low a rate to maintain proper furnace temperature.

Then comes the hazard created by leaving the burner oil valves open too long without purging during unsuccessful attempts to light off.

The last two, in order of importance, are (1) attempts to light burners from hot furnace refractory and (2) not enough air for complete combustion either from fan failure or failure to start draft fans or open dampers.

Recommendations. To avoid furnace explosion, there are a number of practical steps you can take in addition to the five steps already given. For gas-firing they involve: (1) providing a continuous, fixed, premixed pilot for each main burner (2) installing, where a boiler has more than one burner or burner unit, an approved cock and gas safety-control system (3) protecting against low gas pressure on single-burner boilers (4) interlocking the main safety shutoff gas valve with electrically driven draft fan (5) adding approved flame-rod combustion safeguards on 8,000-20,000 lb per hr boilers where practical.

For oil firing, these recommendations hold: (1) thermometers in the oil piping close to burners using heavy oil (2) an oil-pressure gage on the oil-header supplying pressure-atomizing or steam-atomizing burners (3) photoelectric combustion safeguards and interlocks with furnaces having more than two of the six walls as waterwalls or any other boiler where damage would badly cripple plant production no matter what its construction.

Further, in oil firing operate soot blowers only when the burners are firing at high ratings. Then open wide the off-take damper or induced draft to vent steam. Meanwhile throttle the air flow by means of the forced-air damper to prevent blowing out the fires.

Medicine Man

This fierce looking African medicine man is a jack-of-all-trades. He claims to talk with the gods, placate angry spirits, exorcize the devil, drive away evil spirits, cure complicated diseases, make rain, interpret dreams—and, for good measure, divine the future.

You don't believe any one person—especially an ignorant savage—can do all these things? Well, maybe you don't, but I can show you a lot of gullible plant owners, power engineers and operators who swallow bunkum just as impossible.

For example, I've operated in, or visited plants where good hard American dollars are shelled out, year after year, for all kinds of fancy-sounding "cure all" boiler-water treatment.

Engineers who would laugh at a medicine man, go right ahead and let his kind take care of their highly sensitive and expensive boilers.

While most firms selling boiler-water treatment chemicals and apparatus are highly reliable, some outfit is always trying to crash the party with a fancy sounding, but nearly worthless product. It's your job to make him give reasonable proof before acting as his guinea pig. A reliable product thrives on intelligent investigation. A cure-all shrivels up and fades away.

Only witch doctors and medicine men claim to master magic. Let them do their stuff in the jungle—your boilers are too vital for peak production demands just now.

3
Steam engines, steam turbines

Setting Piston Valves

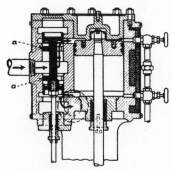

1 Steam engine's piston valve admits steam at center, exhausts over ends. Inside rings (a) are guides. These valves can be set with sticks instead of using old marks, which are usually unreliable

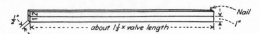

2 One stick for valve and another for ports are needed. Soft wood of smooth grain is best. Use sticks about ½ in. by 1 in. and about 1½ times length of valve

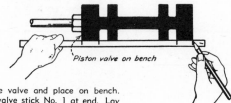

Piston valve on bench

3 Remove valve and place on bench. Label valve stick No. 1 at end. Lay alongside valve about two in. from numbered end. Use a sharp pencil or pen knife for marking the edges of valve

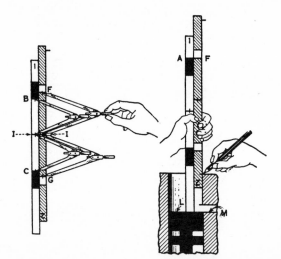

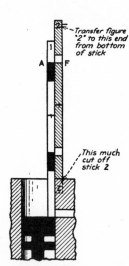

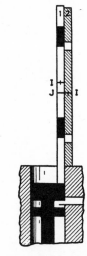

7 With dividers find center between B C on stick 1 and F G on stick 2. Use square to line up centers. With dividers on stick 1 find center of B C and mark across with **square**. Identify with X mark

8 Turn engine crank until top of valve L is flush with port edge M. Hold **sticks in** corresponding position, with A and F in line. Place the sticks on valve and mark stick 2 at cylinder head with pencil

9 Mark nail end of stick 2 with number 2. Saw off bottom end you marked at cylinder head. With stick 1 resting on valve, and stick 2 resting on cylinder block, A and F should now be in line

10 Hold sticks together and turn crank until valve reaches it's highest point of travel. Place mark J on stick 1 opposite center mark I of stick 2. Always turn the engine's crank in same direction

With Sticks

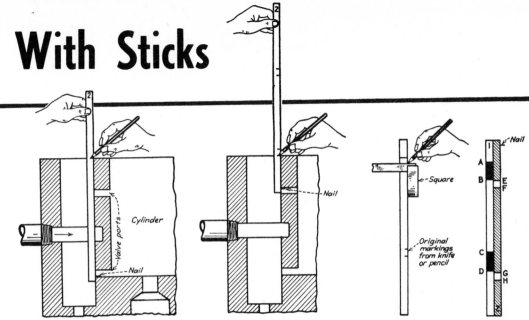

4 Mark port stick No. 2 at one end and drive tiny nail in the other end. Rest nail on bottom of the steam port in valve cylinder and make mark at cylinder head with a pencil or knife

5 Complete marking ports by moving stick to top of bottom port, then bottom of top port and finally the top of top part. Hold stick firmly against cylinder wall and make marks sharp

6 **(A)** Complete port and valve markings with square. **(B)** Turn stick 2 upside down. Shade sections AB and CD on stick 1 and outside sections EF and GH on stick 2, to show valves and ports

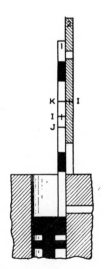

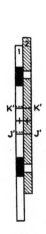

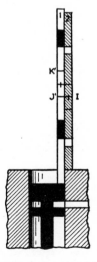

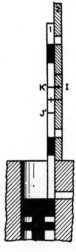

11 Turn crank until valve reaches its lowest position. Make mark K on stick 1 opposite center line I on stick 2. Valve is not set properly because center mark I is not in center of marks K and J

12 Distance K J is valve travel. Find half of this with dividers. Mark K' and J' on either side of center line. Hold sticks with center lines together and draw lines K' and J' across both, using the square

13 Distance from dotted line to K' in Fig. 12 is amount valve is to be lowered. Screw valve stem out of crosshead this amount. Place sticks as shown and turn crank. J' should line up with I

14 Turn crank until valve is at travels bottom. K' should be in line with I. Sticks show what's going on inside engine. Date sticks and keep them for your future reference for checking setting next time

SETTING ECCENTRIC "ON CENTER"

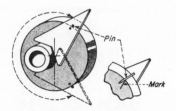
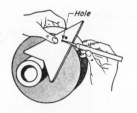
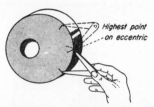

1 To find high point of eccentric throw: use template with pin through it; mark point where pin touches eccentric; swing template around, mark other side

2 Keep template firmly on hub or shaft when marking. One method of marking is back of eccentric where pin hits. Other is with scriber thru hole in template

3 Take pair of dividers and find point on eccentric edge halfway between two points of contact already marked. This is highest point of eccentric throw

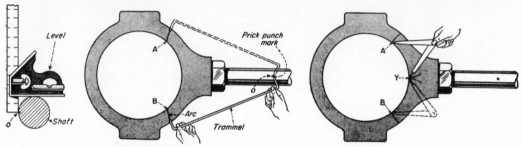

4 Next set machinist's square to half shaft dia. Hold square level, mark **O**. With one end of trammel at **O**, mark points **A** and **B** anywhere on shaft side of eccentric strap

5 Use marks **A** and **B** to find center mark **Y** on eccentric strap with dividers. Mark **Y** is important for future reference. Make it permanent by using small chisel to stamp into strap

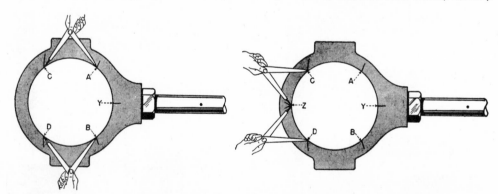

6 Next use mark **A** to make mark **C**. Then from **B** make mark **D**. Keep dividers near inside edge for accuracy

7 Use mark **C** and mark **D** to find center **Z**. **Y** and **Z** are at extreme opposites. Make **Z** permanent with chisel

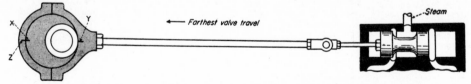

8 When the eccentric sheave mark **X** is opposite mark **Z** or **Y** on eccentric strap, the engine's valve is known to be "on center". That means valve is at extreme position of valve travel, regardless of crank's position. These two reference marks are useful in case eccentric slips, and for finding valve's position without opening steam-valve chest

Bumping a Piston

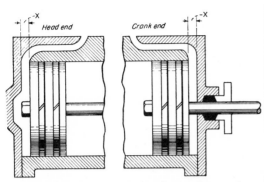

1 "Bumping piston" means finding clearance between steam-engine cylinder head and piston. Clearances **X** should be equal. Wear in crank, main bearing and crosshead pin changes clearance. So center piston between ends before it cracks a head

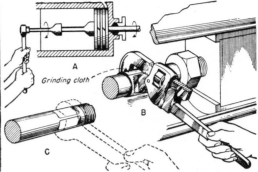

4 Next, remove cylinder head and screw piston against crank-end cylinder head, **A**. On smaller engines, head doesn't have to be removed. Instead turn piston rod with pipe wrench over grinding cloth, **B**, or with open wrench on rod flats, **C**

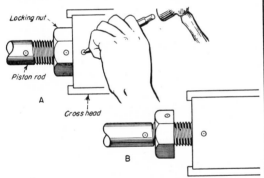

2 **A,** Centerpunch rod, nut and crosshead. When job is done, marks will show how far clearance had been out. (Crosshead mark used in Fig. 6 if scale isn't handy.) **B,** Loosen piston-rod locknut and run nut back to the end of threads

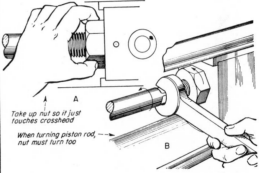

5 Wiggle crank on dead center and adjust piston rod so piston barely touches cylinder head. Then screw nut to crosshead, **A**. Replace cover. Turn engine to head end. Wiggle crank, turn piston rod, **B**, till piston just touches head end

3 Turn engine with bar or with flywheel so piston is on crank end. If engine is large and has jacking engine, use it. This is not accurate operation, but dead center can be guessed by eyeing position of connecting rod at crank-disk center

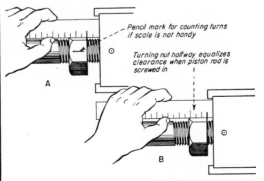

6 Caution: When adjusting piston rod at head end, nut must move with rod. Measure from crosshead to nut, **A**. Screw nut in half that distance, **B**. Without a scale, count nut turns. Turn rod so nut is against crosshead. Tighten nut

Neck-Bushing And

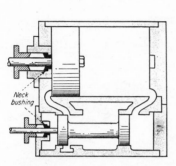

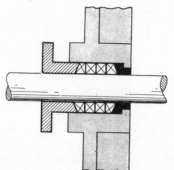

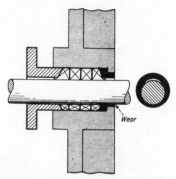

1 Neck bushings are in many types of reciprocating machinery. Here they are shown in steam-engine stuffing boxes

2 Neck bushing is usually made of bronze. It helps seal stuffing box and provide rod bearing when piston wears

3 Worn cylinder and piston lowers rod. Neck bushing and gland wear at bottom. Packing squeezes in opening at top

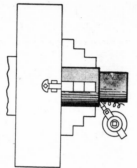

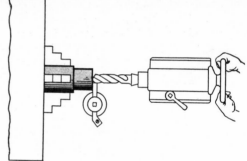

7 Neck bushing can be made of babbitt or bronze. Cast plug by filling a can with metal

8 Chuck babbitt plug or bronze casting in lathe. Turn to snug stuffing-box-diameter fit

9 Then drill a hole slightly smaller than the piston rod diameter. Best way to center any drill in lathe is to run tool holder against drill. This steadies it for center

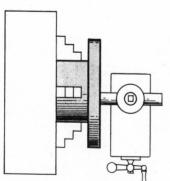

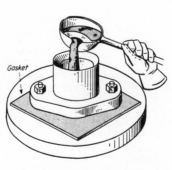

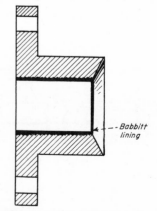

14 Now chuck the gland and rough-bore about ¼ in. out of hole. Rough tool marks help hold the babbitt metal best

15 Bolt gland to flange over gasket, to seal bottom. Then heat gland and pour it full of babbitt metal as shown

16 Chuck gland and drill babbitt. Bore for sliding fit over rod. Neck bushing and gland are ready to install now

Gland Tips

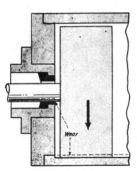

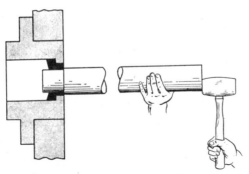

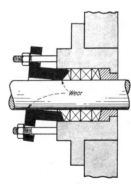

4 Piston and rod have worn cylinder and neck-bushing bottom, made packing **wear**

5 Neck bushing is held in place by making light press fit in cylinder opening through stuffing box. To remove **bushing, turn down round stock, drive out from cylinder end**

6 Cocked gland wears fast. Packing has hard time seal**ing extra** clearance around rod

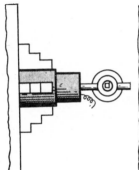

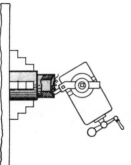

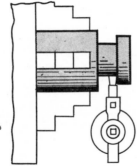

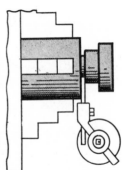

10 Bore smooth hole a few thousandths larger than rod, depending on engine service

11 Set the compound rest for 30 degrees and bevel casting's end as shown

12 Turn diameter for cylinder hole same size as hole. Only one chucking needed here

13 Cut with cutting-off tool and press into machine through stuffing box

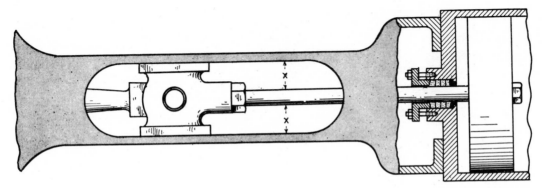

17 With close fitting neck bushing and gland lined with babbitt metal, soft packing in stuffing box has easy job. Rod lasts much longer if packing is not tight. With neck bush-ing supporting piston, cylinder wears less. CAUTION! Be sure X equals X between crosshead guides. Take lower guide wear into account. Renew glands and neck bushings often

Today It Pays to Make

WEER

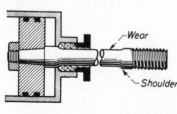

ALL RODS WEAR from reciprocating in stuffing box. Shoulders wedge through the packing, making sealing a tough job

REMOVING PISTON

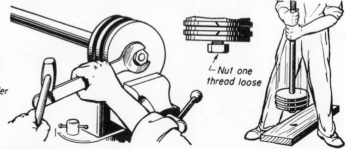

Nut one thread loose

FIRST STEP in removing old rod is to clamp piston in vise. (Never clamp rod in vise.) Loosen nut one turn only. Then smack piston and rod assembly down hard against wooden plank. Nut protects threads and keeps rod from digging into wood

CUTTING TAPER

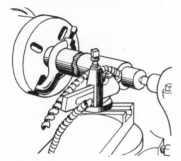

TURN DOWN shaft's end for small thread diameter, after both ends are centered and held between centers with lathe dog

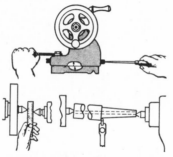

MOVING TAILSTOCK half of total taper is one way to cut. Loosen setscrews, and nut, then jack over distance needed

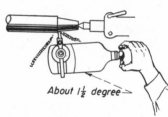

About 1½ degree

COMPOUND REST is handy for cutting short tapers. Find right angle from the old taper, then feed the tool by hand

CUTTING THREADS

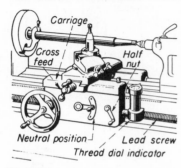

Carriage
Cross feed
Half nut
Neutral position
Lead screw
Thread dial indicator

SET GEARS for right thread, place feed change lever in neutral, engage thread dial indicator against the lead screw

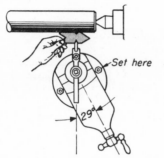

Set here
29°

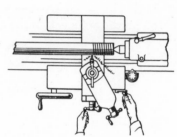

GRIND TOOL to 60 degrees to fit center gage. Place gage against straight shaft and set compound rest to 29 degrees. Cut thread by throwing in half-nut lever at any line on dial (for even threads). Draw the tool away at end with cross feed

New Piston Rods

CENTERING

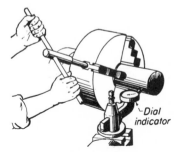

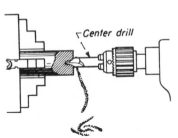

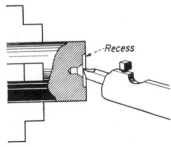

TRUE new shaft with dial indicator by loosening and tightening the chuck jaws until dial needle shows zero all around

NEXT, run center drill against end of shaft slowly or drill tip will break. Be sure to oil the drill before using it

RECESS around center with lathe tool to protect center-hole edge from damage. Shaft will then be true between centers

FITTING

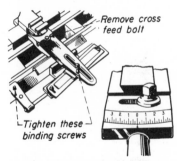

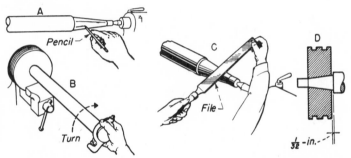

TAPER ATTACHMENT is the third method. Remove cross feed screw, tighten screws shown. Adjust to right inches-per-ft

NEVER FINISH taper with lathe tool only. *A:* First mark taper with pencil. *B:* Then insert taper into position and turn quarter turn. *C:* File end with fine mill file where the pencil marks are rubbed off the taper. *D:* Shoulder should not touch the piston

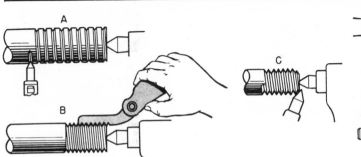

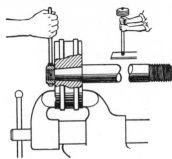

RUN CARRIAGE back by hand. *A:* Feed tool in desired amount by compound rest only. Always use cross feed to remove tool and return to same point. *B:* Check threads with gage. *C:* Cut chamfer at end by placing the tool at an angle, as shown here

BUMP PISTON onto new rod against wood. Then tighten nut. Perfect taper will hold piston on without loading thread

Dead Centering An Engine

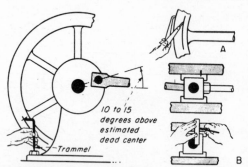

1 Before setting steam valves, engine must be on dead center. Put crank 10 to 15 deg above estimated dead center. Scribe line on flywheel with hermaphrodite, **A.** Make trammel and mark flywheel as shown. Mark the crosshead and guide **B**

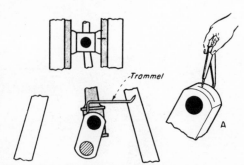

2 In placing some vertical engines on dead center, crank web may be better for markings. After placing crank past dead center, scribe line on web, **A.** Make trammel to reach from web to engine frame. Mark crosshead and guide

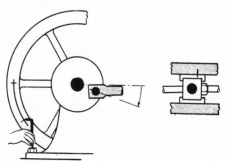

3 Turn engine over dead center so mark on crosshead again lines up with mark on guide. Now mark flywheel with trammel as shown. During all these operations flywheel must be turned in opposite direction engine runs, as explained in Fig. 5

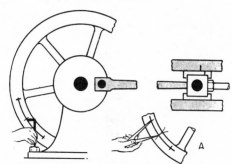

4 Split distance between two trammel marks on flywheel with dividers, **A.** Turn crank until center mark on flywheel comes to trammel. When using trammel always place bottom end in same center-punch mark on engine housing or bolt

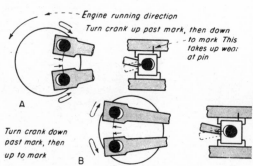

5 Most text books tell when finding dead center to turn the engine in direction it runs. That's wrong. Turn flywheel in the direction of rotation past crosshead guide mark, then back to it to take up bearing wear in the running position

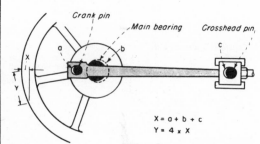

$$X = a + b + c$$
$$Y = 4 \times X$$

6 Unless main-bearing, crankpin and crosshead-pin wear is taken up by turning wheel in opposite direction, wear of **a**, **b** and **c**, or **X**, add up to four times **X**, (**Y**) on flywheel. Then trammel marks false dead center; valves are set wrong

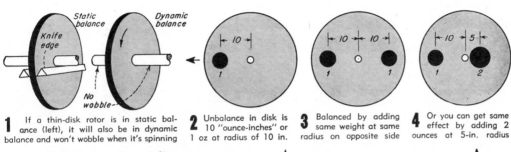

1 If a thin-disk rotor is in static balance (left), it will also be in dynamic balance and won't wobble when it's spinning

2 Unbalance in disk is 10 "ounce-inches" or 1 oz at radius of 10 in.

3 Balanced by adding same weight at same radius on opposite side

4 Or you can get same effect by adding 2 ounces at 5-in. radius

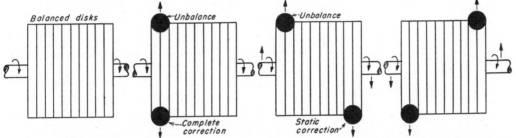

5 If you stack balanced disks to make a rotor, it will be in perfect dynamic balance

6 In this case, weight added opposite unbalance gives static and dynamic balance

7 But placing a weight this way leaves a bad wobble despite perfect static balance

8 Half turn later pressures on both bearings are reversed, meaning bad vibration

How to Balance

▶ IF A ROTOR WILL rest in any position on level knife edges (Fig. 1, left) it's in *static* balance. If it will also spin fast without shaft wobble, it's in *dynamic* balance.

Any rotor in dynamic balance is automatically in static balance, but the converse doesn't hold except for a thin disk. So don't assume a rotor is in running balance just because it balances perfectly on knife edges.

Fig. 2, 3, 4 show how to balance a disk. Find low point on knife edges. Add weight on light side, or drill metal away on heavy side until rotor balances in any position.

Stack of disks, each separately balanced in this way, makes a perfectly balanced rotor.

Fig. 7 and 8 shows that balancing a rotor statically won't balance it for running unless weights are at right point along length of shaft.

To balance a wobbling rotor, correct each end separately according to the wobble at that end as explained in Fig. 11.

A & B = unbalance
a & b = corresponding corrections

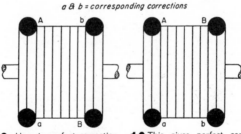

9 Here is perfect correction for two ends in unbalance on opposite sides of the rotor

10 This gives perfect correction where both ends of rotor unbalanced same side

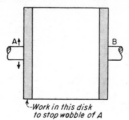

Work in this disk to stop wobble of A

11 Spin the rotor in working bearings and pencil mark "High" side of shaft A. Add weight opposite this mark at left end of rotor until left-end wobble almost disappears. Then do same for right end. Repeat for both ends. Make a third round if it's necessary

Aligning And Boring

1 First step is to stretch piano wire from aligning targets set on approximate center at both ends of the engine

2 Mechanic of Edw Purvis & Son, old time Brooklyn, N. Y. engine firm, mikes distance from 0.020-in. piano wire to machined boss inside crosshead-guide housing. He centers by adjusting the screws on alignment target at either end of long wire

5 Measuring cylinder all around **before** stretching wire tells mechanic cylinder's condition and how deep to set bore

6 After wire is shifted to exact center of engine, measurements are recorded from various points to wire. Then wire is removed and boring bar lined up with mike at same points. Allowance is made for wire size and boring-bar diameter in figuring

Engine Cylinder

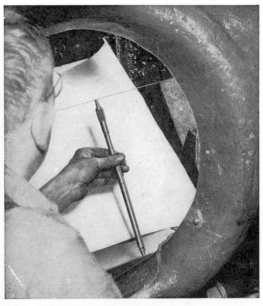

3 Years of experience are needed for right "feel" in this delicate measuring step. Wire must be taut (it broke once on job). Measure from machined bosses on sides and from unworn crosshead-guide edges at top and bottom

4 Inside micrometer rests on edge of crosshead guide and is adjusted until it barely touches wire. Paper behind helps to see wire

7 Pipe supports boring-bar feedscrew handle from studs. String is weak and breaks if tool catches in cylinder

8 Blocks support small boring bar for corliss valve ports. Clamps hold bar in position at both ends of port like this

9 "T" slot in corliss valve is remachined for snug fit with valve stem to reduce lost motion in valve gear

Fix Those Seals

HOW GLANDS WORK

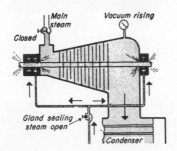

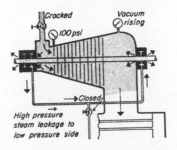

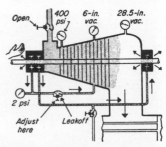

TO START, open gland steam. Steam blowing: out around shaft, seals it; through turbine, is condensed by cooling water

WITH MAIN STEAM CRACKED, leakage past shaft blows back in sealing line to end. No sealing steam needed.

WITH FULL PRESSURE on turbine chest, more steam's at hp end than needed. Excess leakage helps l-p turbine stages

GLAND TROUBLES AND CURES

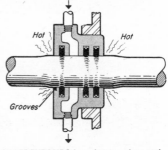

TURBINE VIBRATES and screeches when carbon sealing steam pressure's too low. Then rings ride shaft; hug, overheat it

GIVE FULL STEAM BLAST to carbon rings to 'lubricate' shaft and lift rings. If vibration doesn't stop, slow turbine fast

SHAFT EXPANDS from friction when sealing steam fails. Carbon rings wear grooves fast and shaft glows with heat

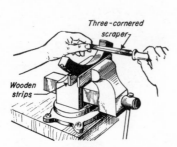

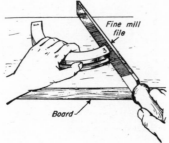

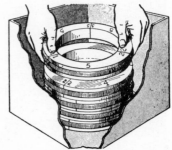

PLACE EACH SEGMENT in vise between wood strips. With 3-cornered scraper, carefully remove shiny inside surface

FILE from one end of each segment with fine mill file until assembled ring has 0.001 in. clearance, for each in. shaft dia

PLACE RINGS in box, exactly as they were in gland (on big job). Ring sides are also sealing joints; watch assembly

Before They Knock Out Turbine

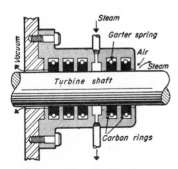

CARBON RINGS in gland restrict steam leakage out and air leakage in. Rings are in segments, held by garter springs

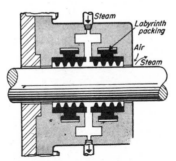

LABYRINTH GLAND has soft white-metal segments in dovetailed grooves around shaft. These restrict leakage at shaft

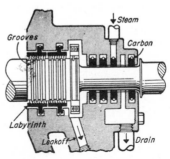

COMBINATION labyrinth with leakoff, plus steam-sealed carbon-ring gland. Grooves at labyrinth break steam path.

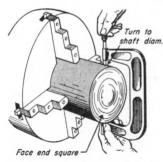

OVERHAUL CARBON RINGS at set periods. First step is to mike turbine shaft and turn short casting to exact shaft dia

REMOVE RING SEGMENTS from gland. Segments are numbered at each end and each segment has a number. Watch this

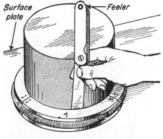

LAY SQUARE END of casting on surface plate. Assemble one carbon ring against it, and check clearance with feeler gage

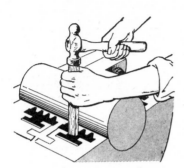

REMOVE LABYRINTH SEGMENTS for inspection by light taps against wood strips. If they do not slide out easily, stop

RUST in gland housing often holds labyrinth segments. Fill dovetail slot with penetrating oil and let set overnight

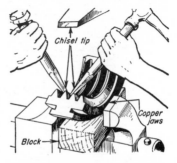

REST SEGMENT in vise on wooden block. Upset sealing edges with flat-tipped chisels so metal almost touches shaft

Solid-Wheel Turbine

How Turbine Works

"SOLID WHEEL" TURBINE is an impulse, helical-flow type, used mostly for noncondensing service. Steam issues from expanding nozzle, at high velocity and enters side of wheel bucket. There steam direction is reversed 180 degrees. Stationary reversing chamber returns steam to moving wheel, where process is repeated until most of steam's energy is spent.

These machines are sturdy, compact, and can take a lot of abuse. Blades and wheel are one piece, and blades have large clearances, with double-rim protection. End play won't damage turbine, and blades won't foul.

Close blade clearance isn't too important because power-producing action of steam acts on curved surfaces at back of buckets. Blade wear has little effect on performance. Steam enters and leaves at right angles to shaft, so there is almost no end thrust. Rotor can be removed from shaft because it's not shrunk on.

If not loaded, turbine can be started up cold in a few seconds. (But always warm up, if possible.) With large wheel clearance and little load on thrust, danger of blades rubbing is small. But because of all these features, this turbine is often neglected and abused. For best results, give it the attention shown here.

Look Inside

STURDY machine stands punishment, but don't be afraid to open for needed maintenance. Check shaft bearings for wear. Allow clearance recommended by maker for your machine. Main bearings also take thrust wear. If over 0.025-in. thrust clearance, renew bearings. Check shaft to see if it's bent, worn at bearings and at carbon rings. If shaft must be replaced, loosen shaft nut on side of wheel, and press wheel off shaft.

Don't wait until the wheel rides on the bottom reversing chamber. With shaft and wheel removed, inspect nozzle, then run wire through it. Also, inspect reversing chamber for wear.

When starting up turbine, first open exhaust so machine warms up gradually. Make sure drain is open so condensate is removed as fast as it forms. Drain should be left open when machine is not running. After machine is warm, crack throttle and turn over slowly until all water works out. Make sure oil rings in bearings turn before leaving machine.

Carbon Seals

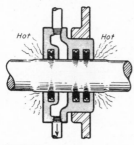

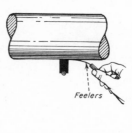

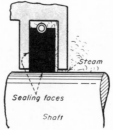

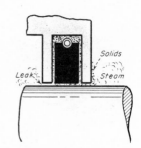

FRICTION expands shaft if sealing steam fails, grooves

CLEARANCE between shaft, carbon rings must be right

SEAL against casing—shaft, same as with a piston ring

CARRYOVER from boiler deposits solids, then rings stick

Is Tough Work Horse

Watch These Details

Dia wheel	Lap
18 in.	1/16 in.
24 in.	1/16 in.
36 in.	3/32 in.

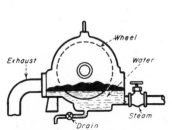
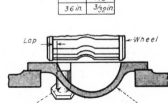
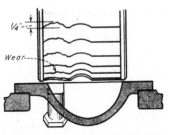

CORROSION is tough if drain is closed on idle turbine. Water floods casing

LAP changes with thrust wear. Check every time turbine is open—inspected

WEAR of 1/8 in. to 1/4 in. on bucket lips opposite the nozzle will down efficiency

GOVERNOR is constant-speed fly-ball type, spring opposes centrifugal force of governor weights. As speed rises, weights move out, closing governor valve. When speed drops, weights move in, allow valve to open. Look for knife edge on weight lever's ends—so that governor will not stick. Slide contact and slide shoe wear. Governor yoke wears into shaft.

Seat of governor valve wears—avoid this by adjusting steam with nozzle so valve does not throttle. That also prevents chatter. If governor hunts, look for improper valve setting, sticking or lost motion in mechanism. Find lost motion: move governor lever by hand, observe all parts.

Common fault is tight valve-stem packing that doesn't allow valve to work freely. Loosen packing or repack. Sticking is from bent or bruised valve stem—or valve wings rusted or scaled. Molded sleeve metallic-packing used needs no lubrication. Just keep packing tight enough so it doesn't leak.

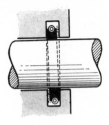
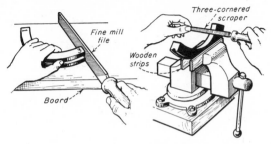

PARTICLES of wear—carryover wedge ring ends open

RINGS COCK from sediment, then they break very easily

FILE ends until clearance is right between shaft—ring

SCRAPE uneven surfaces to fit shaft for a perfect fit

STEAM TURBINE OVERHAUL is a major job requiring skilled technicians, proper tools and a check-off list of repairs and checks

Let's Inspect and Overhaul Your Steam

▶ TURBINES ARE STURDY, need little attention. But you must open them for a general check-up every so often, depending on service and operating conditions.

Check-Up. A machine running at constant load with few shutdowns need not be opened as often as one on start-and-stop operation. That kind of service causes high-temperature stresses from quick-starting, sudden load changes.

If you're not prepared to do the job yourself, be sure to have turbine firm supervise this important work. An experienced man will know what telltale signs to look for and how to correct them so machine won't have to be opened again until next scheduled shutdown.

Bearings. Turbine shown has ball-bearing thrust. Most operators find it good insurance to remove thrust and replace with new every time turbine opening is scheduled, regardless of wear. Reason is that thrust bearing's failure can wreck internal parts by rubbing them against each other.

Flexible Coupling. If you have this type of coupling, dismantle and check for wear, shoulders and galled surfaces. If coupling does not work right, it may transmit thrust from connected machine to thrust bearing. Then overloading

may cause failure. Replace worn parts, clean and assemble coupling.

Right way to check clearance between split bearing and journal is with inside and outside micrometers. Mike shaft with outside micrometers, then mike assembled bearing with inside mikes at various points. Take the difference, and that's the clearance.

Rotor, Stationary Elements. Turbine's efficiency depends on high-velocity steam. So slightest internal defect can be costly. First, keep steam clean so there is no deposit on blades. Soluble deposits usually form at earlier turbine stages, insoluble deposits on middle and later stages. Whatever kind, deposit restricts steam flow, cuts efficiency.

Remove deposits, inspect moving and stationary blades carefully for erosion, corrosion and cracks. Cracks form around blades and spread out from

to be made under supervision

Turbine Now

the jacking-holes in the turbine rotors.

Corrosion can be serious if valves don't seat properly in sealing, drain, throttle and header valves during idle periods. Then steam condenses in turbine, doing damage. Answer is to regrind valves and keep them tight.

It's good practice to indicate rims of turbine wheels while machine is open. Reason is that years of operation and temperature changes cause growth and distort turbine casing, rotor and stationary elements. Turn rotor slowly while indicating wheel rims. Also, take clearances between each wheel at rim —check with previous figures.

Safety. Don't take chances. Use right lifting gear and tools. Have only responsible men on job. Clean all parts thoroughly and remove dirt with vacuum hose. Check everything yourself to make sure no slipup increases outage.

OPEN TURBINE CASING, remove rotor. In photo, hardwood spacer slab prevents sling's wire cables damaging the blades

INSPECT wheel jacking-holes for cracks; all moving and stationary blades and nozzles, for cracks, erosion, corrosion

QUICK clearance check of main bearings without taking leads is to use feelers as shown before taking bearing off shaft

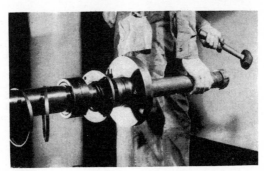

DRIVE new thrust bearings on shaft if old are worn. It's good insurance to install new ball-thrust at each overhaul

Delegate to each man the work he knows best so machine is put back in operation in the least possible time

RETAP all threaded holes so screws enter easily when parts are assembled

CARBON RINGS are used here for both interstage and casing-end sealing. Rings are made in segments, held together with garter springs, and kept from turning by metal stop on spring. These rings work like piston rings—sealing against shaft and side of groove they are housed in.

Clearance between shaft and ring must be right. Keep to turbine maker's recommendations. Before checking clearance, polish shaft section that rings seal, so surface is right. If too much clearance, polish ends of segments with fine abrasive cloth. But this job must be done right or ring ends will leak steam.

If there is much refitting, best practice is to turn piece of short stock same diameter as shaft and use it on work bench to fit rings onto. Carbon dust or solids from boiler carryover often cock rings in groove, causing them to break. Or impurities work into segment ends, causing leakage.

Usually, you have only to clean carbon dust from sides and ends of ring, clean sides of grooves and reassemble rings around shaft. Rings are precision made, so give job to good mechanics.

POLISH shaft sections with a fine abrasive strip where the carbon rings seal

RUB the ends of carbon rings lightly against fine abrasive cloth to clean them

Carbon seals prevent leakage between the various stages

ASSEMBLE the ring segments on shaft and hold together with garter springs

CHECK clearance against manufacturer's recommendation. Sand ring ends if needed

Right clearances do better sealing job, save shaft

LABEL all segments so they go back into the right places, to avoid trouble

REPLACE ring segments in stationary diaphragms after cleaning with vacuum

CLEAN groove sides of carbon-ring housing as sides are also the sealing surfaces

SMEAR sealing compound in turbine housing to seal the lower half of gland box

Carbon rings seal against groove side and shaft, like piston ring

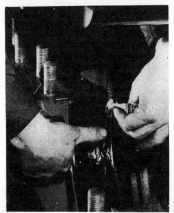

High - pressure steam leaks under carbon-ring housing if it's not sealed

Use vacuum cleaner to remove all dust and foreign particles. Machine's inside must be kept clean

PLACE three strips of light condenser cord on sealing compound as shown here

LOWER half of carbon-ring casing is in place here and sealed against h-p steam

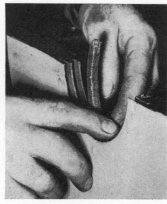

REPLACE soft packing at housing end that keeps oil from traveling on shaft

LOWER rotor assembly into place after removing the wooden blocks and the tools

WITH rotor in place, insert the garter springs in groove, fasten around the shaft

METAL stops prevent carbon rings turning. Move rings so the stops fit in slots

WHEN RINGS are all assembled, move metal stops down into position before closing up the carbon-ring casing. Rub light coating of sealing compound on casing joint, but be careful not to smear on the carbon rings or on the garter springs

Careful work, with every man doing his job right until machine is "buttoned up," will keep this unit running many months until the next major overhaul

PLACE top half of carbon-ring casing carefully in place over coated joint

POSITION carbon-ring segments into the diaphragms against the spring as shown

COAT diaphragm's joint face with graphite and cylinder oil before lowering in place

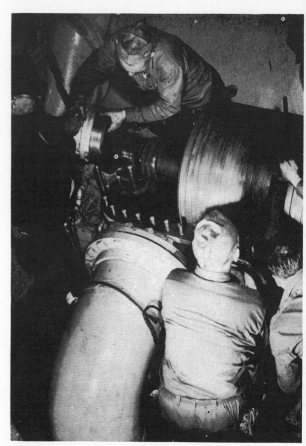

DIAPHRAGMS are all in place between these moving blades

CLOSING the turbine means an anxious period because many things can go wrong to cause delay. From around casing, clear all tools and small parts that might fall in, making you lift cover again—or, worse yet, that might go unnoticed until turbine is turned over with steam. That's why you should use only responsible mechanics.

Before spreading sealing compound on casing joint, again vacuum the joint surface so tiny foreign particles dropping on joint will be removed, instead of shoved inside. Compound used here on joint is Copaltite. Apply only when casing is ready to be lowered after asbestos string is placed as shown. On very large turbines, compound is usually applied with a short, stiff paint brush. Here, it was spread thin by hand to cover surfaces evenly and without smearing on

Use cylinder oil and graphite sparingly or it will end up in the boilers

COAT outside of all diaphragms with graphite and cylinder oil as shown here

RUN this fine asbestos string, three turns, all around the coated casing

CASING joint is well-coated with sealing compound—asbestos string placed

"LOWER AWAY, and here's smooth sailing because this turbine is shipshape"

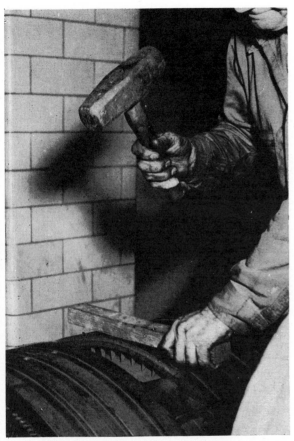

HEALTHY SMACK against wood block helps seat diaphragms

the other parts, such as studs, carbon rings, rotors, etc.

Asbestos twine gives body to compound, and makes a triple seal for steam to work through if there are low spots in the joint surface.

After casing is lowered, take up on all nuts evenly, only hand-tight. Then tighten nuts by staggering from one side of the casing to the other, instead of starting with one nut and going all around.

Scheduled maintenance, as shown in these photos, is the only answer to preventive maintenance. But, again, I should remind you that unless you are set up to do the work, it will pay to have one of the manufacturer's field engineers supervise the job. Before opening a turbine, always have on hand spare parts that might be needed—in good shape.

NOW you can 'shoot' your bearings in line

At long last, here's a telescope that aligns your bearings within a thousandth of an inch or two, and does away with the ancient time-consuming tight-wire job we all dreaded

These targets with mikes are heart of scheme...

MICROMETER target holder is a steel plate shaped to hold three inside micrometers. With target usually placed in first and last bearing bore of machinery to be aligned, micrometers are adjusted to position the cross-wires in exact center of bearing bores. Telescope is set on zero, instrument adjusted horizontally and vertically so target cross-wires and telescope reticle cross-hairs are in same line of sight.

This line of sight then becomes a reference line for the entire length of machine. Targets are centered with micrometers in all other bores of turbine and generator stator. Then measure center-line displacement of these other bores from reference line on micrometer and record. From these readings, plot curve on cross-section paper showing "as found" alignment of the bearings.

INSIDE MICROMETERS center the targets, the telescope then establishes a reference line . . .

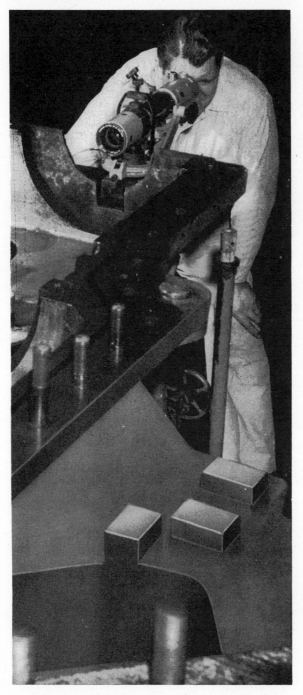

...you're ready to align other bearings

▶ NOT WIRE, but an alignment telescope is used to align turbine-generators at Consolidated Edison Company of New York. This is more accurate, easier to use, and faster than the old tight-wire method, which has been standard for years.

Alignment telescope has been used in the aircraft industry to position in space fuselage and airfoil wing sections without needing large steel jigs.

Problems. There are two basic problems in aligning a turbine-generator: (1) positioning the three to six bearing supports so turbine spindle or spindles are in line with generator rotor shaft at coupling or couplings. (2) allowing the stationary parts (generator stator, turbine cylinders and all fixed internals of turbine—diaphragms, blade rings, dummy rings, etc) proper clearance for moving parts of machine.

Tight-wire method is to stretch a small-diameter wire through turbine-generator with wire ends positioned on proposed center line of machine. Once tight wire is in place, the various parts are aligned to wire with inside micrometers. You measure from wire to round openings or bores in parts to be positioned. By comparing micrometer measurements and doing a lot of arithmetic, relative position of parts to wire are found and parts repositioned accordingly.

This may sound simple if you haven't tried to align with wire, but it's a time-consuming process. Wire is subject to vibration, and is a hazard and obstruction to work going on. Besides, the many micrometer measurements and involved arithmetic often lead to human errors.

Telescope Method. We selected a 60-power Farrand Alignment Telescope, capable of focusing from 18 inches to infinity. In the eye piece is a reticle, horizontal and vertical micrometer scales, and a precision level. Both the large range of displacement and level are desirable for turbine-generator alignment.

Telescope's optical center line replaces tight wire as center line is projected throughout machine's length. Optical center line is set level with the built-in precision level. This optical center line can be displaced vertically and horizontally, plus or minus 0.120 in. Telescope is sighted on cross-wire targets mounted in a target holder, fitted with standard inside micrometers at sides and bottom. Targets fit into bore diameters of 12 to 40 inches. Telescope may be mounted on an end-bearing pedestal or on a separate stand, apart from turbine, *photo.*

Results. Alignment-telescope method is more accurate, less costly, takes less time. Besides, it permits other work to go on in area, even installing or removing of internal turbine parts.

We developed telescope method for heavy machinery on two different turbine-generators. A third unit (53,000-kw topping job) was then completely realigned this new way. Existing alignment was found by centering optical center line through No. 2 and No. 4 bearings. For final realignment, we established a level optical center line through No. 6 bearing. We aligned other bearings, generator stator, and rotating and fixed turbine parts to this center line. As a result, we had to realign and regrout three bearing pedestals and turbine sole plates.

On installing the turbine spindle and generator rotor, all coupling checks showed satisfactory alignment.

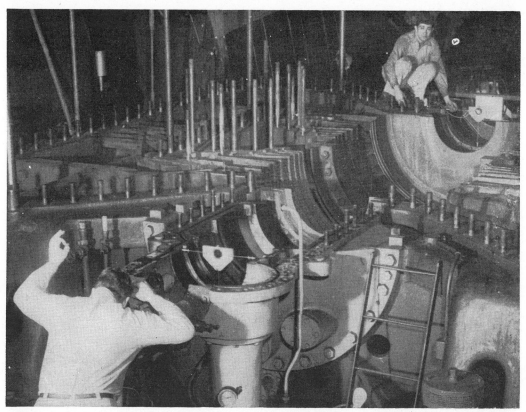

REFERENCE LINE is established with two targets at extreme ends of machine and telescope on stand. By placing the targets in other bearing bores to be aligned, as here, exact amount of misalignment in thousandths of an inch is quickly known

Plot curves of both plan and elevation alignment, *sketch*, as you line up the machine. To do this, set up telescope, as explained before. Center a target with inside micrometers in bearing bore at each end of machine for finding the reference line. With this line established by the telescope, you are now ready to center targets in all other bores to be aligned—or in bores you want information on, whether or not it is practical to align at this time.

From these readings, plot plan and elevation curves on cross-section paper showing the "as found" alignment. Indicate misalignment in thousandths (mils), using enough units to each square to have exaggerated lines far enough apart for easy reading. Also, mark off approximate positions of bearings being aligned with vertical lines.

After making needed shim changes under parts, such as generator stator and bearing pedestal sole-plates, use the telescope again to check and verify the adjustments. After all adjustments are complete, plot the "as left" curve for future reference. Result will be an easy-to-read chart similar to the one shown here.

While no time studies have been made either with tight wire or telescope, personnel using both feel that this method has saved much time and certainly has done a better job.

Use of telescope is naturally not confined to turbine-generator sets. No doubt you will find many uses as time goes on.

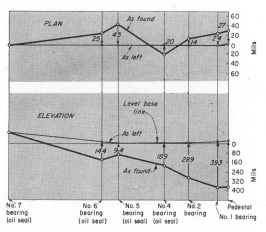

CURVES show reference line, "as found" and "as left" lines. Mark off approximate positions of the bearing bores being aligned. Cross-section paper is handy for this use, roughed out on job as each bearing is being aligned. Manufacturer's data furnish much of the information needed for this important work

If you can't make use of this new method, you may want to align with tight wire

MEASURING distance from bore to wire calls for very fine sense of feel. If there are vibrations or kinks in wire, chances are that readings will not be very accurate

WEIGHTS hung on wire, at both ends of machine, and wire length affect wire sag

GOOD MECHANIC is needed to work with tight wire. Work is slow, holds up job

Tight-wire has been used for years to align bearings of heavy machinery. Wire acts as a datum (something actual or assumed) from which measurements are taken with inside micrometers. In this way parts, such as bearing pedestals, turbine cylinders, blade rings, dummy rings, diaphragms, etc, are aligned.

Wire is stretched tight through center of bearing spaces in machine, with each end of wire fixed in space, according to bench marks established by the surveyor. Micrometer measurements are taken from various bores against wire throughout length of machine.

By adding, subtracting and dividing the micrometer readings and allowing for tight-wire sag, you can find where the center of bore is in relation to where it should be. Then various parts found out of line are moved or adjusted into correct position.

Disadvantages. You can see that this tight-wire method has several disadvantages. One is that it's time-consuming. The job is tedious and must be done by a careful mechanic with the right feel for the job. Then, again, the wire sag varies on each job, depending on its length, diameter and tension. (Tension is obtained with either a spring or a weight at one end.) If any of these factors is not correct, wire sag is naturally inaccurate. Kinks in the wire also lead to inaccuracy.

Also, if the tight wire is displaced from center of bore, angularity of micrometer to wire to a true radius leads to inaccuracy. Errors are also made when manipulating the micrometer readings, together with sag of wire to find true position of a particular part of turbine with respect to datum line.

At times, the wire is bumped, and then it must be reset again. Or the wire vibrates, giving false micrometer measurements. Another disadvantage is that it's impossible to remove a cylinder, blade ring, diaphragm, bearing pedestal, etc, to make repairs or adjustments for alignment purposes while there is a tight wire through the machine's bore. These are some of the reasons that make optical tools a welcome newcomer to the modern maintenance man.

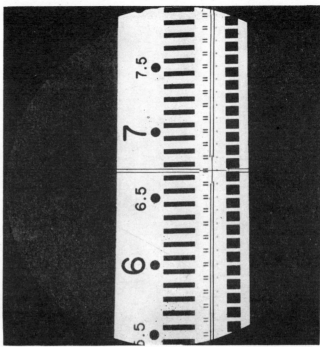

Now, you can <u>see</u> your bearings move with this . . .

Telescope-mirror alignment method

At long last, you can tell misalignment within thousandths of an inch in machines that are running and under full load. Compare this "hot" data with misalignment when machine is cold, then, for best results, realign bearings to favor hot machine

▶ KNOWING EXACTLY how far bearings are thrown out of line on a hot machine has been a lively plant topic for years. But there was no practical way to get accurate answers. Now we've licked this ancient problem with an alignment telescope and mirrors.

Cold alignment. A few years ago we first used a telescope instead of the tight-wire method to align our large turbine-generator units during major overhauls as explained in detail in

"Now you can shoot your bearings in line," on previous pages. Here the telescope is aimed at targets centered in each bearing. But, as is the case with the old tight-wire method, the rotor must be removed. This can only be done while machine is open.

Once bearings were so aligned, question was how much they'd be out of line after machine was started and put under full load. We knew that some bearings get dangerously out of line

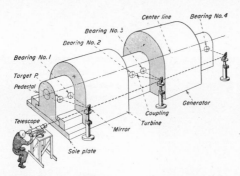

HOT-ALIGNMENT SETUP is to aim telescope parallel to center line of bearings while mirrors reflect bearing targets

TELESCOPE has tilt of one second of arc (about 0.0015-in. at 25 ft), shows the exact position of bearings in the machine

during normal running conditions. If we knew exactly how much, then each bearing could be aligned so it would be in the best possible position during normal operation.

Hot alignment. Sketch, above, shows how bearings of one of our large turbine-generator units were checked by this new method while completely assembled and hot. Targets, or check points, were installed on every bearing pedestal of the turbine-generator during a recent overhaul. These targets were placed on the side of the pedestals to transfer the inside alignment position to the outside. We then mounted a second target on the forward end of No. 1 pedestal, sketch, which we shall call target *P*. This target was used to measure pedestal tilt between the relatively cool end at target *P* and the hotter end at second target on the same bearing.

Next, we mounted our telescope a few feet in front of No. 1 bearing pedestal and 15 feet to the right of the turbine's center line. This gave us an unobstructed line of sight parallel to both the turbine and generator. Stanchions were fastened to the floor alongside the bearing targets and in the telescope's line of sight.

Each stanchion was equipped with a cup mount to receive a mirror. In this way, all targets could be observed by the telescope without disturbing the telescope once it was set. All we had to do was move the mirror from stanchion to stanchion as we looked 'around' the two machines at every target from one spot.

Our optical mirror is ground to a quarter of a wave length and fitted with a 10-sec levelling device. This assures an accuracy of 1.2 mils in 10-ft

distance. Telescope has an accuracy of 6 mils at 120-ft distance.

First test. All targets were set to zero on Jan 22, 1978. We observed them while machine was cold on Jan 22, 23 and 24. No alignment changes took place until the turbine was heated through the exhaust line. That was on Jan 24 at 9:00 am. See elevation chart on following page.

Turbine was started at 4:00 pm against 285-psi backpressure. We slowly brought turbine up to full speed of 3600 rpm at 9:00 pm. At that time we made some valve adjustments and tested the overspeed trip. Generator was phased in at 10:15 pm and load slowly increased to 60 mw. Load was reduced to 52 mw on Jan 25 at 12:30 pm and turbine backpressure was increased to 420 psi. These moves caused the bearing elevation changes that you can easily trace on the elevation chart.

Highest bearing elevations were reached on Feb 4, giving a definitive misalignment, as chart shows. From that point, bearing elevations started going down. We reasoned that ambient temperature was the cause. Reason for our thinking was that all elevation readings were based on No. 4 bearing. So when ambient temperature raised No. 4, that gave impression that the bearing elevations went down. Because we worked from No. 4 bearing, which was our fixed position, we had no way to know its actual rise.

Tilting. No. 1 bearing pedestal tilted between the two targets placed on it from time turbine started. It reached a maximum tilt of 103 mils on Jan 29. From then on it gradually declined to 65 mils. This tilt is represented on the elevation chart by the distance between No. 1 bearing line and heavy *P* line.

We had anticipated that No. 1 pedestal would tilt. That's why we left the sole plate level during the overhaul and installed micrometer check points on this plate in order to measure tilt. But we couldn't detect more than a few thousandths from these points.

Sole plate was checked with a sensitive level at that time. Results were same as with the telescope. Then we inspected the eccentricity coil mounted in No. 1 bearing pedestal. Wear was 85 mils.

Next, we plotted bearing elevation changes from data obtained by telescope. As chart shows, a definite trend was established. No. 1 bearing reached its highest elevation on Feb 4 with 175 mils over No. 4 bearing. After that it dropped back to 102 mils on Feb. 23.

Another interesting point was the tilt of No. 1 pedestal. On Feb 26 we measured a tilt of 85 mils. Unit was then shut down for balancing. During the 12-hr shutdown period, pedestal tilt decreased to 45 mils. This was caused by decrease in elevation of No. 1 bearing during this time. Fifteen hours after unit was started we measured a tilt of 80 mils. During April 28 shutdown, same tilt was noticed. At that time a 70-mil tilt forward reversed itself to a 10-mil backward tilt during a 48-hr period.

Backward tilt. Cause of this reversal was interpreted as being from expansion on the turbine side of foundation under No. 1 pedestal. Expansion itself was caused by steam piping and temperature changes. A second contributing factor was difference in foundation expansion between top and base. Here a 120-mil expansion was measured on the turbine floor level in front of No. 1 pedestal, which is considerable.

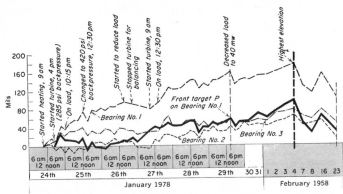

MOVEMENT of each bearing is clearly charted through every stage of warmup period, then during several weeks under full load. Max movement reached 160 mils

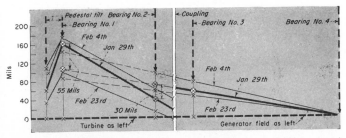

OPTICAL MIRROR on stanchion, target mounted on machine's bearing pedestal

COLD, HOT ALIGNMENT elevation shows pedestal tilt of over 0.060-in. caused by steam lines beneath pedestal. Lines were later insulated to correct the condition

We made several balance moves during Feb, Mar and April to correct excessive shaft vibrations. But the turbine spindle didn't respond to these balancing moves. So we decided cause was something else.

On April 28 we set up the telescope again to make another hot alignment check. We learned that No. 1 bearing was 15 mils high, while No. 2 bearing was 20 mils high. Unit was removed from service for a coupling check on April 29. Coupling check showed the same results as our telescope. So we lowered No. 1 bearing 20 mils and No. 2 bearing 24 mils to correct the misalignment. From experience with this unit, we know that our telescope is thoroughly reliable. We *can* accurately check alignment of our hot machinery while running and under full load.

Telescope used for hot alignment mentioned in this article is a Keuffel & Esser, known as a tilting dumpy level. It has four-screw leveling, an erecting telescope, coincidence reading bubble. telescope axle at mechanical center, and large precise tilting wheel.

Aligning bearings of cold machine...

with a telescope is now doing away with the ancient tight-wire method. Here, a micrometer target is centered in bearing bore. Telescope is set on zero, instrument adjusted horizontally and vertically so that target cross wires and telescope reticle cross hairs are in same line of sight. This line of sight becomes a reference line for the machine's entire length. Micrometers are used to center targets in all other bores of turbine and generator stator. Center-line displacement of these other bores is then measured from reference line on micrometers and recorded. From these readings, you can plot curve on cross-section paper showing "as found" alignment of bearings. Once machines are aligned 'cold,' close them up and take 'hot' readings with mirror method as explained.

The Alchemist

ENGINEERS VISITING the chemistry building of the University of Pennsylvania stop dead in their tracks on seeing this bas relief of the "Alchemist." Created by sculptor Donald De Lee, this work is highly symbolic of chemistry's dubious beginning.

The ancient alchemist tried without success to change base metals into gold, platinum, silver and other so-called "noble" metals. But where the alchemist failed, I figure that the modern metallurgist has succeeded.

Metallurgists have not only produced a metal alloy with noble properties, but at a cost far below those of the traditional noble metals. I'm talking about the familiar high-nickel-chromium steels, created by alloying iron with high percentages of nickel and chromium. Every power engineer knows how vital these metals are to our modern civilization, especially in the power plant.

My point is that many of our highly efficient tools and materials for turning out today's power at greatly lower rates had beginnings about as discouraging as the alchemists' experiments. But like the alchemist, there always were men who didn't know when they were licked.

All this proves that so long as you work toward improvement, you have a good chance of succeeding—if you stick to it.

4
Valves,
steam traps

Look inside your jet-flow

How valve works

▶ THREE TYPES of high-pressure steam-drum safety valves in general use are: (1) jet flow (2) huddling chamber (3) nozzle reaction.

Jet flow makes use of both reaction and velocity of escaping steam. Static pressure on disk overcomes spring tension, causing steam flow. Escaping steam strikes against piston and is deflected downward against nozzle-ring, discharging into body. This reactive force lifts disk higher, thereby increasing area of flow, therefore velocity.

As velocity increases, partial steam discharge takes place through the controlled orifices in guide assembly. Because of its confined velocity, the steam creates an upward force, lifting the valve still higher. Position of disk at "pop" lift is such that area of discharge is greater than area of nozzle; excess accumulation of pressure needed for full lift in huddling chamber and nozzle-reaction types is unnecessary for full capacity in jet-flow valve.

As pressure in boiler falls, velocity through orifices is reduced and valve's disk drops to intermediate lift momentarily. After that, diminishing reactive force controls further reseating, the same as in a reactive valve.

Blowdown adjustment is by an entirely different method from that in other type valves. In this jet-flow-type relief valve, the blowdown is controlled by the location of the exhaust belt reaction-lip, which is varied by adjusting the guide vertically.

Adjusting. If valve opens cleanly but does not seat sharply, nozzle ring is up too high. Lower by removing adjusting ring pin and turning nozzle adjusting ring to left one notch at a time between trials until valve seats sharply with a metallic thud. Adjust ring with screwdriver.

If blowdown is too long, nozzle adjusting ring is too high and must be lowered. If valve warns before popping, nozzle adjusting ring is down too low and must be raised. Turn ring to right one notch at a time until the warning stops completely.

Caution: When raising nozzle-adjusting ring, keep boiler pressure well below popping pressure or gag the valve while making adjustments. To protect personnel, always gag safety valves when adjusting. This is important.

Photo, courtesy Foster Engineering Co

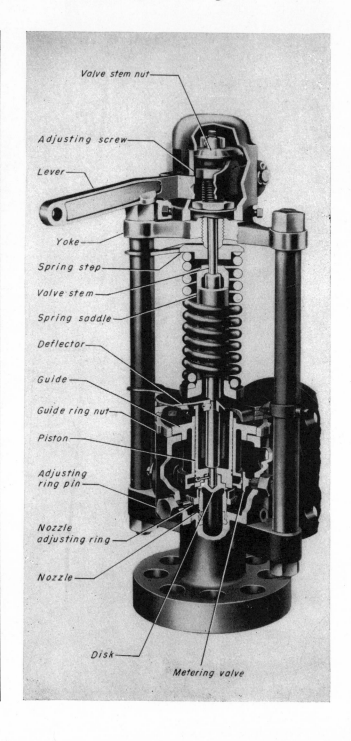

Valve stem nut
Adjusting screw
Lever
Yoke
Spring step
Valve stem
Spring saddle
Deflector
Guide
Guide ring nut
Piston
Adjusting ring pin
Nozzle adjusting ring
Nozzle
Disk
Metering valve

safety valve, keep it working

How to overhaul

1 Remove lifting lever and screw down valve-stem nut. Spring assembly can be removed without changing the setting

2 Now lift out the valve assembly in one piece and turn upside down on work bench so you can lap in the valve disk

3 Lap disk with lapping ring and paste to remove slight depressions. Give it oil rub; clean by dipping in solution

4 Unscrew adjusting ring from nozzle, examine nozzle seat that contacts valve disk. Only perfect seat is steam tight

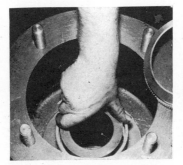

5 Lap in nozzle seat with lapping ring and paste. Again, give oil rub after lapping and rinse in cleaning solution

6 Replace adjusting ring and measure specified distance for that size from straightedge to lip of the adjusting ring

7 With adjusting ring locked in place and all the internal parts thoroughly cleaned, replace guide for disk piston

8 Valve spring assembly is now ready to go into place. Spring adjustment has not been disturbed and valve is set

9 With valve assembled, check the piston by moving with fingers. Piston must be free for valve to work properly

Here's information on

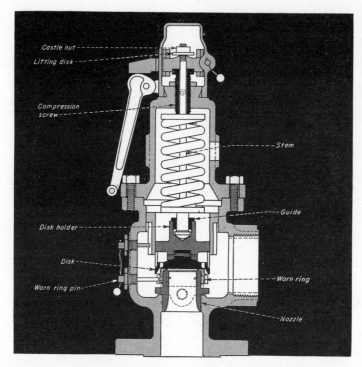

CLOSED. Pressure is contained in throat chamber, no pressure on conical baffle

How valve works

▶ THREE TYPES of high-pressure steam-drum safety valves in general use are: (1) jet flow (2) nozzle reaction (3) huddling chamber. Jet-flow type was covered in detail in other pages in this chapter.

Nozzle-reaction type. In the opening phase of any safety valve, spring force keeping valve closed must be counteracted by a greater force on pressure side. But as disk rises, spring load increases from compression. So total force needed to obtain high lift must also increase. In nozzle reaction valve

How to Overhaul

1 Remove seal, yoke pin, cap, yoke, castle nut and lifting disk from valve

2 Measure compression screw height to save time when reassembling valve

7 Measure distance from top of disk-holder guide to top of the valve body

3 Loosen the locknut, screw the compression screw out of the valve bonnet

8 Remove warn-ring pin and **also** the disk-holder guide pin from **valve** body

nozzle-reaction safety valves

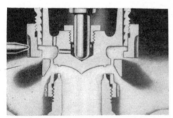

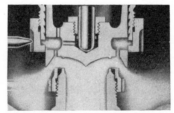

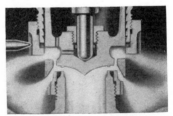

THREE-QUARTERS OPEN while popping. Partial pressure is on the conical baffle

FULL OPEN. Full pressure against enlarged baffle area, full reactive forces

HALF CLOSED. Pressure is against conical baffle on a rapidly diminishing area

shown, energy is produced by a conical baffle skirting the disk. This baffle causes lifting energy to be greatest at or near end of lift, thus giving full-bore lifts or greater at low accumulations.

Opening. As valve opens, steam escaping across seats impinges against conical baffle—then steam also acts against additional in-between areas. This large increase in head-pressure area gives sudden upward force, creating "pop" action. Initial force is increased by dynamic force of jet from nozzle—and impinging against disk face. This force increases rapidly with the flow.

Disk rises quickly from these forces,

which constantly increase from increased velocity and with help of enlarging baffle area. Thus port orifice closes almost instantly.

At the same time, another lifting energy source comes into play. Conical baffle turns fluid jet downward, giving thrust-effect as in steam-turbine blades. This reactive force takes up rising compression load of spring. At usual overpressures of 3% for steam, 5% for gases, 25% for liquids, reactive force quickly lifts disk full-open.

Closing. System pressure-drop is usually gradual, with disk falling slowly at first. Jet tends to hold disk open much longer than needed. In this valve,

a new control principle comes into play.

As disk falls in response to system pressure-drop to about 75% of full-open position, ports begin to vent pressure under the baffle. This cancels part of reactive force holding valve open. At same time, decreasing pressure under baffle works against a constantly reducing baffle area. Finally, total upstream forces become less than spring force—and disk closes sharply from about 50% of its rated lift.

When this or any safety valve needs overhauling, it's best to return it to manufacturer. If you must repair valve yourself, follow instructions carefully.

Photos, Courtesy J E Lonergan Co

4 Mark warning ring with chalk so you can establish same setting later on

5 Bonnet is not under spring tension now; remove body bolts, lift bonnet

6 Remove stem with disk and holder Disk floats in holder, no misalignment

9 Screw out disk-holder guide. With special tool, remove the steam nozzle

10 If light imperfections on seat or disk, tap in both with lapping block, shown. Remove disk from holder for this job. Never lap disk against seat; use the block

Get acquainted with your

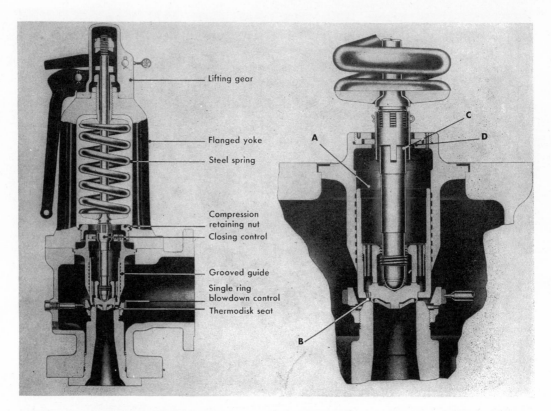

Lifting gear

Flanged yoke

Steel spring

Compression
retaining nut

Closing control

Grooved guide

Single ring
blowdown control

Thermodisk seat

How valve works and other facts you should know

► THREE TYPES of high-pressure steam-drum safety valves in general use are: (1) jet flow (2) nozzle reaction (3) huddling chamber.

► TWO OF THESE types of safety valves are covered in detail in this chapter. Make sure you know what types you have and keep them in working condition.

Huddling-chamber type. In this valve the static pressure acting on the disk area causes initial opening. As valve pops, steam space within huddling chamber between seat and blowdown ring fills with steam and builds up more pressure on roof of disk holder. This temporary pressure increases the upward thrust against spring, causing disk and its holder to lift to full pop opening. After a predetermined pressure drop (blowdown), valve closes with a positive action by trapping steam on top of disk holder.

Blowdown is adjusted by raising or lowering the blowdown adjusting ring. Raising this ring increases blowdown; lowering decreases blowdown. Capacity figure as stamped on name plate and as required by the ASME Code for steam is 90% of valve's actual capacity when flowing at 3% overpressure.

Huddling-chamber-type valve is intended primarily for steam service but will work with gases. However, it is rarely used for such services as it doesn't have a closed bonnet. It's never used for liquid service.

Precision-closing control. With valve in open position and discharging, steam is bled into chamber A through three bleed holes B in roof of disk holder. Likewise, spindle overlap C rises to a predetermined position above the floating washer D. Area between floating washer and spindle is thereby increased by difference in two spindle diameters.

Under this condition, steam in chamber A escapes to the atmosphere through the secondary area formed by floating washer D and the spindle.

At instant of closing, spindle overlap C is so adjusted that it moves down into floating washer D, thereby effectively reducing escape of steam from chamber A.

Resulting momentary pressure buildup in chamber A produces a downward thrust in direction of spring loading. Combined thrust of pressure and spring loading results in tight, positive precision closing without wire drawing or dragging of seat surfaces.

—*Courtesy: Manning, Maxwell & Moore, Inc*

huddling-chamber safety valve

HOW TO OVERHAUL

1 Remove cap and lever assembly, screw compression retainer against spring

2 Remove bonnet, then lift the spindle guide and feather assembly from valve

3 Unscrew the feather (valve disk) from spindle. Hold spindle; turn the disk

4 Check blowdown's ring's position by screwing the ring up against disk

5 Insert power-driven abrasive-cloth recutting tool through feather guide

6 Regrind seat if necessary, using an electric drill, as shown, for speed

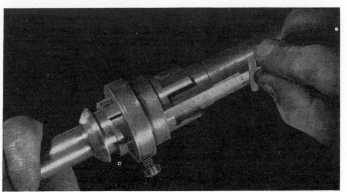

7 After grinding disk on seat, adjust the spindle overlap to keep the right back pressure on the disk for positive closing. Use the scale as illustrated here

COMPRESSION NUT on large reheat safety valve is tightened to increase the popping pressure. Valve on right has been set, gagged

MOVING INNER ADJUSTING RING with screwdriver adjusts the amount of blowdown. This valve is on the hot reheat steam line

When is the best time to set safety valves?

"After the yearly overhaul and just before placing the boilers back into service," says one group. "No," says another, "today we set boiler safety valves with the unit on the line."

Safety valve setting is, and has always been, one of the necessary jobs associated with boiler operation. It is a very important annual chore which must be done correctly, since these devices are often called on (between overhauls) to perform their protective function.

In general, boiler valves operating below 1200 psi are not too difficult to set. But even here, it's not unusual to find an obstinate valve that tries ones patience.

When to set? For several years there have been two schools of thought on the proper time for checking and setting safeties. One group believes in checking settings *after* the yearly overhaul, and just before placing the boiler back in service. This practice was followed either by those who had very little trouble with safety valves, or by those who completely dismantled all valves during the outage. These are both good reasons.

In utility plants, this routine did save some time at the start of the outage. Reason was that the boiler pressure could be lowered with turbine load. Because this practice cooled the turbine and stop valves somewhat, work on the unit could begin as much as a day early. This was very good practice if the valves were generally good performers. Also, if all valves were dismantled yearly, they would have to be checked before going back into service anyway. So this method is often followed.

The other practice is to check valve settings and general performance immediately *after* the boiler has been removed from service for the yearly overhaul. In utility plants, this means holding boiler pressure at rating during the turbine's entire load rejection period.

This practice does have merit. It allows a selective overhaul of valves which (1) cannot be made to operate correctly, or (2) those which leak after popping. But it does require a second session of checks on dismantled valves.

GIANT SAFETY VALVE needs attention as smaller ones. Over-gagging during setting is most common source of valve trouble

10 advantages of setting all boiler safety valves with unit on the line

1—Superheater sections are protected with adequate steam flow at all times

2—Adequate steam flow assures a reliable check of superheater tube metal temperature

3—Valves are checked under more normal operating conditions

4—Pressure increase to popping can be controlled with load and/or a minimum of overfiring

5—Following the pop, pressure can be reduced more rapidly with load to the reheat point

6—There are shorter periods of above normal pressure on the unit

7—It's not necessary to subject the stop valve to cycles of above normal pressure on one side and atmospheric pressure on the other

8—Time for checking the valves is reduced to a minimum

9—Safety valves can be set at a convenient time—two or three weeks prior to overhaul. This allows scheduling of work for faulty valves during overhaul

10—No need to guess at temperature corrections for reheat and superheat valves

Also, there is often a question in one's mind about the correct operation of a valve on a unit which has been cooled down, then heated up again. Does this affect the spring or seating surface? Maybe it doesn't, but I'm not sure.

The advent of installing reheat turbine generator units in electric generating plants brought more safety valve check-out headaches. In general, boilers for these units operate at 1500 psi and above. This means higher pressure to contend with on both the drum and superheat safety valves. Thus conditions here are different.

Reheat valves are generally lower pressure, ranging from 500 to 600 psi. They require closer setting tolerances. But to keep the number of valves per boiler within reasonable limits, and still have adequate relieving capacities, large valves are used. It is normal to find 4- or 6-in. diameter valves on reheat systems. The larger sizes certainly don't ease checking the performance or the maintenance of the valves.

Problem. Reheat valves present a problem in checking operation and performance. During initial startup, the temporary steam-line blowout piping should be arranged so that the safety valves can be set immediately after the steam line blows and before turbine operation.

During initial operation, this gives you proven overpressure protection for the reheater and high-pressure turbine. It also eliminates load swings associated with setting valves with the unit on the line. This is an important consideration since it comes before the unit is completely *wrung out*.

While it is true that subsequent yearly checks must be made with the unit on the line, this is not a big problem. Reason is that, normally, by now all equipment and controls have been put through their paces and should be functioning reliably. But, even with an experienced man operating the intercept valves, a momentary load swing of from 10 to 20% is normal.

Many people, before operating reheat units, would never have considered setting safety valves with the unit on the line. This is understandable since many of them grew up with header system plants where this method of setting safeties was nearly impossible.

But with the unit systems, and a little experience with reheat valves, it became a natural thing to set all boiler safeties with the unit on the line. This holds true even if it is a straight superheat unit. This is one of the less painful ways to perform a distasteful but necessary job. Besides, it offers the 10 advantages shown in above box.

Precautions. Of course it is necessary to use some operating precautions. For one thing, steam flow from the boiler should be limited to the total relieving capacity of ungagged valves. At one point in the operation, all but one valve in either the main boiler or reheat system will be gagged. And steam flow from the system should be limited to a value *below* the relieving capacity of that particular valve.

Most boilers are equipped with devices that will establish a positive limit to the amount of fuel entering a furnace. This may be through pneumatic or electronic control stops or with the fuel valve handjack. In the case of solid fuels such as pulverized coal, normally all but one pulverizer can be taken out of service. If this does not make a positive fuel limit, feeder stops can easily be installed. So limiting the fuel is not a problem.

While setting valves, the entire unit should be under constant supervision by phone, or other reliable contacts between the control room, turbine and safety valve area. Thus abnormal over-pressure situations can be relieved quickly by picking up load, or tripping all fires.

1—FIRST FIND THE EXPANSION FACTOR

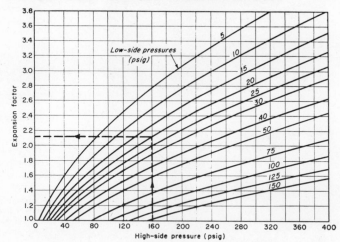

Find your high-side pressure at bottom of chart, draw line straight up until you meet your low-side steam-pressure line, then read expansion factor on left-hand scale

To protect the low side of a reducing valve ..

... How Big a Safety Valve?

The old rule of "safety valve same size as low-pressure piping" is all wet according to Mutual Boiler and Machinery Insurance Co engineers. Their tests show you need more relieving capacity than you can get with a single safety valve same size as piping*

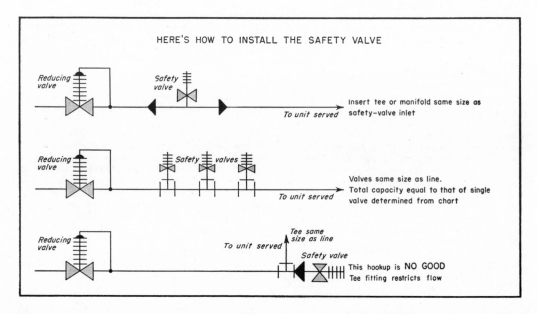

HERE'S HOW TO INSTALL THE SAFETY VALVE

Insert tee or manifold same size as safety-valve inlet

Valves same size as line. Total capacity equal to that of single valve determined from chart

This hookup is NO GOOD
Tee fitting restricts flow

2—PICK OFF THE RIGHT VALVE SIZE

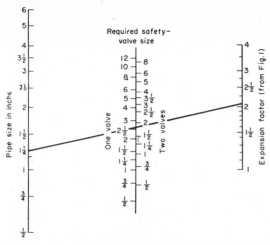

Draw line between expansion factor (from Chart 1) and pipe size of low-side piping to find required safety-valve size

3—THEN CHECK VALVE'S CAPACITY (LB STEAM PER HOUR)

Safety-Valve Setting (Pounds per Square Inch)

Safety-Valve Size	5	10	15	20	25	30	35	40	45	50	60	80	100	120	140
½ in.	60	68	75	80	85	90	94	99	102	105	113	126	140	150	155
¾ in.	112	125	136	147	156	165	174	181	189	195	209	234	258	279	292
1 in.	198	219	237	255	271	287	301	315	328	341	366	410	452	486	507
1¼ in.	389	431	470	505	536	566	594	621	647	672	721	809	889	956	997
1½ in.	540	600	650	700	750	800	845	885	920	955	1030	1150	1260	1350	1430
2 in.	980	1065	1150	1240	1310	1395	1455	1530	1595	1655	1780	2000	2195	2365	2470
2½ in.	1435	1560	1705	1830	1945	2055	2160	2255	2355	2450	2640	2955	3250	3500	3650
3 in.	2375	2625	2855	3055	3255	3445	3620	3790	3940	4105	4395	4945	5440	5840	6080
4 in.	4350	4840	5250	5625	5985	6310	6625	6935	7230	7525	8070	9060	9850	10,500	11,025
6 in.	10,900	12,000	13,050	14,000	14,950	15,800	16,550	17,400	18,050	18,800	20,050	22,500	24,500	26,400	28,000
8 in.	18,900	21,000	23,000	24,750	26,500	28,050	29,050	31,000	32,400	33,550	36,100	40,500	44,350	47,700	50,950

▶ Suppose the diaphragm of a pressure-reducing valve ruptured, leaving the valve wide open! What would happen on the low-pressure side? Would a conventionally installed safety valve keep the low-side pressure down?

A near explosion in a policyholder's plant led Mutual Boiler and Machinery Insurance Co engineers to have some doubt. They started looking into the problem and came up with some startling facts.

For years it's been common practice to install safety valves, on the low-pressure side of a reducing valve, of the same size as the low-pressure piping, set at the allowable pressure of the equipment to be protected. To find out if this was safe enough, Mutual's engineers ran the following test.

The Test. They took a 2-in. reducing valve with 2-in. piping on the low-pressure side, protected by a conventional 2-in. safety valve set at 30 psi.

Pressure on the high side was 143 psi. Then the reducing valve was held open (to simulate a ruptured diaphragm). The safety valve let go okay, but the low-side pressure rose to 105 psi, even with the valve blowing.

The engineers repeated the test with *two* 2-in. safety valves, and the low-side pressure rose to 70 psi with both valves blowing.

So they tested safety valves, same size, of another reputable make. Same story. With one safety valve the pressure shot to 120 psi; with two valves blowing the low-side pressure hit 87 psi.

Conclusion was that safety valves same size as piping did not have sufficient relieving capacity for safe use on the low side of pressure-reducing valves.

Charts. To find out just what capacity was required, they continued their tests and came up with the above charts. Box (above right) tells how to use these charts to find the proper-size valve for most installations and the required relieving capacity. You need to know both size and capacity because a valve the right size may not have the required relieving capacity. In this case you need either a larger safety valve or two or more valves whose total capacity equals the required.

Hooking Up Valve. Box (facing page) shows two approved methods for hooking up safety valves and one method that's NO GOOD. The bottom hookup is n.g. because the tee fitting slows down flow and sometimes lets pressure build up in the up-stream piping faster than safety valve can relieve.

Exceptions. However it's not always necessary to go to a safety valve larger than the piping. One manufacturer has developed a special set of orifice nipples to go on the down-stream side of the reducing valve to limit the maximum possible flow through the reducing valve to a rate that can be handled by the safety valve. Or you can install a rupture diaphragm same size as piping that will let go at a pressure not greater than 20% above safety-valve setting.

NONRETURN valve on boiler is opened by maintenance man at Eastman Kodak Co's Hawkeye plant. Experience shows that it pays to instruct all personnel in the operation and maintenance of valves because their upkeep can run into big money

'KNOW COMMON VALVES,'

CLASS of power-plant youngsters at Dow Chemical Co soak up valve info to help pass upgrading exams. Testing relief valve (right) assures operation when installed

▶ METALS for valves are usually brass, bronze, iron or steel. When selecting valves for steam, cold water, oil or gas, almost any of these metals is good enough. But each has certain limits. Never use brass above 550 F. Bronze can stand 350-psi pressure and temperature up to 700 F. Bronze and brass valves usually come in smaller sizes, up to and including 3 in. Above this size, use iron up to 250 psi and 500 F. Steel is for temperatures above 550 F, and is also used when valve must withstand high internal or external pressures or shock.

Markings. Look at the valve body. A

Each Valve Is Designed to Do a Specific Job. Your Operators and . . .

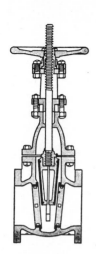

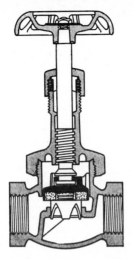

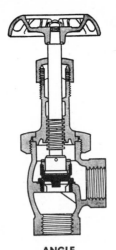

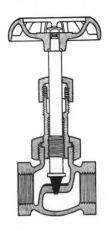

GATE

Outside screw & yoke (os&y) has a rising spindle and double disk taper-seat for sealing

GLOBE

One-piece bonnet, renewable composition disk helps this valve stay tight a long time

ANGLE

Union bonnet and regrinding bevel-disk. Save fittings by putting these at elbow spots

NEEDLE

Union bonnet with a stainless steel disk assures a tight closing and long wearing seat

"2," "4," etc. on the body gives smallest pipe size the valve should be used on. Body marked 200 WOG means valve is safe for no more than 200 psi, whether it's cold water, oil or gas.

Instead of 200 WOG, valve may be marked 125 or 125S. "S" means that this valve is for 125-psi saturated-steam pressure at its equivalent temperature of 344 F. But for "cold" service same valve may be used up to 300 psi. So temperature and pressure are important. Of course, it's dangerous to use a "cold" valve for a "hot" application.

Type and Use. Gate valve is about the most common type. It is named for wedge (or gate) that's lifted for full, free flow. But it's not designed for throttling, close-control, or quick-closing operation. One reason is disk must be opened full diameter of valve, making operation slow. Throttling also causes severe wiredrawing or wear, thus damaging seating surfaces.

Globe valve is so named because of its body shape. This valve is very popular for flow control. Reason is disk travels only a short distance from full-open to closed position, saving operator's time. Throttling also causes wear caused by throttling is easily remedied by replacing or regrinding disk and seat. But trouble is that it restricts or offers more resistance to flow.

Angle valve is same as globe, but body is built on an angle. Use for a 90-

deg turn in the line, thus reducing number of fittings needed. Then there is also less resistance to flow in line.

Y or blowoff valve has all the good features of a globe valve, but it offers less resistance to flow. It is very popular for blowing down boilers. If this or any other valve has a replaceable composition disk, it gives a little. That allows disk to adjust itself to scale and other particles. So tightness is maintained and seat protected against damage.

Needle valve resembles globe valve but has a tapered disk. Use where extremely close regulation is required. Taper fits accurately in its seat, giving drop-tight closure with very little effort.

Check valve is needed for many services. Use as a safety precaution to prevent any back flow. Or use to keep liquid in a line when pump is shut down —then it is called a **foot valve.** Check valve operates automatically, flow keeps it open, and gravity closes it. Two basic types are swing and lift. Use swing check in series with gate valve and operate it either vertically or horizontally. Lift check valve teams up with a globe or angle valve. It is made in three types, angle, horizontal and vertical.

Special type of lift check is called a stop check. Use on parallel boiler installations to prevent back flow, and so cut out boiler. Stem is not connected to disk and acts only as a wedge to hold

disk down. Or stem allows pressure to raise the disk.

Plug valve or cock has a tapered plug with hole for liquid to run through. Plug is turned to line up with pipe openings. Plug can be opened or closed quickly. It handles thick or dirty fluids easily. But keeping it tight may be a problem.

Plug valve comes in various designs. But basically all types are the same, except for variations in stem, bonnet or disk.

Operation of all valves is based on the principle of a nut and bolt.

Rising stem is the simplest and most common type, used for many globe and angle valves. Also for gate valves with split wedge, allowing stem to turn.

Nonrising Stem. Here threaded disk travels up and down on stem. Because of this feature, this design is used only on gate valves.

If a standard right-hand thread were used, turning stem clockwise would allow disk to "run" up, making it come off the seat. To have conventional operation, a left-hand thread is used. Then turning the stem clockwise causes disk to go down, closing the opening. Thread makes the difference between a right- and left-hand valve.

Outside screw & yoke. Here nut is turned, but it can't move up or down so stem moves without turning. Used on solid-wedge gate valves. In this type,

... Maintenance Men Should Know What They Are Made of, Where to Use

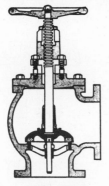

STOP CHECK

With stem open, disk opens by itself to equalize pressure, closes as check

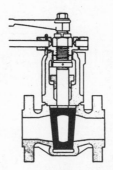

PLUG

Stick-proof plug valve is loosened before opening or closing, then tightened

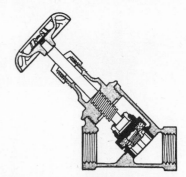

Y TYPE

This has all good features of a globe valve but offers less resistance to flow

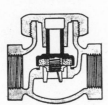

CHECK

One-piece cap and renewable composition disk prevent back flow in the pipeline

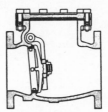

SWING CHECK

Bolted cover, requiring disk in larger type, is used in a horizontal pipeline

BYPASS

Prevent wiredrawing the large valves by running small equalizing line, as shown

threaded section of stem is entirely out of valve body. That makes it easy to lubricate, and it is free from exposure to line conditions. Besides, you can see if valve is open or closed.

Plug valve often sticks because of very tight fit between body and plug. To eliminate this, some plugs can be raised, giving more clearance and making it easy to turn. Cone checks or very large plug valves have this feature. Besides manual operation, valves can be operated with levers, motors, gears or pistons.

Bonnet Types. Screwed-in is the simplest of the bonnet types, being of one-piece construction. Used on small, low-pressure valves, it is not good when frequent opening of the body is necessary, or where shock or vibration is common.

Union-ring bonnet is a type with internal parts that are easy to inspect, repair or service. This construction gives added strength to body against internal pressure and distortion; it's practical only on valves up to a certain size.

Bolted bonnet is most rugged type,

and so is used on almost all large valves. Gasket between body and yoke enables it to withstand high pressures and temperatures.

Type of Ends. Brass, iron and stainless-steel valves in smaller sizes usually have screwed ends. Larger sizes have flanged ends. Solder ends are used when valve is to be connected to copper pipe and tubing. Steel valves may have screwed, flanged or welded ends.

Bypasses are used on steam lines to prevent water hammer by equalizing pressure and temperature on both sides of valve. Water in steam or in "dead" line causes damage if a valve is opened too quickly. Opening valve slowly to prevent this injury may damage seat and disk from wiredrawing. So open bypass valve slowly while main valve is still closed. Because bypass valve has only a small area compared to a large valve, don't worry about wiredrawing.

Maintenance. Preventive maintenance for valves simply means lubricating stem threads, if exposed, and bearing, if there is one; also, tightening down packing nut or gland flange to prevent

needless leakage around the valve stem.

Extensive maintenance depends on service. Repacking may be needed while valve is under pressure. Some valves may be repacked under pressure with valve wide open.

If a valve leaks at the seat, repair it at once. If neglected, seating surfaces may be so badly wiredrawn that valve must be scrapped. Gate valve is repaired by lapping in disk and seats with fine emery cloth.

Metal disks of globe and angle valves are repaired by grinding in with an abrasive compound. Use slug, coin or pin to take up play between disk and stem, and thus keep disk from turning on stem. Disk must turn with stem to allow oscillating motion needed for grinding.

Valves with composition disks may be serviced by replacing disk insert. Right disk is needed for each job. Asbestos composition is often used for saturated steam, air, gas, hot water or condensate. Leather or rubber disks do good job for cold water or other liquids. Special soft metal is for air, gasoline or oil. You must know service for each.

RIGHT	WRONG	
1		Use a wrench on the hexagonal connection near the joint. This gives support to the valve, prevents possible damage.
2	Closed / Open	When installing a valve, keep it in the closed position. It's more rigid and less likely to be twisted.
3		Use good supports, expansion bands or joints. Don't force a valve to carry the weight, sag or expansion of piping.
4		Over-threading may let pipe shoulder against valve seat. Besides injuring valve this keeps joint from making up.
5		Pipe dope on end threads of valve means grit will appear on valve seat sooner or later. Put compound on the pipe threads.
6		Old pipe for a new or altered job is all right if the pipe is *clean.* Dirt or scale left in pipe results in valve trouble.
7		Leverage can murder a threaded joint. A short lever and a few hammer taps is more effective and won't twist the valve.
8		If valve is leaking, don't try to force it shut. This ruins more valves than anything else. Try to flush obstruction.
9		Gate valves and cocks may wiredraw if used for throttling purposes. Use globe or angle valves. These give a balanced flow.
10		Automatic stop and check or lift check needs upright positioning. Disk and piston must have a satisfactory gravity drop.

Here's your pressure-pilot

Applications are many and varied

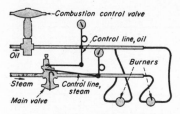

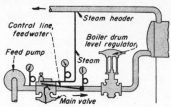

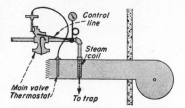

Oil burner. Atomizing steam pressure is regulated to oil burner at preset amount, above the oil pressure. Steam pressure under diaphragm is balanced by oil pressure and adjusting spring on top. Once set, adjusting spring force is constant. As combustion-control valve raises oil pressure, regulator raises steam pressure until pilot diaphragm is again balanced. Improves combustion.

Boiler feedwater. Valve controls pressure drop across boiler level-regulator when boiler is fed by motor-driven centrifugal pump. As demand for feed decreases and level-regulator throttles, main regulator also throttles, offsetting increased pump-discharge pressure. It maintains inlet pressure to level-regulator at constant differential over boiler pressure. Self-operated, no air needed.

Makeup air. Here we provide control of makeup-air temperature with varying outdoor-air temperatures. A maximum coil pressure is set on pressure control of pilot to heat incoming air to needed temperature at lowest outdoor temperatures. Thermostat element then throttles the pilot (and therefore main valve) to maintain constant outlet-air temperature needed. Only valve needed.

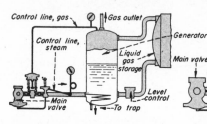

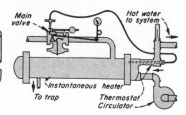

Gas-pressure control. Here main pressure is constant from propane-gas generator; valve limits steam pressure in the generator. Usually pilot double spring controls main valve, admitting just enough steam to maintain set gas pressure. When gas pressure drops and calls for more steam, the pilot (small) limits steam pressure on exchanger to the pressure for which pilot is set.

Remote regulator. Hookup regulates pressure at a distant point from regulator itself. Slight increase in controlled pressure throttles the pilot, which in turn throttles the main valve. One big advantage is there's no limitation on distance so long as control pipe is solidly filled with water. Use hookup where noise or appearance isn't too important at accessible locations.

Hot water. Valve controls converters for hot-water heating systems. Pressure element of pilot limits heater pressure on start-up with cold water. As return temperatures increase, thermostat throttles the pilot and main valve. That maintains a constant converter outlet temperature. Operation of this hookup does not depend on outside power source. Modulating control saves steam.

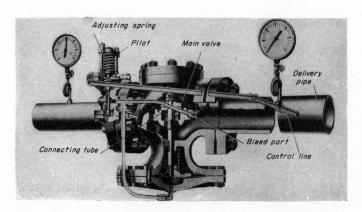

Now let's look

One big difference in this regulator from those featured in the following pages is that its pilot valve is on outside, not in housing. Here, pilot can be removed from valve in a few minutes, replaced if needed. Maintenance men can thus have several cleaned pilots on a shelf, ready to service a large number of pressure regulators throughout

single-seated packless regulator

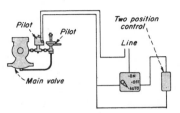

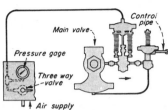

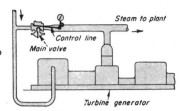

Electric shutoff. Remote shutoff is by electric control where frequent interruption is needed. When electric power is supplied to the pilot, it opens. That admits steam to the pilot and allows it to control the main valve as a simple pressure regulator. Here interrupting current flow allows pilot to close. That closes main valve. Costs less than separate electrically operated shutoff.

Pressure shutoff. Hook-up provides remote shutoff by fluid pressure where frequent interruption is needed. When pressure is applied to diaphragm (double spring), pilot closes. That denies pilot (single spring) steam needed to open main valve. Pilot, when unloaded, opens, which allows it to control main valve as a simple pressure regulator. Can operate a distance from regulator.

Makeup steam. Valve supplies steam to turbine exhaust main. When turbine load decreases and turbine exhaust is not enough for steam load, very slight drop in exhaust pressure causes pressure regulator to feed correct amount of steam to meet demand. Here pilot-operated accuracy avoids undue pressure drop before makeup starts, which makes steam flow to plant very constant.

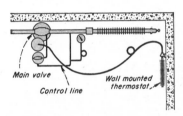

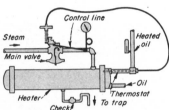

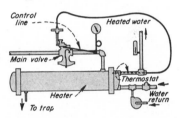

Temperature control. This is economical hookup for warehouses, shops, kilns and other spaces where appearance is not too important. Pilot controls steam flow to the installed radiation that's indicated by the temperature at thermostat bulb. Pilot controls the pressure on the heating surface when the temperature pilot opens wide, as at start-up, for example. No air or electricity needed.

Oil preheater. Here's way to control fuel-oil temperature from an oil preheater. When the oil temperature is near the setting of thermostat element, the element begins to remove pressure loading (imposed by the pressure-loading spring) from the steam-pressure control diaphragm. That allows the valve to throttle and control. Limiting pressure minimizes carbonization here.

Domestic hot water. Valve controls hot-water temperature from a tankless heater as shown. Rising water temperature resets pressure loading. This reduces the heater steam pressure to that needed to maintain the load. This hookup eliminates need for a second regulator of the pressure-reducing type. Thermal element is built to withstand accidental over-temperatures.

inside this valve and see how it works

plant. Then pilot can be cleaned in leisure time, placed back on shelf.

Operating cycle. Main valve is normally closed. Placed in service, pressure acts on under side of small pilot valve. Control line then connects pilot diaphragm chamber to delivery piping. Now compressed adjusting-spring forces pilot valve open, see details in drawing.

Connecting tubing conducts fluid from pilot to main valve diaphragm and bleed port. When pilot opens, fluid flows through pilot faster than it can escape at bleed port, creating a loading pressure that forces main valve open.

Delivery pipe and control line are now being filled with fluid flowing

through main valve. As delivery pressure rises, it overcomes force exerted by adjusting spring and pilot throttles. This, in turn, allows main valve to throttle just enough to maintain the set delivery pressure. When demand stops, pilot closes, allowing main valve to close. *Courtesy: Spence Engineering Co, Inc, Walden, New York.*

For best results know

How valve works

▶ PRESSURE-REDUCING VALVE is an automatic regulating valve for controlling the downstream pressure of a fluid from a higher-pressure source. Three general types are: (1) self-contained internal-pilot piston-operated (2) self-contained external-pilot operated (3) spring- or weight-loaded direct-operated, with diaphragm, bellows or piston. A self-contained or self-operated valve uses fluid being controlled to operate its main valve.

Valve shown is for steam, air or gas service, with initial pressures from 25 to 1500 psi. This type is built for temperature up to 1000 F, for reduced pressures from 2 to 600 psi.

Principle is simple. Screw down on handwheel to compress the adjusting spring against the metallic diaphragm. That opens controlling valve, which admits high pressure from inlet-body port to top of piston. Piston opens main valve and admits reduced pressure to outlet piping. This reduced pressure acts through the outlet body port (or through an external control pipe on some valves) on the underside of diaphragm to balance compressed spring.

Action caused will throttle the controlling valve and control the reduced pressure at the set value. Any load change immediately reacts with a pressure change on diaphragm, instantly repositioning the main valve. That maintains the constant reduced-pressure setting. Change in reduced pressure affects the valve action.

Don't guess at upstream pressure. Use gage at control point to indicate line losses due to friction, which may result in undersizing a valve. Be sure that valve you select is suitable for minimum and maximum load and for superheat conditions.

Never buy a reducing valve to match the pipeline size without checking flow requirement as the line may not be right size to start with. Get information from manufacturer's capacity tables to show size for a given flow and valve design. When reducing to 25% of inlet pressure or less, steam expands in volume two and one half to five times. If you don't know the flow, find it with a flowmeter. Or estimate on basis of equipment performance that valve will service.

Photo, courtesy, Leslie Company

*self-contained internal-pilot piston-operated

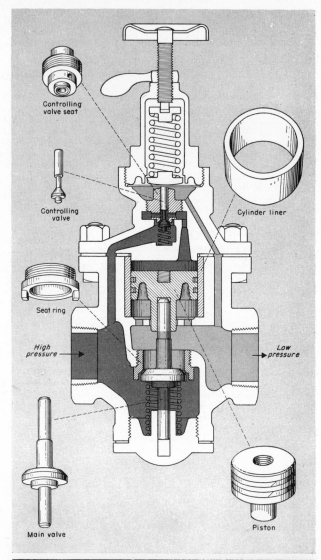

Controlling valve seat

Controlling valve

Cylinder liner

Seat ring

High pressure

Low pressure

Main valve

Piston

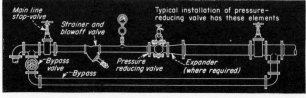

Main line stop-valve

Strainer and blowoff valve

Typical installation of pressure-reducing valve has these elements

Bypass valve

Bypass

Pressure reducing valve

Expander (where required)

128

your s-c i-p p-o* reducing valve

How to keep it working

1 Close stop valves on each side of reducing valve, release compression on adjusting spring by unscrewing handwheel

2 Unscrew adjusting-spring case with wrench. Remove case, take out spring. Remove diaphragm with a suction cup

3 Unscrew controlling valve-seat with socket wrench provided for purpose. Remove valve spring and the tiny valve

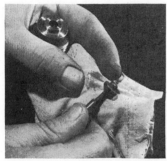

4 Examine seat and valve. If scored or cut, regrind with fine grinding compound. Clean by dipping in kerosene

5 Check clearance between control valve and diaphragm with straightedge and feelers, against maker's recommendations

6 After assembling spring with valve and screwing controlling valve-seat into top cap, remove the cap and gasket

7 Use bolt to remove piston. Examine rings for wear and fit. Remove cylinder liner, inspect the inside for wear

8 Unscrew bottom cap and take out with main valve and spring. Examine for cuts or scored seat, for free movement

9 If needed, grind in valve with liner and piston in place. Clean parts, check for free movement, then reassemble

Here's your air-operated

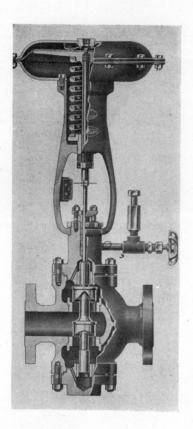

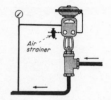

HIGH-PRESSURE GAS to transmission line is usually reduced. The angle valve shown is recommended for this service

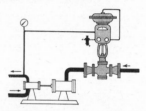

DISCHARGE PRESSURE on steam-driven reciprocating pump may be held closely by controlling the steam to the pump

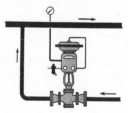

STEAM MAKEUP to low-pressure process or heating. A second controller can relieve the excess on the light loads

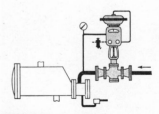

HEATING-STEAM PRESSURE to reboilers and heat exchangers is controlled to meet the exact process requirements

How the control valve works . . .

AIR PUSHING DOWN on the diaphragm provides power for working this control valve. Valve shown has direct-spring return; also made with reverse-spring return, where air pushes up under diaphragm. Increase in operating air pressure above the initial spring setting, of 3 psi for 3- to 15-psi range, pushes motor stem down. This motion is opposed by the spring compression.

Resulting valve action is known as air-to-close when valve plug is moved toward valve seat; air-to-open when plug is moved away from seat. Either action is from direct motor where design permits inverting valve body and plug. Where design doesn't permit air-to-open action with direct motor, use the reverse-motor type (not shown).

Stop in upper case holds initial diaphragm position. Spring is normally set so motor stem starts to move when air pressure to diaphragm is equal to the minimum spring range. Compression can be changed with external adjusting screw.

Spring load is carried by a ball-thrust bearing in motor having more than 100 sq in. effective diaphragm area. Ball reduces force needed to adjust. It also minimizes torsional stress imposed on spring under compression. Travel indicator shows exact plug position between full open and full closed.

Plug design is important. It directly governs flow through valve. Two most common types are double-seated top-and-bottom guided V-port (shown) and parabolic plug (not shown).

diaphragm-controlled valve

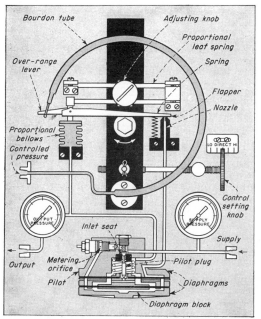

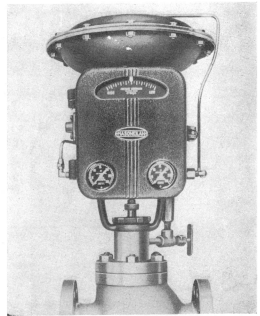

Pressure controller is this valve's brains . . .

In previous pages of this book we learned that a control valve has a pilot valve. Small pilot is either built inside the larger regulator body or is placed on outside. Pilot of this control valve is part of the pressure controller placed on outside of valve; it is connected to valve's diaphragm.

Supply air passes to under side of pilot plug. A little air passes through metering orifice and serves as nozzle supply. When nozzle is covered by flapper, these things happen: (1) Nozzle pressure builds up on top of upper diaphragm. (2) Diaphragm block moves downward. (3) Pilot plug seals exhaust seat in diaphragm block and opens inlet seat. (4) Output pressure builds up under lower diaphragm. (5) Diaphragm block moves upward.

Inlet seat remains open until equilibrium is reached between nozzle pressure and output. When thus balanced, there's no flow of air into or out of pilot—theoretically. But small bleed between supply and outlet increases

pilot responsiveness when at equilibrium.

Decrease in nozzle pressure opens exhaust seat, allowing output air to escape between two diaphragms—through horizontal ports in intermediate plate. Ratio of effective areas between upper and lower diaphragm is such that a change of about 3.5-psi nozzle pressure causes output pressure change of 3 to 15 psi.

Flapper pivots on pin at movable end of proportional bellows. Bellows is connected to output pressure from pilot. Travel of this bellows is proportional to output pressure.

Increase in controlled pressure causes free end of bourdon tube to move flapper clockwise, tending to cover nozzle and increase output pressure. Proportional (feed-back) bellows tends to lower flapper bearing, causing flapper to cover nozzle. In normal operation, reaction of pressure in proportional bellows keeps flapper in position to throttle a steady output pressure.

Movement of feed-back bellows is proportional to: (1) movement of free end of bourdon tube (2) output pressure of pilot. So output pressure must be exactly proportional to movement of free end of bourdon tube.

Function of proportional band-adjustment knob is to vary effective length of proportional flexure rod.

Control-setting knob is for setting controlled pressure. Index setting scale directly above knob has graduated reference marks between Low and High. Direction knob is moved along scale; indicates effect of adjustment.

Range of pressure is for 0 to 10.000 psi, but all bourdons furnished will withstand 25% overpressure without damage, according to manufacturer. Standard controller takes clean dry air or noncorrosive gas at 20 psi. Air set of reducing-relief valve with filter-dripwell is good insurance in air-supply line as shown in application sketches. *Courtesy: Mason-Neilan Regulator Co, Boston, Mass.*

Water Hammer.

RUNAWAY WATER

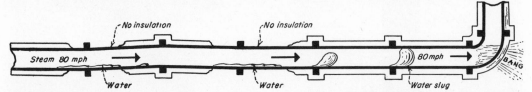

CONDENSATE in steam lines cuts down steam's efficiency besides being harmful to system. Water collects from boiler carry-over or from condensing in poorly insulated lines. If lines sag like these or are not drained properly, steam blows small pools of water together, collecting them into waves, then slugs. Because steam travels fast, slug is carried at that speed until it reaches turn in line. Then incompressible water slug hits piping with loud hammer. This condition is dangerous

DAMAGES

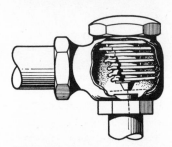

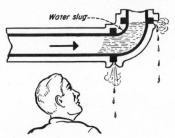

THERMOSTAT BELLOWS is knocked out quickly by badly pounding steam lines. With water slug slammed against delicate bellows, it's deformed so valve won't seat, then steam blows through trap

GASKET JOINTS are usually weakest part, especially if piping is old and gasket material is other than metallic. Water hammering at joints soon forces water through or blows out gasket

DAMAGED PIPING can be costly, dangerous. A high-speed slug of water often shoots through elbow or tee at end of long line, like a cannon ball. Or fitting may crack, making it necessary to replace

CAUSES — CURES

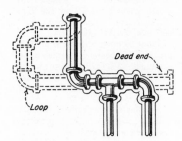

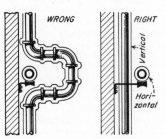

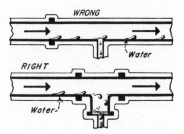

LOOPS AND DEAD ENDS trap water. Then water's picked up by steam for knockout wallop. Go over old system and eliminate all useless fittings or hookups that condense steam — collect water

VERTICAL LOOP around horizontal line is bad. This 4-in.-dia loop has flow resistance of 32 ft straight piping. Try to run vertical lines against wall, with horizontal lines a foot away from wall

RUNNING DRAIN into large pipe does not do good job. Water shoots over small opening. Right way is to place fitting in line that has large "pot" for collecting water from pipe's lower half

. . Damage, Causes & Cures

TRAP WATER BEFORE IT DOES DAMAGE

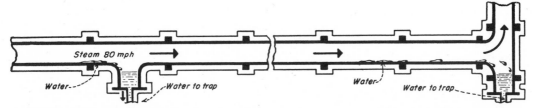

STOP WATER HAMMER completely if from boiler carryover, water condensing in long lines, lines not sloped downward, or if poorly insulated. First step is to place tees in line as shown, with drain line to traps from flanged bottom of tees. Make sure lines have slight drop toward discharge end if carryover problem is tough. Next, insulate lines completely and of right material for steam temperature. Piping layout shown here will cure old system, save money and prevent outage in long run

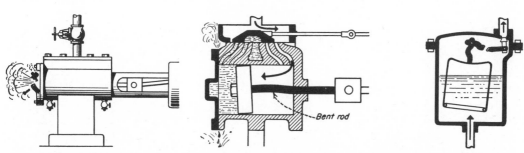

STEAM-ENGINE CYLINDER HEADS crack or let go completely from a large slug of water. Water can't leave cylinder fast enough in a high rpm engine. Steam separator above engine gets flooded

BENT PISTON RODS are common when water carries over into steam engines or into steam reciprocating pumps. This is more common when pump looses prime and races, then gets slug of water

DAMAGED TRAPS are common when a heavy shot of water is forced against working parts. And it's no fault of trap's construction. Steady pounding by water hammer distorts or deforms vital parts

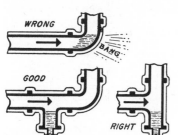

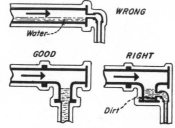

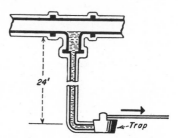

DRAINING AT ELBOW of vertical line is important. Tee with drain line before elbow is good. But simpler and most effective drain is to replace the el with a tee and hook to steam trap

DEAD END of headers should be drained. But drain must be at lowest part so water won't collect. Right way is to use tee and tap into side of bottom leg. Then dirt is kept out of the trap

PLACE TRAP well below steam line. At 24 ft shown, pressure of water leg above trap is 10 psi, in addition to steam pressure above water. The lower trap is the higher head pressure is on trap

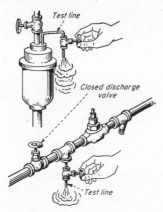

1 **TEST VALVE,** trap discharge section or line, allows tightness check

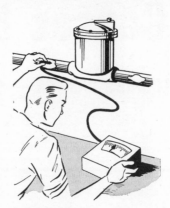

2 **PYROMETER** checks inlet, discharge temperature. File two spots

3 **STEEL ROD** held against trap body helps you to hear discharge action

4 **GLOVE TEST** checks inlet, outlet temperature; it is easy to make

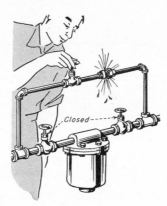

5 **LEAKY BYPASS** valve wastes plenty steam. Check tightness regularly

6 **CLEAN TRAP** parts in suitable fluid for good system operation

Know-How Keeps Steam Traps Healthy

Your surest bet for long trap life and better plant efficiency is right maintenance. Here are solid tips for most of the traps you will use, also a trouble-shooting chart for easy checks

► IF YOU WANT TO LIVE happily with your steam traps you have to treat them right. They're just like any other piece of mechanical equipment—they need regular inspection and maintenance.

Trap Tests. Make these regularly. Check large high-pressure traps daily. Do the same for units handling dirty or corrosive condensate. Look over medium-pressure traps once a week. Low-pressure systems need a monthly check.

Best way to test a trap is with test-valve hookup, Fig. 1. Some traps have a tapping for test connection. On others you have to fit a tee in return line. Be sure there's a tight valve beyond test connection to isolate trap from return line.

If steam comes from test line, look for trouble. Some traps give off "lazy" flash when air is mixed with steam, or condensate load is light. Others discharge almost continuously when there's too much air in system. Traps that have lost their prime blow steam at high velocity. Close inlet valve a few minutes, then open slowly. If trap catches prime, its mechanism is probably OK.

Pyrometer Method. File clean spots on inlet and discharge lines. Take temperature readings of incoming steam,

TROUBLE-SHOOTING CHART FOR STEAM TRAPS

TROUBLE	POSSIBLE CAUSE AND CURE

TRAP DOESN'T DISCHARGE

1 Steam pressure too high, pressure-regulating out of order, boiler pressure gage reads low, steam pressure raised without altering or adjusting trap. On the last item consult trap maker. He can supply parts for higher pressure or tell you how to adjust trap
2 Plugged strainer, valve or fitting ahead of trap; clean
3 Internal parts of trap plugged with dirt or scale; take trap apart and clean. Fit strainer ahead of trap
4 Bypass open or leaking; close or repair
5 Internal parts damaged or broken; dismantle trap, repair

TRAP WON'T SHUT OFF

1 Trap too small for load; figure condensate quantity to be handled and put in correct-size trap
2 Defective mechanism holds trap open; repair
3 Larger condensate load from (a) boiler foaming or priming, leaky steam coils, kettles or other units, or (b) greater process load; find cause of increased condensate flow and cure, or install larger trap

Note: Traps made to discharge continuously won't show these symptoms. Instead, the condensate line to trap overloads; water backs up

TRAP BLOWS STEAM

1 Open or leaky bypass valve; close or repair
2 Trap has lost prime; check for sudden or frequent drops in steam pressure
3 Dirt or scale in trap; take apart and clean
4 Inverted bucket trap too large, blows out seal; use smaller orifice or replace with smaller trap

TRAP CAPACITY SUDDENLY FALLS

1 Inlet pressure too low; raise to trap rating, fit larger trap, change pressure parts or setting
2 Backpressure too high; look for plugged return line, traps blowing steam into return, open bypass or plugged vent in return line
3 Backpressure too low; raise

CONDENSATE WON'T DRAIN FROM SYSTEM

1 System is air-bound; fit suitable vent or trap with larger air capacity to get rid of the air

TROUBLE	POSSIBLE CAUSE AND CURE

2 Steam pressure low; raise to the right value
3 Condensate short-circuits; use a trap for each unit

NOT ENOUGH STEAM HEAT

1 Defective thermostatic elements in radiator traps; remove, test and replace damaged elements
2 Boiler priming; reduce boiler-water level. If boiler foams, check fires and feed with fresh water while blowing down boiler at quarter-minute intervals
3 Scored or out-of-round valve seat in trap; grind seat or replace old trap body with new one
4 Vacuum pump runs continuously; look for a cracked radiator, split-return main, cracked pipe fitting or a loose union connection. Or pump shaft's packing may leak
5 Too much water hammer in system; check drip-trap size. Undersized drip traps can't handle all condensate formed during warm-up so hammering results. Fit larger trap if drip lines are clean and scale-free. Size for warm-up load, not for load with mains hot
6 System run down; older heating plants are sometimes troublesome because a large number of trap elements are defective. Easiest cure is replacement of all thermostatic elements in the radiators. This is low-cost, sure

TRAPS FREEZE IN WINTER

1 Discharge line has long horizontal run where water collects; make discharge line as short as possible and pitch away from trap
2 Trap and piping not insulated; fit insulation to outdoor traps and piping connected to them

BACK FLOW IN RETURN LINE

1 Trap below return main doesn't have right fittings; use check valve and a water seal, or both, depending on what the trap maker recommends
2 High-pressure traps discharge into a low-pressure return; flashing may cause high backpressure. Change piping to prevent return pressure from exceeding trap rating
3 No cooling leg ahead of a thermostatic trap that drips a main; condensate may be too hot to allow trap to open right. Use a 4- to 6-ft cooling leg ahead of thermostatic traps on this service. Fit strainer in cooling leg to keep solids out of trap

---

outgoing condensate, Fig. 2. If inlet temperature is about same as that of steam, and outlet is at or near that of steam in common return, trap is good.

When pyrometer shows rise and fall in discharge temperature, trap is working intermittently. If outlet temperature is nearly the same as inlet steam temperature in bucket traps, steam is blowing through. Impulse traps often have higher outlet temperature than other types, discharge continuously on many loads.

Use Your Ears. Hold one end of steel rod against trap, Fig. 3, and listen for discharge. If innards rattle much, trap may have lost prime. When you make this test don't be fooled by other sounds "telegraphed" along pipe.

Canvas Gloves. Put on a pair of these and grab inlet line with one hand, discharge with other, Fig. 4. You should be able to feel a distinct temperature difference with most traps. Without a

difference, the trap has lost its prime.

Bypasses. Traps are easier to live with if fitted with a bypass. But if you neglect bypass valve, it may leak more steam than you'd ever think. So it's smart to disconnect bypass line once a year and test bypass valve with full working steam pressure, Fig. 5. Replace disk or seat at once if valve leaks.

Strainers. These are low-cost and save much misery. But don't forget that a strainer doing its job gets dirty. A dirty strainer ahead of a trap is almost worse than none at all. So clean your strainers regularly. Easiest way is to install a permanent blowoff line. If you have to remove the screen from a strainer in a vertical line, be careful not to allow dirt to fall into pipe.

Freezing. Sometimes unprotected traps freeze in extremely cold weather. With traps that are not non-freeze, keep discharge line short without any horizontal runs. Some trap makers have automatic drain valves that allow condensate below a certain temperature to leave trap and inlet line after steam is shut off.

Remember, some traps don't need freeze protection. But play safe and insulate outdoor traps. Heat savings make it worthwhile.

Dirt in Trap. If much sediment and scale collect in a trap, you'll probably have to take the unit apart. Wash parts in kerosene, rust-penetrating oil, or other cleaning fluid. Traps in lines serving reciprocating engines or pumps may have a coating of oily paste if too much cylinder oil is used. Always check this when you have trap apart.

Trouble-shooting chart, above, gives some hints on trap care. Remember. it is general, and some traps may react a little differently, depending on their design. If you ever have any doubts call in trap manufacturer.

I Must'a Opened the Wrong Valve

OH NO, NOT *again*, BRUNO! *That's the second time you flooded that machinery room this week. And we just got through drying out all those motors and putting them back into service this morning.*

Does this picture bring back unhappy memories to you readers? Sure it does. This kind of thing is repeated almost daily—in some plant or another around the country. The wrong valve is opened or closed. The wrong switch is thrown or pulled. Equipment isn't lubricated, or it's lubricated with the wrong oil. So, slowing down our power output goes on and on, like "Ole Man River."

So what? Just this: all these expensive boners can be slashed way down by using check-off sheets for starting up and shutting down equipment. Label all your valves and controls to make them foolproof. Then employ well trained, competent help. And I've found the more modern and automatic the plant, the more need for well rounded men to keep it running. That also means training each new man before turning him loose to gum up your expensive works.

Careful screening will find men who intend to make plant operation a career, men who take pride in their profession. With no licensing required of operating engineers in most of our states, it just doesn't pay to hire the first jack-of-all-trades who comes along and is willing to work for less.

Mechanics of Bruno's caliber might cut down on the payroll but you'll surely pay through the nose before they are through with you.

5
Refrigeration

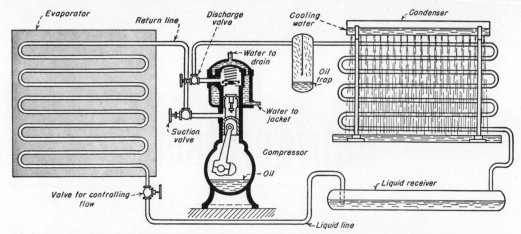

1 A simple refrigerating system consists of a cooling coil where the liquid boils; a compressor that compresses the vapor to a higher pressure, a condenser that cools the vapor and changes it back to a liquid and a liquid-storage vessel

How Machines Make Cold

Knowing how liquids act at different temperatures and pressures helps you spot refrigerating troubles in a jiffy.

▶ REFRIGERATION is nothing more than removing heat from a substance we wish to cool. All we have to do to coax heat out of a warm object is to place it near or in a cold substance. But we can't get very far with natural cooling in the summer because we don't have enough nature-made cold substances. So the only answer is—make your own cold with a mechanical system, Fig. 1. Knowing *what* has to be done, let's see how we go about building such a system.

Energy Needed. To make artificial cold, we'll have to get heat to flow "uphill" from a cold body or area to a hot one (relatively speaking). This means working against nature and it's going to take energy, somewhere along the line, to haul the heat away. This costs money.

For the time being let's forget the energy part of our problem and, start by letting the heat move downhill, naturally. We can do this by taking advantage of (1) what happens when liquids change to vapors and back and (2) the way these changes are tied to the pressure acting on the liquids.

How Liquids Act. To get our thinking straight about the characteristics of liquids, let's start with a few we know because we can see and feel them. Water is certainly a familiar one and we know it is still liquid right up to 212 F as long as we keep it in a vessel open to atmospheric pressure. We also know that if we keep on adding heat to the water it will boil and turn to vapor at this temperature and pressure, Fig. 2A.

Other substances that we can see as liquids open to the atmosphere are chloroform, which boils at 140 F, and ether, which boils at 95 F.

Now that we have a few boiling points off our chest, let's see what they have to do with moving heat. Take water for example; if we want it to boil, we add heat. For every pound we have to add one Btu to raise the temperature one degree. Starting with water at 32 F we need only 180 Btu to get one pound to 212 F. But, to turn a pound into steam we must add 970 Btu (latent heat of vaporization).

Effects of Pressure. So far we have been talking about how a few liquids

act at atmospheric pressure and we can see that the temperatures involved will not do much in the way of keeping things cold. Fortunately we can take advantage of a liquid's high latent heat of vaporization and get it to boil at the temperature we need by varying the pressure acting on it. Increasing the pressure raises the boiling temperature and decreasing it lowers the boiling point.

But to do this we must use a closed container which puts the liquid out of sight. A good example is a boiler feeding a condensing turbine. Here the water does not boil until it is heated considerably above 212 F, Fig. 2B, because the boiler confines the vapor given off as the water gets hot so pressure is much higher than atmospheric.

Cool Steam. Steam from the boiler drops in pressure as it passes through the turbine to enter the condenser, which may be under a 29-in. vacuum (a pressure much lower than atmospheric). You can prove that steam leaving the turbine is considerably below 212 F by laying your hand on the exhaust hood. And it can't be a liquid

138

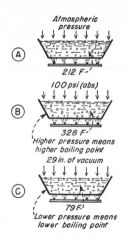

A Atmospheric pressure

212 F

B 100 psi (abs)

328 F
Higher pressure means higher boiling point

C 29 in. of vacuum

79 F
Lower pressure means lower boiling point

2 Pans above show how pressure acting on water surface affects boiling

Vent

Atmospheric pressure

Insulation

Ammonia

−27 F

3 Cylinder of liquid ammonia vented to atmosphere holds −27 F in box

at this point because the turbine blades wouldn't last long if water passed through them.

Last, but not least, you can refer to the steam tables and find that water can exist as a vapor at the low-temperature and low-pressure conditions of the turbine exhaust.

Now we're getting somewhere. If water can exist as a vapor in a vacuum of 29 in., and we know that it is cool enough for us to lay our hand on the turbine exhaust, we have proof that water will boil into steam at a temperature below 212 F if we keep the pressure below atmospheric and add heat, Fig. 2C. Another thing in our favor, from a refrigerating standpoint, is that it still takes about the same number of Btu to vaporize the water at this lower temperature and pressure.

If we have to talk about liquids penned up out of sight, let's see how a common refrigerant like ammonia acts. Checking the tables of ammonia characteristics we find that it is a gas at atmospheric pressure and temperature; so we have to confine it in a closed system to keep it liquid permanently if the temperature is higher than its boiling point of −27 F. By the way, we have to expend energy to get it penned up.

Looking at it another way, suppose we have a cylinder of ammonia whose temperature is 80 F. A pressure gage connected to it will read 138 psi. The more we cool the cylinder (at 70 F pressure is 113 psi) the less the gage will read until we get it down to −27 F. Then the gage reads zero.

How Ammonia Cools. Suppose we put the cylinder inside a heavily insulated box and open the valve so ammonia can escape, Fig. 3. As the pressure falls, the ammonia starts to boil furiously because it is at 80 F. The only source of heat for boiling is from the liquid and its metal container. So what happens. The large body of liquid cools rapidly, and as it cools its vapor pressure keeps in step with the temperature until the tank and liquid reach −27 F. Pressure gage then reads zero and only a small whiff of ammonia gas comes out occasionally as heat leaks in from outside the insulation. We now have a tank of liquid ammonia open to the atmosphere, but we lost about 19% as vapor or gas in the boiling process to cool the cylinder down to −27 F. As soon as we opened the ammonia valve and released the pent-up pressure, the liquid started to boil because it contained too much heat to remain a liquid at the resulting pressure.

Heat Haulers. Boiling absorbs heat, in this case from the liquid itself, so the liquid ammonia soon gets so cold it no longer contains an excess of heat units (vapor pressure and temperature are in balance) and boiling stops. The same thing holds true with other refrigerants except they have different boiling temperatures at atmospheric pressure. For example sulphur dioxide boils at 14 F, methyl chloride at −10 F, Freon 12 at −20 F, Freon 22 at −44 F and carbon dioxide at −109 F.

Real Refrigeration. Suppose we need a lot of room. We insulate the space, cover one or more walls with pipe coils,

to get lots of cooling area, and connect them to a large cylinder of ammonia. Open the valve, feed liquid ammonia into the pipe coils where heat picked up from the cooling area causes it to boil rapidly, and what happens. The resulting gas escapes to atmosphere and we have a refrigerated space at −27 F. If this is too cold, put a valve on the vent to atmosphere. Then throttle the discharge and build up pressure in the coils until you get the temperature you want.

Fine, you say, but that ammonia blowing to waste costs money. We can't afford to get our refrigeration by wasting ammonia but saving and using it again requires investing a lot more money than our simple gadget cost.

Starting Uphill. So far we have been letting heat flow downhill in its natural way. But reclaiming the ammonia is another matter. Here is where it takes energy, because we have to buck nature to get the heat out of that −27-F ammonia vapor and turn it into a liquid. And we have to remove the heat with some substance available to us at normal temperature—say water at 70 F. Let's follow through without worrying about pressure drop caused by pipe friction.

From a practical standpoint the best we can hope to do with 70-F water is to cool the gas to 80 F. Now, how are we going to cool gas that has a temperature of about −27 F, with 70-F water? Maybe the ammonia tables will give us the answer; so let's see what they say about 80-F ammonia. Well liquid ammonia at 80 F has a vapor pressure of about 138 psi gage. But our gas is at atmospheric pressure. Sure, but you said you wanted to reclaim it, so dig down and buy yourself a compressor. And while you're at it, get one rated for at least 150 psi to cover yourself if the cooling water should push well above 70 F on some hot summer day.

While in the buying mood get a good condenser and some pipe and you're ready to build a closed refrigerating system. So you want to hook it up like Fig. 1. OK, now vapor from the cooled space (evaporator) enters the compressor, which has to do work on the gas to get it up to 138 psi. The gas gets hot—in fact it jumps to about 280 F in the compressor cylinder. We get the pressure we need all right but end up with a lot of very hot gas that has to be cooled down to 80 F before it will start condensing back to a liquid. With the correct size condenser our 70-F water should do the job nicely and prepare the ammonia for another journey around the circuit.

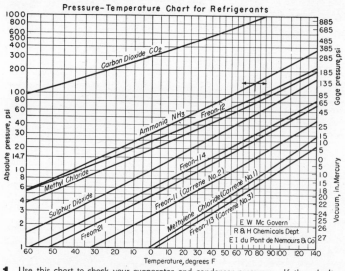

Pressure–Temperature Chart for Refrigerants

E W Mc Govern
R & H Chemicals Dept.
E I du Pont de Nemours & Co

1 Use this chart to check your evaporator and condenser pressures. If they don't match the respective temperatures, check your instruments before looking further

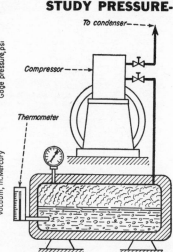

2 Compressor takes vapor away and raises it to the condensing pressure

How Machines Make Cold

Here we learn how many Btu a pound of refrigerant absorbs when changing to vapor, Part 2. A further step compares heat content per cu ft of vapor, which determines compressor capacity

► ON PAGES 134 and 135, we learned how the boiling temperature of a liquid depends on the pressure acting upon it; so let's see how this affects the various common refrigerants. But first let's learn how refrigeration is measured and machines rated.

Standard Ton. Large pieces of refrigerating equipment are rated in tons. Now just what is a ton of refrigeration? *It is the cooling effect obtained by melting a ton of ice in 24 hr.* Notice that the capacity is based both on the amount of heat absorbed and time—a ton of refrigeration is not just a quantity, it is a rate of cooling.

Btu Per Ton. Let's see what one ton of refrigeration means in Btu. The latent heat of fusion (freezing) of ice is 144 Btu per pound; the amount of heat that must be removed from 32-F water to turn it into ice at 32 F. In other words the process takes place without a change in temperature. So if we have to take 144 Btu out of every pound of water to freeze it into ice, the ice is

going to absorb 144 Btu when melting to water.

A ton of ice, therefore, must absorb $2000 \times 144 = 288{,}000$ Btu in melting. And if the process *must* take 24 hr to complete, then it occurs at $288{,}000 \div 24 = 12{,}000$ Btu per hour and $12{,}000 \div 60 = 200$ Btu per minute.

Now we can define a ton of refrigeration as the removal of 200 Btu each and every minute in a 24-hr period. But don't jump to conclusions and say one ton of refrigeration will produce a ton of ice in 24 hr.

This may sound confusing after using a ton of ice to define a ton of refrigeration, but remember that a ton of ice absorbs a constant and never varying number of Btu as it melts. But if we attempt to freeze this ton of ice in 24 hr, we have mechanical losses and heat flow into the system from the surrounding atmosphere. Both of these use up refrigerating capacity. It takes about 1½ tons of refrigerating capacity to freeze a ton of ice in 24 hr.

Equally important when figuring capacity is the rate at which the refrigeration must be done. A one-ton compressor (rated on 24-hr basis, remember) is not large enough to handle a 288,000-Btu load if the cooling job has to be done in 5 or 6 hours.

Machine Rating. Just stating that a one-ton compressor will produce one ton of refrigeration in 24 hr is not the whole story. We have to be more specific about the operating conditions to be able to compare machine performance. These conditions are defined as (1) an evaporator pressure corresponding to a saturated-vapor temperature of 5 F with a 9-F allowable superheat at the compressor intake and (2) a condensing pressure corresponding to a saturated-vapor temperature of 86 F with an allowable subcooling of 9 F at the evaporator expansion valve.

Now that we have the standard ton streamlined down to 200 Btu per min at 5- and 86-F pressure conditions, we can compare the refrigerants.

TEMPERATURE RELATION AND VAPOR VOLUME

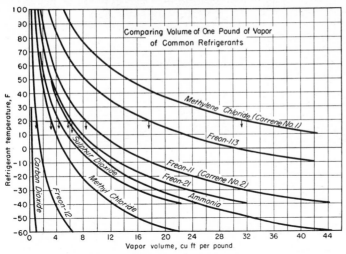

Comparing Volume of One Pound of Vapor of Common Refrigerants

3 Here is how the vapor volumes of the common refrigerants compare. Before using this information to select compressor capacity, figure Btu content per cu ft

BTU NEEDED TO BOIL ONE POUND OF LIQUID REFRIGERANT

(Latent heat of vaporization)

Temp. deg F	Ammonia	Methyl Chloride	Sulphur Dioxide	Carbon Dioxide	Freon– 2I	Freon– II	Freon– I2	Freon– II3
−40	598	190	178	137	114	87	73	. .
−30	591	189	177	134	113	87	72	72
−20	585	186	175	129	112	86	71	72
−10	576	184	173	125	111	85	71	71
0	569	182	171	120	110	85	70	71
10	561	179	168	114	109	84	69	70
20	553	177	165	109	107	83	68	69
30	545	175	163	102	106	82	67	69
40	536	172	159	95	105	82	65	68
50	523	169	156	87	104	81	64	68
60	518	167	152	77	103	80	63	67
70	509	164	149	64	101	78	62	67
80	498	161	145	45	99	77	60	66
90	489	158	141	. . .	99	75	59	65
100	478	154	137	. . .	97	76	57	65

Boiling Temperatures. Take a look at the simple curves in Fig. I. They represent vapor pressures of the common refrigerants at different temperatures. If you prefer gage pressures, use the right-hand scale, otherwise use the absolute scale at left. Refrigerants all act pretty much alike although each one has its own pressure-temperature range. To explain the curves, suppose we use the small compressor setup in Fig. 2 with its suction line connected to an insulated drum of ammonia. To change drum pressure we can run the compressor at any speed we like.

Suppose the ammonia has a temperature of 75 F. What should the pressure gage read? Tracing up the 75-F line to the ammonia curve, follow the arrow to the right and read 135 psi. Let's run the compressor at full speed for several minutes. A quick look at the gage shows the pressure to be 65 psi, and we can hear the ammonia boiling furiously. What happened? Well we took away part of the ammonia's vapor pressure, so the liquid gets busy trying to restore conditions to a balance.

Temperature Lag. At this stage of the game the thermometer may not have changed, but if we watch it closely it will show a drop in temperature by the time the liquid stops boiling. The boiling uses up some heat in the liquid and it has to be a little cooler as explained last month. If pressure rises, say to 85 psi, and remains steady, the curve shows that thermometer should read about 58 F.

To get the temperature you need, run the compressor and pump the ammonia pressure down to the value corresponding to the temperature shown on chart. Then run it just fast enough to carry the vapor away as it forms and still hold the pressure steady. This is the principle of any refrigeration system regardless of what refrigerant is used. Now let's see how refrigerants differ when they boil.

Vapor Volume. The curves in Fig. 3 show the volume of one pound of the different refrigerants after they have been changed to vapor at the temperatures given on left-hand scale. The vapor in each case has a saturation pressure corresponding to the temperature so we can compare the refrigerants in terms of volume per pound.

For example, suppose we need a room or space held at 20 F. We find (arrows) that carbon dioxide has the smallest volume. Freon-12 is 1 cu ft, methyl chloride is almost 3½, sulphur dioxide 4½, ammonia almost 6, Freon-21 over 6, Freon-II is 8½, Freon-113 almost 18 and methylene chloride 31½. On this basis alone it looks like compressor size, needed to do a given refrigeration job, varies all over the lot.

But just a minute. Suppose we examine things a little closer. Remember we learned that cooling and freezing equipment is rated on a ton-per-24-hr basis, equal to 200 Btu per min. Fig. 3 shows how much vapor a pound of each liquid forms. But take a look at the table. At 20 F ammonia latent heat of vaporization per pound is 553 Btu. Even though its vapor volume is almost 6 cu ft per lb, the ammonia vapor does contain 553 ÷ 6 = 92 Btu per cu ft.

Compare this with Freon-12 having a volume of 1 cu ft and heat content of 68, which means that it contains 68 Btu per cu ft. So you see compressor capacities for the two refrigerants do not vary much for the same cooling job at 20 F. On the other hand there is quite a difference between ammonia and Freon-113 capacity requirements.

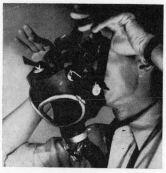 

USE THE RIGHT GAS mask for your refrigerant and keep it near the shop door

THIS ALL-PURPOSE mask is good for ammonia saturations only up to 3 per cent

Know Your Refrigerants

Okay! So you were never knocked out by a bad refrigerant leak. But don't get cocky.

▶ EVERY NOW AND THEN it happens: A line ruptures, a packing gland lets go, and the refrigerant fills the machine space. Your first impulse is to plug the leak quickly. But stop . . . think of the danger you're headed for.

First, imagine your refrigerant is ammonia. Now ammonia gas in itself leaking out isn't bad. But when ammonia comes in contact with moisture it forms *aqua ammonia* (ammonium hydroxide). THAT'S DANGEROUS, especially if a lot of these two get mixed.

In your haste to stop that leak you rush into the room without a mask. Then you inhale ammonia gas. The moisture in your throat and lungs is all the water needed to form aqua ammonia in your respiratory system. And depending on the air's ammonia saturation, your lungs get irritated. If it's bad, you might not get out of that room alive.

What To Do. Now let's see the right thing to do. In the first place, if the leak is bad, don't take chances. Before tackling it put on the all-purpose gas mask. It should be near the door for just that purpose. Then spray the room with water. Water absorbs most of the ammonia quickly and washes it to the floor.

Even then, exposed parts of your body which come in contact with ammonia are irritated. Therefore, use leather gloves along with the gas mask. But the all-purpose gas mask shouldn't be used if air contains more than three percent ammonia. That's stated clearly on the gas mask's cannister. How do you know when saturation gets that high? By ammonia coming through the mask's filter. If that happens, use an oxygen breathing apparatus or an air line mask.

Small Leaks. Now let's find a minor leak. First, an ammonia leak makes itself known by its odor. Ammonia is a pungent gas.

CAUTION! Whether ammonia or any other refrigerant, keep your eyes away from refrigerant lines. That's especially important on system's high pressure side. If liquid refrigerant hits your eyeball, it freezes immediately. This has happened many times. If it does, do *not* rub your eyes. Take the following first-aid treatment. Drop sterile mineral oil into the eyes, as an eye wash; follow with a weak solution of boric acid or sterile salt solution— not over 2 percent table salt.

Repairing Leaks. Copper, brass or bronze corrode quickly in ammonia systems. Ammonia itself doesn't affect these materials, but moisture gets into system, producing aqua ammonia. So use iron or steel.

Since ammonia is harder to seal than steam or air, make joint right. Use either screwed fittings with litharge and glycerine, or solder. Tongue and groove flange joints with lead or soft gaskets will also work.

Ammonia and Fire. Ammonia gas itself is not explosive and doesn't burn easily. At high temperature it burns with a greenish-yellow flame. But it decomposes around 1800 F into its constituents parts of nitrogen and hydrogen. Hydrogen burns. When mixed with air in right proportions, it EXPLODES!

Carbon Dioxide. When carbon dioxide leaks, it mixes with air. There's nothing to worry about if leak is small. When air's mixed with more than one half of one percent you get headaches and feel drowsy. In larger doses, you *pass out.* When air contains about 8 percent carbon dioxide, you *suffocate.*

So if leak is large, use an oxygen breathing apparatus or air line mask when entering room. Don't use all purpose gas mask because it relies on atmospheric condition in room. Human beings suffocate with less than seven percent oxygen in air. The flame of a flame safety lamp goes out at 16 per-

ROUND UP OF COMMON REFRIGERANTS

	AMMONIA (NH₃)	CARBON DIOXIDE (CO₂)	SULPHUR DIOXODE (SO₂)	FREON—12 (CCL₂ F.₂)
EFFECT ON PEOPLE	When in contact with water forms aqua ammonia, which is dangerous to life; can ruin foodstuffs	Non-irritating and non-toxic. A mixture of more than 1½% causes drowsiness and headaches, and possibly insensibility. An 8% mixture with air causes suffocation	Relatively harmless. If inhaled causes sore throat. When mixed with water forms sulphurous (H_2SO_3) acid. Will produce itching if body is wetted by it	Non-poisonous, non-irritating. Concentration above 20% has anaesthetic properties. Above 1000 F, decomposes, forming extremely irritating products
EFFECT ON METALS	Iron and steel aren't affected by aqua ammonia. Copper, brass or bronze corrode quickly. Copper alloy won't work for compressor bearings	Materials won't react with CO_2 when in pure form. In presence of moisture forms carbonic acid which attacks iron and steel. Copper and brass may be used	Pure sulphur dioxide is harmless to all metals. In presence of moisture forms sulphurous acid that attacks metals rapidly	Non-corrosive to all metals used in refrigeration machines
EFFECT ON LUBRICANTS	Solubility in oil very low. If water is present in ammonia, oil emulsifies. Ammonia is lighter than oil. Use mineral oils that don't contain soaps or fats	Has no effect on oils and greases. Oil must be suitable for low temperatures, however. Glycerine is preferred since it's adaptable to low temperatures	Has tendency to absorb lube oil. Is heavier than oil. Light colored oils containing no water are best	Freon-12 is miscible with lube oil in all proportions. Use straight run mineral oils. Oil must not contain water, sediment, acid, soap or resin
TESTING FOR LEAKS	Colorless, pungent, suffocating gas. Use litmus paper, sulphur sticks or sulphur candle	Colorless, odorless and has acid taste. Soap suds or lime water turns cloudy contacting CO_2. Peppermint added to CO_2 detect small leaks by smell	Detected by smell, litmus paper, or soap suds. When aqua ammonia is applied a dense white smoke results if there is a leak	Odorless in concentrations less than 20%. Detect by soap suds, or a Halide torch
FLAMMABILITY	Does not burn readily in air unless heated. Above 1600 F it decomposes and forms explosive mixture	Non-inflammable, non-explosive. (May be used as a fire extinguisher)	Non-inflammable, although under certain conditions may reduce to sulphur trioxide which is inflammable	Non-inflammable at ordinary temperatures. When Freon-12 breaks down, the constituent gases do not burn or form explosive mixtures
BOILING TEMP AT ATMOSPHERIC PRESSURE ★	—28 F	—109.3 F	14 F	21.6 F
PRESSURE RANGE AT 86 F IN COND AND 5 F in EVAP ★	(Low Side) 19.6 psig (High Side) 154.5 psig	316.8 psig 1028.3 psig	5.9 in. vacuum 51.8 psig	11.8 psig 93.2 psig
NET REFRIGERATING EFFECT OF LIQUID 86 F and 5 F ✱	474.6 Btu	55.5 Btu	142.8 Btu	51.1 Btu
WEIGHT COMPARISON	Lighter than air	Heavier than air	Heavier than air	Heavier than air

★ ASRE Refrigerating Data Book, 1949

cent. Don't use all purpose gas mask when flame won't burn.

Carbon Dioxide. This has been popular in refrigeration of food. One reason is, leak won't spoil food. That's not so with ammonia.

Freon. Freon is non-poisonous and non-irritating. But a large freon leak causing over 20 percent concentration with air knocks you out.

Freon itself won't burn. In the presence of flame or hot surfaces above 1000 F it decomposes. Then it forms toxic products that are extremely irritating. So take all precautions when welding. But unlike ammonia, when Freon-12 breaks down into its constituent parts, it won't burn or form explosive mixtures.

Sulphur Dioxide. This is a relatively harmless gas. It produces some throat irritation if breathed freely.

Sulphur dioxide mixing with water forms sulphurous acid (H_2SO_3) which is relatively harmless. This solution produces an itching where it touches your body. But continued exposure, strangely enough, tends to make you immune to its action. But does have a corrosive effect on plant equipment. Don't confuse this with sulphuric acid (H_2SO_4) which is both dangerous and destructive.

If you want to reach a ripe old age working with refrigerators, better soak in this information right now for good.

Purging your refrigeration system

The knowledge is common among refrigeration engineers and system operators that air in the condenser causes a rise in head pressure. But what they're apt to overlook is *why* air's presence increases head pressure and *how*, in many ways, this increase can add significantly to overall costs.

First of all, let's briefly review air's effect on condenser performance, then see how this adverse effect makes itself felt throughout the system. When air molecules spread themselves across the condenser's heat-transfer surfaces it becomes extremely difficult for the refrigerant-gas molecules to reach these surfaces to give up their heat energy. Thus air's presence has the effect of reducing amount of heat-transfer surface available to the refrigerant gas. The only way to offset this "smaller-condenser" effect is by increasing temperature—and pressure —of the refrigerant gas so that total amount discharged by the compressor will be condensed.

As discharge pressure rises, a number of effects are apt to follow. One, certainly, is an increase in power input. Accepted rule: for each 4-lb increase in head pressure—due to air or other noncondensable gases in a refrigeration system — compressor power costs go up 2% and compressor capacity drops 1%. This 4-2-1 formula for increased power cost and decreased capacity applies to all common refrigerants, under wide differences in suction and head pressure. It's obvious that such power-cost increase means a profit decrease some-

where along the line. Too, decrease in compressor capacity means operating the unit more hours to get the same refrigeration effect. In many cases, other losses caused by lowered compressor capacity may add up to far more than the extra cost involved in operating the compressor against the higher head pressure.

Added cost factors: Cooling-water use increases and heat-transfer efficiency drops off. Using more cooling water may mean much or little to the individual plant, depending upon its location, pumping system, water source and cost, recovery factor and so on. However, it's a safe bet that unnecessary water used in any plant ups operating costs to some degree.

Excess head pressure takes its toll from the compressor itself. Bearings (this applies to the driving motor also) are subjected to greater loading. Higher temperatures accompanying increased discharge pressures tend to shorten valve life, promote lubricating-oil breakdown. And gasket failure often goes hand-in-hand with excessive head pressure.

From a practical standpoint, it's impossible to keep air and other noncondensable gases from sneaking into the majority of refrigeration systems. If system operates with suction pressure below atmospheric, air leakage is most likely through valve packing and compressor seals. But, whether suction pressure is above or below atmospheric, air will find additional ways of getting into the system: (1) when system is opened for cleaning

or repair (2) when you're making or breaking hose connections to discharge refrigerated cars or trucks (3) when you're adding compressor lube oil (4) by refrigerant or lube-oil breakdown (5) from impurities in the refrigerant itself.

A sure sign that air or other noncondensables have crept into your system is a discharge-pressure-gage reading higher than normal for any given refrigerant temperature leaving the compressor. Table 1 lists temperature-pressure relationships for three widely-used refrigerants: saturated ammonia, Freon 12 and Freon 22.

Example: If ammonia temperature leaving the compressor is 84 F, theoretical condenser pressure is 149 psig. If the discharge-pressure gage reads 169 psi, 20-psi excess pressure is increasing power costs by 10%.

Liquid temperatures are easy to read if the liquid line from the condenser is equipped with a mercury well. Without the well, reasonably accurate readings can be obtained by taping the bulb of an immersion thermometer to the liquid line. Cover the bulb with about 1-in. of asphalt sealing compound or similar protective substance, then insulate with suitably heavy cloth. After a lapse of 3 or 4 minutes it should be possible to take a reading accurate to within $\frac{1}{2}$ to $\frac{3}{4}$ of a degree.

Remember, however, that air and other noncondensables are not the only causes of excessive condenser pressure. If condenser simply isn't big enough for the job, or if it's permitted to operate with fouled or scaled tubes, head pressure will go up. But in most cases air is likely to be the culprit and must be purged before head pressure can be reduced to the proper level.

Purging, like everything else, can be done in more than one way. To be effective it must be done slowly, giving air molecules time to migrate to purge connections.

It's possible, of course, to purge manually, though it isn't practical from an expense standpoint except on very small units. Table 2, for instance, shows that manually purging 1 cu ft of air from an ammonia system at 90 F and only 4-psi excess pressure due to air wastes 51 cu ft of ammonia. Or, from Table 3, manually purging 1 cu ft of air from an F-12 system at similar temperature and excess-pressure conditions allows 32.3 cu ft of Freon to escape.

At best, then, manual purging either takes a lot of time or not all the air is removed. In the latter case, adverse effects of air in the condenser continue, though at reduced rate. Summing up the case against manual purging: (1) It wastes refrigerant. (2) It takes a man's time which might be invested to better advantage elsewhere. (3) Air is almost never completely eliminated. (4) Released refrigerant gas may be disagreeable,

could be dangerous. (5) Purging may be postponed or ignored because of time and trouble involved in doing it.

Automatic purging, then, appears to be the logical way of handling the problem. Let's refer again to Table 2 and contrast the 4216 cu ft of ammonia lost by manual purging at 80 F and 4-psi excess head pressure due to air, with the 0.25 cu ft of refrigerant lost at a temperature of 0 F in an automatic purger with head pressure of 157 psia. Under these conditions automatic purging is 42.6/0.25, or 170 times, as efficient from a refrigerant-loss standpoint. Checking Table 3, we note that automatic purging of a Freon-12 system with same temperature and excess head pressure is 26.3/0.31, or 84.8 times, as efficient as manual purging.

Automatic system purging, commonly referred to as "refrigerated" purging, is applicable to practically any system at relatively low cost. It involves nothing more than installing commercially available, simple, dependable refrigerated devices. These automatically and completely separate air by condensing the refrigerant gas. Then they release the air to the atmosphere. Advantages derived from these devices, which require almost zero maintenance, are obvious in view of today's constantly rising operating costs and dependability requirements.

1—Too-high gage reading means air in the system

If temperature, F, is:	Gage pressure should be:		
	Ammonia	Freon 12	Freon 22
44	64.7	40.70	75.04
46	67.9	42.65	78.18
48	71.1	44.65	81.40
50	74.5	46.69	84.70
52	78.0	48.79	88.10
54	81.5	50.93	91.5
56	85.2	53.14	95.1
58	89.0	55.40	98.8
60	92.9	57.71	102.5
62	96.9	60.07	106.3
64	101.0	62.50	110.2
66	105.3	64.97	114.2
68	109.6	67.54	118.3
70	114.1	70.12	122.5
72	118.7	72.80	126.8
74	123.4	75.50	131.2
76	128.3	78.30	135.7
78	133.2	81.15	140.3
80	138.3	84.06	145.0
82	143.6	87.00	149.8
84	149.0	90.1	154.7
86	154.5	93.2	159.8
88	160.1	96.4	164.9
90	165.9	99.6	170.1
92	171.9	103.0	175.4
94	178.0	106.3	180.9
96	184.2	109.8	186.5
98	190.6	113.3	192.1
100	197.2	116.9	197.9
102	203.9	120.6	203.8
104	210.7	124.3	209.9
106	217.8	128.1	216.0
108	225.0	132.1	222.3
110	232.3	136.0	228.7

2—Ammonia lost through manual purging

Temp, F	NH₃, psia	Air, psia	Total psia	Cu ft lost per cu ft air purged
90	180.6	4	184.6	51.0
		8	188.6	25.5
		14.7	195.3	13.9
80	153.0	4	157	42.6
		8	161	21.3
		14.7	167.7	11.6
70	128.8	4	132.8	35.3
		8	136.8	17.7
		14.7	143.5	9.6
30	59.74	97.26	157	0.64
20	48.21	108.79	157	0.45
10	38.51	118.49	157	0.33
0	30.42	126.58	157	0.25
—10	23.74	133.26	157	0.18
—20	18.30	138.70	157	0.13

3—Freon 12 lost through manual purging

Freon, psia	Air, psia	Total psia	Cu ft lost per cu ft air purged
114.3	4	118.3	32.3
	8	122.3	16.1
	14.7	129	8.8
98.76	4	102.76	26.3
	8	106.76	13.1
	14.7	113.46	7.15
84.82	4	88.82	23.2
	8	92.82	11.6
	14.7	99.52	6.3
43.16	59.60	102.76	0.77
35.75	67.01	102.76	0.55
29.35	73.41	102.76	0.41
23.87	78.89	102.76	0.31
19.20	83.56	102.76	0.23
15.28	87.48	102.76	0.18

INDICATING REFRIGERATOR compressor shows what's going on inside machine

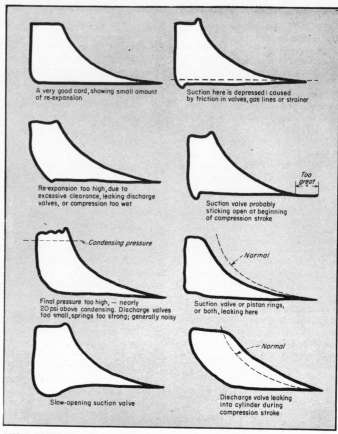

A very good card, showing small amount of re-expansion

Suction here is depressed; caused by friction in valves, gas lines or strainer

Re-expansion too high, due to excessive clearance, leaking discharge valves, or compression too wet

Too great

Suction valve probably sticking open at beginning of compression stroke

Condensing pressure

Final pressure too high, — nearly 20 psi above condensing. Discharge valves too small, springs too strong; generally noisy

Normal

Suction valve or piston rings, or both, leaking here

Slow-opening suction valve

Normal

Discharge valve leaking into cylinder during compression stroke

THESE CARDS show the experienced engineer that his compressor is either in first-class shape (top left) or that a number of things are wrong, and exactly what they are

The Indicator-Card Story

▶ You may not know it, but James Watt invented the steam-engine indicator, as well as the double-acting steam engine. As a young man, Watt was an instrument maker. One day he repaired a small model of the Newcomen-type atmospheric engine for the University of Glasgow. The model fascinated him so much that it awakened his first interest in steam power.

After building the first steam-powered double-acting engine, Watt developed the indicator. He needed it to find out what went on inside the cylinder at all times.

High-Speed Engine. Modern high-speed steam engines don't need to be indicated as often as old slow-speed corliss machines, with their many adjustable parts. The more valves and valve linkage, the more chances for engine to get out of adjustment.

In the same way the enclosed-type refrigerating machine of today generally operates satisfactorily for years, without needing to be indicated. For this reason most compressors today are no longer fitted with indicator connections, unless so ordered.

But the indicator does provide a check on what is occurring within the machine. Taking indicator readings may prevent serious waste in your plant. This is especially true if the

compressors are opened for inspection only after long intervals.

Large Compressors. Large ammonia and Freon compressors seldom run at speeds above 360 rpm. Taking an indicator card under such conditions is not difficult. Easiest way is to fit a home-made eccentric to the shaft, between stuffing box and belt wheel or motor, to reproduce the motion of the pistons.

Caution. *On ammonia machines, use an indicator with steel cylinder and piston. Ammonia corrodes brass.*

Set the eccentric so it is at top of its stroke when piston of cylinder being indicated is at top of its stroke.

KEEP THESE SAMPLE CARDS FOR REFERENCE
TO COMPARE CARDS FROM YOUR COMPRESSORS

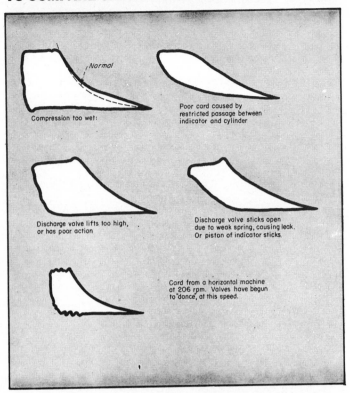

Normal

Compression too wet

Poor card caused by
restricted passage between
indicator and cylinder

Discharge valve lifts too high,
or has poor action

Discharge valve sticks open
due to weak spring, causing leak.
Or piston of indicator sticks.

Card from a horizontal machine
at 206 rpm. Valves have begun
to "dance" at this speed.

VALVE AND OTHER mechanical faults in refrigerating compressors are often not known without opening machine, so cards are taken. Here's what they tell you

You can use a steam-engine indicator to check refrigerating-compressor performance if it can be connected to the cylinders. Here is how to interpret the more common cards taken

Best way to do this accurately is to set crank on dead center with trammel

Then set eccentric with trammel. You don't need to cut a keyway on shaft or keyseat in eccentric. Just hold in right position with setscrew. Then place permanent guide mark on shaft and eccentric for future settings.

What Cards Tell. The refrigerating work being done by each cylinder and the indicated horsepower required to drive the machine, can be found from the cards. Any other method of measuring the cooling load is likely to be much harder than with cards.

The cards shown here carry notations that tell their own story. They will help any engineer find out how to fix what's wrong. It pays refrigerating plants with large slow-moving compressors to take these cards at set periods. It's the only way to know what's going on inside the machine.

Then operating engineer can place his finger on the fault the first time without having to guess at trouble. I've known of compressors that ran with major efficiency losses for years before trouble was discovered accidentally.

Cards not only help to make adjustments but keep engineers on their toes.

The attention you give your new refrigeration system just before putting it into operation may make all the difference between a lemon and a peach.

These easy-to-follow steps tell how to make . . .

START AND STOP the compressor several times at 10-second intervals before running in. Read compressor pressure gages. Correct operating difficulties before proceeding

Final checks and adjustments on

▶ THE BEST WAY to avoid costly and annoying shutdowns after a refrigeration system is operating is to go over it with a fine-tooth comb before starting. Then check it again after running 72 hours and once more after a week.

Before operation:

Before starting your new refrigeration compressor for the first time, check the entire system. Put all controls through their complete cycle of operation.

Adjust the low-pressure switch so it starts the compressor whenever suction pressure rises above the desired setting. Adjust the differential or cutout points as low as the job will stand to prevent short-cycling.

Check cutout point for the high-pressure switch, too. This is generally 175 psi.

See that all interlocks work right. For example, the compressor should be "locked out" if the evaporative condenser fails to run properly.

Check the thermostatic expansion valve for proper superheat adjustment. Put a thermometer on the suction line near the remote bulb of the thermostatic valve (sketches, facing page). Read refrigerant temperature at a point as near the remote bulb as possible, but on the compressor side of both the remote bulb and the external-equalizer-line connection.

You get the most accurate temperature reading by using a thermometer well. If no well is provided, attach thermometer with duct seal or use a clip-on-type thermometer. In these last two methods, reading accuracy depends on the bond between thermometer and pipe.

Read the suction pressure at the gage installed in the back-seat port of the compressor suction valve. Difference between thermometer reading on the suction line and the temperature calculated from the suction pressure is the superheat.

This method does not consider any

pressure drop in the low side of the system. If you suspect abnormal pressure drop, take steps to measure it and determine its effect.

One method for measuring pressure drop is to install a tee at the connection between the thermostatic expansion-valve external-equalizer line and the suction line proper. Connect a refrigeration compound gage to this tee. This gives the true pressure reading at the evaporator outlet. Gage readings can then be used for obtaining superheat measurements.

Most air-conditioning installations have superheat settings of about 10 degrees.

Also, check head pressure. If it is too high, adjust the water-regulating valve for increased flow. If head pressure is too low, adjust valve so that the water flow will be reduced.

Set thermostat to maintain desired temperature in the area to be cooled.

Just before starting the system, check for the following conditions:

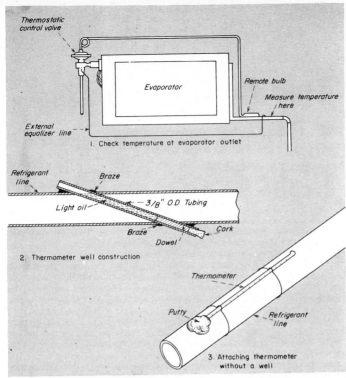

1. Check temperature at evaporator outlet

2. Thermometer well construction

3. Attaching thermometer without a well

USE THERMOMETER WELL to get most accurate reading of refrigerant temperature. Otherwise, tie thermometer on line, use putty to insulate bulb from the atmosphere

refrigeration systems

1. Shutoff valve on water line to shell-and-tube condenser open.

2. Oil in compressor at or above center of sight glass.

3. Compressor suction and discharge valves open.

4. All moving parts free to move. (Check by turning compressor over by hand several times from the drive coupling.)

5. All shutoff valves open except by-pass valves installed for other purposes.

6. Solenoid stop valve on magnetic coil control.

7. All compressor suction, discharge and oil-pressure gages connected and open to read operating pressures.

8. Readings on compressor pressure gages OK when compressor is started and stopped several times at 10-second intervals. (Correct any operating difficulties before proceeding.)

After making these checks, let system run normally for at least 72 hours before making a final check.

During 72-hour run-in:

Mechanic should stay on the job continuously for at least 48 hours of the 72-hour run-in period. Here are the things he should do:

1. Keep a close watch on oil level in the compressor, never permitting unit to run short of oil.

2. Immediately after starting, take amperage and voltage readings on all motors to determine that they are running at nameplate conditions.

3. Check entire system for leaks several times during the 72-hour run-in.

4. Check all controls for proper operation; reset if necessary; see that they are calibrated.

5. Observe and record all operating pressures and temperatures.

6. Check superheat setting of the thermostatic expansion valve; adjust if necessary.

7. Adjust and set compressor capacity control system. Even though this may be preset at the factory it should be reset if necessary.

After 72 hours:

Here are the points to check after system has run 72 hours:

1. Compressor oil level. If low, do not add oil immediately, but operate the system three or four hours and check frequently to see that returning oil does not restore the proper level. If oil has not returned at the end of four hours, add some. If oil is at the proper level on starting, check frequently during the 3- or 4-hour run to see that level stays constant.

2. Refrigerant flow at the sight glass. Flow should be solid with no bubbles. These indicate a refrigerant shortage. If bubbles are present now, there is probably a leak, and system should be checked completely with a halide torch.

3. Liquid line from receiver to expansion valve. See that there is no appreciable temperature change over the length of the line. If a strainer or stop valve has a warm inlet and cold outlet, there is restricted flow or inadequate subcooling. If liquid is properly subcooled, there is an obstruction. Remove and clean the obstructed part.

4. Superheat setting of the expansion valve.

5. Adjustment of the evaporator pressure regulator valve.

6. Head and suction pressure. If correct, the suction and discharge shut-off valves should be back-seated. Remove test gages (unless they are permanent) and plug gage tappings.

7. Compressor seal. Stop system and test seal with a halide torch to be sure it is holding tight.

8. Motor alignment. Check compressor motor for alignment and adequate lubrication. Avoid over-lubrication.

9. Clean the water strainer ahead of the water-regulating valve.

10. Air-handling units. Adjust belts for proper tension and check alignment. Check for adequate lubrication. Avoid over-lubrication. Check air filters to be sure they are clean.

After one week:

1. Drain compressor crankcase. Remove inspection plate and thoroughly clean crankcase, swab out with carbon tetrachloride or similar cleaner. Then recharge with clean, new compressor oil, as specified by the manufacturer.

2. Tighten bolts in the coupling between compressor and motor.

3. Clean all filters in the entire system both in the compressor and liquid line.

Do not take these steps at the time of the 72-hour check. More regular running time is needed to clear the system of foreign material.

—*Courtesy, The Trane Company*

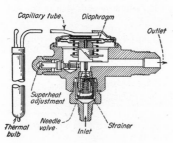

1 With a thermostatic expansion valve gas leaving evaporator is superheated

Step Up Refrigerating-System Capacity: Use Right Expansion Valve

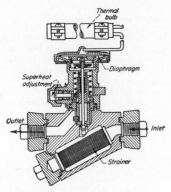

2 Theromostatic expansion valve for ammonia system; 1- to 19-ton rating

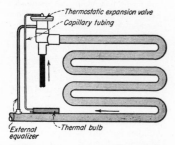

3 Freon-12 thermostatic valve has ⅛- to 1-ton capacity, angle construction

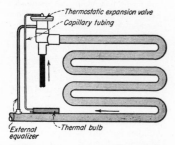

4 External equalizer on thermostatic valve offsets pressure-loss effects

▶ IF SOMEONE ASKED what you consider the most important valve in your refrigeration system you'd probably say the expansion valve. It's one valve you can't do without. If it works right everything is OK; when it goes on the blink you hear about it fast.

What Valve Does. Main job is to control refrigerant flow to evaporator. If too little liquid refrigerant enters your evaporator, it quickly flashes to a gas without absorbing much heat. When too much liquid enters coil, not all of it is vaporized and some floods back to compressor suction.

Either way you run into trouble. With not enough liquid, cooling capacity of system falls off. Too much liquid can wreck compressor valves and pistons because cylinder clearance is very small and liquid refrigerant is almost incompressible.

Thermostatic Valves. Today many expansion valves are the thermostatic type. Superheat of gas leaving evaporator, Fig. 1, changes the pressure of liquid in the thermal bulb. Pressure rise or fall is carried through capillary tube to a diaphragm in the expansion valve, opening or closing the valve.

Fig. 2 shows a typical ammonia thermostatic expansion valve, its thermal bulb and capillary tube. Pressure from thermal bulb opens valve against spring. Evaporator pressure and spring force tend to close valve. With a steady load on evaporator the three forces are balanced and valve stays open in one position. When load goes up, thermal bulb pressure rises from higher suction-gas temperature and valve opens wider to admit more liquid. Opposite happens when load goes down. Liquid in thermal bulb is usually the same as the refrigerant in system.

Gas Superheat. To remove most heat with a pound of your refrigerant it should enter evaporator as a liquid, leave as a gas. With this setup, gas leaving coil will be slightly superheated.

Thermal bulb is connected at end of coil, Fig. 1.

Suction-gas superheat depends on amount of refrigerant fed coil and how fast it is evaporated by heat from load. If your valve is set for 10-F superheat, then gas leaving evaporator will be 10 F warmer than the evaporating liquid. So at some point in coil, say *A*, saturated gas starts to absorb heat to become superheated.

Between point *A* and the thermal bulb, evaporator coil is used only for superheating. When load falls there isn't enough heat absorbed between *A* and bulb to give 10 F desired. Bulb cools, closing valve to cut liquid flow to coil. Point *A* moves nearer liquid line so there's more coil length for superheating.

When load goes up, superheat temperature rises. Thermal bulb gets warmer, opens expansion valve wider to admit more liquid. Point *A* moves nearer bulb. So you can easily see that as load changes, point *A* moves to give right coil length for superheat needed.

What Superheat? Your refrigerant picks up far more heat when it is vaporized from liquid to gas than when just dry gas is superheated. So the section of coil used to superheat the gas does little cooling. For this reason it is wise to keep superheat low. Then you can use more of the coil length for useful cooling.

But too low a superheat may cause liquid to flood suction line. Too high a superheat cuts capacity. You'll find many valves with a superheat adjustment range of 5 to 25 F. With a load that changes quickly a superheat of 10 F is often used. Manufacturer will set superheat before shipping valve if you ask him to do so. Some valves also have a suction-pressure limit adjustment to prevent system overload.

External Equalizer. Large pressure drop through system low side cuts compressor capacity and can also reduce

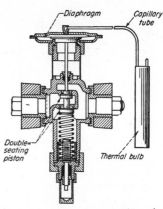

Modern-design expansion valves are neat little packages that do a big job. No matter what kind of system you run you must know what makes them tick. Here's practical know-how that's a real help in large or small plants

5 Double-seating piston in this valve controls F-12, F-22, methyl chloride

expansion-valve capacity. As we saw above, pressure in thermal bulb regulates valve opening by acting against spring and evaporator pressure. With large pressure drop through coil, saturation temperature of gas near outlet is lower because gas pressure is lower.

With lower saturation temperature more superheating of suction gas is needed to hold expansion valve open. This means you are using less of your evaporator to take large amounts of heat from load. System capacity falls.

One way to step up evaporator capacity when pressure drop is large is to use an external equalizer, Fig. 4. This is a tube connected between diaphragm underside and evaporator outlet. Average pressure in evaporator is now on diaphragm underside.

With an external equalizer, valve is almost completely free of pressure-drop effect and responds only to superheat in suction gas. Valve gives full capacity at low superheat. System capacity goes up.

Hookup. Some valve makers recommend that you never use a thermostatic expansion valve without an external equalizer when pressure drop through evaporator is more than 2 psi. Exact location of evaporator end of equalizer depends on evaporator, but usually it is tied in just beyond the point of greatest pressure drop.

With a long single-pass evaporator where all pressure drop is in coil, hookup in Fig. 4 is good. For coils fed from a liquid distributor head, see valve maker. He'll tell you the best way to hook in the valve.

Thermal Bulb. Both internal and external thermal bulbs are used. But since external bulbs are good for most jobs they're more common.

Install external bulb on horizontal run of suction line near evaporator outlet, Fig. 6. Don't put bulb on a length of pipe in which liquid can collect and be trapped. This can give false valve

response. System capacity may fall.

Where suction line is ⅝-in. OD or less, bulb is usually clamped to top, Fig. 6. Clamp bulb on side of line ⅞-in. OD and larger. Have bulb just above horizontal center line of pipe. Clean pipe and apply a coat of aluminum paint before clamping bulb to it.

Internal Bulbs. Use these where you need quick response with low superheat. They are also good when suction line has many fittings near the point where bulb should be clamped to line. Don't install near point where liquid can collect.

Automatic Valves. With a fairly constant evaporator load you can use an automatic expansion valve, Fig. 7. It maintains constant suction pressure in your system. Where two or more evaporators are connected to one compressor or where load changes much, a thermostatic valve is usually recommended.

Bottom spring, plus evaporator suction pressure, acts against top springs, Fig. 7. Top-spring tension is controlled by stem at top of valve and may be set for desired suction pressure. When evaporator pressure gets too high, excess pressure is transmitted to bottom of valve diaphragm through equalizing port. Top spring compresses, allowing bottom spring to move pin toward seat. Suction pressure is cut to point where top and bottom pressures balance.

When suction pressure goes below desired point, reduced pressure in equalizing port unbalances diaphragm pressures. Top spring pushes pin away from seat, raising suction pressure to point where top and bottom pressures again balance. When compressor stops, pressure increase in evaporator closes valve tightly.

Keep Valves on Job. This article gives only basic facts to guide you. Valve selection and location are other problems you'll run into. Your best bet for help on these is the valve maker. He'll be glad to help.

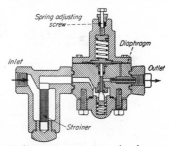

6 Top, external thermal bulb clamps to pipe. Bottom, an internal bulb

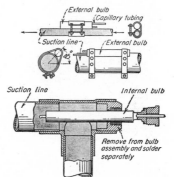

7 Automatic expansion valve for ammonia systems; externally adjustable

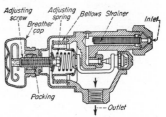

8 Automatic expansion valve for Freon, sulphur dioxide and methyl chloride

NEAT PIPING shows good planning to insure efficient running at all times. You can make your plant both attractive and efficient by working closely with manufacturer

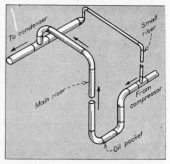

1 Use double risers when gas velocity goes below 1500 fpm at low capacity

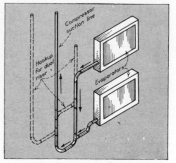

5 Loop suction line to form a seal for oil with evaporators below compressor

Eleven Piping Tips for Smart-Running Small Refrigeration Systems

Pipe your small refrigeration units right and you'll get top-notch year-round output. These pointers can save you time, money

▶ NOT ALL small compressor builders hook up their units the same way. But there are some setups many follow. Here are eleven for Freon and methyl chloride.

Hot-Gas Pipes. When compressor load changes much and condenser is high above compressor, double riser, Fig. 1, is good. At low load, oil collects in pocket, seals large riser. Gas passes through small riser at 1500 ft per min, the velocity needed to carry oil. At max-

imum load, gas flows through pocket, carries oil with it at 1500 ft per min.

Oil trap, Fig. 2, does same job as double riser. Size riser for normal pressure drop at maximum load. Oil goes direct from trap to compressor.

Where condenser is above compressor and has a bottom inlet, loop hot-gas pipe so it forms a seal, Fig. 3. Have loop top six feet above condenser liquid level or connect vent line as in Fig. 3. Close vent valve after pumping down.

Suction Pipes. With two or more evaporators above or at same level as compressor, loop lower suction line, Fig. 4. This stops oil from top evaporator from collecting in lower while it is shut down.

With two or more evaporators below compressor, run suction pipe as solid lines in Fig. 5 show. When dual riser is needed, run as shown in dotted and solid lines of Fig. 5. Lower evaporator would then be above upper one shown.

Expansion Valves Some evaporators

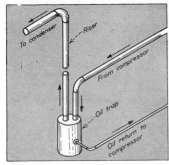

2 Oil trap may be used instead of a double riser. Trap returns the oil

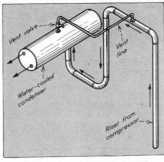

3 Loop hot-gas line before entering condenser. Vent cuts siphoning action

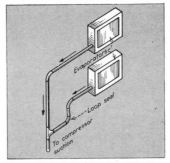

4 Connect suction line like this when evaporators are above the compressor

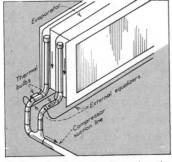

6 Use separate suction lines when the evaporator has two expansion valves

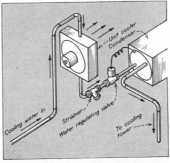

7 Cool the compressor room with a unit cooler hooked up as shown in sketch

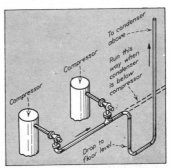

8 Compressors in parallel can use the horizontal discharge line shown here

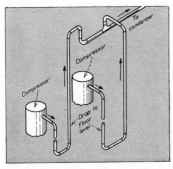

9 Run loops to floor when compressors in parallel discharge to high main

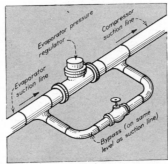

10 Bypass on same level as evaporator pressure regulator saves you time

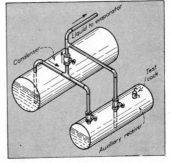

11 Auxiliary receiver holds pumped-down liquid; don't fill to the top

have two or more expansion valves. To stop one valve circuit from affecting the other, run separate suction lines, Fig. 6. Run each line several feet before connecting to common suction pipe.

Room Cooling. Where compressor room is hot or local code requires room ventilation, hookup in Fig. 7 is good. Cooling water on way to condenser passes through unit cooler fitted with fan that runs when compressor does.

Parallel Operation. When compres-

sors are close together and they discharge to one condenser above them, run discharge line as in Fig. 8 if line can be close to floor. Where line can't be close to floor, run as in Fig. 9. Loop is used where condenser is above compressors.

Backpressure Regulator. One-valve bypass, Fig. 10, lets you work on vital parts of backpressure regulator without shutting system down. Have bypass line on same level as suction line.

Receiver. Most small systems don't need a separate receiver, but where there's a large charge, a receiver is a big help. Hookup in Fig. 11 is good because receiver is valved off at all times except when system is pumped down.

Remember that hookups we've shown here are general. Maker of your compressor may prefer different ones for his unit. So be sure to check with him before changing your old piping or running new lines. Sketches are guides.

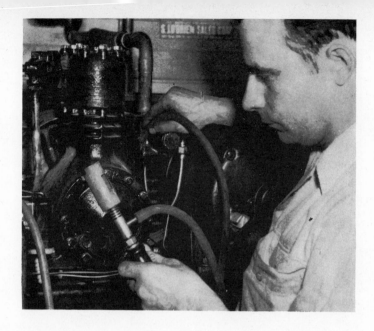

11 steps to foolproof refrigerant-leak testing

▶ PLANNING TO INSTALL more refrigeration capacity to handle summer air-conditioning loads? Having trouble playing detective with small leaks in your existing system? Then the latest guide used by The Trane Company is for you. It's your road map to a refrigerant system that's free of even the smallest leaks. Here, in 11 easy-to-follow steps, is the gist of it.

1 If test pressures exceed relief-valve setting or safe-pressure limits on bellows used in pressure controls, remove these items before building up pressure. (*Test before covering pipes or turning on condenser water.*)

2 Plug the thermostatic-expansion-valve inlet to make sure both high- and low-pressure sides of the system are fully tested. Common way to do this: Remove cap screws and insert a plug in valve's flange body. Hold plug in place with a solid diaphragm over the inlet. And don't forget to disconnect the external equalizer line from the expansion valve.

Remove the thermostatic expansion valve's power assembly during leak testing to avoid damaging it.

3 For high-side test connection: Close compressor suction, discharge valves to make sure you won't be testing the compressor itself. Connect a drum of dry CO_2 to the liquid charging valve. A word of caution: Don't use oxygen or acetylene instead of CO_2. You'll be inviting a possible explosion. And be sure to have a gage and regulator on the CO_2 drum, to control pressure.

CO_2 pressure can build up to 1000 psi or more if it isn't watched. Second gage, between regulator and liquid charging valve, reads test pressure. Install this gage with a flare tee.

Admit CO_2 to system's high side until gage reads 225 psig or pressure specified by local codes. Close regulator and charging valves. Then disconnect drum just ahead of the tee. To read high-side test pressure, cap tee and open the charging valve.

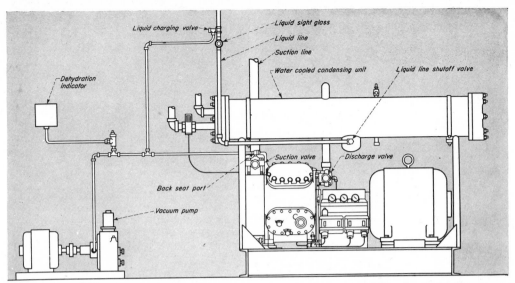

RECOMMENDED METHOD of connecting your vacuum pump to the refrigeration system before charging it with refrigerant

4 For low-side test connection: Admit CO_2 to the suction line through the external equalizing pipe that was disconnected from the expansion valve (see step 2). Drum should be upright and fitted with standard gage and regulator. Gage, installed with flare tee, between regulator and equalizer line reads test pressure. To isolate drum from system, install a packless diaphragm valve between regulator and test gage.

After admitting CO_2 slowly until test gage reads 125 psig (for Freon-12 systems), close regulator and packless diaphragm valve. Then disconnect drum at the packless valve and cap the valve's open side.

Don't forget to check with local codes for test pressures used with different refrigerants.

5 With system under test pressures, tap the sweat or solder connections with a rubber or rawhide mallet. Hit hard enough to start faulty connections leaking, but not hard enough to break a sound joint.

6 A drop in gage pressure shows up a major leak. You can spot such leaks by tracking down escaping gas noise.

To find smaller leaks brush all possible leakage points with soap solution and watch for bubbles. A few drops of glycerin in the soap solution insures good bubbling. Using distilled or rain water for the solution yields best results.

7 After bubble-testing of high- and low-side systems, break charging-valve and external equalizer-line connections; allow CO_2 to escape. Repair all leaks spotted by gage pressure loss or by bubble testing. A word to the wise: Don't try to repair leaks while the system is under pressure. Also, take bad joints apart, clean and remake them as a new joint. Don't add more brazing material over a poor connection.

8 After repairing leaks, charge high- and low-side of the system with a small quantity of Freon-12. Use five pounds of F-12 for each ten tons of refrigeration, adding larger part of the F-12 to the low side. (For charging, follow the same steps used for CO_2, above). Next, disconnect refrigerant tank, hook up CO_2 drum and build up to original test pressures.

9 Check the entire system with a halide leak detector, photo on facing page. Make sure you run the exploring tube over all parts of the system under pressure. Small amount of escaping refrigerant colors the flame green. Big leak may put out the flame or turn it dense blue with a reddish tip. If F-12 in the air colors the flame regardless of the searching tube's position, ventilate the room thoroughly.

If an open flame can't be used, re-run the bubble test where leaks have been repaired. If additional leaks show up, you'll have to run through repairing and testing steps again.

10 After you've tracked down all system leaks, let the equipment stand 24 hours under test pressures. If gages show pressure changes—taking into account room-temperature changes—you're up against more leak testing. But remember: Pressure will rise or drop about three psig for each 10-F rise or fall in temperature.

11 After final leak testing and repairing, release pressure, remove all expansion-valve plugs and reassemble the equipment for normal operating conditions. You're now set to pull vacuum on the system, see drawing at top, to prepare it for charging.

▶ UNLESS A NEW air-conditioning refrigeration system is charged right, it will work poorly. When there isn't enough charge in system, high- and low-side pressures drop below normal. Then system capacity drops too. If too much Freon is added, high head pressures result. This also cuts the systems capacity for cooling.

Charging the new system for the first time is important. Here's where the operating engineer first gets acquainted with his new unit—while it's warming up.

As Freon first enters system, check for leaks. Read "How To Test New Freon System" (following page) for complete details on finding leaks. If you find major leaks, pump out system right away and repair before going on.

When charging, go slow so you can check for the minor leaks. Feed Freon gradually. Periodically, stop compressor and weigh Freon drum to see how much charge you're feeding in. Mark minor leaks as you find them. No new system can be expected to be leakproof. It's best to find all leaks now.

After charging to full pressure, check again for leaks. Mark them. Then pump out system, fix all leaks and recharge.

Check liquid strainer. Freon-12 is a good cleaner and loosens scale and dirt in system during first charging.

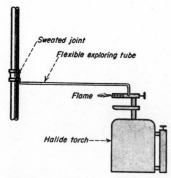

1 First step in charging new refrigeration system is attaching an accurate compound gage to the suction-gage connection. Gage should show pressure above atmosphere as well as inches vacuum. Attach pressure gage to discharge connection. Install these gages permanently

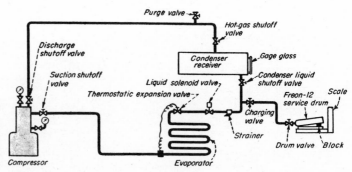

2 Place Freon-12 service drum on scale and connect to charging valve on liquid line. Use ⅜-in. flexible copper tubing or a flexible metallic hose. This is important, because solid piping strains as scale lowers and causes leaks or a break. Open the drum valve slightly to test charging connections for leaks. Leaks bubble when connections are coated with soap-and-water solution. Open compressor discharge and suction valves, condenser hot-gas and liquid valves and liquid solenoid valve. Then open charging valve, and crack drum valve enough to raise 40 psi in entire system

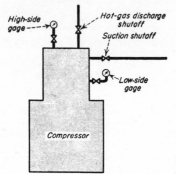

3 Test every joint in system with exploring tube of a halide torch. Pay particular attention to all connections at compressor. Slightest trace of Freon-12 turns torch flame from blue to a green color. Repair leaks and recheck with torch to make sure entire system is tight

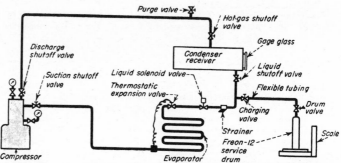

4 Turn on air-conditioning fan to put load on evaporator. Close condenser liquid valve. Note weight of service drum. Lay drum on scale and block so valve end is lower. If scale won't handle drum, lay it on floor and weigh from time to time. Open charging valve and drum valve. Circulate water through condenser and run compressor intermittently until system is charged. Freon should show in bottom of condenser gage glass. Close charging valve and open condenser liquid valve. For air conditioning, keep suction pressures 30 to 45 psi and head pressures 115 to 140 psi

TEST NEW FREON SYSTEM

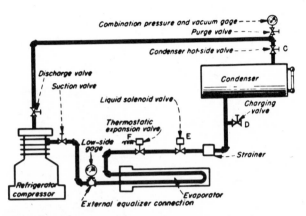

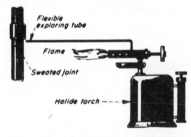

▶ BEFORE PLACING a new air-conditioning plant in service, the refrigerating-system piping must be thoroughly checked for leaks. It makes no difference how well the piping joints have been made up or how reliable the workmen are, the only sure way to know if the system is tight is to test for leaks. What do leaks mean to the operating engineer? Just this; he will lose his Freon charge and the system won't work right. Even the tiniest leaks will cause him headaches day after day. Every joint in the system is a possible source of leaks. The contractor should make a thorough search for leaks before turning the plant over to you. Since it's your baby after it's put in, you'd better know how to make this test right so you can check the work while it's being done. The following steps show the right way to test for Freon-12 leaks in any refrigeration system.

1 Close the discharge and suction valves to the compressor. Open the condenser hot-side valve **C**, charging valve **D**, liquid solenoid valve **E** and thermostatic-expansion valve **F**. Disconnect the external equalizer line to the thermostatic-expansion valve to keep it from closing. Then attach pressure gages to the high and low side of the system. The high-side gage can be attached to the purge valve. The low-side gage can be attached to the external equalizer connection after the piping is removed

2 Connect a drum of CO_2 or dry nitrogen to the charging valve. Raise pressure in system to about 15 psi. Check suspected joints by coating with solution of soap and water with a few drops of glycerine. If no bubbles show, raise pressure to 50 psi and soap again. Resweat defective joints and test again until no bubbles appear and pressure in system is maintained

3 Replace the CO_2 drum with a Freon-12-service drum. Raise the pressure in system to about 10 psi. Test all joints with the exploring tube of a halide torch. The slightest trace of Freon-12 will turn the torch flame from its normal blue color to a characteristic green color. A gasoline torch can be used for this—also denatured alcohol or acetylene gas

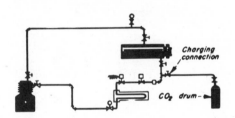

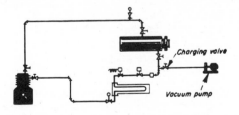

4 Disconnect the Freon-12-service drum and reconnect the CO_2 cylinder. Remove any relief valves that might raise and cap openings. About 10 psi Freon is still in system. Boost this to 175 psi with CO_2. Check all joints with the halide torch. Repair any leaks and repeat procedure until leaks are stopped. Keep 175 psi for 24 hours. If pressure holds, system is tight

5 Release pressure from system. Connect the suction side of a vacuum pump to charging valve. Discharge pump to atmosphere, running it for 24 hours. If the system does not lose more than about 1-in. of vacuum, it can be considered dehydrated and free from leaks. That's because this is a triple check for leaks and the only method to use before accepting the job

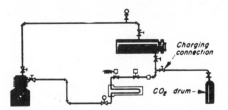

How to Charge an Ammonia System

▶ IF LARGE AMMONIA SYSTEMS were like sealed units kitchen refrigerators, operating engineers wouldn't have to worry about charging them. But their life isn't that simple. Leaks waste ammonia, or oil and water find their way in, so the operator must constantly check quantity and quality. When ammonia gets low, he must add it. When charge gets too impure, he must draw it off and replace with fresh ammonia.

Like all other jobs around the plant, there are right ways and wrong ways to charge ammonia systems. Here are pointers on the right way.

How Shipped. Manufacturers ship dry (anhydrous) ammonia in steel cylinders holding 150, 100, 50 and 25 lb net. Regardless of size, liquid ammonia fills only 85% of the cylinder at 70 F. Extra space allows for expansion of ammonia at higher temperatures, say, up to 125 F.

Cylinders will stand normal expansion of ammonia, but not the powerful forces built up when exposed to high temperature. Keep cylinders away from steam lines, hot exhaust pipes and heating places—and never apply a torch.

Take Care of Drums. Ammonia manufacturers test their cylinders regularly to be sure they can stand high pressure, and that rough handling and rust haven't made them too weak for safety. Plant operators can help by emptying drums as soon as received and returning them promptly. If necessary to store drums, keep them in doors, never outside in direct sunlight or on ground where moisture starts rusting.

When a cylinder reaches the plant, weigh it first, then test. Makers of ammonia guarantee both quality and quantity; to be sure, check both—it's easy and worth while. Weigh cylinders with protecting cap off before and after emptying. Check gross and empty weights against shipping charge.

Quality Test. It's lucky that the test for dryness and purity of ammonia is also the simplest. With reasonable care, nothing more than a small flask (4–8 oz) one old stopper and sampling pipe, you can duplicate the accuracy of the manufacturer's test.

The sketches show the details. Sampling pipe can be tubing, as shown, or ordinary black pipe, but must be clean. Fitting a shield (of metal, wood or cardboard) near the end will keep im-

These clear and detailed practical instructions are based on information from National Ammonia Division, E I du Pont de Nemours Co, and Henry Bower Chemical Mfg Co

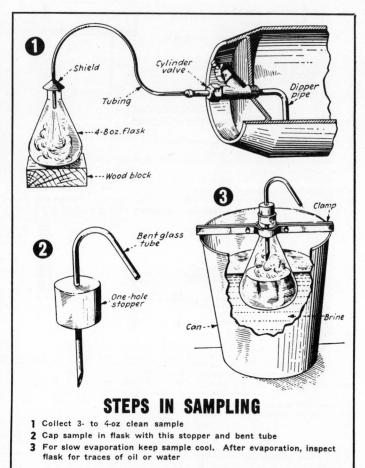

STEPS IN SAMPLING

1 Collect 3- to 4-oz clean sample
2 Cap sample in flask with this stopper and bent tube
3 For slow evaporation keep sample cool. After evaporation, inspect flask for traces of oil or water

purities out of the flash when filling.

Place cylinder in horizontal position with dipper pipe pointing down (valve outlet or valve stem will point upward, depending on type of valve). Open cylinder valve slowly to allow first gas, then liquid, to flow, sweeping impurities out of sampling tube.

Stand flask under tube with mouth against shield as shown, and let in about 3- to 4-oz of ammonia. Then remove flask quickly and cap with a one-hole stopper, fitted with glass tube bent as shown in Fig. 2.

Lipson Sampling. The chief trouble with tests may come from ammonia's

CHARGING

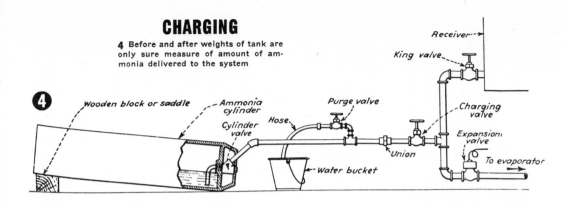

4 Before and after weights of tank are only sure measure of amount of ammonia delivered to the system

Labels on figure: Receiver, King valve, Wooden block or saddle, Ammonia cylinder, Cylinder valve, Hose, Purge valve, Charging valve, Expansion valve, Union, Water bucket, To evaporator

habit of picking up moisture. Leaving the flask open, even for a short time, picks up a surprising amount, so be sure to close the flask quickly.

The tube in the stopper carries away gas during evaporation and points down, so that froth collecting at the mouth won't fall back into the flask. Better still, avoid testing on humid days; in such weather the amount of water in the flask at the start may be enough to make the test worthless.

Ammonia will evaporate with the flask standing in the open air, but frost collecting on the glass will slow down evaporation. For a better job, hang flask in can or bucket of brine by some simple clamp, Fig. 3. Or hold it under a stream of tap water. A low-temperature bath gives more accurate results—shouldn't be much above 32F.

After the ammonia has evaporated, dry the outside of the flask and look at what's left inside, if anything. Water will show as a single drop at the bottom, or as a lot of tiny drops. Oil may show as a film or a small puddle. Reject the cylinder if you find the slightest trace of oil or water.

After checking for oil and water, wave the flask gently through the air to sweep out ammonia gas; then cautiously smell the flask. A nauseating odor points to pyridine, an organic impurity.

How to Charge. Ninety-nine times out of a hundred or better the ammonia will be found to be pure, and you're ready to charge. Most large plants take the ammonia charge to a connection between "king" valve and expansion valve. Fig, 4 shows how to connect the cylinders at this point. The purge valve helps, but isn't absolutely needed. Raise the bottom end of the cylinder about three-quarters inch and be sure the dipper pipe points down.

If charging line and connections are

tight, close king valve and open cylinder valve and charging valve. After king valve is closed, pressure between it and expansion valve will drop to suction pressure, allowing liquid ammonia to flow into system. To be sure cylinder is discharging into system, put it on scale and watch change in weight. As cylinder empties it gets cold near valve from evaporation of last few ounces of liquid.

If no scales are available, this cooling effect will show emptying. But don't depend on it; always weigh cylinder after charging to be sure it is empty.

Sometimes a mistake in hooking up equipment, or some other trouble, causes reversed flow—from system to cylinder, overfilling it. Exposing such an overfilled cylinder to even slight overheating might bring on a dangerous explosion —better be safe than sorry.

When cylinder is empty, close cylinder and charging valve and open king valve. Lead pressure from charging line to purge valve. If no purge valve is installed, crack the union carefully.

Charging With Gas. To charge a system with gas, instead of liquid ammonia, turn cylinder over so inside dipper points up. Ammonia may be taken from the cylinder at the rate of 1½ to 2 lb per hr; heating cylinder with warm water will speed up the flow, but is not recommended.

Always check to make sure that gas flows out of cylinder and does not condense into it. Again, weighing the cylinder before and after will show for sure.

How much ammonia should a system contain? For the first charge or a complete recharge, best answer comes from the engineer who designed and installed the system. If the figure isn't handy, you can calculate amount needed from data given in refrigerating handbooks.

In a running plant, the gage glass on

receiver tells the story; receiver should be about one-half full. Without gage glass, you have to fall back on watching operation. If only lower coils frost when expansion valve is wide open and plant is not overloaded, the system needs ammonia. If a clicking sound is heard in expansion valves, it checks the lack of ammonia.

HOW TO PREVENT ICE ON

Induced Draft Towers

1—Run two-speed motors at half speed to retard ice formation. Fans use 15% full-speed power.

2—Shut down some fans (if multi-cell unit).

3—Cover louvered area's upper portion to limit air flow.

4—Reduce tower's water flow and stop one or more fans (if multi-cell unit).

5—Bypass water to part of cooling tower and stop one or more fans (if multi-cell unit).

Louvers and Filling

1—Reverse fan's motor (for limited time only) to blow warm air out louvers. *Caution:* Some speed reducers don't supply oil to upper bearing when reversed.

2—Stop some fans temporarily, but don't stop water. When cells thaw out, start fans and repeat process in other cells.

3—Cover upper portion of louvered area to limit air flow.

CENTRIFUGAL REFRIGERATION LOG

JOB NAME: XYZ Company   JOB No. 18   MACHINE SER. No. 0000   SIZE 17M43-7-6

TIME	VACUUM	LIQUOR TEMPERATURE	WATER TEMP IN	WATER TEMP OUT	CHILLED WATER GPM	VACUUM-OR PRESSURE-A	AIR INDICATOR	WATER TEMP IN	WATER TEMP OUT	CONDENSING WATER GPM	DAMPER POSITION	CONTROLLER OPERATING POINT	TURBINE RPM	MOTOR RPM GEAR RATIO	BEARING TEMP THRUST END	BEARING TEMP SEAL END	OIL PRESSURE BEFORE SEAL	OIL PRESSURE AFTER SEAL	PURGE SUCTION	PURGE DISCHARGE
1	2	3	4	5	6	7	8	9	10	11	12	13	14	15	16	17	18	19	20	21
AM																				
9	—	—	—	—	—						8			—	—	—	—	—	5*	60#
10	13.5	46	60	51	7#		80	89			8	4		4100	110	130	17.5	14	5*	60#
11	15	42	56	48	8#		82	93			8	7		4500	110	132	17	13	—	
12	15.5	41	56	48	9#		85	95			8	7		4500	112	132	17	13	—	
PM																				
1	15.5	41	56	48	9#-		84	93			8	7		4500	115	135	17	13	—	
2	15.5	41	55	47	9#		84	94			8	7		4500	118	135	17	13	—	
3	15.5	41	57	48	9#+		85	95			8	7		4500	116	136	17	13	5*	60#
4	15.5	41	58	48+	9#+		85	95			8	7		4500	115	135	17	13	5*	60#
5	15	42	53+	49	10#		85	95			8	7		4500	115	135	17	13	—	
6	14	45	57	50	8#		85	95			8	4		4100	116	135	16	12	—	

STOPPED — LIQUOR LEVEL: 2/3 glass   FULL SPEED: ___
MOTOR — FULL LOAD AMPS 240   TOT. PTS. ON CONTRLR 11
MACHINE ROOM TEMP: 90 °F AT 10 AM   92 °F AT 3 PM
DATE 4-1-51   ENGINEER SLM

CENTRIFUGAL refrigeration systems like this are easy to operate and maintain

TYPICAL LOG SHEET for centrifugal refrigeration system has spaces for all important entries. Carefully kept logs like this one make trouble-shooting easier, surer

Don't Baby Centrifugals...

... they're rugged, dependable, have few troubles. Just follow these solid tips on refrigeration jobs for easy running. Then when trouble pops up use the 60 time-tested remedies given

Carrier Corporation

▶ CENTRIFUGAL compressors are plenty reliable—you don't often have operating troubles with them. But if trouble shows up you want to cure it fast. To do that you must know trouble signs, causes and cures. Table II, *facing page*, gives you dope on operating troubles and what to do about each. Let's see how we can best use it in our everyday work in the plant.

Operation. If you've run reciprocating compressors on refrigeration jobs you know that high head pressure cuts output. Same can be true for centrifugals. Head pressure increase, gradual or sudden, means there's air in condenser if load, speed and other conditions remain the same. Check temperature difference between condenser-water discharge and condenser gas. Increase, with fixed load, may mean air in condenser. Rise of cooler pressure can

mean air, too. Run purge unit to remove air from machine.

Other troubles outside condenser cause head-pressure increase. Table shows what they are, how to correct. Moisture removal is as important as removing air. Check machine for tube leaks if you get much water from purge unit each day. The water separates fast when machine is stopped. So if you're getting water, run purge unit awhile after machine stops, before starting.

Typical Controls. Low compressor oil pressure, refrigerant and brine temp automatically shut down machine, as does high condenser pressure. Settings of safety controls for Freon-11 vary, but 6 psi cut out for low oil pressure and 12 psi cut in are usual. Refrigerant-temperature cutout is usually set a few points below design temperature, depending on brine used.

Chilled-water cutout works a few degrees below minimum leaving temperature selected. Condenser-pressure cutout goes out at 15 psi, in at 8 psi for F-11. Only necessary difference in

motor and turbine safety controls is in method used to shut down.

Your safety controls should stop machine whenever you have troubles listed above. When this happens look over table for causes, check machine and correct.

Drive. Table lists operating troubles for motors only. That's because turbine troubles can't be generalized too easily. If you have a turbine drive consult instruction book supplied with machine. It lists major troubles and what to do.

Compressor Log. Good log helps spot troubles fast. Fig. 2 shows typical sheet good for most refrigeration jobs using centrifugals. Table I, *facing page*, gives points to check during running and best intervals between checks.

Remember, not all refrigeration jobs are the same. So you·may have to change your operating procedure a little to suit the job. Main idea to keep in mind is that you should regularly inspect and care for your compressors during all operations and shutdowns. Don't baby 'em—just treat 'em right.

I: FOR BEST RESULTS RUN YOUR CENTRIFUGAL BY THIS SCHEDULE

COMPRESSOR _____

Oil Pressure	Hourly
Bearing temperatures	Hourly
Change oil	Yearly
Pump-chamber oil level	Hourly
Atmospheric float-chamber oil level	Hourly
Replace oil filter	Yearly
Clean oil strainer	Yearly

COOLER _____

Clean tubes	When needed
Refrigerant temperature	Hourly
Refrigerant level	Daily
Refrigerant pressure	Hourly

CONDENSER _____

Clean tubes	When needed

Refrigerant pressure	Hourly
Safety controls	Six months

PURGE SYSTEM _____

Compressor oil	Daily
Motor lubrication	Monthly
Clean condenser	Monthly
Oil level	Hourly
Change oil	Yearly
Relief valve	Daily
Water removal	Daily
Oil separator	Yearly
Compressor valves	Yearly

COUPLINGS _____

Lubrication—oil type	Weekly
Lubrication—fluid-lubricant type	Six months

Alignment	Yearly

GEAR _____

Oil level	Hourly
Oil pressure	Hourly
Change oil	Yearly

MOTOR _____

Oil level	Hourly
Change oil	Yearly
Brushes	Yearly

TURBINE _____

Oil pressure	Hourly
Oil level	Hourly
Emergency trip	Daily

SPARE PARTS _____

Spares on hand	Yearly

II: USE THIS CHECK CHART TO CURE SYSTEM TROUBLES FASTER

SYMPTOM	POSSIBLE CAUSES OF TROUBLE; CURES
CONDENSER PRESS. HIGH	1 Not enough condenser water 2 Water valves throttled 3 Water pump not running right 4 Clogged strainer or tubes 5 Cooling tower not operating right; fan shut down; sprays or pump clogged 6 Scaled condenser tubes 7 Air in condenser; find and repair leak 8 Condenser float valve stuck closed; adjust float
CONDENSER PRESS. LOW	1 Water inlet temperature low; throttle flow 2 Refrigerant charge too small; add refrigerant if needed
SUCTION DAMPER CONTROL NOT RIGHT	1 Control air off; check air filter and controller 2 Damper links or blades binding; move damper through entire range by hand 3 Controller doesn't operate right; check and adjust
F-11 LEVEL TOO LOW	1 Charge lost or trapped in condenser or economizer by the floats; find leak; check floats; add refrigerant 2 Purge leaking; check relief valve
LOSS OF CAPACITY	1 Condenser not transferring enough heat; see first item in this table 2 Hot-gas bypass valve partly open; close valve 3 Gradual contamination of refrigerant by oil; rectify refrigerant 4 Sudden increase in difference between refrigerant and water temperature; check division plates and gaskets in cooler waterbox for breakage 5 Gradual increase in water and refrigerant temperature; clean cooler tubes, spray deck; remove excess oil from refrigerant, regulate speed to load, check division plates and gaskets
COMPRESSOR SURGES	1 Load too light; open bypass valve 2 Air leak; run purge and repair leak 3 High condenser pressure; see first item in this table
COMPRESSOR SHUTS DOWN	1 High condensing pressure; check water quantity, condenser float valve; purge air 2 Low oil pressure; check strainer cleanliness and reducing-valve setting 3 Low refrigerant or chilled-water temperature; check refrigerant level, chilled-water flow; regulate load
COMPRESSOR 2ND STAGE FROSTS	1 Economizer float valve blocked; check float operation 2 Light load with high compressor speed, cold condenser water; increase condenser pressure, reduce speed 3 Suction-damper closure causes circulation of condensate from economizer

SYMPTOM	POSSIBLE CAUSES OF TROUBLE; CURES
THRUST END OVERHEATS	1 Bearing worn from lack of lubrication; replace 2 Excess pressure from gear or couplings; check for mechanical interferences
HIGH OR LOW OIL PRESSURE	1 Too much oil dilution of refrigerant or foaming on start; pump out by starting compressor momentarily several times; remove diluted oil and replace 2 Oil pressurestat out of adjustment; check setting, readjust 3 Compressor speed too low (turbine)
COMPRESSOR BEARING TEMPERATURE HIGH	1 Not enough cooling water flowing through oil cooler; open throttling cock to oil cooler 2 Too much oil dilution of refrigerant on starting or electric oil heater off during shutdown 3 Pump-chamber temperature above 150 F on start; turn on cooling water earlier during start; check for clogged cooling coil 4 Thrust or shaft bearing scored; check bearing and shaft-end clearance
COMPRESSOR LOSES OIL	1 Worn labyrinths; check rotor balance and bearing wear 2 Oil system leaks; check oil piping and correct 3 Oil level too high; drain oil
PURGE UNIT DOESN'T RUN RIGHT	1 Motor thermal overload burned out; replace 2 High-pressure cutout-switch contacts open; check setting and operation; reset if needed 3 Air-cooled condenser fouled with lint and dirt; clean with compressed air 4 Shutoff valves closed 5 No power; check switch, contacts and connections 6 Broken belt; replace 7 Separator float valve stuck; valve off unit, evaporate F-11, open tank and fix float 8 Air relief valve stuck; stop purge unit, remove valve and repair or replace 9 Suction-line reducing valve stuck; valve off purge unit at condenser, remove valve and repair or replace
COMPRESSOR COUPLINGS NOISY	1 Misalignment; check at operating temp; realign 2 Not enough lubrication; add oil or grease 3 Excessive wear; replace
DRIVE MOTOR OVERLOADED	1 High chilled-water temperature or flow; regulate 2 Damper open; regulate as needed 3 Refrigerant level too high; remove refrigerant
GEAR OVERHEATS	1 Oil level too high; drain oil 2 Fouled cooler; clean 3 Not enough cooling water; increase flow 4 Misaligned coupling; check

HOW SOON THIS SYSTEM LOSES ITS EFFICIENCY DEPENDS ON HOW MANY OF THESE MISTAKES OPERATORS MAKE

19 Ways Operators Ruin Refrigeration Equipment

Check your practices against these operating mistakes.

▶ To LIST ALL the ways operating engineers can add to the normal wear and tear of refrigerating equipment would be to recite the history of refrigeration. But over the years, the commonest mistakes I've seen pulled are listed here. And they're all a result of either incompetence, negligence, carelessness, or just downright misuse. None of them is a major problem to beat at the start, but they can all cause extra expense and maintenance time lost if neglected.

1 WRONG LUBE OIL. Using oil with low flash point for lubricating compressors. While not likely to flash in the system, the wrong oil will break down under pressure, wear and heat of compression. Then it deposits carbon or sludge which scores poppet valve stems, dampener pistons, piston rings, cylinder walls, etc. If pour point is too high, any carryover may congeal in the evaporator.

Using wrong lube oil makes compressor "boil"

2 CONTAMINATED OIL. Leaving oil in the system too long without either filtering, blowing down to remove sludge, or renewing, will lead to improper lubrication and scored or carbonized parts.

3 DIRTY FILTERS. Forgetting to clean the oil filters before charging new oil to crankcase. Entrained impurities work loose from the filters and circulate through the whole system.

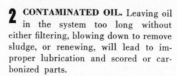

Neglected oil filter puts plant out of kilter

4 UNFILTERED BLOWDOWN. Another common error is neglecting to filter the dirty blowdown oil from crankcase before recharging to system.

5 DIRTY OIL CAN. Ruining otherwise clean oil by using a dirty can to charge it into compressor crankcase.

6 FORGETTING TO OIL or grease separately lubricated bearings.

7 CARRYOVER OIL. Some ammonia systems don't have separators at compressor discharge. Oil carryover to condenser or evaporator causes drop in efficiency and greater wear to compressor when operator forgets to remove from system.

8 WOOLEN WICKS. Using woolen wicks for capillary feed lubricators. Ammonia is death to wool even though soaked in oil. Bits of wool get into system and clog oil passages. Use cotton and renew more often.

9 FORGETTING TO PURGE. Air and other non-condensable gases get into every refrigeration system, reducing efficiency. Compressor has to work against higher head pressure to carry load and so wears out faster.

10 CLOSED DISCHARGE. Many a man has tried to start a compressor against a closed discharge valve. If motor is improperly protected and fails to cut out, untold damage can result . . . including explosions.

11 FROST ON EVAPORATOR acts like an insulator causing lower heat transfer, lower suction pressure, higher compression ratio, higher discharge-gas temperature, longer running time, possibility of lube-oil breakdown, greater wear. Defrost periodically.

12 LEAKS NOT REPAIRED. Danger from direct loss of refrigerant into air, water and/or brine

Plant owner gets hooked when grease gun's overlooked

Compressor is abused if purger isn't used

Compressor discharge blocked; whole plant may soon get rocked

Evaporator frost makes higher power cost

with indirect losses from corrosion by chemically-active vapor.

13 V-BELT TENSION. If belt is too loose, driving pully slips on starting and may burn belt. If belt is too tight belt will go in no time, also put wear on bearings.

14 IGNORING MOTORS. Allowing sliprings and motors to get and stay dirty; windings full of lint, dust and even oil, can lead to motor's over-heating and wearing out faster.

15 INSTRUMENTS. Many operators don't have enough instruments. Others have enough but assume they're always working perfectly and forget them. Then they operate blindfolded and don't know what's going on inside. Might as well not have them.

16 DIRTY CONDENSER TUBES. Unless they are cleaned periodically, higher head pressures result, calling for more compressor power and cooling water.

17 WRONG GASKETS. Using rubber gasket material on ammonia equipment is asking for trouble.

18 MISTREATING EXPANSION VALVE. Using a pipe wrench on the rim to unscrew the diaphragm bonnet of a thermoexpansion valve may save time but it doesn't save valves.

19 BLOCKING SAFETY VALVES. Make's one wonder how some men get licenses.

When the V-belt sags, output power lags

Meter on the blink; efficiency will sink

On the rim he put a wrench; now the valve is on the bench

Of course, not all mistakes are operating mistakes.

THIS OLD MACHINE IS A MONUMENT TO GOOD OPERATION. ABOVE MISTAKES COULD HAVE SHORTENED ITS LIFE

Refrigeration Oil Level

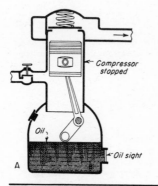

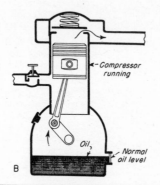

1 Oil level shows incorrectly in refrigeration compressor after system's been shut down for a long time, *A*. That's because there's no way to know how much liquid refrigerant is mixed with the oil. After a long shutdown, oil contains a lot of liquid refrigerant, so the level naturally is higher.

The best time to check oil level is after the machine has been running for some time, *B*, then there is only a little refrigerant mixed with the oil. While compressor is running, refrigerant is evaporated out until only normal quantity remains in solution. Then oil level shown in the sight glass is correct.

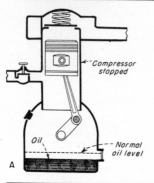

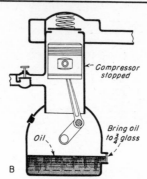

2 If oil level is lower than normal after long shutdown, *A*, it's a sure thing that actual running oil level is much too low. If compressor is started without adding oil, it won't be long before refrigerant is pumped out of oil and machine may be ruined, or at best the bearings may be harmed.

In this case, be sure to add oil until level reaches ¾ full in sight glass, *B*, before starting compressor. Then after machine runs a while, check level again. It should be normal after refrigerant is pumped out of oil. Never let oil get out of sight in glass, or you won't know how high it is. Keep crankcase oil at right level.

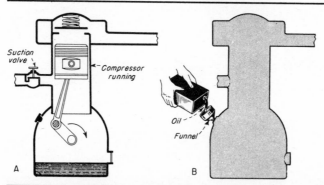

3 To add oil to crankcase, close suction valve, *A*, and open discharge valve. Then run machine a few revolutions. This will reduce crankcase pressure to a few pounds above atmosphere. *Don't pump down to vacuum.* Idea is to have slight pressure so moist air does not enter the refrigerating system.

Then remove oil filling plug slowly to bleed off the slight pressure in crankcase. Add oil through a clean, dry funnel, *B*. Pour oil steadily so no air enters through funnel. Screw in plug as soon as funnel is empty. Caution: Never leave a Freon system open for long, close it as soon as possible. This goes for other refrigerants also.

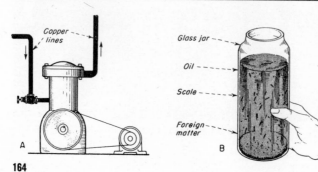

4 Most Freon-12 systems use copper tubing, *A*. Usually great care is taken when installing tubing to keep dirt out of system. If system is originally clean, chances are crankcase oil will not be contaminated enough to need removal for filtering or replacing with new after the breaking-in run.

When steel or iron piping and fittings are used with Freon-12, be sure to draw a sample of oil from crankcase every six months, *B*. Check oil in glass jar. If tiny particles of foreign material show up after foam has settled, replace all oil. It is a quick way to test oil. Be sure to check and renew the oil every six months.

6
Air conditioning, heating, cooling towers

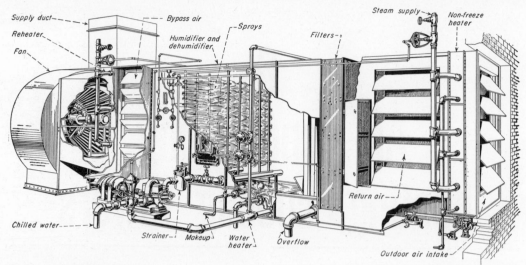

1 Typical air-conditioning equipment for a centrally located compressor and water cooler. Outdoor air is heated by steam coils in air stream, then is humidified or dehumidified as needed. After air is washed, it's ready for distribution

Ways to Air Condition

Outside air is now "tamed" for year-round indoor use to serve any purpose.

▶ MOST PEOPLE THINK air conditioning is for summer only, but it's used all seasons. To do a complete job, conditioning equipment must control these four properties of air—temperature, humidity, cleanliness and air motion.

Temperature. Temperature is kept at a set comfortable level by adding or removing heat. To do this, one heat exchanger for heating in winter and one for cooling in summer are used. Heating medium may be flue gas, steam or hot water.

Steam or hot water is circulated in copper coils through air to be heated. Fins bonded to coils outside increase the heating surface, Fig. 3.

Temperature is usually regulated by controlling steam or water temperature in the coil with a thermostat. Another way is a bypass damper, mixing heated and unheated air. The same coil or unit is often used for summer cooling. Cooling medium is chilled water, brine, or direct-expanding refrigerant. Cooling is also done by passing unconditioned air through cold water sprays.

Humidity. Moisture content of air should be controlled for health and comfort. Engineers speak of moisture content as "relative humidity," which is the air's degree of saturation, expressed in percent. Heated air's moisture-carrying capacity is increased and its relative humidity decreased, if no

moisture is added. Air is "drier."

Too-low relative humidity makes a space feel colder than its temperature indicates. That's because moisture is evaporated from persons in room, lowering their skin temperatures. To avoid this, water vapor is added to air after heating.

Many types of humidifiers are used —steam, air atomization, impact, hydraulic separation, forced evaporation (pan), and air washer. The first two are too noisy for most offices. The common impact humidifier uses a high-velocity water jet against a target, breaking water into a fine spray. Then room air is blown past targets to pick up the spray as it evaporates. Humidistat turns jet off and on automatically.

The hydraulic-separation humidifier has a slotted nozzle and whirling chamber, similar to oil-burning atomizers. Whirling chamber throws fine water spray through nozzle slots into heated air stream. Control is by turning water off or on.

Mechanical separation uses mechanical means for a fine spray in stream. Forced-evaporation (pan) humidifiers are wide shallow containers of water, placed in air stream, and heated. Evaporation is controlled by water temperatures. This is simplest humidification method. It's used in small units, but can't be controlled within narrow limits.

In air washers, air stream is passed through three or more banks of water sprays, some directed with air flow and some against it, Fig. 4. Eliminator plates after the sprays remove entrained water droplets.

The water is usually in a closed system with reservoir pump and float valve for adding makeup as water is evaporated. A steam coil can be placed in reservoir to control the spray's temperature. This type controls humidity closely. It cleans air and cools it during cooling season with cold water, thus eliminating cooling coils.

Dehumidification. Moisture is taken from air at most times during cooling season, because the moisture in the air

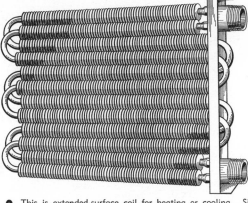

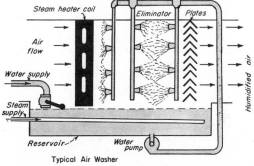

3 This is extended-surface coil for heating or cooling. Six to eight fins per inch, bonded to outside, increase surface

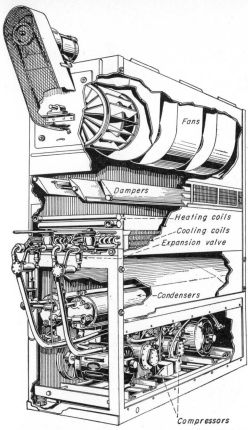

2 Packaged unit conditions air and has advantage of being easy to install and with all equipment in one small space

Typical Air Washer

4 Air flows from filters separating dust from air stream, then to water spray removing remaining impurities by washing

contains heat. Condensing some of air's moisture by cooling below its dew-point (air's saturation temperature) with a cooling surface or spray is one way. Another way is absorbing some moisture in chemical desiccant beds, over which air is passed. Many installations have two beds of chemicals (such as silica gel) for continuous use. Then one is reactivated while other is in service.

Dehumidifying equipment selected for any job depends on amount of air to be handled, desired temperature of leaving air, relative humidity, water conditions, and other variables.

Cleaning. Conditioned air must be cleaned. Either disposable or non-disposable filters are usually used. Operation is the same for both types.

There's a baffled passage through the filter. It's coarser on the inlet than outlet end, so dust and lint particles in air strike filter divisions. Most filters are coated with a light odorless, non-inflammable oil to make baffles sticky and hold dust better.

Disposable filters are made of cardboard, glass wool, steel wool, or a similar low-cost substance in cardboard frames. These are usually thrown away when dirty.

Non-disposable filters are built of copper mesh or grill in a metal frame. These are cleaned with gasoline or a caustic solution, then dipped in light oil and allowed to drain before re-installing. Air washers remove most dust from air, but oily particles sometimes pass through them. A filter is best ahead of washer.

Place filters so both outside and re-circulated air pass through. Two banks of filters installed in one duct insure complete air cleaning.

Air Motion. Air motion is necessary for proper ventilation of all parts of conditioned spaces—also for same temperature throughout. Fans take suction over the coils and discharge to conditioned spaces through outlets.

Outlets are placed to avoid drafts and dead air pockets in rooms. Outside-air inlet is usually placed in fan's suc-

tion so fresh air mixes with recirculated room air before passing through coils. This makes a slight positive pressure in conditioned spaces. Then air leakage is outward, avoiding outside uncondi-tioned air leaking into spaces.

Types of Units. Not all air condition-ing units do both heating and cooling. Summer air conditioners don't heat and winter conditioners don't cool. How-ever, both features are included in a unit for year-round service.

Packaged Units. Packaged units come with all needed equipment in a cabinet, Fig. 2. In smaller sizes, cabinets are often in harmony with office or store furnishings. Then they can be placed inside room to be conditioned. Larger packaged units have duct-work for fully automatic room conditioning that is too large for one outlet, Fig. 1, or where a number of rooms are to be served.

Packaged units need little mainte-nance for long periods, are easily in-stalled and moved.

Units over fifteen tons are usually custom built for each job.

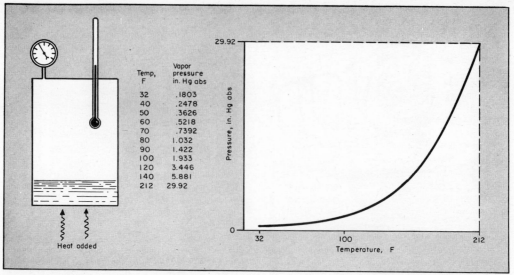

Temp, F	Vapor pressure in. Hg abs
32	.1803
40	.2478
50	.3626
60	.5218
70	.7392
80	1.032
90	1.422
100	1.933
120	3.446
140	5.881
212	29.92

Heat added

RELATIVE HUMIDITY is actual amount of moisture in air at any given temperature, divided by the greatest amount of moisture the same air could hold without condensation. A convenient way to look at this is in terms of pressures. That is, actual pressure exerted by moisture (*vapor pressure*), divided by highest vapor pressure possible under the same conditions, gives relative humidity. Let's look at a system that gives us the relation between this highest possible vapor pressure—the *saturation condition*—and temperature.

At left above, we have a sealed chamber with all air evac-

uated and water in the bottom. Now we get the system at exactly 32 F and let it settle. The pressure gage then shows 0.18 in. Hg abs. If we do the same for 40 F, the gage reads 0.25 in., and so on up to 212 F. Precise values of these gage readings for various temperatures are tabulated.

If we plot these pressures and temperatures, we get a *vapor-pressure curve, above*. For atmospheric air, this curve represents highest pressure that water vapor can have without condensation—it's 100% relative humidity. This pressure, plus that of dry air, makes up atmospheric pressure.

We often say, 'It isn't the heat, it's the

But what is humidity and how do we measure it? That's where the often misunderstood term "relative humidity" comes into the picture. To get it clear, start with explanations in the diagrams and captions, then read text to see how we find it

▶ IN THE DIAGRAMS and captions we've got a line on what relative humidity is, and on one of the items we need to measure it—saturation vapor pressure. We've also seen what wet-bulb temperature is. Now we need to know how to use wet-bulb temperature to get the other item—actual vapor pressure—involved in relative humidity.

If the air were saturated (100% relative humidity), we could measure vapor pressure directly below the saturation curve. But in our example, air

isn't saturated, so its actual vapor pressure is something less. As shown in the last diagram, actual vapor pressure is found under the curve at a point called *saturation temperature*. This is the temperature to which our air mixture would have to be reduced to bring it to 100% relative humidity.

To find the actual vapor pressure, we recall that at the wet-bulb temperature (*see captions*) the heat from the air to a water drop equals the heat taken from the drop to vaporize water. So we set

up heat-transfer relations for each condition and put them equal to each other.

For temperatures from 30 to 130 F, this gives a simple and fairly accurate rule: Actual vapor pressure equals saturation vapor pressure at wet-bulb temperature minus 0.11 times difference in dry- and wet-bulb temperatures. Then actual vapor pressure divided by saturation vapor pressure at dry-bulb temperature gives relative humidity. We usually multiply by 100 to get relative humidity as a percentage.

Let's see how this works for the diagram figures, taking wet-bulb temperature as 58.3 F. From the table, saturation vapor pressure at wet-bulb is 0.491 in. Hg abs. Then 70 minus 58.3 gives 11.7 as difference between dry and wet bulbs. Multiplying by 0.11 yields 0.129. Subtracting this from 0.491 we get 0.362 as actual vapor pressure.

Saturation vapor pressure at 70 F is 0.739 in. Now if we divide actual va-

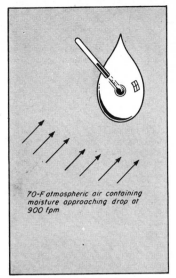

70-F atmospheric air containing
moisture approaching drop at
900 fpm

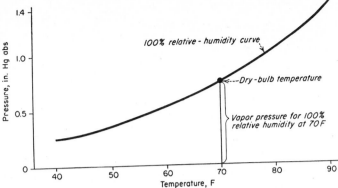

DRY-BULB TEMPERATURE. Curve shown here is a blownup reproduction of vapor-pressure curve on facing page. Water drop we've just been talking about is at 70 F, so it shows as a point on saturation curve at that temperature. Such a temperature, measured by ordinary thermometer, is called *dry-bulb temperature*

ACTUAL VAPOR PRESSURE. We now know highest vapor pressure possible for air at various temperatures *(table, curve, p 118)*. Let's call it *saturation vapor pressure*. For relative humidity, we also need actual vapor pressure.

We measure this by a *psychrometer*. To see how this gadget works, let's look at a water drop at 70 F, with moist air, also at 70 F, blowing on it at 900 fpm.

humidity'

por pressure by this, and multiply by 100, we get 49% relative humidity.

Wet-Bulb Thermometer. We've seen what relative humidity is and how we can figure it if we know dry- and wet-bulb temperatures. Dry-bulb comes from an ordinary thermometer, but how do we get wet-bulb? Even if you've never seen a wet-bulb thermometer, you've probably guessed by now that it's an ordinary one with its bulb surrounded by a wick wetted with water.

In what is known as a *sling psychrometer*, dry- and wet-bulb thermometers are mounted side by side. The instrument can be twirled around like a charm on the end of a watch chain to get air moving past the wet bulb at 900 fpm or so. This simulates the conditions of that drop of water we talked about in the diagrams and captions. Many psychrometers have a slide-rule gadget on which you set the two temperatures and read relative humidity directly.

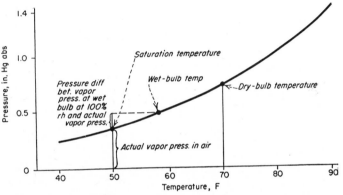

WET-BULB TEMPERATURE. Now let's see what happens when air at 900 fpm starts blowing on the drop. Water at surface begins to evaporate. This vaporizing takes heat—latent heat of vaporization—so the drop is cooled. Temperature the drop finally reaches when cooling stops is what we call *wet-bulb temperature*

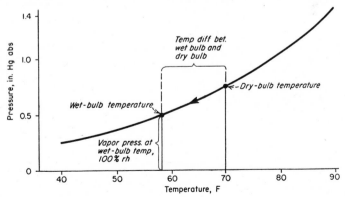

COOLING OF THE DROP stops at wet-bulb temperature because heat coming from the air to the drop eventually equals heat given up by the drop. As we'll see in more detail in the text, we can find actual vapor pressure of the air, and thus its relative humidity, once we've established the wet-bulb temperature

Here's how to figure humidity, wet-bulb temperature, air mixtures,

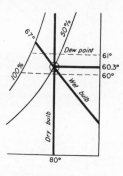

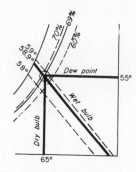

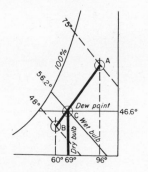

Humidity...

Let's assume you know dry-bulb temperature is 80 F and wet-bulb temperature is 67 F. You want to find dew-point temperature (temperature at which moisture will condense out of the air for a given specific humidity and pressure, as air-moisture temperature is reduced) and % humidity.

Find 67 F on the 100% humidity curve and follow the line parallel to constant wet-bulb lines. Then trace the line from 80 F on the dry-bulb scale. Intersection shows dew-point temperature is 60.3 F, humidity is 50%. This point represents condition of the air.

Wet bulb...

Now suppose you know dry-bulb temperature is 65 F and dew-point temperature is 55 F. You want to determine wet-bulb temperature and the % humidity. Approach is much the same as in our first example.

Follow the 65-F constant dry-bulb temperature line. Then, moving over to the vertical scale, locate 55 F. Intersection pinpoints air-moisture condition. Since chart doesn't have a fine enough grid, you'll have to estimate 58.9-F wet-bulb temp along the constant 100% relative humidity curve. The 69% humidity is also an estimate.

Air mixtures...

Let's say 25% (point *A*) of a mixture of two air streams is at 96-F dry bulb and 75-F wet bulb. The balance (point *B*) is at 60-F dry bulb and 48-F wet bulb. Then, 0.25 × 96 = 24; 0.75 × 60 = 45. Resultant dry bulb of the mixture is 24 + 45 = 69 F.

Now draw a line between *A* and *B*. Intersection of 69-F dry bulb with the line between *A* and *B* establishes point *C* for the mixture. Wet-bulb and dew-point lines, running through point *C*, are conditions of the mixture. Its wet-bulb temperature is 56.2. Dew-point temperature is 46.6 F.

Get to know air-moisture data

It sets the groundwork for well-planned air conditioning

Psychrometric chart shown here in skeleton form is nothing more than a graphic portrayal of various air-moisture mixtures.

Lines of constant dry-bulb temperature are almost vertical. Constant dew-point temperature lines are horizontal. Constant wet-bulb temperatures slope downward to right. Constant % humidity lines are curved. Before using chart you must know any two values. You can then pinpoint the air-moisture condition. —*Courtesy The Trane Co*

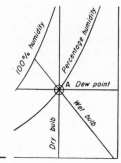

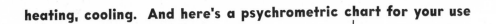

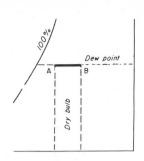

Heating, cooling ...

The chart can also trace air-conditioning cycles. Remember, in tracing any cycle, dew-point temperature is constant as long as there's no change in moisture content of the air.

Thus heating air without changing its moisture content would take place along a horizontal line of constant dew point. The dew-point line would, naturally, be set by the initial condition of the air. Initial air at point *A* would be heated to point *B* along horizontal line *A-B*. If you cool air without condensing moisture, the condition of the air moves from point *B* to *A*.

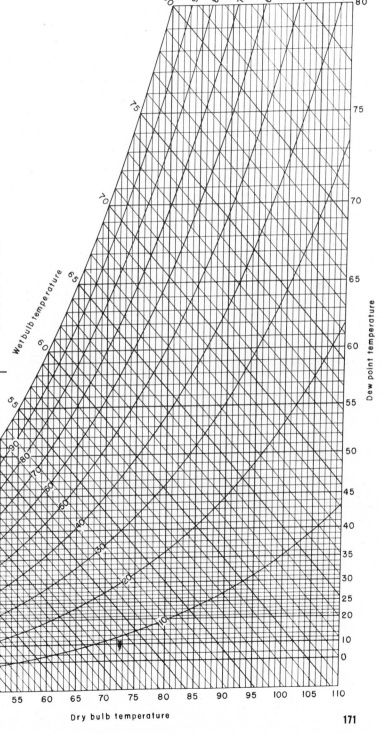

Dry bulb temperature

Cool Your Building Roof

TYPES OF SPRAYS IN ACTION

STATIONARY streams from jet *(below)* shower fine droplets. Roof thermostat controls water, keeps roof temp right

ROTARY cooler revolves slowly through gearing as wheel breaks up thin spray, uses no thermostat or electrical wiring

WHIRL JET throws fine sprays over roof, runs continually. Excess water, if any, runs into drain sump for recirculation

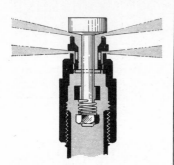

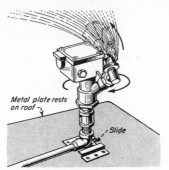

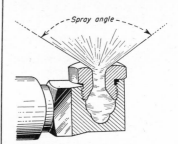

DOUBLE row spray head throws upper and lower streams. Upper sprays wide outer circle; lower inner area 10-ft diameter

JET of fine spray is thrown over roof at 45 deg. Cooler runs continuously, revolves just enough to moisten roof

NOZZLE has no moving parts, but is designed so water whirls in jet, throwing hollow spray cone at very wide angle

CONTROLLING ROOF TEMPERATURE

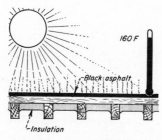

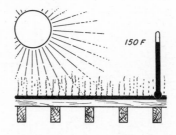

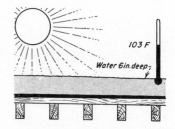

INSULATED roof's temperature gets high as 160 F on hot day because heat can't get in. Roof deteriorates fast

UNINSULATED roof of same construction reached 150 F on same day. The spaces under this roof are hot, need cooling

FLOODED roof is cooler, but water stores heat, gets stagnant, mosquitoes breed—so must be treated. Roof may also leak

By Making It Sweat

THIS MUCH HEAT GOES IN

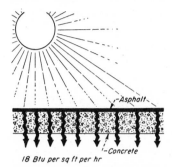

18 Btu per sq ft per hr

ASPHALT over concrete let in 18 Btu one hot day in August. Black roof absorbs most heat from hot rays of sun

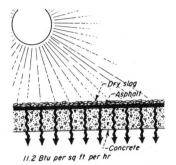

11.2 Btu per sq ft per hr

DRY SLAG over asphalt of same concrete roof on same day absorbed 11.2 Btu. This is less heat, but too much for comfort

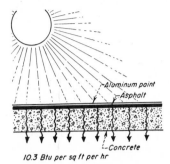

10.3 Btu per sq ft per hr

ALUMINUM paint over asphalt of same concrete roof on same day let in 10.3 Btu because light color reflects rays

ADVANTAGES OF SPRAYED ROOF

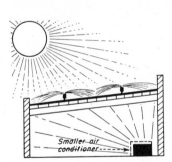

LESS air conditioning for top floor by 20%. Top floor cooler than floors below, smaller power bills, less upkeep

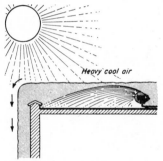

WALLS are cooler because heavy roof air spills down along walls, cooling them, so help make inside spaces comfortable

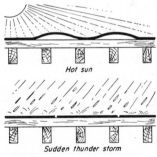

SAVES ROOF as volatile oils vaporize less. There's no thermal shock from sudden thunder storms because roof is cool

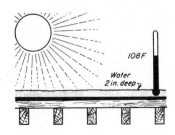

LESS WATER raises temperature slightly. But 108-F water blanket won't draw heat up through roof as will sweating roof

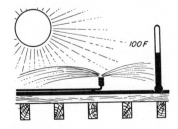

SPRAYED roof is cooler—spaces below are about 85 F—water evaporates fast as it hits roof, pulling heat from inside

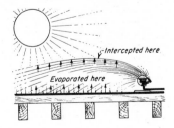

SPRAYED roof cools in two ways. Water droplets intercept heat before hitting roof; roof film evaporates, pulling heat

Air Conditioning

1—NOT ENOUGH COOLING . . . INSULATE DUCTS AND CHECK AUTOMATIC DAMPERS

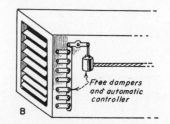

INSUFFICIENT COOLING in spaces when refrigerating system works right calls for a checkup outside the machinery room. Don't overlook anything

DUCTS WITH NO INSULATION often pass through hot areas, **A.** Quite often these areas are normally cool, but occasionally enough heat reaches ducts to transfer heat to conditioned air inside. Standard duct lagging is good investment. Also look for automatic dampers being out of order, **B.** Oil moving parts and work by hand to free

3—ODORS ARE BAD . . . CLEAN DIRT FROM INSIDE DUCTS AND CHECK DRAINS

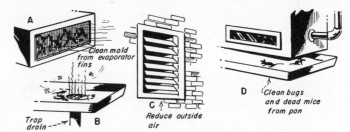

ODORS are disagreeable when carried with conditioned air. They make the operating engineer unpopular and defeat purpose of supplying treated air

MOLD OR DIRT accumulations on evaporator fins, **A,** spoil air passing through them. If drain lines are poorly trapped or water in gooseneck dries up, **B,** sewer odors escape. Outside air improperly dampered, **C,** mixes with inside air, contaminates air supply. Look out for foreign matter such as bugs and dead mice in drain pan, **D.** Keep clean

5—LOW HUMIDITY . . . INSPECT SPRAY NOZZLES, STRAINER AND SOLENOID VALVE

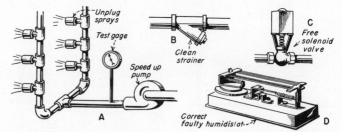

LOW HUMIDITY is uncomfortable, it dries mucous membranes, causes headaches. Modern systems control humidity, keeping air's moisture right

SPRAY NOZZLES get plugged, **A,** or water pressure to nozzles is low or even shut off. Gage may not show right reading. Strainer, **B,** might be plugged. Check solenoid water control valve, **C.** It might be sticking or coils may be burned out. Inspect humidistat, **D.** Sensitive element is delicate, if insulated with dust it shows faulty reading

Trouble Shooting

2—MECHANICAL NOISES . . . TIGHTEN CABINETS AND FAN BELTS, SUPPORT PIPES

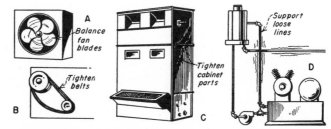

MECHANICAL NOISES in spaces are not only hard on nerves; but indicate that something is radically wrong with equipment. Get busy finding cause

FANS GET OUT OF BALANCE at times, **A.** Shut down and check. Rebalance in workshop. Fan belts slip, **B,** sending irritating noises through ducts. Belts are easy to tighten. Loose cabinet parts vibrate, **C.** It's good practice to tighten all screws at set periods. Pipes to refrigerating compressor vibrate, **D.** Bracket pipes right

4—NOT ENOUGH HEAT . . . VENT HEATING COILS, CHECK TRAP, BUY CIRCULATOR

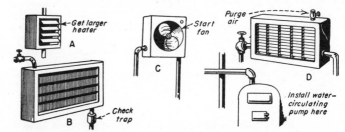

LOW TEMPERATURES from insufficient heat call for action. Discomfort is hard to take and doesn't help production or operator's reputation

HEATER, A, may be too small for job or not spotted right. Drain lines from heater, **B,** are trapped improperly or trap doesn't work. Often fan, **C,** isn't running and no one checked it before complaining. Hot-water coils, **D,** get air bound. Open vent before looking further. Circulating pump is needed in line, **E,** or piping is poorly arranged

6—FLOODED FLOOR . . . BAFFLE AND INSULATE CHILL-WATER PIPING FROM AIR

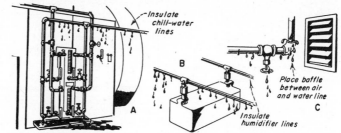

WATER ON FLOOR is not only messy but creates expensive problems by soaking into floors, shorting electric equipment and even causing accidents

CHILL-WATER COILS leading into conditioning equipment, **A,** should be insulated to prevent sweating. Humidifier piping in cold fresh-air current, **B,** should be baffled or insulated from air. Water supply line, **C,** is in cold air stream or valves' packing leaks. Drip pans are usually a nuisance. Insulate lines and keep after all leaks

temperature and relative humidity. The lower the temperature and relative humidity the lower the vapor pressure. Water vapor is present in air as a gas and will penetrate any material that absorbs air. When an insulated body is at temperatures above ambient, vapor pressure of water is higher at the insulation's inner surface than in the surrounding air. This pressure differential drives vapor outward from the inner surface (drawing below), through the insulation into the outside air.

Under these conditions insulation need only be protected from water impingement on the cold outer surface with some form of weather barrier. Heat of the insulated body keeps insulation fairly free of water vapor and moisture.

When temperature of the insulated body is lower than ambient, vapor pressure at the inner surface is less than that of air at the outer edge. So direction of water vapor flow reverses bringing water vapor into the insulation. Amount of vapor flow varies with the difference in vapor pressure. This in turn is proportional to temperature difference between the insulated body and the outside air.

Sketch below shows what happens to water vapor inside insulation. At dew point temperature the water vapor will condense. Actual formation of water takes place at the first cooler vapor resistant surface within the insulation. In some cases this is at the cold surface itself. If the insulation absorbs and retains water, its effectiveness as a heat barrier is soon lost. If temperatures within the insulation are below 32 F, ice forms to buckle the material; allow still more water vapor to enter and cause additional destruction.

Icing and water damage don't show up immediately with some of the water impervious insulation used for cold lines. Water vapor in these cases may enter through cracks or unsealed joints. Any resulting localized freezing causes only minor damage. But during shutdown all accumulated ice in the system thaws and runs along the piping to collect at some low point in the system. When low-temperature

Vapor barriers combat leakage: Achilles' heel of low-temperature systems

Insulation for temperatures below 32 F is only as good as its vapor barrier. A vapor soaked material has very little insulating value. First let's see how water vapor gets into insulation and then examine various ways to keep it out.

Air has a measurable vapor pressure governed by its

Moisture is the culprit that destroys low temperature insulation . . .

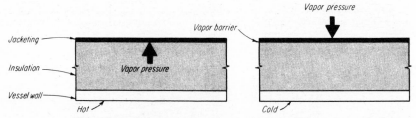

When the inside of a wall is insulated and interior space is held at a low temperature, moisture present will condense. This happens at the dew point — the critical temperature at which invisible vapor condenses to water. Dew point temperature varies with temperature differential and amount of humidity present. If insulation absorbs and retains moisture it soon loses insulating effectiveness. And if temperatures within the material are below freezing ice forms to destroy insulation. Solution: vapor barrier on warm side to stop moisture.

Higher vapor pressure at warm insulation surface drives out potential vapor leakage

Higher temperature at outer surface forces vapor laden air into the insulation

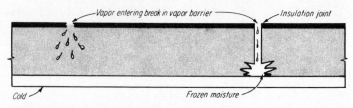

Trouble starts when vapor leaks in through damaged vapor barrier or through joints not properly sealed. Moisture travels along inner surface, freezes and destroys insulation

operation is resumed this large accumulation of water then freezes to severly damage insulation at this point.

What happens in those parts of the low-temperature system operating at temperatures above 32 F? Of course, the moisture won't freeze and visible harm may not occur. But water does collect in the insulation next to the inner surface with the attendant danger of severe metal corrosion.

Vapor barriers are the only practical answer to moisture problems. Aim is to stop movement of vapor at as high a temperature as possible so that dew point temperature will not occur where moisture is present or where condensation can take place within the insulation. Completely effective vapor barriers do not exist in practice either because of material's properties or defective application.

Since it is practically certain some moisture will pass the vapor barrier, what other provision can be made to keep the insulation dry and prevent its eventual destruction? In the case of refrigerated spaces this can be done by using insulation that readily permits water vapor flow plus a vapor-porous interior finish. Any small amount of moisture that gets by the vapor barrier then migrates through the insulation and deposits at the coldest spot inside, usually the refrigerant coils.

But when insulation is applied to cold tanks, pipes and similar equipment, there is no chance for moisture to escape on the cold side. So the vapor barrier must keep out moisture or the insulation will eventually fail.

Materials used include: (1) structural barriers (2) membrane barriers and (3) coating or surface barriers.

Structural barriers are the rigid sheets of reinforced plastics, aluminum and stainless steel jacketing. These come either as flat, corrugated or embossed sheets, often fabricated to exact dimensions and ready to install. They are usually fastened with bands, clasps or some other device. Joints may also be sealed with a plastic or tape.

Membrane barriers are the metal foils, laminated foil and treated papers, coated felts and papers and plastic films or sheets. One example is a new asbestos fiber coated with a polyvinyl choloride film. Today many of these barriers come attached to the insulation. Others are available in roll form. The joints of molded pipe insulation often come with flaps that can be sealed with an adhesive or tape.

Coating barriers come in fluid form as paint or in semi-fluid state. These are also known as surface coatings and often are the hot melt type used over insulation of cold storage spaces, piping systems and equipment. Material may be asphaltic, resinous or polymeric type. Pigments and solvents naturally depend on application and design. These are applied by brush, roller, trowel, spray or mop.

Rating of vapor barriers is according to vapor transmission rate expressed in perms. A perm of one, for example, means that only one grain of water will pass through one sq ft of material in one hour for every inch of mercury pressure differential.

In low-temperature insulation, a better vapor barrier performance than one perm is needed. For coolers operating between 32 and 50 F, 0.2 perm rating is sufficient. Most asphalt-saturated, shiny glazed papers qualify.

At still lower temperatures even greater efficiency is demanded. Practical rating in the freezing range below 32 F is about 0.01 perm. Laminated paper and aluminum foil products or aluminum foil alone, are used.

Great care must be taken in application to avoid damaging the vapor barrier material. Joints between sheets must be sealed and care taken to avoid rupture due to expansion and contraction of the structure itself. When the vapor barrier is punctured, the resulting hole must be sealed.

Adhesives used to laminate some vapor barriers of the membrane type; paper to paper, foil to foil, foil to paper, paper to plastic, etc, often have flame extinguishing properties. When material reaches combustion stage, gases or vapors released tend to smother the flame.

... but correctly applied vapor barriers effectively keep most of it out

Jacketing sealing these insulation joints is a 15-lb felt applied directly over the surface. In addition it serves as a protective cover

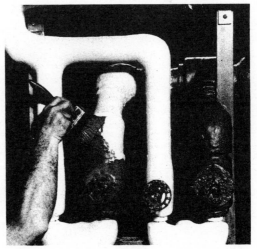

Plastic insulation finish being applied here is a combination vapor barrier and paint, used for cold and hot lines and equipment

1 This canvas water bag exposed to hot desert sun is cooler than the surrounding atmosphere because of evaporation

2 Electric fan evaporating perspiration carries away excess heat, cooling body to normal temperature despite the heat

Why Cooling Towers Cool

►IN OPERATING refrigerating plants or condensing turbines, an important problem is removing and dissipating heat from the compressed refrigerant or the exhaust steam. This heat is usually removed from the "gas" by transferring it to water in a heat exchanger.

From there it may be dissipated in a number of ways. If the equipment is on the bank of a river or lake, the intake and discharge connections can be piped to prevent mixing the heated discharge water with the cooler inlet water. If the source of cooling water is a well or city supply the discharge may be piped back into the ground or into the sewer or open waterway. But such outright waste of this heated water is usually costly and in some localities illegal.

There are a number of reasons why cooling water should be reused. The most important is that very few plants are lucky enough to be near a limitless water supply. Another is that using city water for cooling is costly unless it's reused. The third is that most water contains dissolved salts, and using a continuous supply of such raw water quickly coats the heat exchanger with scale. In a cooling tower the water is cooled by exposure to air after each passage or cycle, allowing the same water to be reused a number of times.

Atmospheric Water-Cooling. Water, by having its waste-heat load transferred to the atmosphere, can be reused in a continuous cycle and thus conserved. The heat can be transferred by bringing the water and air together

Here's the lowdown on why cooling towers cool below atmospheric temperatures.

indirectly (non-contact) as in an automobile radiator. Another way is with evaporative-cooling equipment such as atmospheric spray ponds, cooling towers, etc.

Water, when cooled by the evaporative method loses about 1000 Btu (heat units) for each pound of water evaporated. This heat taken away in the water vapor produced is called the *latent heat of vaporization.* When air removes heat (in vapor) from water in this way, it can cool the water below atmospheric temperature.

This fact is important in transferring heat from water to air. It allows the water cooled by evaporation to serve plants having a great variety of temperature needs. It also allows a little water to cool a much greater heat load than if it were not below atmospheric temperature.

Cooling by Evaporation. Surface evaporation keeps water cool in the old-fashioned porous canvas water bags, Fig. 1. Likewise evaporating perspiration helps the human body, Fig. 2, to keep its normal 98½ F temperature in places where the surrounding atmosphere is much hotter.

This latent heat of evaporation is the

primary cooling effect produced by blowing air over wet surfaces or through sheets of falling water in a cooling tower.

Cooling-tower engineers also refer to "sensible heat"—heat you can *feel*—as temperature. The higher the temperature of a substance, the greater its sensible heat. Where the air is cooler than the water there is some tendency (not counting evaporation) for the air to cool the water—that is for the air to get hotter (gain sensible heat) while the water gets cooler (loses sensible heat). On the average about 75% of the total heat removed is by evaporation (latent heat).

Wet- and Dry-Bulb. Cooling-tower engineers, like air-conditioning engineers, often speak of *dry-bulb* and *wet-bulb* temperatures. The temperature of air as read on the ordinary thermometer is called the dry-bulb temperature. Wet-bulb temperature is the final reading when the bulb of a thermometer is covered with wet cloth and the instrument is whirled around in a sling.

Wet-Bulb Is Lower. When the humidity is 100%, the wet-bulb temperature will be the same as the dry-bulb (ordinary thermometer). For any lower

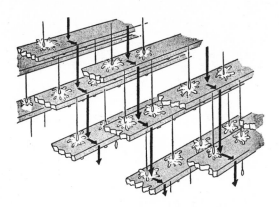

3 Cooling tower lowers water's temperature because water drops spreading over laths are exposed to atmosphere. Slight evaporation carries away heat, cooling water

4 Schematic definition explains terms "cooling range" and "approach"

humidity some of the water will evaporate to cool the bulb, so the wet-bulb reading will fall below the dry-bulb. The drier the air the greater will be the difference.

Knowing the wet- and dry-bulb readings you can always find the relative humidity from a table or chart.

Other things being equal, cool air is better than warm air for a cooling tower. Most important of all is a low wet-bulb temperature, indicating either very cool air, very low humidity, or a combination of the two.

Here's a scientific fact worth remembering: No cooling tower, even in theory, can ever cool water below the wet-bulb temperature of the inlet air. In actual towers, of course, the final temperature will always be a few degrees above the wet-bulb temperature.

Fig. 3 pictures water falling from one level to another in a cooling tower. The splashing droplets wet the wood laths giving a big exposure of water surface to the upmoving air.

Cooling effect can be speeded by (1) increasing air velocity over the wet surfaces, (2) increasing exposed wet surface, (3) lower barometric pressure, (4) increasing water temperature to the tower, (5) reducing air's humidity.

Cooling-tower designers must consider these properties of air: (1) wet-bulb temperature, (2) dry-bulb temperature, (3) humidity, (4) total heat in Btu, (5) pressure, (6) weight and (7) velocity.

The wet-bulb temperature assumed in designing evaporative water-cooling equipment is generally close to the average maximum wet-bulb for the summer months at the given location. The

exact wet-bulb temperature selected will depend on operating conditions, temperature limitations and geographical location.

Operating Terms. Those who operate or manufacture specialized mechanical equipment gradually develop a somewhat specialized "glossary" of words and phrases that describe particular related functions, parts, or characteristics of their equipment. Here are some of the most common terms used with atmospheric water-cooling equipment.

Cooling Range is the deg F that water is cooled by the water-cooling equipment, Fig. 4. It is the difference in temperature between the hot water coming to the cooling tower and the temperature of the cold water leaving the tower.

Approach is the difference in deg F between the temperature of the cold water leaving the cooling tower and the surrounding air's wet-bulb temperature, Fig. 4.

Heat Load is the amount of heat thrown away by the cooling tower in Btu per hour (or per minute). It is equal to the pounds of water circulated times the cooling range.

Pumping Head is the pressure required to lift the water from the water's top in the basin to the top of the tower and force it through the water-distribution system, Fig. 5. Pumping head is equal to the static head, plus the friction loss in the distribution system and the velocity head (head required to maintain velocity of water).

Drift is the small amount of unevaporated water lost from atmospheric water-cooling equipment in the form of

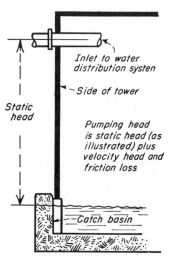

5 The term "pumping head", shown here is used in cooling tower design

mist or fine droplets. It is water entrained by the circulating air. Drift is a water loss independent of the water lost by evaporation. The drift loss, unlike the evaporation loss, can be reduced by good design.

Blowdown is the continuous or intermittent wasting of a small fraction of circulating water to prevent a concentration of chemicals in the water. The purpose of blowdown in water-cooling equipment is to reduce the soluble solids, or hardness. This reduces the scale-forming tendencies of the water.

Makeup is the water required to replace the water that is lost by evaporation, drift, blowdown, and small leaks.

Water Cooling With Free Air

What's the answer to your water-shortage problem? Is it a spray pond? A mechanical-draft tower? A natural-draft tower? OR a combination job? The right system, as explained here, will save on water bills and conserve water supply

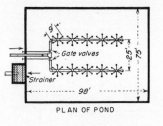

PLAN OF POND

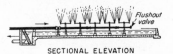

SECTIONAL ELEVATION

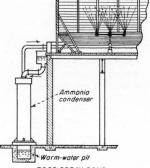

ROOF-SPRAY POND

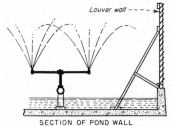

SECTION OF POND WALL

1 Spray pond details here show good practice in pond design. Whether pond is on ground or on roof, consider all features shown here for best service

▶ COOLING WATER available for condensers (steam, ammonia, or other vapors), internal-combustion engines, oil coolers or other heat-exchange apparatus, is often impractical or costly. Water may even be contaminated with sewage or acids.

When water is scarce or expensive, you don't waste it after a single passage through apparatus. Instead, use the limited supply repeatedly, and cool it after each use. Atmospheric air cools three types of cooling systems: cooling ponds, spray ponds, and cooling towers.

Cooling Ponds. Water is economically cooled by circulating through a cooling pond, if ample surface is exposed to the atmosphere. Ponds are best if only a little water is to be cooled, and land is cheap nearby or if there is a natural pond. The pond surface should receive warm water at one end and deliver cooled water from near the bottom at other end.

Cooling rate depends upon ponds surface area, difference in temperature between water and air, velocity and humidity of air, and length of air path per unit area of pond surface. Pond's depth is usually from 2 to 4 ft, but pond's volume should be enough to give storage capacity to meet load variations.

Spray Ponds. Spray ponds, Fig. 1, reduce surface area of cooling ponds. The water is forced through nozzles, which spray it through the air in tiny particles. Spray brings the water particles into intimate contact with air. That greatly increases the surface exposed per unit weight of water cooled.

Evaporation and resulting cooling is fast. For a given duty, the spray pond needs only from 1 to 10 percent as much surface as the simple cooling pond. This is an important advantage when land for pond is expensive or space is restricted.

Nozzles or spray heads are special non-clogging types, operating at from 3 to 15 psi, and usually at 6 psi. They are spaced at 8- to 15-ft intervals, on rows of pipe from 15 to 20 feet apart. Since water must have an intimate contact with air, nozzle arrangement should have adequate air lanes through the spray.

There must be ample pond surface beyond spray to catch drift in high winds. If space is limited (as on building roof) louver fences will prevent excessive water loss. Much as 6 percent of water cooled is lost by evaporation and drift, even with louvers. Total loss depends upon heat removed from circulated water.

When a spray pond is used for cooling condensing water from a steam power plant, energy used in pumping water is from 1 to 2 percent of station's output.

For more cooling, respray the pond water. Here are two ways: 1—Use mixed spraying system, in which pond water is added to hot water in definite proportions before spraying. 2—By independent respraying (double spraying) using a separate distributing system. Either method requires more nozzles, larger or longer pipes, greater pond area, and more pumping energy.

There are no general rules for spray-pond selection or design. Reason is wide range of operating conditions, cost factors, local atmospheric conditions, etc. All this makes each problem an individual one.

Cooling Towers. A cooling tower is a wind-braced housing or shell, (wood, concrete, brick, or metal), enclosing a network of obstructions, or filling. Water to be cooled is pumped into a distributing system at tower's top, from

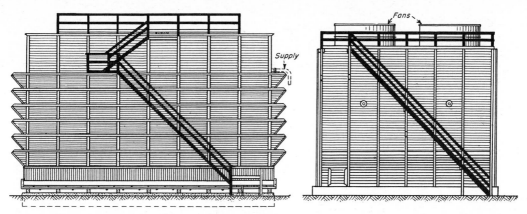

2 Atmospheric cooling tower is simple. Air enters through louvered sides and cools water sprayed through the filling

3 Mechanical-draft tower has one or more fans to give circulation. This tower has two large fans on top

which it drops in thin sheets or sprays to the filling. Filling is arranged so water spreads out to expose new surfaces to air flowing through tower. Cooled water is collected in reservoir or sump.

Air passing through tower becomes partly or completely saturated by evaporation from a portion of the water. This evaporation is mostly what cools the water. With plenty of cooling surface and air, the water can be cooled to wet-bulb temperature of air, as a limit. Depending on conditions, water is usually cooled to within 4 to 40 F of wet-bulb temperature. A cooling tower is often preferred to a pond. Reason is that it gives greater cooling effect per unit of ground area occupied. Also, it gives a greater cooling range because of longer contact time of air with finely divided water. Besides, it's convenient to inspect and repair.

The filling in cooling towers can be made of boards or lath, hollow tiles, metal sheets, wire screen or special grids. Redwood, cypress, and other woods that don't deteriorate rapidly in water, are most common. Filling's evaporating surface exposed, per unit volume of tower, depends on type of tower and type and arrangement of filling. Usually this is from 5 to 20 sq ft per cu ft of space filled. The area free for air flow is from 65 to 85 per cent of total cross-sectional area occupied by filling. Velocity of air flow through tower depends on type of tower and operating conditions. It's usually from 100 to 700 ft per min.

Cooling towers differ principally in their housing arrangement and draft-producing methods. Usual classification as follows: 1—atmospheric towers,

2—chimney-draft towers, 3—mechanical-draft towers, and 4—combined mechanical- and natural-draft towers.

Atmospheric cooling tower, Fig. 2, (or open natural-draft tower) is the simplest tower. Air enters through louvered sides that prevent water being blown out, and flows through in a transverse direction. Air circulation depends on wind velocity. These towers are often designed to cool about 1.5 gallons of water per minute per square foot of active horizontal area, with a 5 mph wind velocity.

For effective cooling, they are limited in width, so won't carry large cooling loads unless they are long. Tests show that atmospheric towers with a deck wider than 12 ft lose efficiency. Many manufacturers make standard towers, with deck widths of 6 ft, 8 ft, 10 ft and 12 ft, for atmospheric units.

Louvers add to overall width of tower but these given are standard deck widths. To prevent ice formation in winter, a secondary distributing system supplies water to only lower portion of filling.

Chimney-draft tower (or closed natural-draft tower) has air inlets at its base. Sides above inlets are closed and extend, chimney-like, from 60 to 90 ft above filling. Air circulation is independent of wind conditions and flow is upward, counter to falling water. Draft is caused by difference between density of high-humidity air column and temperature within tower. Also from density of cooler and drier-atmospheric air column outside. Because of better draft, these towers are well suited for water of high initial temperature. Air velocity through stack is from 100 to 200 ft per min, and the evaporation

water loss and drift is usually small. The stack portion is relatively expensive for small towers, but is a minor item for large tower.

Mechanical-draft cooling tower resembles the chimney tower, but its housing isn't so tall. It has more compact filling, has one or more forced-draft or induced-draft fans, Fig. 3, to give positive circulation and control of air supply. To prevent excessive water loss and nuisance caused by drift, baffles are normally installed at the tower's top.

For a given ground area, mechanical-draft towers may have from two to five times cooling capacity of chimney towers. Velocity of air flow through tower is usually between 300 and 700 ft per min. Loss of water by evaporation and drift are very small.

Combined mechanical- and natural-draft tower permits use of natural draft, with the fans shut off, when atmospheric conditions are favorable or when cooling load is light. Then mechanical draft is always there if needed.

Selecting Systems. Selecting type and size of cooling system for a given duty depends on requirements, first cost and operating costs. Cooling system can seldom be considered by itself because its relation to apparatus served must be studied.

For example, an expensive cooling system will cool condensing water to within a few degrees of air wet-bulb temperature. Whether or not this is economical depends on saving in fuel burned to generate steam for prime mover that reduces condenser pressure. Contrast this with high fixed costs and high operating costs of water-cooling equipment. And after selecting yours, double check—be sure you're right.

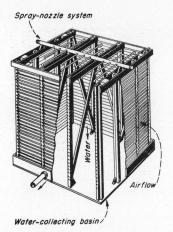

Spray-nozzle system

Water

Airflow

Water-collecting basin

1 **Spray-filled atmospheric** cooling tower is efficient under 30,000 Btu per min.

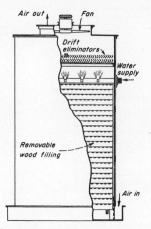

Air out Fan

Drift eliminators

Water supply

Removable wood filling

Air in

2 **Mechanical-draft** towers are wood or spray filled, or a combination

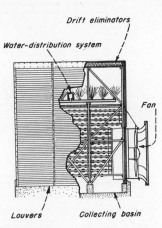

Drift eliminators

Water-distribution system

Fan

Louvers Collecting basin

3 **Forced-draft** towers take corrosive water well. Fan is easily serviced

---

Here's the lowdown on the principal types of cooling-tower designs in use.

Water-Cooling

▶ THE BASIC PRINCIPLE behind all water-cooling towers is that air passing over exposed water surfaces picks up small amounts of water vapor. The little bit of water evaporating takes a lot of heat from water left behind. **This heat is** called latent heat of vaporization.

All cooling towers depend on circulation of air over water. The main difference in various tower designs is the method of air circulation. The two chief methods are natural circulation by wind and mechanical circulation by fans. Other design differences occur in the way water surfaces are exposed to air.

Spray Ponds. A spray pond consists of a number of spray nozzles over a water-collecting basin. Nozzles spray droplets—not mist, because cooled water must drop into pond and not be blown away. Louver fences on pond's leeward side keep water from being carried away. These enclosures are a "must" in restricted areas and on roofs.

Spray ponds are best for large-capacity service where efficiency isn't important and moisture drift isn't objectionable. These ponds are cheaper although basin costs and pumping needs are high.

Atmospheric Towers. For water-cooling needs of less than 30,000 Btu per minute, the spray-filled, atmospheric cooling tower, Fig. 1, is good. This design is a narrow spray pond with nozzles at top and a high louver fence. The nozzles spray downward.

These towers range from 6 to 15 ft. high and are taller than they are wide. Capacity ranges from 0.6 to 1.5 gpm per sq. ft. of tower cross-sectional area, which is about one-fourth that of an equivalent spray-pond area. The louvers are always wet, hence add to the water surface exposed to the cooling air.

Spray-filled towers are used (1) when equipment served can stand a few degrees rise in the cold-water temperature at low- or zero-wind velocities, (2) where drift from the tower isn't objectionable and (3) when tower can be placed so the wind isn't cut off by buildings, trees, etc.

Mechanical-Draft Towers. These towers are usually a vertical shell made of wood, metal, transite, or masonry. Water is distributed near the top and falls to the collecting basin. It passes through air that's circulated from bottom to top by forced- or induced-draft fans, Figs. 2 and 3.

The air contacts the hottest water just before leaving tower. Because it passes against the flow of water, a given quantity of air thus picks up more heat than the average equal quantity of air on natural-draft equipment. So less air is needed to cool the same amount of water. As air is supplied by fans, air quantity must be held to a minimum for low operating cost.

The inside of a mechanical-draft tower may be filled with a spray of water droplets from nozzles or packed with wood filling on which water cascades from top to bottom, Fig. 2. In many cases, a combination spray-filled and wood-filled design is used.

In the spray-filled tower, the area cooled by air is the combined surface area of the water drops present in the tower at any one time. The net, free cross-sectional area of the air spaces in a spray-filled tower is usually greater than that of the wood-filled tower for the same gross area. This means lower air velocities and a longer contact time between air and water in the same size structure for a spray-filled tower. Before discharging into atmosphere, the water-laden exhaust air passes through a drift (spray) eliminator which removes water droplets carried along.

In the wood-filled tower, lumber is

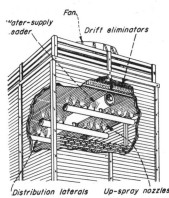

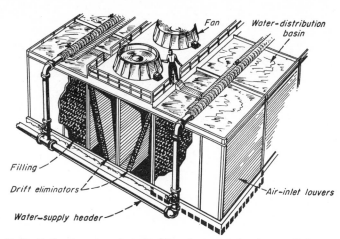

4 **Counterflow** induced-draft tower. High air velocity stops recirculation

5 **Double-flow** tower is packed with filling. Large wetted surface greatly improves cooling capacity. Because filling is horizontal, draft loss is not increased

Tower Designs

laid through the tower both horizontally and vertically. Water is sprayed over the top layer from nozzles, troughs or splash heads. Water drops from piece to piece of wood filling. As air moves upward or across, it strikes a large wetted surface, repeatedly breaking up the falling drops and providing new drop surfaces whose combined areas are several times the wood-fill area.

This tower's efficiency is improved by increasing the filling, height, area, or air quantity. Increasing height increases the time air is in contact with water, without any more fan power.

Increasing area at constant fan power increases air quantity and air-water contact time because of lower velocity. The water-surface area in contact with air is increased in both cases. Increasing the air quantity decreases the time air is in contact with water, but since more air passes through, the average difference between water temperature and air's wet-bulb temperature is increased. This speeds up the heat-transfer rate. More air means increased fan power. Air velocities through wood-filled towers vary from 250 to 400 fpm.

Mechanical-draft towers don't depend on wind velocity, so it's possible to design them for more exacting performance. They require less space and less piping than atmospheric towers. Pumping head varies from 11 to 26 feet. Over-all plant economy from lower water

temperature usually offsets the added operating expense and initial cost compared to atmospheric towers.

Forced-Draft Towers. A mechanical tower with forced draft, Fig. 3, works well with corrosive waters because the fan can be near the ground. This way parts most likely to corrode are easily serviced. Fan maintenance and depreciation are high. Heated air leaving the tower's top at low velocity is an objection. At times the air recirculates to the fan inlet. With unfavorable winds, this reduces efficiency as much as 20%.

During cold weather, recirculation can cause ice formation on nearby equipment and buildings, or in the tower's fan ring, sometimes breaking the fan. Limited fan size (12 ft. or less) means more fans, motors, starters, and wiring in larger towers than needed by induced-draft towers, which use fans up to 18 ft. in diameter.

Fan's location on top minimizes noise. It offers a neater appearance and is more adaptable to architectural surroundings.

A counterflow induced-draft tower has the fan on top, Fig. 4. Air movement is vertical across the filling and upward at a high velocity to prevent recirculation. With small loads, the fan is at one end for horizontal flow.

Double-Flow Towers. A demand for compact design, better construction,

lower cost, larger capacity, more flexible operation, and improved all-around performance has produced the double-flow induced-draft tower, Fig. 5, sometimes called a common-flow tower.

Air flow is horizontal with fans centered along the top. Each fan draws air through two cells paired to a suction chamber that's partitioned midway beneath the fan and fitted with drift eliminators which turn the air upward toward the fan outlet.

Double-flow towers use a low pumping head, varying from 11 to 26 ft. Operating advantages are: (1) Horizontal (crossflow) air movement as water falls in a cascade of small drops over the filling and across the air stream. This offers less resistance to air flow, therefore a lower draft loss. (2) It has longer air travel than conventional design. (3) The open water-distribution basin is accessible for cleaning during operation. (4) It has a close-spaced, wood, diffusion deck under the water basin for uniform water distribution to the wood filling. (5) Water loading in most cooling towers has a maximum of 6 gpm per sq ft caused by the blanketing spray effect. Heavier loadings, up to 10 gpm per sq ft, are possible in double-flow towers for steam condenser service. (6) A modern double-flow tower occupies less than 1/20 the area needed by a spray pond for equivalent service.

Water-cooling towers are dependable and do their job well if they are selected right. Just remember that each type of tower is designed for the exact job it must do. To get the most out of your tower, consult an expert first.

1 All-wood natural-draft spray tower for 5 to 200 tons of refrigeration

2 Steel-cased vertical induced-draft tower for 50-300 tons refrigeration

Water-Cooling Tower

▶ BOOKS CAN BE WRITTEN on starting, operating, maintaining and shutting down water-cooling equipment. But here are important points in nut shell.

Water Treatment. Makeup water required by a cooling tower depends on evaporation loss, drift loss, and blowdown. Evaporation losses average 0.80 percent of water circulated for each 10-F range. Drift loss is water carried away, as droplets or mist, by air.

In properly designed induced-draft towers, drift loss is about one-tenth of one percent. Most manufacturers guarantee a drift loss not over two-tenths of one percent. Amount of blowdown water wasted depends on circulating-water hardness, type water softening used and drift loss. Blowdown is controlled to keep the concentration of soluble and scale-forming solids below the point where scale forms or would be caused by corrosion.

Algae formations plug nozzles and prevent proper water distribution over the tower filling. This slimy growth also collects on equipment served by the cooling tower, reducing the heat-transfer rate. Hold algae at a minimum or eliminate it. This is done by adding a little chlorine, copper sulfate, potassium permanganate, or other chemicals to circulating water.

Most water has some scale-forming materials, but calcium and magnesium carbonates cause most trouble in water-cooling systems. Scale formation on equipment reduces heat-transfer rates. This scale can be reduced or prevented by softening the makeup water with lime and soda ash, zeolite, sulfuric acid or some phosphates.

Water softening or treatment requires close regulation and control by a chemist. Too high a concentration of soluble solids in cooling-tower water raises the effective wet-bulb temperature. This in turn raises the temperature of water leaving the tower. That may cause sludge deposits or corrosion in the system. Concentration of solids is usually controlled by blowing down or by a continuous overflow to the sewer.

Delignification. Sodium carbonate in circulating water causes wood fibers, coming in contact with water, to separate by dissolving lignin which binds fibers together. The technical word for this is "delignification." It leaves the wood surface in a white fibrous condition. Continued exposure reduces wood's structural strength.

Delignification first appears on parts of the tower that are alternately wet and dry. This is because evaporation at these points increases the concentration of dissolved solids quickly. Sodium carbonate in harmful amounts is generally indicated by a high pH of 9 to 11. The effect of sodium carbonate may be neutralized with sulfuric acid. It's best to have water's pH value at 7 to 7.5 (7.2-pH value is neutral for redwood).

Two-Speed Motors. For adapting tower performance to temporary or seasonal decreases in heat load, and especially for winter operation, two-speed motors are best for fan drives. The chief advantage is that at half speed, fans need only about 15 percent of their full-speed power. Especially in multi-fan towers, the flexibility of two-speed motors saves power.

At half fan speed, a cooling tower's hot- and cold-water temperatures increase 6 to 8 F for a given wet-bulb temperature and heat load. When wet-bulb temperature is 10 F below design temperature, fan can be at half speed. Then same cooling range is obtained for a given amount of water.

For same wet-bulb temperature and range, (but 6 to 8 F higher hot- and cold-water temperatures) a tower needs about 60 percent more effective cooling area at half fan speed than at full speed for a desired heat load.

During winter months, when it's not cold enough to stop fans, but still cold enough so ice forms at full fan-speed, run some or all fans at half speed.

Starting Up. Before starting a water-cooling tower for first time, or after a long shut down, clean and inspect it thoroughly. Remove all dirt in the catch basin beneath tower.

Inspect fans to make sure all bolts are tight, fans turn freely and that clearance between blade tip and fan cylinder is right. Check fan's drive shaft to see that all bolts are in place, flexible couplings or universal joints are in good condition and that the shaft guards are in position and properly secured.

Motor, drive shaft and speed reducer should be in line. Make sure speed reducers are filled with right grade and quantity of clean oil. Also open vent

3 Vertical induced-draft wood tower for 150-2000 tons of refrigeration

Upkeep

connection on top of speed reducer and tighten all bolts.

Inspect motors for lubrication. Tighten motor frame and anchor bolts. Inspect nozzles. piping and flumes of distribution system. Replace missing nozzles and stop all leaks. Look over tower framework and tighten loose bolts. Don't tighten bolts too much or wood swell may break them.

Stop catch-basin leaks. Small wood-basin leaks stop when wood is soaked. In concrete basins, check expansion joints in walls and floor. Make sure makeup float valve works freely and doesn't leak when closed. See that overflow is open and working properly. Check sump assembly for screen and cover (or suction screens if concrete basins) being properly installed.

Fans. Large fans draw air through tower and discharge upward to atmosphere. Inspect tower fans when starting to make sure they run freely. It's very important that fans rotate in right direction. Check the fan motor's power input. If these motors aren't loaded to specified horsepower, contact manufacturer for setting the fan-blade pitch. Speed reducers are sometimes noisy when started first time, but generally quiet down when properly broken in. Noise developing in speed reducers after they've been running some time means excessive wear.

Circulating-Water. Adjust the catch-basin float valve to keep water level 5 to 6 in. below wood filling. Keep at least 6 in. of water in redwood or steel basins. Open or close all valves as required, then prime and start pumps.

Adjust water being circulated to tower-design quantity. Water flow can be checked with orifice meter, weir, pitot tube or pump performance curves. Equalize water distribution over all tower parts.

Operation. Keep distribution system, nozzles and catch basin under tower free of dirt, algae and scale. Don't operate fans driven by variable-speed devices above design speed. Concrete basins should have double suction screens. That is, one set of screens in front of another so all water goes through both. When cleaning, first remove screen water passes through.

Inspect oil in speed reducer weekly and add oil for proper level. Grease drive shafts equipped with splines or lubricated universal joints weekly. Drive shafts with disk couplings don't need lubrication. Check cooling tower's operation daily. Because cooling tower is part of a system, it's faulty performance may be a symptom of failure elsewhere in system.

Cooling range drops from light heat load or excess water. Cooling range increases from heavy heat load or insufficient water. When a condenser is badly scaled, circulating water is usually retarded and pumping head increased.

When water quantity is incorrect: (1) Check pump for speed, pressure, air binding; (2) check condenser for scale, air or restrictions and (3) check pipe lines for air, partially closed valves, dirt or restrictions.

Cold-Weather Operation. Extremely cold water doesn't normally increase performance much. It does increase operating hazards considerably. Water-cooling towers operated in sub-freezing weather are subject to ice formation on louvers and outer portion of filling. To prevent icing, keep cold raw-water (tower circulating water) temperature high as practical, considering effect on equipment served. Follow one or more of these procedures:

For induced-draft towers: (1) Run two-speed motors low speed, or shut off some fans. (2) Shut down some cells, and put all water over remaining cells. (3) Reduce water flow to tower, and shut off some cells. (4) Bypass cooling tower with part of water, and shut off some tower fans or cells.

De-Icing. To remove ice after it has formed on the louvers and filling: (1) Reverse direction of fan rotation to blow warm air out through louvers to melt the ice. (2) Shut down some fans temporarily, but don't shut off water. When these cells are thawed out, start

the fans and repeat on the other cells.

Caution! Never operate fan in reverse more than 5 to 10 minutes. Most speed reducers don't supply oil to upper fan-shaft bearing when reversed. Lube-pump may deliver no oil when reversed. So running fan reversed too long may cause bearing failure. Where towers are operated intermittently, protect the exposed piping against freezing. Protection required depends on length of shut-down period and severity of expected weather.

Mechanical Maintenance. Set up a regular schedule for mechanical maintenance of cooling-tower equipment. Grease ball-bearing motors with water-resistant grease every few months.

Overgreasing ball bearings causes failure, as does undergreasing. Check motor's insulation yearly with testing set. Paint motor windings yearly with insulation varnish and the motor frame with rust-resistant paint. Refill speed reducers with clean oil every 3000 hours, or at least once a year. Type oil for speed reducers is usually shown on nameplate.

Excessive clearance or play between pinion and ring gears indicates wear. Pinion-shaft end play or fan-shaft side play shows bearing wear. Oil leaking around pinion shaft or fan shaft indicates worn oil seals.

Paint fan blades and hubs as often as necessary to prevent corrosion. Corrosion depends on water used in tower and on atmosphere location. Tighten blade clamps and hub bolts, and rebalance when necessary.

Structural Maintenance. Remove dirt, scale, bugs and debris from distribution system. Replace or repair damage or missing parts. Clean and paint all metals that may corrode. Redwood doesn't require painting for weather protection. Tighten loose bolts, making allowance for wood swelling when wet.

Clean drift eliminators, because dirty ones reduce air flow. Line up eliminator spacers. When installing eliminators, don't leave holes between bundles as these allow excessive drift. Clean dirty wood-filling slats.

Shutting Down. When shutting down tower, especially in winter, always drain to prevent freezing and corrosion. Leave drain open so rain and melted snow escape. Operate fans for about five minutes, once a week, to keep upper fan-shaft bearing oiled. Protect metal parts from corrosion.

Do maintenance or repair work while cooling tower is completely shut down, if possible. This helps you do thorough job and doesn't restrict operation.

FOG RISES high above towers on calm day. It's nuisance when blown across highways and railroads, especially in winter

DISAPPEARING FOG here absorbed by air quickly because sub-zero outside air condenses moisture, eliminating fog

HOW TO PREVENT ICE ON

Induced Draft Towers

1—Run two-speed motors at half speed to retard ice formation. Fans use 15% full-speed power.

2—Shut down some fans (if multi-cell unit).

3—Cover louvered area's upper portion to limit air flow.

4—Reduce tower's water flow and stop one or more fans (if multi-cell unit).

5—Bypass water to part of cooling tower and stop one or more fans (if multi-cell unit).

Louvers and Filling

1—Reverse fan's motor (for limited time only) to blow warm air out louvers. *Caution:* Some speed reducers don't supply oil to upper bearing when reversed.

2—Stop some fans temporarily, but don't stop water. When cells thaw out, start fans and repeat process in other cells.

3—Cover upper portion of louvered area to limit air flow.

Winter Tips On Cooling Towers

Your water-cooling tower will do a better job this winter if you know how to prevent ice, reduce fog and follow a planned maintenance schedule. You may also save expensive repairs

▶ WATER-COOLING TOWERS can cause grief in winter—unless you watch out. Most important, follow the manufacturer's instructions. Your cooling tower's useful life depends on its built-in qualities, the climate, type of service, severity of operation and your general care and maintenance.

Just remember that well maintained and clean equipment gives best operating results at all times, also cuts down over-all upkeep cost. Set up a regular inspection schedule to insure steady operation, regardless of the type of water-cooling equipment you have. In

most cases a daily inspection is enough. Use maintenance and inspection schedule shown to save you headaches this winter.

Fogging. Fogging occurs under certain atmospheric conditions. By definition, air's dew point is temperature at which air reaches a state of saturation when cooled. When air is cooled to its dew point, moisture begins condensing, and fog results. It's very objectionable next to buildings, highways, or railroad tracks. In one case, fogging took place largely because tower wasn't de-iced enough. Tower was located just north

MAINTENANCE INSPECTION SCHEDULE

General recommendations
(more frequent inspection may be desirable)

	Fan	Motor	Drive shaft	Speed reducer	Eliminators	Filling	Cold-water basin	Distribution system	Float valve	Structural members	Suction screen	Steel casing
Check for unusual noise or vibration	W	W	W	W						Y		
Inspect keys and keyway	S	S	S									
Check oil level, and oil for water and dirt				M								
Make certain vents are open				M								
Inspect for deterioration		S			Y	Y		Y		Y		
Lubricate	S	M	S									
Tighten loose bolts	Y	Y	Y						Y	R		
Repaint	R	R	R				R		R	Y		
Clean							R	R	R	R		
Inspect for clogging						W	M				W	
Check for leakage							M	M				
Change oil (at least)				S								
Check oil seals at speed-reducer shafts				S								

W-weekly; M-monthly; S-semi-yearly; Y-yearly; R-as required

Properties of saturated air

Wet-bulb temp	Btu per lb	Grains of water vapor per lb
90F	56	217
80	43	156
70	34	110
60	26	78
50	20	54
40	15	36
30	11	24
20	7	16

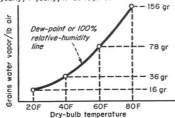

MAINTENANCE AND INSPECTION schedule is reliable guide to prevent trouble this winter. Properties of saturated air and psychrometric graph help study fog formation

of state highway, so the fog was bad for vehicular traffic.

When tower ices up, fans move less air. That means air leaves tower at a "low velocity and high temperature". This causes much fog when warm tower air mixes with cold outside air. Then fog is carried across highway, obstructing traffic, icing highway and windshields.

Fog is airborne droplets formed by vapor condensation. As warm and nearly saturated air leaves cooling tower in cold weather, it mixes with surrounding air, then cools, condensing vapor, and forming fog.

Contrary to general belief, fog is usually worse during mild winter weather of 50 to 60 F than with sub-zero temperatures. From table and psychrometric graph, shown, air at low temperature has small heat and low moisture content. Let's see what causes fog.

Troublesome Fog. If saturated air leaves tower at 80 F and is cooled to 60 F, it will condense: 156 − 78 = 78 grains of moisture per lb. of air, for a heat removal of 17 Btu per lb of air (see table above). In this case, the moisture from tower cannot be absorbed rapidly enough by outside air, so troublesome fog occurs.

Disappearing Fog. For the same heat removal of 17 Btu per lb of air (as before), assume saturated air leaves tower at 57 F and is quickly cooled to 20 F. It then condenses: 71 − 16 = 55 grains of moisture per lb of air. As this saturated air leaves the tower, its vapor rapidly condenses. Heat of air and vapor is quickly absorbed by the cold outside air. The small cloud of fog above the tower soon disappears because it dissipates faster when surrounded by the colder air.

Comparing the second condition with the first we see that there are 30% fewer fog-producing particles and there is a much greater temperature difference between the saturated air and the colder outside air. This bigger temperature difference speeds up the transfer of heat from the vapor in the saturated air to the outside air. Thus with less fog-producing material and faster reactions the cloud is smaller and less troublesome.

With colder air, temperature difference is greater and warm tower air moves upwards into atmospheric air faster (chimney effect) than in denser atmosphere of moderate winter weather. Moderate winter weather and high winds are a bad fogging combination.

Reduce fog trouble like this: (1) Place tower on south side of or away from railroad tracks and highways. (2) Maintain low air temperature, (70 F or less) at tower outlet. (3) Put as much air through tower as operating conditions allow. (4) Maintain high exit air-velocity from tower. (5) Keep minimum ice formation on louvers for continuous maximum air entering tower. Fogging usually comes from atmospheric conditions and not always from tower design.

Ice Prevention. Extremely cold water normally doesn't increase performance much. It does increase operating hazards a lot. Water-cooling towers operating in freezing weather are subject to ice formation at air inlets. Small amounts of water vapor are likely to freeze on inlet louvers and nearby wood-filling. Ice starts to form on the lower louvers section and climbs upward. This restricts inlet area and reduces air flow and increases temperature of water circulated through tower.

Most of the tower circulating water is rarely cooled to freezing temperature. In fact, the cold-water temperature seldom gets lower than 60 F, except in towers used for specific operating conditions. For this reason, ice forms only on parts of tower that are lightly wetted by fine drops which splash toward entering air stream.

To prevent icing during cold-weather, keep tower temperature of circulating water as high as practicable. A temperature-control valve may be used to bypass some hot water to cold-water basin. In general, ice formation in water-cooling towers may be prevented, controlled, and removed by varying tower air flow.

When unit is operated off and on during winter weather, drain water from all exposed piping and basins. This insures protection against freezing and corrosion. Leave basin drains open during winter shutdown to allow rain and melted snow to escape.

Did you know ...

that much of the trouble in removing a nut from a stud is that threads, sticking out of stud rusts, get clogged with paint, etc?

One way to prevent this is to saw off excess portion of stud flush with nut's top after drawing nut down tight. Then dab a coat of paint over stud's cut end and top of nut. Burr caused by hacksawing the stud is easily removed first time nut is backed off.

Original design					Modifications					Improved conditions				
Case No.	Gpm	In, F	Out, F	Wet bulb, F	Fill	Dist system	Mach equip	Fan stack	Frame ht	Gpm	In, F	Out, F	Wet bulb, F	% improvement
1	13,200	120	90	80	•	•				20,000	120	90	80	51.5% increase, gpm
2	1250	125	75	65.2	•	•	•			1850	125	75	65.2	48% increase, gpm
3	183,600	128.8	94.2	75.1	•	•				191,000	122.5	90	77	4.5 F colder water
4	3100	105	85	77	•	•	•			3880	105	85	77	25% increase, gpm
5	4500	110	85	78	•	•	•			6500	103	85	78	40% increase, gpm
6	13,200	120	90	80	•	•	•	•	•	22,200	120	90	80	68% increase, gpm

IMPROVEMENTS: (1) New spray distribution system, new 60-deg drift eliminators and film packing. (2) Larger motor and gear unit, larger spray nozzles and film packing. (3) New spray distribution system, new film packing. (4) Larger motor and gear unit, nozzles with larger orifices and film packing. (5) Motors increased from 40 to 60 hp and larger gear units changed from h-p spray to 1-p downspray type and put in additional splash-type fill. (6) Increased flume height of tower by 16 ft, new spray distribution system, larger motor, more splash-type fill and tip seals placed in fan stack.

Here are six ways you can increase cooling tower capacity

When cooling tower load is exceeded, you may install a second unit, install a larger unit, or increase your tower's capacity. The third alternative may give new life and efficiency without requiring new, costlier equipment

"We need more cooling tower capacity but have no room for a larger unit. What can we do?" "Is there anything we can do to get more capacity out of our present cooling tower?" These are questions asked by anxious tower owners almost daily.

To see how much can be done to improve cooling tower capacity, look at the table above: Capacities have been increased by more than 60%. Savings can be relatively high, when you consider that capacities are usually increased without enlarging the physical size of the present unit, or the amount of space, piping and electrical alterations to the plant area.

Cooling tower owners are often troubled since many units serving power,

You may be able to juggle these six cooling tower variables,

FILL type is important because newer designs may increase the surface area; also, the time the water's surface is exposed to the upcoming air. Up to 20% capacity may be gained

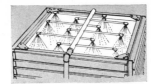

DISTRIBUTION SYSTEM is good for up to 15% improvement. The positive-pressure spray type, for example, remains balanced as the tower ages. The spray patterns may be improved

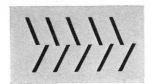

DRIFT ELIMINATORS and inlet louvers may give up to 5% better performance. Some older designs naturally restrict air flow. Staggered blades at 60 deg are much better

Variables to consider before increasing capacity

Tower type	Natural vs mechanical draft
	Forced vs induced draft
	Counterflow vs crossflow
Fill	Open rugged splash vs thinner, dense type
	Splash vs film
	Plastic vs wood
	Honeycomb vs redistribution, spaced
Water distribution system	Closed pressurized vs open pan
	Spray vs through vs open flume
	H-p upspray vs l-p downspray
Fan cell arrangement	Single cell vs multicell
	One large fan per cell vs multiple fan cells
Draft eliminators	One-pass vs two-pass vs three-pass
	Blade angle of 45 deg vs 60-deg angle
	Plastic vs wood vs metal
	Honeycomb vs slats
Louvers	Offset vs flush
	Straight wall vs sloped wall
	Adjustable vs fixed
	Closed spaced vs wide spaced
Fan stack	Conventional vs velocity recovery type
Partitioning	No partitioning vs partitioning

process plants, or air-conditioning systems have been called upon to serve larger cooling loads as plants have expanded. And—sad to say—many towers have not been able to keep up with actual plant operating conditions since startup. This is not surprising when you realize that this type of equipment is often selected and installed solely on a low first-cost basis. As might be expected, a unit thus selected more than likely is not the most efficient, or even the most practical one for the job.

Increasing capacity. Perhaps half of the field-erected

—and a few of the prefabricated—towers installed today can be improved to some degree. This is largely because new technology, more efficient components, and new materials can be designed into a harder working, more rugged and, therefore, more reliable package.

Before trying to upgrade capacity, a complete study must be made of cooling requirements. Then a careful inspection and analysis of the existing tower components is needed. These can be tied together with respect to existing tower size, pressure drop, air flow characteristics, drift loss, fan performance curves, gear selection and service factor and installation problems.

Studies and recommendations should be made by a cooling tower specialist. Once the findings are carried out to a practical conclusion, very substantial savings result which may, and often do, save the cost of a new unit.

Let's look now at the alternatives we have to choose from when planning to increase capacity. By juggling the variables around, the results will vary. The mechanical draft counterflow tower lends itself more readily to capacity increases than does the crossflow design. The reason is that, by modifying counterflow tower components, greater returns are realized than by making adjustments to a crossflow unit.

Another thing to consider is that, in many cases, actual plant operating conditions change over the years. For example, increasing loads usually calls for specifying a new cooling tower. Now let's zero in on each of the seven variables and see specifically what can be done:

Fill configuration. Cooling tower packings have improved greatly over the years. When older towers were originally installed, tower manufacturers designed units to perform according to specified conditions by using the type of packing that they considered most economical for their fabrication facilities to produce. The design was, of course, consistent with performance and test data available at that time. Since then, improvements in design have come a long way, and more field test data is currently available on different fill configurations.

depending on your present design

MECHANICAL EQUIPMENT moves the air through a tower—fan blade angle, speed, size motor — all are variables. Up to 10% more capacity can be obtained on some towers

FAN STACK can up capacity as much as 7%. This can be done without changing motor size. The proper velocity recovery-type stack can deliver more air with the same motor

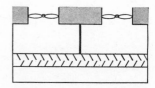

PARTITIONING is very important in larger multicell towers. It's not easy to predict percentage of improvement but it prevents short circuiting of air, if one cell does little work

---

For example, most older counterflow towers have splash-type packing installed. Rows of splash decks are usually spaced about 2 ft apart in the tower. The theory is to break up the water into drops as it cascades through the tower. And each drop's surface is increased by continuous interception of its fall by the splash decks, thus exposing a fresh evaporating surface on every new drop.

Appraisal of the fill arrangement, and its quantity, by an experienced cooling tower specialist will establish if more fill can be added to provide increased capacity. In this case, film surface packing, or special packings that combine *both* film surface as well as additional splash surface, are installed. Some fill designs will increase the surface area and also the time that the water's surface is exposed to air. As a result, the rate of exposure may be high enough to permit the reduction of air required. Therefore, film packing may be added without increasing fan horsepower requirements.

The percentage of the improvement in capacity that results from adding film surface-type fill depends on the severity of duty, performance level and the height of the tower. For example, by adding substantial quantities of film packing for a very low tower that is subjected to very severe duty, a 40 to 50% capacity improvement can be obtained. On the other hand, if a very high tower is used for a very easy duty, the performance could be adversely affected when film-type packing is installed. In this instance, cfm is more critical to performance than is the performance level itself.

However, for a reasonable performance level, a 20% increase in capacity can be achieved for a tower of average design. A 20% increase in capacity may be equated to a 20% increase in gpm at the same temperature level, or approximately a 20% decrease in the approach to the wet bulb at the same flow rate and heat load.

Distribution system. Some older towers have open flume or trough distribution systems. This type of system — especially in multicell installations — is hard to balance. The problem is then compounded when water loadings change in the process. Flooding or dry spots also steal from the tower's effectiveness.

Sometimes the gravity feed clogs at the downspouts. As you know, each downspout has a diffusion deck, or splash plate, under it. If these are broken or out of line, the effective cooling volume is reduced. Here, changing to a positive pressure spray-type distribution system reduces balancing problems. An advanced design header-lateral spray system ensures a good water pattern over the entire fill area and permits the passage of more air through this area of the tower.

Again, an analysis of the tower in operation reveals possible areas for improvement. Water spray may not fully cover upper fill layers. Occasionally the water is half-way down through the fill before it is evenly distributed across the entire plan area of the tower.

The obvious answer to this problem is to modify the nozzles so that the spray pattern is altered, or corrected. This simple improvement often adds capacity to an existing tower.

Some older, high-pressure upspray towers have been converted to low-pressure downspray units with improved results. This process entails raising the distribution system level, which permits additional rows of fill decks to be installed. While this modification may increase capacity by as much as 12 to 15%, there will also be an overall, small decrease in pumping head.

Drift eliminators. Many older towers have extra-heavy drift eliminators. Others have close-spaced eliminator blades of 45 deg angles. Both designs have been found to be too conservative for some requirements. One reason is that these designs naturally restrict the air flow. By replacing these with a staggered drift eliminator blade at a 60-deg instead of a 45-deg angle, more air is permitted through the tower and additional capacity results.

Drift eliminator modifications are usually made when the drift eliminator requires replacement. However, a counterflow tower improved by this simple change can increase capacity by 4 to 5%. On a 10 F approach tower,

Today's new technology allows more effective cooling

this is equivalent to almost 0.5 F of colder water.

In some cases, an extra pass of drift eliminators can be installed to enable the change of fan or gear to draw maximum cfm without causing a drift loss problem.

Inlet louver design improvements can also sometimes increase tower capacity. The older 45-deg closed-spaced louvers are too restrictive. If an off-set louver with return drip panel is used, a greater amount of air enters — and enters more easily. Additional advantages occur during cold weather periods of operation, since splash-out problems are minimized when fans are cycling during light loads, or partial load conditions.

Mechanical equipment. If additional capacity is expected from an existing cooling tower, more air movement is generally needed. But many installations already operate at maximum rated horsepower. In such cases, be sure to take readings at the motor leads to determine how close the motor is running to full load or nameplate amperage. If there is room for additional load, the fan blade-angle can often be increased to draw more cfm. First obtain a set of fan curves from the cooling tower vendor to see what cfm improvement can be expected with a pitch-angle change. Since horsepower varies as the cube of cfm, there isn't much room for movement without requiring a larger motor.

Conversely, an increase in the air rate varies as the cube root of the increase in horsepower. But a substantial increase in horsepower may reflect only a relatively small increase in capacity, as summarized in the table below:

Motor change, hp	Hp change, %	Approx. capacity change, %
25 to 30	20	6
30 to 40	33 1/3	10
40 to 50	25	7½
50 to 60	20	6
60 to 75	25	7½
75 to 100	33 1/3	10

If the unit is installed with the fan pitch angle at maximum efficiency level, a change in fan speed may be in order. This is done by changing the gear ratio. Thus you may help slightly with your gearing service factor, and it's just a case of changing the ring and pinion gear.

A major modification includes increasing the fan size. But this is rarely done, because increasing cfm by increasing fan speeds strains the limits of the tower from another viewpoint: Too high a velocity through the tower and drift eliminators can cause a nuisance problem from excess carryover.

Let's say that full effectiveness can be realized from the larger motor with increased pitch angle or fan rpm. Here a capacity improvement of as much as 10% can put money in the bank.

Fan stack. The importance of fan stack design for top performance cannot be overemphasized. It isn't practical to change or increase the motor size on many existing units. One reason is that any change of the electrical service to the cooling tower site might be unreasonably costly. Here is where the installation of a parabolic venturi-type fan stack might be the answer because, with this velocity

recovery-type fan stack, the fan is capable of delivering 6 to 7% more air through the tower with the same motor. And 7% more air means 7% more tower capacity.

Partitioning. Some larger multicell towers were built for operation only at design load. Some of these units have two fans per cell and there is no partition in the plenum area. If one fan is shut down for repairs, or if the tower is operated with only one fan per cell for any reason, the mechanical draft feature would be rendered practically ineffective. This is because the operating unit would then be drawing most of its air through the adjacent fan opening, thus bypassing the fill area, which is extremely wasteful.

But even if plenum areas are properly partitioned between fan cells (in counterflow towers) a transverse partition should be extended down to the top of the louver level. Only then can the effective counterflow principle be realized. If the unit is installed in a wide open area, be sure to install a longitudinal partition to prevent blowthrough when high winds hit at 90 deg to the longitudinal axis of the tower. When your plant operates at partial load, remember that the proper partitions between fans and cells will greatly improve operating efficiency.

Efficiency from proper operation and maintenance. Your cooling tower will be the focus of more and more attention as increased effort is directed to the more efficient use and conservation of water. (See Water, a special 48-page in-depth that was prepared by POWER for the June 1966 issue.) The operating and maintenance manual for your cooling tower should be studied by operating personnel to make sure that peak efficiency of the equipment is realized. While the cooling tower is manufactured from durable materials, it is probably subjected to one of the most extreme operating conditions of any type of equipment in use today.

A consultation with a trained cooling tower engineer will enable you to properly appraise your tower components. The distribution system (air water pattern) is the key to the tower's operating efficiency. This experienced survey may make the small difference needed to maintain the design capacity level. The dividends obtained by following such a procedure will more than justify the comparatively small amount of time invested.

Summary. We have outlined six possible ways to improve the cooling tower performance. But there are a number of conditions under which modification is not practical: If substantial plant and/or tower expansion is planned, economics generally dictate that added capacity be assumed by a new tower unless: (1) the existing tower is poorly designed, (2) real estate or space for the new unit is at a premium, or (3) repairs are planned and needed on the existing tower.

But your cooling tower does not have to be old or damaged to consider the services of competent cooling tower engineers. With their help, existing towers may be given a new life and greater efficiency without installing costlier new equipment.

Delignification . . .

fungus attack and . . .

salt deposits on wood cooling - tower slats are three good reasons for topnotch water treatment. Fortunately, cooling towers don't usually end up in this condition. But it's good insurance to check for trouble symptoms. To aid you, here are . . .

18 Q & A's on treating cooling water

1 Is there an ideal treatment for all cooling waters?
A Economics plays an important role in both selection of cooling-water systems and methods of water treatment. It's *possible* to prescribe an ideal water for each cooling job, but equipment and chemical cost is out of the question for most installations. So treatment must be tailored to fit the size and type of cooling-water systems currently in use.

COOLING SYSTEMS

2 What are the main types of cooling systems?
A The two main classifications are "once through" and "circulating." In the simple once-through system water passes through the heat-exchange equipment only once. It is increased in temperature about 20 F before going to waste.

Circulating systems are more complex. Cooling water also passes through a heat exchanger or other apparatus where cooling occurs. Temperature increases from 6 to 20 F. Heated water is then cooled in another exchanger using an external cooling medium such as an air-cooled radiator.

3 Is there a trend to the use of circulating systems?
A Since the once-through system calls for an almost unlimited supply of cooling water, today's trend is toward the use of circulating systems where the water is cooled for re-use. Cooling is brought about in the open type by actually evaporating a part of the cooling stream. In a closed system the cooling is usually done in an air-cooled radiator. A separate circulating or once-through setup can also be used.

4 Where are open and closed systems used?
A Open type is most frequently used for cooling water in chemical plants, oil refineries and power-plant condensers. Closed system is usually limited to jobs where the quantity of cooling water isn't large. Typical applications include compressor cooling jackets and diesel engines.

EXCESSIVE SCALING

5 What causes scale formation?
A Most common type scale found in heat-exchange equipment is calcium carbonate formed by the breakdown of calcium bicarbonate in the cooling water at high temperatures. Magnesium compounds and calcium sulfate are rarely the cause of scale in cooling systems. Scaling is affected by temperature, rate of heat transfer, calcium, magnesium, silica, sulfate and alkalinity concentrations and the pH.

6 How is the Langelier index used?
A Langelier index measures the tendency of calcium carbonate to precipitate from water under given conditions of calcium hardness, alkalinity, pH, temperature and total dis-

solved solids. Charts, nomographs and special slide rules can be used to quickly calculate this index for any water supply when the analysis is known. Positive index means water has a tendency to deposit scale while a negative index shows a tendency to dissolve scale and corrode. Note that index does not measure amount or rapidity of scaling. This depends only on calcium carbonate or bicarbonate content.

Index is decreased by reducing calcium hardness, calcium bicarbonate, pH or a combination of all three. Total dissolved solids has a relatively minor effect on the index.

Blowdown can be used to reduce the calcium and alkalinity in the circulating water. But it increases makeup and is seldom used as the only means of control.

Other treatment methods used to adjust the analysis include: cold-lime softening, usually followed by an acid feed for the adjustment of pH and alkalinity; partial zeolite softening with hard water by-pass, also followed by an acid feed if needed. Some moderate-hardness high-alkalinity waters can be handled with an acid feed alone.

7   What is the "controlled scale" method?

A   It is a method of keeping index of the circulating water slightly positive, say in the range of 0.5 to 1.0 at the highest temperature in the system. This is usually enough to insure deposit of a thin impervious layer of calcium-carbonate scale on the surfaces of the cooling-water system. Careful adjustment of pH, alkalinity and calcium is needed to keep the desired index and protect the system against the ravages of excess scaling and corrosion.

8   What is the role of surface-active chemicals?

A   Surface-active materials prevent crystal growth and therefore scale formation. They increase the solubility of scale-forming salts so precipitation doesn't occur when solubility limits are exceeded. Chemicals used for this purpose include special types of phosphates, tannins, lignins, etc. They can handle the job of scale prevention alone and are also helpful in broadening the range of Langelier index control in the controlled-scale treatment.

CORROSION

9   How do you control cooling-water system corrosion?

A   Corrosion inhibitors such as chromates, phosphates, silicates and alkalies are used. Anodic inhibitors of the chromate and phosphate type decrease metal attack. But if the concentrations are too high, pitting and tuberculation result. Use of chromate in combination with phosphate has been successful in controlling pitting, tuberculation and other metal losses from corrosion, and dosages are at economical levels.

Since O_2 and CO_2 are primary cause of corrosion one of the more direct treatments is deaeration. With a vacuum deaerator, oxygen can be reduced to as low as 0.1 ppm depending on water temperature and equipment design.

SLIME AND ALGAE

10   What causes slime and algae to form?

A   Algae growths are composed of millions of tiny plant cells that multiply and produce large masses of plant material in a short time. Slime growths are a gelatinous mass of micro-organisms that cling tightly to secluded surfaces in the system, trapping organic and inorganic matter and debris along with scale-forming materials. Any appreciable buildup seriously interferes with heat-transfer efficiency.

11   What chemicals give effective treatment?

A   Chemicals that prevent slime and algae troubles are those that exert a toxic action on the micro-organisms.

Chlorine is fairly general in toxicity for most bacteria, algae and protozoa, but continuous application in a once-through system is too expensive. Intermittent treatment gives good results at much more reasonable cost.

Copper sulfate is toxic to simple algae, but is less toxic to slimes. Use is limited because copper sulfate precipitates at a pH of 8.5 or higher. Phenolic and other quarternary ammonium compounds are also used. They have the advantage of resisting precipitation at high pH and are not removed from the system by aeration in the cooling tower.

12   Why is an intermittent feed best?

A   Intermittent large doses of chemical rather than a steady feed of a much smaller quantity keep the micro-organisms from getting used to the treatment and building up an immunity to it. It also avoids the danger of a new strain of algicide- or slimicide-resistant bacteria developing.

13   How are accumulated slime and algae removed?

A   Best way to get rid of these growths is by mechanical cleaning. In the case of algae the algicide may only loosen the growth enough to set it free in the system to plug lines and damage pumps. After the system is completely cleaned, start out on the right foot with a good chemical-treatment program to prevent future buildup of these growths.

WOOD DESTRUCTION AND DECAY

14   What causes cooling-tower wood destruction?

A   This destruction is usually caused by one or a combination of: (1) chemical attack commonly known as delignification (2) mechanical rupture of wood cells from salt crystallization and (3) biological attack causing wood decay.

15   What is delignification and when does it take place?

A   It is the removal of tannins, lignins and cellulose products from the lumber, leaving long stringy fibers with greatly reduced strength. It apparently occurs in the presence of high concentrations of sodium carbonate in the circulating water. Oxidizing agents like chlorine speed up the process.

16   Why are salt deposits harmful?

A   Intermittent splashing on parts of the cooling tower produces concentrated salt deposits by evaporation. Salt-saturated water in these areas is absorbed by the wood. Then crystals form beneath the surface of the wood mechanically disintegrating it. Result is a mass of soft loose fibers susceptible to both chemical and biological attack.

17   What is fungus attack and how is it caused?

A   It is a form of biological attack that is caused by micro-organisms that use the wood for food. These wood-destroying fungi reproduce by means of spores that are normally airborne and alight on the surface of the wood. The attack is not confined to any particular section or make of cooling tower. It is most noticeable in the mist sections and takes place on both the surface and the interior of the wood.

18   How do you combat fungus attack?

A   Proper control of alkalinity, pH and solids content of the circulating water will go a long way toward preventing chemical attack on the wood. It will not *prevent* biological attack, but will decrease chances of damage from fungus.

One preventive today is use of properly impregnated cooling-tower lumber. Chromated copper-arsenate salts show much promise for this. A different approach is to try to chemically destroy the fungi in the circulating water by treating with sodium pentachlorophenate, zinc sulfate, and sodium chromate. Use intermittent dosing. It is often effective to spray a concentrated solution of toxicant directly on the cooling-tower wood in those sections most likely to be attacked by the fungus, including tower mist areas.

BRRR R R—avoid heating failures this winter by doing work now . . . Schedule maintenance at once . . .

14 Q&A on preparing your heating system _now_ for winter

1   What's best time to overhaul building heating system?
A   Most building operators do this job when equipment comes off the line in spring. It's best to clean boiler and sooty ducts as soon as system is shut down. Reason is that ashes and soot absorb moisture from humid air. Because of sulfur in these deposits, resulting sulfuric acid rapidly corrodes the metal.

2   Is galvanic action a problem in hot-water heating systems?
A   Galvanic action can be a very serious problem because of dissimilar metals in piping, fittings, heat exchangers, boilers, etc. Galvanic action precipitates iron into water, which lowers pH down toward acid condition.
Water going toward acid increases galvanic action, which in turn precipitates more iron, causes oxidation—corrosion. If water is not treated for five to ten years, look for serious trouble. If organic compounds are used, 1 part in 500 parts of water often prevents trouble.

3   What equipment in heating system should be checked?
A   Everything. Follow system through building and come back to boiler with return system. Take in all piping and check joints for leaks, corrosion and condition of insulation.

4   Should hot-water heating systems be drained in summer?
A   Many large buildings have hot water heated by steam in heat-exchangers. Hot water is then circulated through space

heaters. Many air-conditioning systems having chilled water are now being adapted for hot-water heating in winter. Chilled water is usually at 40 F, while hot water may be anything over 180 F. But slime and algae form in such systems when water is under 140 F. Because oxygen is gradually liberated from system while water is hot in winter, don't drain in summer, but treat water chemically.

5   Radiators: What maintenance is needed?
A   Check for physical defects—usually they need little maintenance. Some steam radiators have orifice plate on inlet side between valve and radiator. In that case they have no trap on outlet side. If it's a high-pressure system, you might take coupling apart once in five years or so and check hole size in orifice for wear. Radiators that don't heat properly usually have fouled valves. Make sure air bleeder valve on radiator works.
When trap is on outlet, renew valve disk if needed, or grind in. If the packing gland of valve is screwed down all the way, remove old packing and put in new.

6   Unit heaters of the ceiling type: What maintenance is needed?
A   Coils of small unit heaters often spring leaks. Check by placing pressure inside coil. Open steam trap and clean out, check valve seat and disk, also bellows for cracks. Check electric motor on unit. Clean out motor and grease bearings. Don't forget that more motor bearings are burned up from

Check returns and all the traps . . .

To keep your building warm on coldest days

overgreasing than from lack of lubrication. If in doubt, leave out plug of bearing and run motor so excess grease can run out. Then screw in plug again. If unit has fan, clean the blades. This allows them to run more freely and also keeps them from getting out of balance, which loads the bearings and shaft.

7 Valves: What attention is needed?
A All valves in system, whether you buy or generate your own steam, should be checked for tightness, packing, good working condition. Best way is to lift valve bonnet. Then you can usually see if valve is tight by observing seat and disk condition. At times a valve needs a new stem, or packing may be hard, then leak steam during heating season.

8 Expansion joints: What attention is needed?
A Expansion joints are often of slip type. Because there is constant movement in this joint when steam is turned off and on throughout zone, you should repack joints each summer. Before doing so, shine sliding part with fine grinding cloth to make sure no burrs or scale cut packing.

9 Reducing valves: What attention is needed?
A Besides large reducing valve that steps down steam from utility firm as it enters building, there may be other reducing valves at various zones of either high or low pressure. One zone may operate on as low as 1 psi while another uses 50 psi. Open each reducing valve, check valve seat and disk, and other internal working parts.

Diaphragm may be critical part, although that depends on design and material. If there are any indications of failure from constant flexing, it's good insurance to replace diaphragm with new.

10 Condensate pumps: What attention is needed?
A Clean electric motor and lubricate same as any motor in system. Because this unit can shut down entire system, have only responsible person do work. Check float for leaks and clean out condensate-return tank if needed. Float may leak, get gradually water logged.

Remove packing from pump shaft and check shaft for grooving. If heavy wear, check pump and motor alignment. Best way is to break coupling (remove bolts) and use dial indicator. Don't assume that equipment stays in line just be-

cause it was in line at one time. Misalignment is costly.

Most pumps of centrifugal type need little maintenance, especially on condensate-return system. But checking wearing rings and impeller often shows that parts need renewing.

11 Return piping: What attention is needed?
A Leaks and corrosion are very common in return lines. Look for corrosion around piping connections, especially at threads of nipples. If portion of your return line is buried, check for leaks as that is where corrosion takes place. Put hydrostatic test on buried section of pipe. If possible, run piping in covered trench to avoid future headaches.

12 Relief valves: What attention is needed?
A New type relief valves on hot-water boilers are tested and approved by National Board of Boiler and Pressure Vessel Inspectors. They are stamped showing they comply with ASME Heating Boiler Code. These valves, rated in Btu discharge capacity of either water or steam, are more reliable than older type valves. Insurance firms advise renewing old valves with this newer type.

13 Controls on boiler: What maintenance is needed?
A Safety valve should lift easily when test lever is raised. Clean water column so water level is easily seen in glass. Experience shows that if automatic water column is attached to boiler and water column, draining weekly during heating season keeps it working right.

Summer is time to check low-water fuel cutout, emergency water feeder, pressure controls and firing controls. This includes any automatic firing device. Take low-water fuel cutouts apart and check carefully. If manufacturer has local service center, let him do this work. Or you might return equipment to factory.

14 What is common cause of radiators or other space heaters, which are supplied by one riser in system, not heating while other units do heat?
A Building sections that support piping and hangers often settle, which may restrict circulation in riser. Heaters on other risers may have good circulation at expense of one that has settled. To avoid, check all possible places where settling might take place. Do this every summer when lines are cold and then elevate the risers to their original positions.

Gas fires

INFRARED HEATERS ON CEILING keep busy factory workers comfortable far below, but don't needlessly heat the air inbetween

MODERN GAS-FIRED INFRARED HEATER, complete with controls, has a parabolic reflector to focus heat where you want it

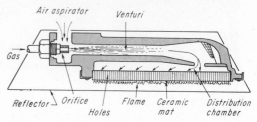

CROSS SECTION of generator shows orifice for gas metering, aspirator, venturi, ceramic mat. Temperature quickly hits 1600 F

What is infrared?

Infrared radiation for space and industrial process heating is catching on rapidly. For one reason, thermal energy is radiated directly to the object to be heated, without dissipating in the intervening air.

Infrared heaters supply controlled heat to people and equipment. Rays of energy from the generators, like the rays of the sun, become heat only when they hit an object. In an infrared-heated building, people are warmed by rays from the generator and by radiation, conduction and convection from the floor, wall and machinery surfaces which, in effect, become radiators themselves.

A basic physical phenomena is electromagnetic radiation. Infrared is one form of the electromagnetic radiation which runs in a spectrum from radio waves, through infrared, through the light we can see and on to X-rays. All electromagnetic radiation spreads in the form of waves. Frequency and wave length are in an inverse relationship: the higher the frequency, the shorter the wave length.

Radiations in the low frequencies are used for broadcast radio; the electromagnetic spectrum is used for radio or radar right up to the edge of the infrared region.

Beyond the infrared spectrum, where wave lengths are measured in microns, about 1/50,000 in. is visible light. Then follow ultraviolet light, soft X-rays, hard X-rays and gamma rays. Infrared, like light, can be focused with mirrors and lenses, yet it can pass through some opaque materials, such as silicon and germanium, in just the same way that radar waves do. Every object at a temperature above absolute zero (−459.6 F) generates infrared waves. For example, all the sun's heat reaches us as infrared radiation—we sense it with our skin.

How heaters work

Low-temperature infrared generators—radiant panels, hot water or steam radiators, electric coils or combustion chambers of heaters—are widely used for heating. But only a very small part of their heat is transferred by radiation, so large surface areas are needed.

High-temperature electrically heated infrared lamps are efficient sources of short-wave infrared, and they're commonly used for industrial processing. But besides the price of the equipment, you also have to figure in terms of cost of power, electric lines, transformers.

Gas is an inexpensive infrared source. The gas is used to heat a ceramic, and when the ceramic gets hot enough it gives off large amounts of infrared radiation.

One generator used by various American manufacturers was developed by Gunther Schwank in Germany. It consists of a gas-fired burner operating under atmospheric pressure. Heaters with burners of this kind are used to warm buildings, thaw frozen cars of coal or ore, heat dies, molds and cores, bake finishes, dry materials, control humidity or supply process heat.

Heart of the Schwank heater is a small chunk of ceramic 3/8-in. thick, each sq in. of it perforated by 200 tiny holes of 0.055-in. diameter. This ceramic is a near-perfect insulator and the most powerful generator of infrared energy developed thus far, it is claimed.

Let's look at a cross section of the generator to see how the device works: Gas is metered through an orifice, located at the opening of the mixer tube or venturi. All air needed for combustion is aspirated through the chamber shown. Gas and air mix in the venturi and feed to the distribution chamber on the inside of the ceramic mat.

these economic infrared space heaters

CERAMIC MAT, developed in Germany, is only ⅜-in. thick, with 200 tiny holes per sq in. It's a near-perfect insulator

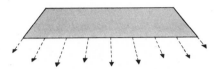

STRAIGHT-SIDED REFLECTORS mount 10 to 20 ft high where a spread pattern is desirable, as around a busy factory floor

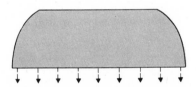

PARABOLIC REFLECTORS are best for mounting heights from 20 to 45 ft. They aim the infrared beams in spotlight fashion

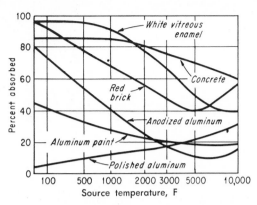

TYPICAL BUILDING MATERIALS absorb varying percentages of the energy that strikes them, depending on wave length

Effect on materials

Curves above show what percentage of energy at each wave length some typical building materials will absorb. Some materials transmit infrared rays in the same way that glass passes light rays. But glass itself won't transmit the longer infrared rays. Because quartz and quartz glass are more transparent they are used as windows for infrared. Sodium-chloride crystals are very transparent, and so they're used as lenses and prisms in spectroscopic apparatus. Other materials are used as filters.

Metallic surfaces reflect most of the infrared radiation striking them, and polished aluminum is about the best of all reflectors. Nonmetals like wood, masonry, concrete and fabrics absorb most of the infrared that strikes them, converting the radiant energy into internal heat.

These curves show that most nonmetallic substances absorb a higher percentage of the longer wave lengths while metals work just the opposite way. So when a painted surface, a concrete floor or some other nonmetallic body is to be radiant-heated, long-wave infrared is more effective than shorter rays close to wave length of light.

Air-gas mixture burns at the outside surface of the ceramic mat—a separate minute flame at each of the tiny holes. After a few moments of operation, the flame is almost invisible, but the surface of the ceramic reaches about 1600 F while its bottom stays at 325 F.

Large port area results in low backpressure in the burner, which makes it possible to introduce all air needed for combustion as primary air. Low thermal conductivity of the ceramic plate reduces flow of heat from the red-hot burner face to burner's interior.

Of course, the surface of the ceramic has a high infrared emissivity. How effectively the heat of combustion converts to infrared radiation depends on surface condition and temperature. The Schwank burner's stated efficiency at 1650 F is about 60%.

Reflectors for infrared heating may be straight-sided or parabolic. The parabolic type, made of polished aluminum, is highly reflective. It's used for heights over 20 ft, since it projects an infrared-heat beam of undiminishing intensity, similar to the beam of a searchlight.

Ventilation for these units meets requirements of the American Gas Assn and Underwriters Labs for indoor use in nonexplosive atmospheres. Manufacturer claims that their carbon-monoxide production is less than that of a domestic gas range. They are vented in the peak of the building's roof, and heat output of the combustion gases is recovered by mixing with air before gases are exhausted through the roof ventilators. These gases may warm the ceiling to help make up heat losses there.

Sources: Hupp Corp and Bettcher Mfg Corp, Cleveland, Ohio; ASME paper No. 58-A-210 by Marc Resek.

Don't wilt when air-conditioning trouble hits you. Just use
this handy chart to . . .

WELL-KEPT air-conditioning units give little trouble; keep 'em fit with right care

Track Down Air-Conditioning Ills Fast

SYMPTOM	POSSIBLE CAUSES OF TROUBLE; CURES
HIGH HEAD PRESSURE	1 Not enough condenser water, or water too warm; check water-regulating valve, look for plugged pipe 2 Noncondensable-gas or air in system; purge until head pressure becomes normal 3 Condenser fouled; clean tubes, tube sheets, heads 4 Too much refrigerant; bleed excess to a drum 5 Evaporative condenser faulty; check and correct air and water flow; clean surfaces 6 Evaporative condenser too small; install larger condenser or use small watercooled unit for peak loads
LOW HEAD PRESSURE	1 Too much or too cold condenser water; adjust water-regulating valve 2 Not enough refrigerant; look for leak in system; repair; charge 3 Leaky compressor valves; check to see if suction pressure rises more than 2 psi per min in Freon units. If it does, overhaul compressor 4 Oil return valve or trap leaks; spot by feeling for hot oil return; repair or replace valve or trap
LOW SUCTION PRESSURE	1 Evaporator load light, causes compressor to short cycle; run at lower speed or fit capacity control 2 Clogged strainer; clean or replace strainer 3 Low-pressure switch setting too low; adjust 4 Compressor oversized; fit with a capacity control or run at lower speed 5 Hiss in expansion valve. This is caused by flash gas in liquid line. Cure by subcooling liquid 6 Evaporator pressure drop too large; fit an external equalizer to stop high superheat 7 Not enough refrigerant; look for leak in system; repair; charge
HIGH SUCTION PRESSURE	1 Too much load on evaporator; check conditioned-space insulation, fresh-air leakage, infiltration 2 Compressor suction valves faulty; overhaul 3 Expansion valve stuck open. With this trouble suction line will be too cold; repair valve 4 Evaporator oversized; run compressor faster or buy a large compressor 5 Low-pressure switch set too high; adjust 6 Expansion valve oversized; use smaller valve

SYMPTOM	POSSIBLE CAUSES OF TROUBLE; CURES
SYSTEM CAPACITY LOW	1 Evaporator pressure drop too large; use an external equalizer on expansion valve 2 Superheat adjustment wrong; fit external equalizer or adjust expansion valve 3 Clogged strainer, valve or pipe; clean or replace 4 Dirty cooling coil; clean dirt off coil; defrost to remove excess ice; check air flow to see it is normal after cleaning and defrosting
COMPRESSOR WON'T STOP	1 Temperature high in conditioned rooms; look for oversized fan, wrong damper setting or too much infiltration. See that rooms are insulated right 2 Discharge pressure high; check cooling water flow and temperature, clean condenser tubes or evaporative-condenser coil, purge excess Freon, air and noncondensable gas 3 Too low a temperature in conditioned rooms; check thermostat settings in the rooms, adjust to needed temperature; look for cold drafts that may be striking thermostat 4 Not enough refrigerant; look for leak in system; repair; charge 5 New heat loads in conditioned spaces; these may come from new or different production methods; increase compressor capacity or reduce heat load
ODORS IN ROOMS	1 Refrigerant leak; find, repair, charge, test 2 Brine leak; find, repair, adjust concentration and test piping 3 No trap in air washer drain to sewer; fit with suitable trap 4 Cooling-coil surface dirty; shut down air washer and clean coil; check intake-air filters for dirt 5 Odor-producing units near air intake; relocate odorous units or air intake
COMPRESSOR OIL LEVEL TOO LOW	1 Dirty strainer, tubing or valves; clean 2 Liquid slugs in suction; reset superheat, check to see thermal bulb is securely fastened 3 Not enough charge; look for leak, repair, charge system, and test with a halide torch 4 Suction line traps oil; repitch and fit loops 5 Oil separator faulty; repair, add oil, test

7

Diesel engines

Let's Take Over Watch

1 — Read logbook carefully to see what happened on the previous watch . . .

YOU ARE RESPONSIBLE for the plant the minute you take over a watch—your reputation as an operator is at stake. Study the logbook, and make sure you know of any unusual conditions that arose on the previous watch. A smart operator can often prevent breakdowns from telltale entries.

Above is a photo of George Green, first assistant engineer, studying the log in a smoothly running plant.

Use selector switch to check each cylinder's exhaust temperature. If temperature to any one cylinder is high, check log sheets to see if it has been high before. If not, adjust fuel oil to that cylinder until temperature is the same as in others.

If that cylinder knocks, it might be overloaded. If all its adjustments are right and it still knocks, the cause might be that the cylinder firing before the knocking cylinder is dead. Then this one knocks because it does the work of both. If dead, check fuel pump, also valve mechanism for that cylinder, starting with camshaft. Check lube-oil temperature going into engine. Adjust cooling water to lube-oil cooler if oil temperature isn't right.

Next, check fuel, lube and cooling-water pressures. Try to keep them as nearly ideal as possible. Lube-discharge temperature from engine often warns of overheated bearing or badly overheated parts inside. Try to locate cause by feel, or by listening carefully with ear to flashlight placed against various parts of engine housing.

Cooling water from engine tells the same story as lube oil from engine. If temperature has been climbing gradually over several weeks, the cause is usually scale building up inside engine jackets. Or it could be a fouled heat exchanger.

Make sure you know where lube oil is in governor by looking at oil-level sight. Keep level right. If it is a self-cleaning lube-oil strainer, give handle one turn as in photo. That scrapes strainer clean. Then check oil pressures to see if they have changed because of clean strainer.

Crankcase oil level should be up to mark. Check it with bayonet, and make sure you read it correctly. Lubrication is the engine's lifeblood. You can't always depend on the engine's tachometer readings. Good investment for your plant is a hand-type revolution indicator. Place it on end of shaft at pedestal bearing. If engine's critical speed is known to be within operating range, make sure engine runs slightly over or below this speed. While at the pedestal bearing, feel the bearing and, if it is oil-ring lubricated, look inside for good measure to see that rings are turning and level is right.

Trouble with modern engines is they run well with little attention for long periods. Then operators get lax, and they forget to make their rounds as they should.

While on watch, keep on your toes during peak-load periods. Get the habit of listening to each engine when taking over watch, and while making each round. Listen at each cylinder until you get so expert that you can detect any unusual noise. After a while you will know what every little sound means. Before turning watch over to your relief, be sure to tell him about anything that might give him trouble.

In a Diesel Plant

2—Check: Cylinder exhausts . . .

3—Lube-oil temperature . . .

4—Fuel, lube, cooling pressure

5—Check: Lube discharge . . .

6—Cooling discharge temp . . .

7—Governor's oil level

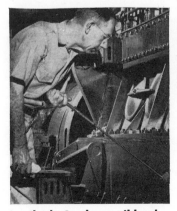
8—Check: Crankcase-oil level

9—Shaft revolutions

10—Clear: Lube-oil strainer

Let's Start Your Diesel

Getting Ready

1 Open the inlet and outlet of engine's jacket and the cylinder-head cooling water, make sure water is flowing through

2 Pump up the lube-oil pressure by hand to about 5-10 psi. This prevents scored bearings caused by oil draining

3 Open fuel valve and pump up fuel by hand to make sure lines are filled, and also to help clear the lines of air

Starting Up

7 Open starting air-valve to crank the engine, close valve as soon as engine is firing and rolling under own power

8 Test cylinders for water by holding a dry rag under open-cylinder blow-off valves. If moist, check for cracks

9 Bring speed up slowly. If cold, wait till lube is 100 F. Observe oil, fuel, cooling pressure, and then throw on load

Keeping on Your Toes

13 Check bearing and other temperatures every hour. Do this by feel as shown, and by reading the various gages

14 Every hour, oil starting air-distributor (this engine). There are various oil cups that need hourly hand oiling

15 Check voltage and load every hour, jot down on log sheets. This practice is often skipped in small plants

Engine and Keep It Rolling

4 Open blowoff valves of each cylinder to avoid any damage if there is water in the cylinder from a cracked liner

5 Open starting-air valve on storage tank if the pressure is right for starting. If not, start the air compressor

6 Set the governor to starting speed. This engine is now ready to start as it has been properly prepared to roll

10 Check exhaust temperatures of all cylinders with selector switch as shown. Do this before leaving engine

11 Exhaust of this cylinder is very low because of air in the fuel-oil line. Crack connections and let air bleed out

12 If venting the fuel-line won't help, shut it down and replace the fuel injector because the unit is probably faulty

16 Every hour, log exhaust temp, cooling-water temp in and out, lube-oil temp in and out, lube-oil pressure in, out

17 Once a day, fill wick oilers for rocker-arm rollers, exhaust and also intake valves, etc, depending on engine

18 Enter time engine was started, or shut down and other important info in log book. Sign when going off watch

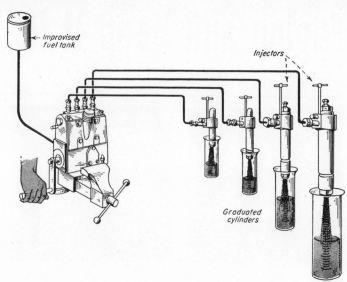

YOU CAN DO ALL THESE THINGS:

1 **Find faulty pumps**

2 **Clean pumps as they should be**

3 **Nail down unexpected troubles**

4 **Grind in pump valve seats**

5 **Learn why pumps stick**

6 **Correct early or late timing**

7 **Stop engine misfiring**

8 **Calibrate pumps in your shop**

1 Hold pump in vise, place graduated glass cylinders under each injector. With temporary handle on pump, crank about two minutes. Compare oil in cylinders

Give 'Zing' to Diesel-Fuel Pumps

▶ CRACKERJACK OPERATORS, who can do almost any diesel job, often shy clear of fuel pumps. They think it takes a genius to overhaul one.

Diesel engine's fuel pump is heart of the system. It takes patience, care, and the maker's maintenance manual. But a manual leaves out plenty that a practical man should know.

Most troubles in barrel- and plunger-type pumps are from damage to these two parts. Symptoms vary, but include low exhaust temperature and low firing pressure, unbalanced pump, misfiring, and engine failure to develop full power.

If engine misfires, first find missing cylinder by exhaust temperature (low reading). Or cut out each cylinder in turn by disconnecting fuel line at nozzle union or opening bleeder screw. When defective cylinder is found (misfiring ceases), remove injector and replace with a good spare. Start engine —if misfiring continues, trouble is in pump.

Clean dirt, grease, oil, etc, from pump housing with diesel fuel or kerosene. Dry. Remove faulty pump of missing cylinder. Tie covers over all open lines.

Soak hands in fuel oil. This prevents corrosion starting from sweaty hands on fine, finished surfaces. Work on linoleum-covered bench. Place all parts

in bath of diesel fuel oil, acetone, carbon tetrachloride, etc, as removed. If more than one pump is being checked, keep all parts in separate, fuel-filled trays or cans. CAUTION: Remove sticking dirt with stiff bristle, NOT wire, brush.

When clean, check these pump parts with hand magnifying glass for: (1) minute lengthways scratches on plunger or barrel. If scratches are deep enough for fingertips to feel, replace barrel and plunger, Fig. 2. (2) Dull patches on bright surfaces. Remove by rubbing lightly with used crocus cloth that has been soaked in lube oil or finest lapping compound (jeweler's rouge) on piece of felt. (3) Tiny pits, rust, especially at plunger ends. Remove by lapping, Fig. 3. (4) Roughness at fuel-helix edges. If nicked, renew, Fig. 2.

These troubles are from dirty fuel, so check all filters. Corroded patches are from water in fuel (centrifuge or settle out), also from sweaty hands handling parts at previous overhaul.

Pitting, etc, can be from heavy-content-sulfur fuel, mixed with water. Resulting acid eats away metal at helix edges. Change to lower-sulfur-content fuel. Ask oil dealer for analyst's report.

When cleaned and polished, some plunger and barrel assemblies are tested

for vacuum by holding finger over open end of barrel (if top isn't closed). Place plunger in barrel and push. Stop pushing. Compressed air should then push plunger out. If no compression, replace pump unit.

If ports in barrel prevent this test, cover ports' top-openings with finger. With plunger in place, invert assembly. With finger removed from top, plunger should start sliding. Cover hole—plunger should stop, Fig. 4. If it keeps sliding, it's badly worn. Smear with lube oil, repeat test. If light lube oil seals plunger enough to stop it falling out, parts are worn but can be used.

Scoring on one side of plunger (on pump with delivery valve holder attached to pump barrel) usually means that holder was overtightened, barrel distorted. Renew parts.

Discoloration, scratches, tiny pits, call for lapping. Use fine compound (American Bosch No. BM 10007 or Carborundum Co's No. H-400). If lapping on surface plate, make sure plate is flat. Plate that is used for years dishes. Level plate by lapping against another plate, or face off in lathe. Plate-to-plate rub once a month keeps it level.

Early or late timing can cause engine to detonate. This means a loss of power. Look for worn pump mechanism, such

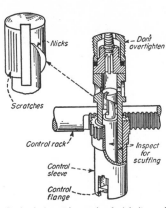

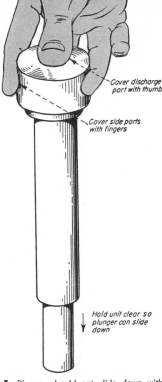

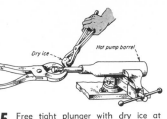

2 Look for nicks at the fuel helix and scuffing on the plunger, inside barrel

Cover discharge port with thumb

Cover side ports with fingers

Hold unit clear so plunger can slide down

5 Free tight plunger with dry ice at one end after barrel is heated in oil

6 Test for sliding fit by pushing on plunger while turning it to and fro

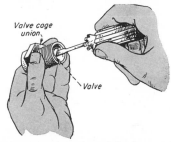

3 Lap plunger ends like this when tiny pits and rust start showing on end

4 Plunger should not slide down with ports covered by fingers like this

7 Lap in this valve with lapping compound while working with screwdriver

as cams. Also look for lost motion between pump and engine crankshaft from gear-train wear. Clean worn cams with crocus cloth or stone. Measure used cam for wear against pump drawings. Then make allowance to pump's stroke to compensate for loss of travel from cam wear.

Broken fuel-pump plunger-spring causes cylinder to misfire. If spring is weak, engine also misfires. Replace broken and weak springs. Touch nicked enamel on spring coils with enamel to prevent corrosion spreading. Don't dip springs in cleaning solution—it softens enamel, breaks off easier.

If no bench-testing equipment is available for calibrating pump, do this: Clamp pump in vise and connect some kind of drive mechanism to pump coupling or shaft. Connect pump's suction line to gravity fuel tank (can with strong flexible connection). Invert can, punch venting—filling holes. Protect hole with wad of screening, hang above pump.

Bleed all air from pump. With nozzle attached, place control rack in delivery position and start pump by turning wheel's crank at least 100 rpm. When all injectors spray fuel—stop. Place

glass cylinder of same size under each injector tip, Fig. 1.

Engine misfiring may be from leaks. Tighten fuel-line union nuts, nozzle bleeder screws, or delivery valve holder. Such leaks are easily found on clean engine. Wipe engine down daily. When tightening won't stop, repair leaks temporarily by chasing crossed threads with file. Then smear with jointing compound.

If misfiring starts, then stops, etc, cause can be plungers, which stick in barrels, breaking free. Run on kerosene lube-oil mixture for short period. Trouble is dirt or gummy deposits, varnishes sticking to sliding surfaces. Check all filter units. You might have to change fuels.

Sticking barrels may not be freed easily from plungers. Soak overnight in carbon tet, acetone or diesel fuel. If this fails, warm barrel in hot oil, place in vise and apply dry ice (with tongs) to plunger, Fig. 5, and pull.

When plunger is out, smear with tallow. Work in and out, turning to left, then right. Continue until plunger slides without pushing, Fig. 6. Pumps are lapped so finely that plunger left on bench in a hot shop (when barrel is in

fuel-oil) may not enter barrel because of expansion. Place both in fuel bath if this happens.

Governor troubles are also from sticky plungers. If speed drops when applying load, or engine races when load is removed, or engine "hunts," (speed rises-falls at same engine load), try control rack by hand (if that type). If you can't move by hand, plunger might be stuck in barrel. Free as above. Dirty rack teeth often act as if plunger is sticking.

Misfiring can also be from dirty, pitted or worn fuel delivery valves. Delivery valves must slide into their seats under own weight. If they don't slide, don't push. Clean with mutton tallow.

With all cylinders in place and no drops at nozzle ends, start pumping at 100 rpm for about two minutes. If fuel in each jar is within 5% pump is fairly well calibrated.

Test two or three times; repeat after every pump adjustment. Try control rack in several positions; if possible, include STOP. Test can also be made with cylinders under injectors while cranking the engine, after first priming system. Test can be made on engines with individual pump near each cylinder.

Loading Effects on

UNDERLOADING

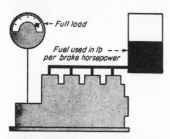

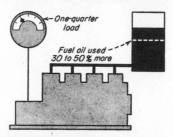

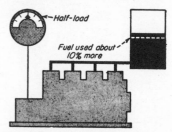

DIESEL ENGINE is usually built for a certain brake-horsepower or load. It does best all-around work when run full load or not less than two-thirds load

ONE-QUARTER-LOAD: Fuel used in lb per brake horsepower hour is much more than at full load. In some cases it's 30 to 50 percent more than at full load

HALF-LOAD: Fuel used in lb per brake hp per hour is only little more than full load consumption. In most cases it's about ten percent more than full load

OVERLOADING

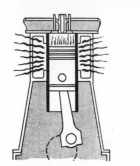

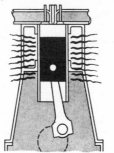

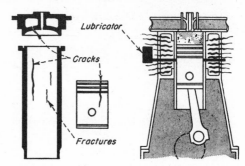

CYLINDER WALLS AND PISTONS become gradually over-heated as engine runs overloaded. That means heavy wear and tear on various engine parts, with possible early engine failure. Watch the cylinder lubricators if engine is overloaded

CYLINDER LINERS, working overloaded are liable to fractures and even cracks. Some cylinder heads crack under overload when head bolts are tightened too much. Cylinder lubrication fails. This will cause you early trouble and expensive repairs

OVERLOAD EFFECTS AND CURES

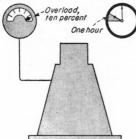

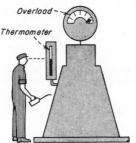

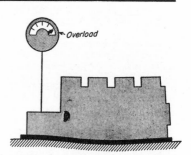

TOTAL TIME engine is overloaded can usually be controlled. Keep down to one hour and not over ten-percent overloaded. During this time, keep checking engine carefully. Pay especially close attention to the engine temperatures and to lubrication. Underloaded engine's only fault is high fuel use. Engine still runs cool

FOUNDATION SETTLING causes slight misalignment with time. It's exaggerted here, but overloaded engine heats up from this condition, causing more wear

Diesel Engines

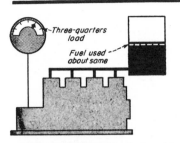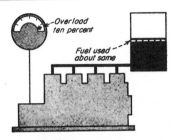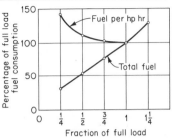

THREE-QUARTERS-LOAD: Fuel used in lb per brake hp per hour is about same as full-load consumption. In most cases the difference is small enough to ignore

AT TEN PERCENT overload, fuel used in lb per brake hp-hr is only little more than at full-load and about same as at three-quarters load. But engine wears fast

TOTAL FUEL USED at various loads increases gradually with the load. But fuel used in lb per brake horsepower hour goes down as load on engine increases

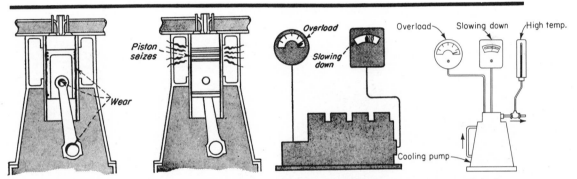

EXCESSIVE FRICTION and wear between piston and cylinders in overloaded engine is certain if overload continues for long. This causes rings to stick and pistons finally seize. To avoid, try to balance load with standby engines during peak loads

REDUCTION IN SPEED often occurs when engine's badly overloaded. If water circulating pump is attached directly to engine's shaft, cooling is reduced just when it's needed most. Then engine's temperatures shoot up, causing plenty trouble

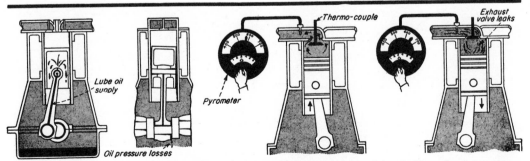

LUBE-OIL SYSTEMS get less efficient with time and poor maintenance. Out-of-line engines have lower lube-oil pressures. Overloading knocks out engine

UNEQUALLY LOADED CYLINDERS are found by checking exhaust temperatures from each cylinder with pyrometer. Highest temperature shows cylinder carrying most load. Adjust fuel so temperatures are within ten degrees. High exhaust temperature can be from leaky exhaust valve. Over- and underloading cause black smoke

Know your diesel cooling-

Types of closed systems

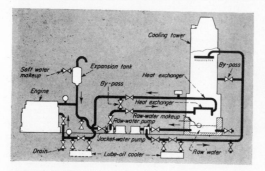

COOLING TOWER of induced-draft type with heat exchanger

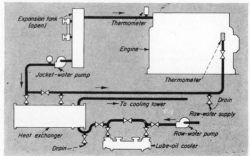

SHELL AND TUBE heat exchanger located inside building

Putting temperature where you need it

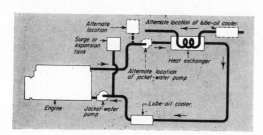

MODERATE TEMPERATURES are held by hookup of this kind

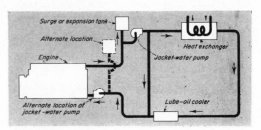

LIGHT TEMPERATURE to engine, lower temperature to lube oil

Emergency supply

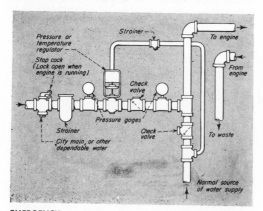

EMERGENCY water supply takes over when other system fails

What you must know about diesel

► AMOUNT OF HEAT absorbed by cooling water varies with type of engine and design of cylinders, exhaust manifold, pistons, lube-oil system and anything cooled by the system. Most engines are designed for minimum temperature difference between cooling-water inlet and outlet (from 15 to 20 F). This temperature difference is ideal.

Don't overlook seasonal temperature variations that may freeze your cooling water and congeal the lube oil. Hook up system so all cooling water flows through engine continuously. Never throttle flow to raise outlet temperature. Use only soft or treated water for uniform heat transfer and high efficiency of cooling through engine jackets. A value of about 8 pH is best to prevent deposits in water spaces. Keep system vented so water spaces are always completely full.

Selecting types of cooling systems shown depends on many factors. One engine maker may prefer the circulating pump with suction from engine, another insists on installing discharge to engine. Good rule to remember: Always install heat exchangers with outlet at top, unless they are vented.

water systems

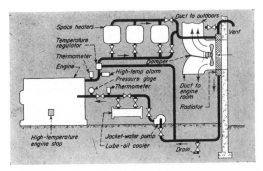

RADIATOR uses extracted heat for heating building spaces

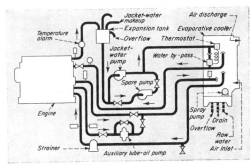

EVAPORATIVE COOLER has raw-water spray pump in its base

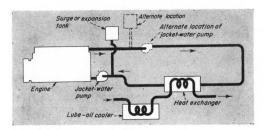

HIGH ENGINE TEMP from lube-oil cooler in separate circuit

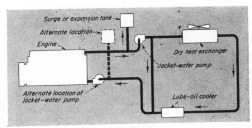

HIGHER ENGINE TEMP from some water by-passing exchanger

cooling systems

Open system has water flowing through the engine jackets, then wasted or recirculated through a cooling tower. *Remember:* Continuous evaporation in cooling tower or spray pond increases hardness and impurities. This accumulation doesn't remain in solution, but deposits on engine jackets and retards heat transfer. This is the main trouble with open system for diesel engines.

Closed system is best, may have one or two complete circulating systems. *Example:* If a radiator is used with air as cooling medium, use single system with water flowing through engine jackets, then through radiator. Jacket-water outlet temperature and rate of flow water must be specified by engine builder. Keep in mind that raw-water temperatures, rate of flow and heat load govern the type of cooling apparatus you should use.

Cooling tower may be one of two types—atmospheric or induced-draft. Atmospheric type depends on natural draft or air movement. Induced-draft type has a fan. In some places it's best to install heat exchanger inside the building.

If so, use a shell and tube type exchanger. Then circulate raw water over the tower and through the heat exchanger.

Emergency water connection, sketch, bottom of page 210, assures flow of jacket water if regular system fails. Use this hookup where city water or other source is available at all times. Make sure this water enters system at point where it will be tempered before going to engine jackets.

Alarms warn you when temperatures or pressures go haywire. Automatic shutdown devices stop engine at set temperature or pressure. Use thermostatic controls to automatically adjust flow of jacket or raw water. You can connect them so engine always operates with a fixed outlet temperature from engine or lube-oil cooler, without reducing volume of flow through your engine.

If possible, bring cooling water to normal operating temperature before starting engine. Do this by using heat from another source or by-passing cooling tower, reactor or evaporative cooler until normal temperature is reached.

Courtesy, Diesel Engine Manufacturers Association.

Your Liners Can Crack

SLOW CRACKS

LINERS USUALLY CRACK a few inches from top—run circumferentially around

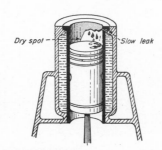

Dry spot — — Slow leak

SLOW CRACK allows water to leak into cylinder. Water washes off cylinder oil

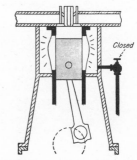

Closed

DRY SPOTS rust when engines shut down. Then liner heats fast, seizing the piston

HOW TO PREVENT Continued

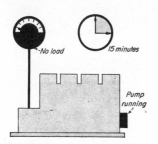

No load   15 minutes

Pump running

RUN ENGINE for 15 minutes with no load if the cooling pump is on the engine

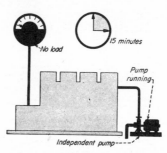

No load   15 minutes

Pump running

Independent pump---

ALWAYS COOL SLOWLY by unloading for 15 min; then run pump another 15 min

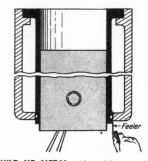

←Feeler

BUILD UP METAL with welding torch if no metal-to-metal contact at the bottom

CAUSES

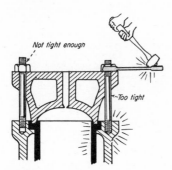

Not tight enough

←Too tight

UNEVEN TIGHTENING or too much tightening of cylinder head bolts is bad

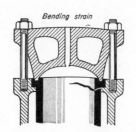

Bending strain

TOO TIGHT bolts cause bending strain on cylinder head—transmitted to liner

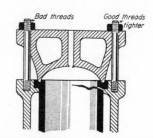

Bad threads   Good threads tighter

GOOD THREADS help nut tighten more than bad ones—torque wrench reads same

Unless You Stop Them

HOW TO PREVENT——————————————————

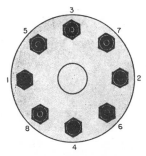

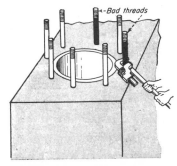

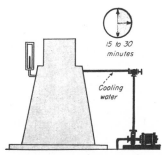

TIGHTEN NUTS in sequence shown. Take up slack on all, then drive them tight

REMOVE BOLTS and nuts with damaged or worn threads and replace with new

RUN COOLING WATER 15 to 30 minutes after shutting down to cool the engine

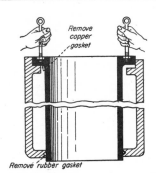

PRUSSIAN BLUE under collar, and turn the liner slightly to mark the high spots

SCRAPE HIGH SPOTS shown by prussian blue until the bearing surfaces are level

GRIND BEARING smooth with cast-iron plate. Use new copper and rubber gasket

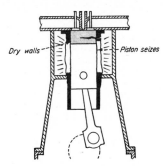

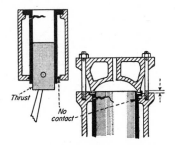

CUTTING OFF cooling water soon as the engines shut down overheats the liners

PISTON RESTS in same spot each shutdown with water off, overheats the liner

NO CONTACT between liner and block at the top or bottom from warped casting

BEARING FAILURES TELL OPERATOR ABOUT ENGINE'S CONDITION

1 Tiny particles of cast iron are embedded in the bearing surface here

2 Foreign matter caked to bearing-shell back caused high pressure and heat

3 High pressure from foreign matter back of thin bearing caused failure

4 Fatigue and corrosion teamed up in high-speed high-pressure bearing

5 Bond failure is from faulty tinning or no tinning. Lining breaks away

6 Ridge around crankpin or journal may spoil bearing if not stoned down

Your Bearings Talk . . .

▶ YOUR DIESEL BEARINGS can tell you a lot about the engine's condition and operation, if you understand their 'language'.

To interpret the bearings' story, you should know lining materials, their characteristics, main causes of bearing failure and how to identify them, bearing-design features and lubrication.

Various lining materials are used in heavily loaded bearings for resistance to fatigue corrosion and for greater life.

Babbitt Linings. Babbitt makes fine bearing linings within its load limitations. Its frictional coefficient is low; melting point is low and it's soft and easily worked; it bonds well; and conforms to shaft. Also, foreign particles embed into it. Big advantage is that it's less abrasive and works well with unhardened crankshafts.

WHY BEARINGS FAIL

Dirt and other abrasive materials cause most trouble. They scratch and score surfaces.

Particles are often embedded in the babbitt. With a harder material, these particles might remain on the surface to score the crankshaft. If particles removed by light scraping are metallic and ferrous (shown by a magnet), look for excessive engine-parts wear, Fig. 1, such as liners, pistons and rings. If dirt, check air filters.

Mechanical conditions cause failures also. Bearing misalignment is common. Then crankshaft is distorted. Such distortion deforms the bearing surfaces and causes localized high pressure areas so bearings get hot. Bearing-shell distortion is from foreign matter on shell's back, Fig 2. This causes high pressures and temperatures, but only in bearing effected, Fig. 3.

Modern engines with higher cylinder pressures and higher speeds have two common bearing failures. These are fatigue and corrosion of lining material.

Cracks become deeper, until they alter direction and break parallel to the bearing backs just above the bond layer, Fig. 4. Then as load increases on surrounding areas and oil fails, bearing burns up completely.

OTHER FAILURES

Corrosion. Corrosion is from acid components of combustion, together with condensed water vapor, collecting in lube oil. Build-up of harmful acids is speeded by excessive oil temperatures. Acids etch bearing surfaces. Then area is reduced, until in extreme

* Photos, Federal-Mogul Corp.

cases results are similar to fatigue. At times it's hard to tell if failure is from fatigue or corrosion, and not only because appearance is often similar. But both causes may have caused final failure.

Most babbitts resist corrosion well, but resistance varies with different types. With tin-based babbitts, for example, the high oil temperatures promote corrosion, cause fatigue failure before corrosion is serious. Lead-based babbitts with over five percent tin or one-half percent silver aren't effected much by corrosion. The SAE formulas and a number of proprietary metals resist corrosion well.

Bond Failure. This has been largely lessened by modern manufacturing techniques. The bond layer or "tinning" is very thin and inclined to be brittle. But metal loss occurs if entire area isn't properly bonded. This condition is easy to tell because the roughened portion of bearing back is visible, either with no tinning or faulty tinning apparent, Fig. 5.

Now, let's roll out the bearings and examine all of them. By checking a complete set, it's often possible to tell if a condition is general or localized. A look at the engine log and any recorded abnormal events may help.

Fatigue. Let's say evidence of a bearing fatigue condition is found in all bearings. Cause may be high peak cylinder pressures, possibly from long periods of overload (found from engine's log) or bad adjustment, (such as early timing).

Fatigue of only one or two bearings might mean unequal fuel delivery to cylinders. Then overworked bearings fail. Check this with pyrometer readings. If readings are OK, check the instruments and gages.

Bearings failing near flywheel are often from special causes. Consider method of power takeoff. Weight attached to already overhanging flywheel, radial loading from belting or chains, or a poorly aligned outboard bearing is straw that breaks the camel's back.

Obviously, all engine parts are interrelated and so failure's 'cause and effect' will be related. With tips given here, your trouble shooting success depends on you.

TROUBLE SHOOTING

Before replacing bearing shells, note all faults and make repairs. For example, suppose there's evidence of misalignment and this is confirmed by crankshaft strain-gauge readings (this chapter). Condition may be from sprung shaft or poor bearing alignment. Remove suspected shaft, place between lathe centers or on roller supports at ends. Then check with a dial indicator while turning.

Forged shafts can be straightened but it's a tough job and might need a light regrind for final truing. If bearing mountings aren't right, maybe bedplate wasn't originally level. Or maybe it settled from cracked foundation. Re-leveling bedplate should take out sags or warps.

If fault lies in bearing supports, problem isn't so easy. But this is uncommon. Two courses are open, the simplest is using oversize or semi-finished bearings that are align bored. This has disadvantage of making the bearings non-standard. Next best is to remove some metal from cap faces, replace them and align-bore bearing supports.

This takes good equipment for boring and expert workmanship. Remove only enough metal to clean it up. And metal should come from bore's top and bottom, leaving parting lines untouched. If shaft's axis is slightly changed, check drive gears for meshing.

Before installing bearings, check the crankshaft. Mike journals and crankpins for being out-of-round or taper. If these conditions are found in excess of maker's limits, replace or regrind and use oversize bearing shells.

Another condition, especially with softer shafts, is wear on each side of oil hole. Many bearings have fairly long centrally located oil grooves lining up with oil hole. Area around hole is more heavily loaded, so expect greater wear. This might not effect old bearing since the two surfaces have conformed to each other. But such a ridge may rapidly destroy a new bearing shell, Fig. 6. Remove the ridge by carefully stoning journal.

REPLACING SHELL

Outer diameters of "thick wall" bearings are ground. This insures concentricity and close fits with saddle and caps, both for support and for maximum heat conductivity. Good contact is further insured by providing "crush". The shell ends should extend beyond the faces of saddles and caps, so there is an interference fit. The pressure at the parting line forces the bearing into close contact with the supporting surfaces. The resulting friction also holds the bearing against rotation. Tangs or dowels are also used.

See that this crush exists and that there are no burrs or high spots to cause excessive interference at the parting line. There's usually a relief at this point and on the flanges' faces. It takes care of any swelling when compressed and acts as an oil reservoir.

You can usually tell shell looseness by carbon deposits on their backs. Such looseness may result in bearing turning. Then bearing will fail from oil holes misaligning.

ASSEMBLING

Take great care during installation to exclude dirt. A sizeable particle behind bearing's back will raise lining. That reduces clearances and creates a highly loaded area. Fatigue and corrosion follow rapidly, as well as actual melting and metal displacement. Dirt on bearing surfaces causes immediate scoring and abrasion.

Make sure bearings are not reversed and that all tangs, dowels, etc are properly located. Then there won't be distortion. When shims separate bearing halves, install them so that they prevent excessive oil escape without galling the shaft.

Clearances. Clearances should never be less than recommended by the manufacturer. At times they must be somewhat larger, depending on mechanical conditions. Journal's surface condition is important. Success of present close tolerances are partly due to improved shaft finishes.

Scratched or roughened surfaces have tiny sharp crests and valleys. Crests protrude through oil film and thus reduce actual effective oil-film thickness. Then more clearance is needed to form a thicker film.

Carefully stoning or polishing journals with fine emery and crocus cloth helps bring good finish. Be very careful to remove all abrasive particles afterwards.

Some large engine bearings can't be taken up for wear. This is justified by long bearing life. In one case bearing lasted 30,000 hr before clearances reached their limit. By then fatigue failure comes in and replacement is best solution anyway.

If bearings can be taken up, restore clearances close to minimum practical recommendations. Clearance of 0.003 in. is equivalent to same amount of wear. Then bearings will have to be overhauled sooner.

When wear approaches halfway mark, decide whether to continue shells in service. Bearings' overall condition, cost of replacement and expected remaining hours of operation must balance cost of early replacement job and engine down time.

With this information, you're all set to do a little trouble-shooting.

DIRTY?_____ CHECK STRAINERS AND WEAR_____

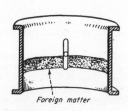

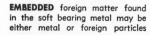

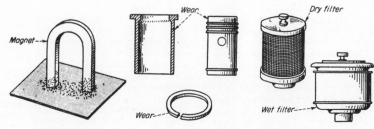

EMBEDDED foreign matter found in the soft bearing metal may be either metal or foreign particles

MAGNET tells if matter is metallic. If so, check liners, pistons, rings for excessive wear. If not metallic, check·air filters for foreign particles being drawn into cylinders. Grit in bearing metal scores shafts, increases bearing clearance, causes overheating

CORRODED?_____ STOP OIL CONTAMINATION_____

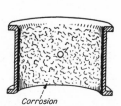

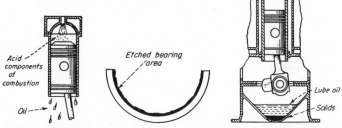

FATIGUE and corrosion are often difficult to tell apart. Both are very hard on the bearing lining

ACID COMPONENTS of combustion and condensed water vapor collect in lube oil, cause corrosion. Affect all bearings alike. Etched bearing surface reduces effective bearing area, causes failure from high unit pressure if not corrected soon

LOOSE?_____ HAVE GOOD 'CRUSH' FIT_____

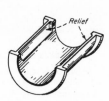

CARBON deposit on back of shell advertises that the bearing shell is loose and in need of correction

CHECK by placing straightedge on bearing, insert feelers at 1, 2, 3. Shell must have "crush" fit at these points. Loose shell may turn, cut off oil supply. Relief at parting edges cares for swelling; they give oil storage for good operation

Diesel Bearing Troubles

MISALIGNED?_____ MAKE STRAIN GAGE TEST_____

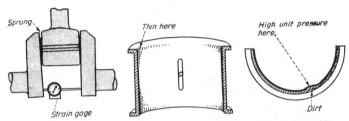

MECHANICAL cause is often from misalignment. That deforms bearings, causes high-pressure areas

CRANK BEARINGS' misalignment is found with strain gage (see pp 206–207). Distorted crankshaft may break if not found in time. Bearing-shell distortion is also from foreign particles under thin bearing metal; then heating affects bearing

BOND FAILED?_____ TIN AND BABBITT CORRECTLY_____

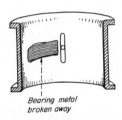

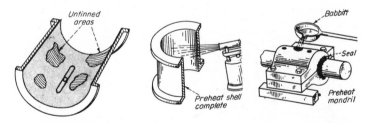

BOND FAILURE causes the thin pieces of the soft bearing metal to break away from the bearing shell

UNTINNED shell areas before babbitting are chief bond-failure cause. Improved methods make this fault uncommon in modern bearings. When tinning shell, preheat carefully and tin completely. Also preheat mandril and shell to right temperature

OIL PRESSURE LOW? TAKE UP ON CLEARANCES_____

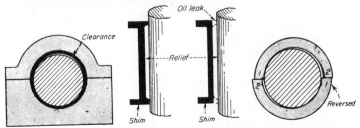

LOW OIL pressure will be hard on the bearings. Keep the pressures that the manufacturer recommends

EXCESSIVE bearing clearance oozes badly needed oil, lowers pressure, makes engine noisy. Heavily loaded bearings may even fail. Shims must seal bearing halves at relief ends, but not touch to scrape oil off shaft. Reversed shells may leak oil

Next Shutdown,

DAMAGES

CRANKCASE wreck ($4771 here) is most common damage when bolts fail

PISTON SKIRT AND ROD break from smack by crank when bolts let go

CRANKSHAFT jamming loose parts when bolts fail can cause this

CAUSES

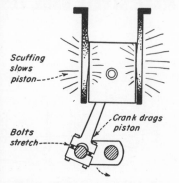

Scuffing slows piston

Crank drags piston

Bolts stretch

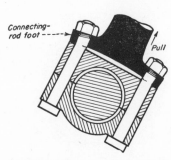

Connecting-rod foot

Pull

PISTON SEIZING or scuffing stretches crankpin bolts in a multi-cylinder engine as crank drags piston along

CONNECTING-ROD foot flexing bends bolts. This also happens from piston scuffing. Check foot with straightedge

PREVENTION

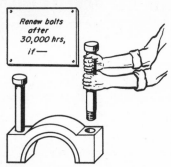

Renew bolts after 30,000 hrs, if —

LOG BOOK

RENEW BOLTS after 30,000 hours of engine operation if engine has run smoothly with no scuffing or seizing

CHECK LOG for engine's history since last bolt change to see if conditions existed that could cause bolt elongation

ELONGATION LIMIT

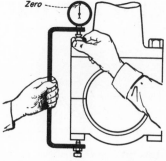

Zero

DON'T EXCEED bolt's elongation limit from engine instruction book. Reaching limit means bolt reached yield point

BEFORE TIGHTENING new bolts with wrench, use your stretch gage like this and set the dial so it registers zero

Check Crankpin Bolts

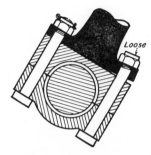

NUT COMING LOOSE strains other bolt. Tap all nuts every few months with light hammer. Sound shows loose nut

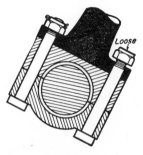

NUT NOT TIGHTENED is next. Or one nut pulled up tighter. Both bolts must have exact strain to avoid trouble

STRAINING bolt from over-tightening is common. Torque wrench is NOT answer as thread friction varies on both nuts

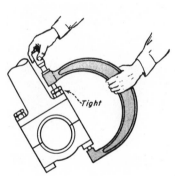

BEST CHECK for bolt elongation or stretch is to mike the bolt **before** loosening nuts, with bolts strained

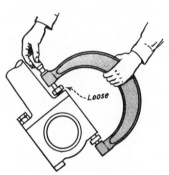

THEN MIKE the bolt **after** loosening nuts. Compare measurement with that of tight bolt for amount of stretch

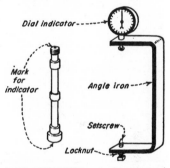

IF NO LARGE MIKE, make 'stretch gage' from strap iron and dial indicator. Always read at same spots on crank bolt

TIGHTEN BOTH BOLTS until stretch indicated on dial is exactly same for both. Record bolt's stretch in the log

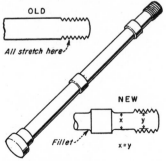

CHROME MOLYBDENUM steel SAE-41-30, ASA 3125 and 3130, with collars, stretch full bolt length, relieve threads

WHITEWASH TEST shows hidden crack. Clean the bolt, rub with kerosene, wipe dry, then whitewash. Crack will show

Deflection-Test Your

DAMAGE

SHAFT BROKE at web from fatigue as a high bearing bent shaft with each rev.

HOW TO TEST

1 Strain gage set measures misalignment without taking crankshaft out

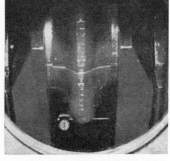

2 Place gage between webs. Spring tension against points holds it in place

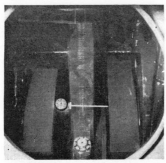

6 Set dial to zero. This gage balances itself so face is always to operator

7 Jack the crank to quarter position. Needle hasn't moved from zero mark

8 Jack to top, then quarter position. Needle stays on zero; this crank's OK

12 Lift assembly a few thousandths by screwjack, enough to remove shim

13 Remove 0.015-in. shim. Then take away jack and tighten down bolts

14 Now gage readings in four spots on generator-end webs read zero

Crankshaft Once a Year

3 Spin balanced gage around a few times after setting the dial to zero

4 0.003 in. reading shows that points have imbedded into metal this much

5 Jack the crank over so that it is near the bottom dead-center position

9 Gage on generator end crank shows top dc out 0.003 in. from bottom dc

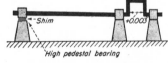
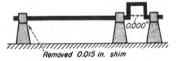

10 High pedestal bearing is answer. Removing 0.015 in. should correct

11 This pedestal bearing wore less than crank bearing. Loosen nuts

15 Check air gap all around on generator with feelers. Adjust if uneven

HOOKE'S FORMULA

$$\text{Stress} = \frac{43{,}500 \times d \times t}{L^3}$$

43,500 = a constant
d = distortion in mills (not a decimal)
t = crankweb thickness, inches
L = ½ engine stroke in inches

If greatest difference between any two readings is 0.003, the web thickness 3 in., and stroke 14 in. we get: (43,500×3×3 ÷ (7×7) = 7989 psi. The answer is strain in psi on webs.

16 Calculate stress on webs from distortion to see if limit is exceeded

17 If stress exceeds 20,000 psi, whitewash test to find cracks like this

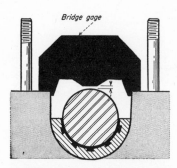

1 Bridge gage with feelers tells of wear in lower half of crank bearing

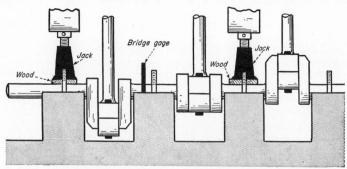

2 Jack-testing main bearings forces crankshaft down against lower bearing halves. Then bridge-gage readings show any spring in crankshaft when engine is loaded

4 Easy Ways To

Don't wait for that crankshaft to break

▶ WE WERE YAPPING about crankshaft failures during a bull session. The boys agreed that any fracture of journal, web or pin could be blamed on the manufacturer. Because a "blanket" explanation of "fatigue failure," would transfer blame from operating engineer to engine builder.

"And why should a manufacturer be blamed for what's often the operating engineer's fault?" I asked.

"Look here, Jones," yelled the super, "in nine cases out of ten the bearing of any machine with a broken crankshaft is well lubricated. The bearing metal's in good condition, the engine runs at the right speed and not overloaded. That's my experience and I've worked with diesels a long time."

That's a general statement, so I asked how he measured these apparently perfect main-bearing surfaces for wear. He explained the usual method of removing the shells and measuring the metal thickness on precision bearings. On older babbitt-metal bearings, he took leads.

"Then how do you know the crankshaft is resting on bottom halves of all main bearings when engine's running. And how can you tell if they all take their share of the load?" I asked.

"If an engine's well balanced, as shown by indicator cards and temper-atures, then wear must be equal on all bearings. That shows the shaft is resting on bottom halves of main bearings at all times," he came back.

I asked if he thought the crankshaft bridge gage measurements wouldn't tell exactly what was happening to these bearings. He said he hadn't time to bother with 'em. So I stopped arguing. It was time to explain a few things.

Using Bridge Gage. Test bearings for wear with a bridge gage each time main bearing caps are removed.

First remove all main-bearing caps and measure each with engine manufacturer's bridge gage, Fig. 1. If no gage, make one and calibrate it first time engine is opened. Find wear by inserting feeler gages between gage anvil and journal's top. Repeat at each main bearing. This checks bearing wear under crankshaft.

Next roll out lower half of No. 2 bearing. Place jacks over No. 1 and No. 3 bearings with jacks' tops against "A" frame or crankcase webbing. Place wood under jack, Fig. 2. Force crankshaft against No. 1 and No. 3 bearings. When jacks are tight, measure distance between gage and journal with the bridge gage at No. 2 journal.

If reading is greater than first one, No. 2 bearing is higher than neighbors. Now remove bottom halves of

No. 1 and 3. Replace lower half of No. 2. Measure No. 1 and 3. If reading is lower than before, No. 2 is high and adjoining bearings don't carry their share of load. If this wear is great, it causes shaft to whip. That hammers and distorts bearing metal, bends the shaft, which finally fractures. Repeat this process at all main bearings.

Finding Crack. If difference is over a few thousandths, inspect crankshaft closely with a magnifying glass for fracture signs. Or sprinkle magnetic powder on shaft.

If no powder, clean suspected fracture areas (at fillets where crankpin meets web and web journals) with kerosene. Clean as if trying to force kerosene into metal.

Dry rub the surface with rags. Then chalk entire area with school chalk. Kerosene seeps out of hidden fracture and shows up crack. To make sure, repeat process. If chalk is again stained with kerosene, check with a strong magnifying glass.

Another way is to make a light paste of powdered chalk and alcohol. Alcohol evaporates fast, leaving a film of chalk. An oil-filled crack is outlined in dry-chalked surface.

Warp Test. If no cracks are shown, give crankshaft a deflection or strain

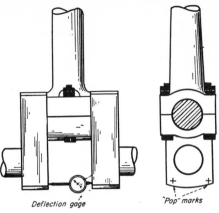

3 Strain-gage shows if there's deflection in crank webs at each quarter position when shaft is jacked slowly around

4 Bend test is made by attaching dial gage to main bearing. Remove the main-bearing caps, then revolve shaft slowly

Check Crankshafts!

THESE TESTS HOLD GOOD FOR ALL ENGINES

1 BRIDGE-GAGE TEST for wear of lower main bearing.

2 JACK TEST to find spring in crankshaft when loaded.

3 DEFLECTION TEST to see if crankshaft is sprung.

4 BEND TEST to find if crankshaft has whip action.

test. This tells if crankshaft should be scrapped. It also shows stresses crankshaft has been working under. A deflection gage looks like inside micrometer with a dial gage.

Place gage between crank webs and rest against "pop" marks. If no pop marks, make them exactly opposite each other and near web's edge, Fig. 3.

Tighten all bearing caps until engine can just be barred over after removing a few shims. If shell-type bearings, place shim metal between cap and journal before tightening cap nuts.

Gage in position holds itself between webs. Don't touch after it's in place. Take readings every 90 degrees. Next take difference between any two readings and apply Hooks' formula. That tells if crankshaft is safe.

Hook's Formula. The formula is 43,-500 multiplied by the number of thousandths of in. maximum difference between any two readings times thickness of crank web and divided by the square of half length of stroke.

So if greatest difference between any two readings is 0.003 in., the web thickness 2 in., and the stroke 10 in. we get: $(43,500 \times 3 \times 2) \div (5 \times 5) =$ 10,440 psi.

The answer is strain in psi exerted on webs. The maximum allowable strain is 20,000 psi on any crankshaft.

Using Micrometer. If no crankshaft deflection gage, take measurements with inside micrometer. Set micrometer so you just feel cheeks at first reading.

At second reading, take a regular "feel" and note whether you have to withdraw or extend spindle to get same setting. These readings are approximate, but give some idea of crankshaft's condition.

Suppose after testing all throws like this, only trouble spot seems to be in crank throw next to coupling. Cause may be not in bearing but from faulty alignment between diesel and whatever it's coupled to. Then realign diesel to driven machine. Play safe and take deflection readings at major overhauls.

Bend Test. Remove bearing caps and attach a dial gage to a stud. Have gage's "leg" against top of journal, Fig. 4. Slowly rotate crankshaft. Any whip or bend shows by gage-needle movement.

If crankshaft fractures at fillet between web and pin, examine area closely. If there's a further crack at right angles to break, it shows trouble was caused by torsional vibration.

Remember that cracks due to fatigue only, take a long time to develop. If a crankshaft and bearings are checked regularly, condition causing failures can be prevented.

To make a quick main-bearing test, stop engine, whip off crankcase doors quickly and feel each bearing. Bearing cooler than others is not carrying its load. Inspect it for excessive wear at first chance.

Cause of crankshaft failure is usually journal-bearing failure from operating engine at critical speeds, overloading or overspeeding, fatigue failure or operating engine with torsional vibration damper (if fitted) out of commission (usually loose on the shaft).

Temporary Repairs. Cracked webs can be repaired in an emergency, usually in isolated parts of the world or out at sea. A ½-in. steel strap is placed around the web. It is fastened to the web by cap screws and should be as wide as the web.

Cracks in solid crankpins (those not drilled with oil passages) can be checked by drilling out the pin. This is not an easy job. The crankpin is drilled to about two-fifths of it's diameter. Then a steel pin is driven in to strengthen it. This is at best a "kill-or-cure method".

To avoid having to make these repairs, why not check your crankshaft at the next overhaul. Then you'll know.

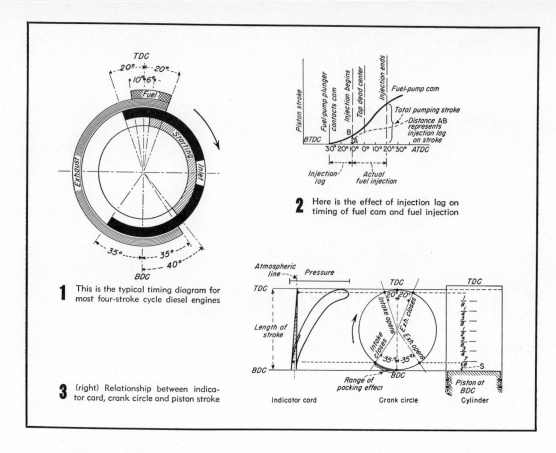

1 This is the typical timing diagram for most four-stroke cycle diesel engines

2 Here is the effect of injection lag on timing of fuel cam and fuel injection

3 (right) Relationship between indicator card, crank circle and piston stroke

Right Timing Saves Fuel

Packing effect, ignition and injection lag, phases of exhaust, and valve overlap of diesels

▶ Good DIESEL ENGINE performance hinges on its timing. As an operating engineer, you should not only know the timing diagram of your engine, but why it's so timed. All this gives you a better background for trouble shooting. Diesel engine timing is varied to suit the engine design. It depends on revolutions per minute, piston speed,

connecting rod length, manifold and valve design and other factors. In this article, we will consider only the timing diagram of a typical medium-speed, 4-stroke-cycle diesel engine.

Our timing diagram in Fig. 1 pictures these events:

1. Inlet valve opens at 20 deg before top dead center (TDC). Closes at 35 deg after bottom dead center (BDC).

2. Compression starts at 35 deg after bottom dead center. Ends at top dead center.

3. Fuel injection starts at 10 deg before top dead center. Ends at 20 deg after top dead center.

4. Expansion starts at 20 deg after top dead center. Ends at 35 deg before bottom dead center.

5. Exhaust valve opens at 35 deg before bottom dead center. Closes at 20 deg after top dead center.

Packing Effect. On the first downward piston movement, with the inlet valve open, air rushes into the cylinder to fill the partial vacuum caused by the downward moving piston. When the piston reaches bottom dead center, it would seem the inlet valve should be closed. That's because an upward movement of the piston would tend to force air from the cylinder. However, according to the timing diagram, the inlet valve remains open until 35 deg after bottom dead center.

This is because the inertia of the incoming air keeps rushing into the cylinder even though the piston is now on the upstroke. This is called the *packing effect*. When finally the piston reaches a point on the upstroke where any further upward movement would back the air out the inlet valve, we time the inlet valve to close.

By taking advantage of this *packing effect,* we increase the volumetric efficiency. This is simply the ratio of the volume of fresh charge taken in during the suction stroke to the full piston displacement. We also increase the potential horsepower of the engine, since we have more air available to burn additional fuel. Because of a greater volume of air at the beginning of compression, our compression ratio can be so much less in arriving at a normal final compression pressure and temperature, Fig. 4.

Ignition Lag. We know that fuel must be injected into the cylinder before top dead center, because of the time required to heat the fuel to its auto-ignition temperature. This time is known as *ignition lag.* The amount of this advance of fuel injection varies with: (1) speed of the engine, (2) quality of the fuel, (3) compression ratio, (4) degree of atomization and (5) other design features.

Injection Lag. But there's another factor. It's *injection lag.* On our timing diagram fuel is actually injected into the cylinder 10 deg before top dead center. This is the *arc of injection.*

By noting the position of the fuel cam at the point where the pump plunger is starting to pump fuel, with all bypass ports closed, we find the cam timed about 30 deg before top dead center, yet fuel injection does not occur until 10 deg before top dead center. The difference between these crank angles is 20 deg, and is known as injection lag, Fig. 2.

Injection lag is caused by:

1. Expansion of fuel-oil discharge lines under high pressure, resulting in

the need for additional volume of fuel to be displaced.

2. Compressibility of the fuel. When fuel is near 3000 psi, it is compressed about 1% of its volume.

3. Leakage past the fuel-oil plunger.

It is more appropriate to express the injection lag as "percent of the pumping stroke." Fig. 2 shows how great the injection lag can be in a pumping stroke. It's often as high as 20%. That's why it's mighty important to position the fuel cam correctly on the cam shaft.

Phases Of Exhaust. We're now at the final phase of the 4-stroke cycle. At the working stroke's end, the exhaust valve opens. The opening of this exhaust valve is usually set to open earlier on high-speed engines. From this we see that timing the exhaust valve determines the length of the working stroke. It's desirable to have as long a working stroke as possible. That's to get the most work from the expanding gas in the cylinder. Yet, the exhaust valve is opened 35 deg before bottom dead center (BBDC), even though a pressure as high as 50 psi is in the cylinder just before the exhaust valve opens.

This timing is justified because at 35 deg before bottom dead center the piston is practically at bottom dead center. This is shown by the distance *S* in Fig 3, and represents less than ⅛th of the stroke. Therefore, little power has been lost. Also, the useful push on the crank is small here.

By releasing the gas before bottom dead center we reduce the back pressure at the beginning of the return movement of the piston. In this way, the exhaust of the spent gas may be divided into two phases.

First phase is due to the difference in

pressure between the cylinder and the outside atmosphere, starting at the point of release and ending when the piston has traveled through bottom dead center and back to about the same point. In this brief period over half the exhaust gas escapes.

This early exhaust opening reduces the pressure in the cylinder and therefore reduces the pumping loss that would otherwise occur in the second phase of the exhaust. Second phase makes up about 90% of the exhaust stroke and results from displacing the gas by the upward moving piston.

By permitting the exhaust gas to escape in these phases, the gas velocities are reduced and their flow resistance lessened in passing through the exhaust manifold and piping. Back pressure during exhaust in the engine is usually between ½ to 1 psi. If the average mean indicated pressure of an engine is 100 psi, then each psi back pressure means a decrease of 1% of the mean pressure and power of the engine.

Overlap. We notice on our timing diagram that the exhaust valve closes at 20 deg after top dead center (ATDC). Before it closed, the intake valve opened at 20 deg before top dead center (BTDC). There is then a period where both the intake and exhaust valves are open at the same time. This is often termed the *scavenging effect* of a 4-stroke-cycle diesel engine.

In this scavenging effect the velocity of the escaping gas to the exhaust creates a partial vacuum in the cylinder at the end of the stroke, which permits the entry of fresh air through the now open intake valve. This scavenges the remaining burned gas.

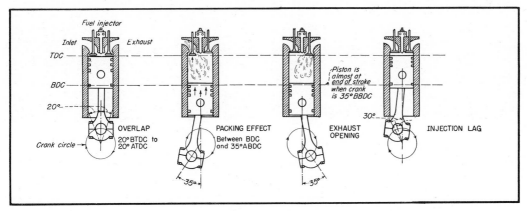

4 These diagrams show the piston and crank relations during overlap, the period when both inlet and exhaust valves are open; packing effect, inertia of the air rushing into the cylinder; exhaust-valve opening and start of injection lag

Oscilloscope spots

How can piston-ring blowby and cylinder scoring be located while machinery is operating? Our solution might prevent unexpected breakdown in remote areas.

We assembled an instrument package and bought an ultrasonic translator detector to inspect valve defects on compressors. Because the unit has recorder and earphone output jacks, we married the unit with an engine/compressor analyzer. Providing a direct readout of horsepower, capacity and power loss, the analyzer has a rotary motion transducer, a power and logic circuitry unit, and

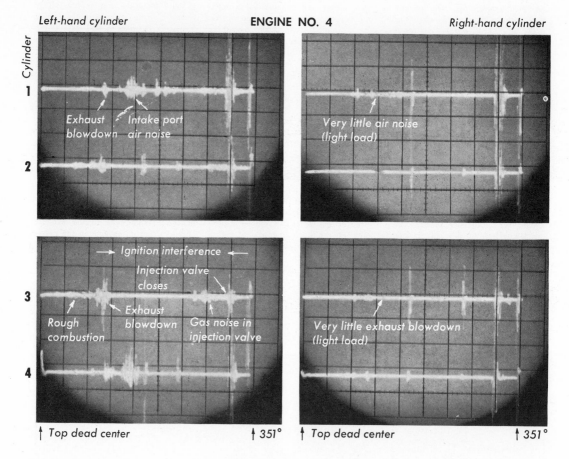

Left-hand cylinder   **ENGINE NO. 4**   **Right-hand cylinder**

Cylinder

1

Exhaust blowdown — Intake port air noise

Very little air noise (light load)

2

Ignition interference

Injection valve closes

3

Rough combustion — Exhaust blowdown — Gas noise in injection valve

Very little exhaust blowdown (light load)

4

↑ Top dead center   ↑ 351°   ↑ Top dead center   ↑ 351°

Normal engine...

Ultrasonic diagrams above were made on engine No. 4, which is identical to engine No. 5 (right-hand page).

By comparing the two sets of diagrams, you can see the dramatic difference in the relative condition of the two engines. When the above diagrams were made, this engine was operating at full speed, but at a very light load. Under these conditions, a 2-cycle engine often misfires. That is why some of the cylinders, indicated above, show very little exhaust blowdown and intake air port noise, while other cylinders very definitely show these conditions.

Fortunately, the device's sensitivity is such that a minimum gain provides enough signal amplitude and a constant reference for cylinder and engine comparisons as well. Thus many faults may be found while equipment is operating.

piston blowby while engine operates

an oscilloscope to display various engine parameters as a function of crankshaft cycle time.

Starting with the rotary transducer, synchronized with top dead center flywheel markings for cylinder No. 1, each cylinder is analyzed in turn. This is done by adjusting the transducer's selector in accordance with the engine's crank-angle data. Thus the horizontal oscilloscope sweep displays 360 crankshaft degrees.

Ultrasonic energy released by the interaction of gas molecules—bypassing compression at a ring or cylinder wall fault—is converted into an electrical signal by the ultrasonic translater detector. This 10-milliwatt signal

from the unit's phone output provides the vertical display on the analyzer oscilloscope.

Sensitivity of the device is such that a minimum gain provides enough signal amplitude, as well as a constant reference for cylinder and engine comparison. The scope's vertical gain controls the presentation.

The probe microphone is held against any flat cylinder head surface (photo at left) and the detection unit's intensity meter observed. The unit being tested in the photograph is an 8-cylinder, V-type engine with a 2-cycle engine compressor and four integrally housed compressor cylinders. Engine bore is 9¾ in. by 10½-in. stroke.■

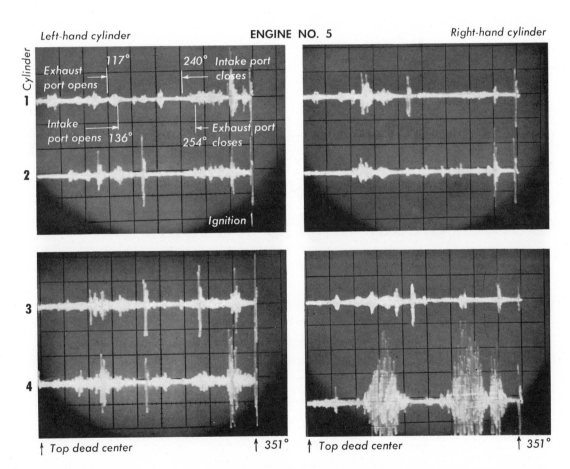

Faulty engine . . .

In operation only nine months, this unit gave no outward indication that its mechanical condition was faulty. But as the trace photos above reveal, all cylinders **actually had** severe piston ring blowby. Five cylinders have pronounced vertical rises which indicate scoring. Right-hand cylinder No. 4 shows severe scoring. The culprit? Actually, the wrong piston rings had been installed.

Troubleshooting involves comparing the oscilloscope photographs—each horizontal grid equals 40 crankshaft degrees—with a timing diagram of the engine. We can then determine what mechanical operation is occurring at each maximum vertical display. Most malfunctions are found by comparing traces from each cylinder. But the operator must know both the engine (or compressor) and the instrumentation.

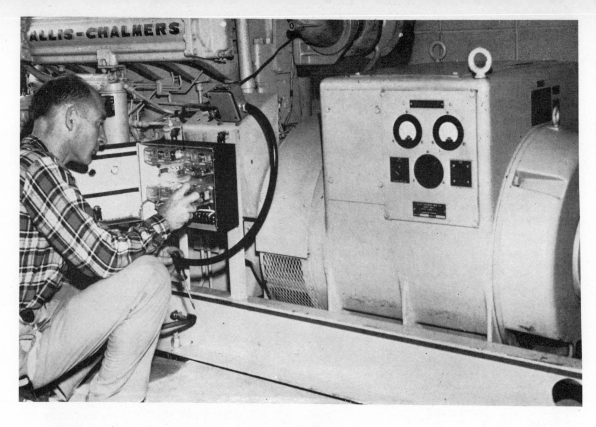

You can train your senses to help . . .

Detect diesel engine troubles

- LOOK for fuel leaks, exhaust smoke color, bent or crimped fuel lines, or any restrictions in the air intake system
- LISTEN for noises produced by engine loping, misfiring, metallic rubbing, or knocking
- TOUCH to check on temperature, linkage binding, abnormal vibration, loose fittings, condition of the air system hoses, and crimped lines that are not readily visible
- SNIFF for odors from leaking fuel, burning materials, and any unusual exhaust smoke

Allis-Chalmers

Our service engineers tell us that it is good practice to use four of the five senses when checking out diesel engines. But don't forget to use *common* sense as well as touch, hearing, sight and smell. Of course, tools must be on hand to locate the source of trouble, analyze the cause, and then help do the repair work.

If using your senses instead of instrumentation in today's world of automation sounds like a throwback to the days of the cave man, don't be alarmed. It isn't. Remember that almost every plant has smaller diesel engines, such as emergency units for generating electricity, and engines for driving pumps, compressors, etc. With these

units, complete automation is not practical.

How do you check out smaller engines? Here is where the senses—including the common sense of the operator—enter the picture. Once the source of the malfunction has been identified by your senses, it is then only a matter of obtaining the right instruments and tools to correct the fault. To help you nail down the exact fault, a trouble-shooting chart of 48 items is given on the facing page. It shows the possible causes of each malfunction.

About everything that can happen to a diesel engine is listed. Keep this chart near your diesel engine and use it to keep your engine running smoothly.

DIESEL ENGINE TROUBLE SHOOTING CHART

SYMPTOMS

CAUSE	Poor Performance, Surge, Erratic Action, Low Power	Rough Idle, Engine Vibrating	Excessive Engine Smoke, Black	Excessive Engine Smoke, White-Blue	High Fuel Consumption	Engine Overspeeds	Hard Starting	Engine Will Not Start
1. Throttle linkage adjustment	■	·						
2. Throttle linkage sticking, binding	■							
3. Incorrect governor setting (exterior)	■					■		
4. Incorrect governor setting (interior)	■					■		
5. Air cleaner restriction	■		■				■	■
6. Excessive lube oil in air cleaner				■	■			
7. Incorrect fuel delivery	■		■		■		■	
8. Low fuel supply pressure	■						■	■
9. Overflow valve leaking or stuck open	■						■	
10. Fuel filter clogged, restricted	■						■	■
11. Air leaks in fuel supply system	■						■	■
12. Incorrect bleeding, fuel tank	■						■	■
13. Fuel return line restricted	■							
14. High-pressure tubings, restricted	■						■	
15. Injection pump to engine timing	■		■	■			■	■
16. Engine valve timing	■		■				■	
17. Sticking of fuel control rack or injection pump	■					■	■	■
18. Nozzles defective, leaking, worn	■	■	■	■	■		■	
19. Incorrect nozzle opening pressure	■		■				■	
20. Nozzle incorrectly torqued	■		■				■	
21. Nozzle valve sticking	■		■				■	■
22. Delivery valve sticking or leaking	■						■	
23. Fuel transfer pump inoperative	■							■
24. Exhaust pipe or muffler clogged or pinched	■		■					
25. Engine-pump drive worn	■						■	
26. Incorrect timing, advanced-retarded	■		■				■	
27. Incorrect fuel oil, water contamination, wrong grade or type	■	■	■				■	
28. Lube oil pump dirty or clogged				■				
29. Lube oil system restricted				■				
30. Lube oil level too high		■	·	■				
31. Lube oil level too low	■	·						
32. Lube oil too heavy, contaminated	■						■	
33. Engine mount deflection		■	·					
34. Engine running cold				■			■	
35. Engine overheating	■	·		■			■	
36. Poor compression	■			■			■	
37. Engine head gasket leaking	■						■	
38. Excessive carbon deposits in combustion chamber	■		■					
39. Engine valve guides worn or gummed				■				
40. Pistons or rings stuck, worn or broken	■		■	■			■	
41. Plunger sticking							■	
42. Injection pump plungers worn			■					
43. Injection pump rollers or cams worn				■				■
44. Fuel tank vent plugged								■
45. Fuel tank valve closed; fuel tank empty								■
46. Fuel lines or filters plugged	■							■
47. Incorrect nozzle popping pressure		■						
48. Pump drive broken								■

227

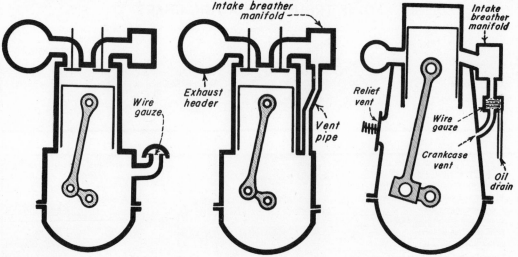

1 Engines with an outside vent should be screened to keep fire and sparks out

2 An open vent pipe from crankcase to intake manifold is very dangerous

3 This crankcase vent keeps flame from passing down, vents and drains oil

What About Crankcase Explosions?

Diesel engines are one of the safest prime movers, but there have been crankcase explosions.

▶ EVER HEAR of a diesel crankcase explosion? Now don't go deserting your diesel, or quitting your job until you read this through. Experience shows that diesels are as safe or safer than steam engines if you compare the total horsepower of each in use. But just the same, there have been explosions. These accidents haven't been confined to any one make or type of engine, nor to any one country. Usually, the operating engineer is to blame because he fails to take proper precautions when the warning sounds. Let's see what causes these rare accidents, and why.

Explosion Causes. First there must be an explosive mixture in the crankcase. That means hot oil and air, vaporized to the right mixture. How does this happen?

The motion of the crank, connecting rod and crankpin, all spraying lube oil through the crankcase air, does the

vaporizing. The lube oil usually enters the bearings at 20 to 30 psi and is sprayed out at 130 F to 140 F. This warm oil being sprayed around is a blue haze in appearance. Stick a flashlight to the glass port of any diesel and see for yourself.

The lube oil in the sump can't explode. It's flash point is from 375 F to 400 F, but it isn't vaporized. That leaves the oil vapor the real danger and then only if its ignited.

Ignition. Lets see how this vapor can be ignited in your crankcase. It might be set off by a spark, by redhot metal or from piston blowby. Then again it might be from an outside flame getting into the crankcase.

A spark is caused when bare metal strikes against bare metal. This happens if the indicator gear comes adrift. The telescope piping or a broken lube-oil line strikes another metal part.

Piston blowby is a common occurrence, but isn't dangerous unless the rings are broken and the condition so bad that red-hot gasses blow past the piston into the vaporized oil.

Blowby can be dangerous if caused by stuck rings or rings that don't seal the hot gases. Eventually it will blow all the cylinder lubricating oil off the cylinder walls, causing the bare metal of rings and piston to rub against bare cylinder walls. The resulting sparks from this hot metal can ignite the gases and this is dangerous.

Red-hot metal in the crankcase is caused by two moving parts rubbing against each other without lubrication. This happens if the wrist-pin-bearing, main- or crankpin-bearing lubrication fails. Then the rotating parts tend to seize, causing sparks before they fail.

The chance of crankcase vapors igniting from an outside source is very slim in most engines. This can only happen with an unscreened, outside vent pipe, Fig. 1, found usually on small engines or older large units.

Some operating engineers get a brainstorm and hook up a suction line from the intake breather manifold to the

228

crankcase for removing crankcase vapors, Fig. 2. This idea is OK, but if not done right will work in reverse. Then the combustion flame passes from the cylinder through this pipe to the crankcase. How can this happen? From a leaking intake valve, after-burning in the cylinder, or from backfiring.

Explosions. Investigation shows that when a damaging explosion does happen, it is always a double one. The first explosion is very mild. It may rupture the crankcase or blow off a crankcase door. That's when trouble starts. Air rushes into the crankcase through this opening and combines with the hot oil vapor. That causes a really first-class explosion—no fooling.

The diesel manufacturers of today keep their engines safer than most prime movers by installing a line between the crankcase and the intake-breather manifold or scavenging air-pump suction side. This removes some of the vapor in the crankcase.

They place screens and an oil drain in this line to prevent lube oil from passing up. This also keeps flames from passing down, Fig. 3.

Some engines carry a slight vacuum in the crankcase. This is caused by a suction to the scavenging blower, which removes the vapors and air needed for supporting an explosion. Then if the remaining vapors are ignited, their explosive force is slight.

Manufacturers take another precaution by fitting relief doors or ports to the crankcase. This relieves the crankcase pressure in case there is a slight explosion. Because the first explosion is always slight, these spring-loaded doors open and relieve the pressure, causing no damage. After opening, they close immediately, allowing no air to enter the crankcase for supplying the needed air for the secondary explosion, which is the dangerous one.

To prevent outside flames reaching the crankcase vapors, manufacturers place a flame proof screen gauze on the vent pipe, Fig. 1.

As for metal-to-metal contact caused from loss of lube oil to bearings or wrist pin, don't worry about it too much. This isn't likely if the engine has proper care. If this condition should develop, it is because the operator isn't watching his lube oil gages. They show a drop in pressure before damage can be done.

Lube-oil alarms fitted on some engines sound a warning when there is a slight drop in pressure. Then again, all mechanical cylinder lubricators are fitted with sightfeeds so the engineer can check his cylinder lubrication.

When it comes to broken piston rings causing blowby, this is always indicated by a loss of compression and lowered combustion pressures. Most engines have an indicator or pressure gage to warn the engineer in plenty of time.

Recent Explosion. One unusual explosion happened recently in a two-cycle, solid-injection marine engine. A cracked injection nozzle was the cause. That allowed a stream of fuel oil to be pumped into the cylinder without being atomized.

Fuel oil flowing down the cylinder wall washed off the lubricating cylinder oil. The engineer on watch knew something was wrong, but waited too long before stopping the engine. That gave the piston time to become over-heated from friction.

When the chief engineer was finally called, he ordered the engine started without investigating the trouble properly. Then it happened. Three men were killed instantly.

What actually caused the explosion? First of all, stopping and restarting the engine made a perfect condition for this casualty. The partial vacuum in the crankcase was broken when the engine was first stopped. By the time it was started, the crankcase had filled with fresh air. The first few revolutions provided the heat for igniting.

Avoiding Dangerous Conditions. When stopping an engine because of unusual noises, loss of power or low lube-oil pressure, always let it cool for at least

ten minutes. Make sure the cooling water and lube oil is circulating during this time. Then the crankcase doors can be opened with safety.

Adding excess lube oil to the crankcase will also cause trouble. The oil is whipped up by the cranks and heated to a high temperature because it's splashed on the hot cylinder walls. This condition caused a minor explosion in a four-cycle engine.

An oiler put too much oil into the crankcase. The relieving oiler removed the sounding rod to check the sump. That caused an explosion. Why? Because the crankcase was under a partial vacuum with the rod in place. The air needed to support combustion was admitted through the sounding tube.

Common Sense Prevents Trouble. To prevent these casualties, just use common sense. Keep the right temperatures and pressures and check the engine for unusual noises. Don't overfill the crankcase and make sure only responsible assistants add oil to it.

It is only when the first warning signs are ignored or improperly investigated that conditions are right for an explosion.

In case there are abnormal noises, temperatures or pressures or too much oil in the crankcase, stop the diesel as soon as possible. Keep the cooling water and lube oil running. Then allow it to cool for at least ten minutes before opening the crankcase doors.

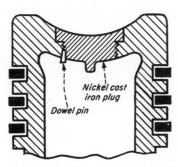

Nickel cast iron plug

Dowel pin

Piston-crown Plug

▶ BECAUSE THE CROWN of diesel pistons takes the greatest heat punishment, pistons sometimes crack at this point. To

do away with scrapping otherwise good pistons, some engineers fit their piston crowns with nickel-cast-iron plugs shown in sketch. If this plug cracks, it's renewed with a new plug and the piston is saved.

The plug is threaded and screwed into the crown. To make sure it doesn't work loose and wreck the engine, a dowel pin holds it in place.

While plugs add to the original cost, they have proved efficient in some cases. They have the added advantage of standing more heat than the piston, which reduces the chance of cracking in the first place.

In machining them, it's well to make threads and joint a perfect fit, especially if the piston is water or oil cooled, for leakage won't help. This idea might be applied to other uses by operating engineers with cracking problems.

HOW TO

Check For Eight Diesel Faults

Courtesy, Kent-Moore Organization, Inc

FAULTS are easy to diagnose if you have the right instruments for doing the job

1 **RESTRICTION** to blower intake is found with vacuum side of this gage

2 **DEPTH** of cylinder liner is found here by using dial indicator on gage block

3 **BACKPRESSURE** on engine is read at a glance from gage hooked to exhaust

4 **RPM** of engine is read from tachometer attached by flexible extension

5 **TIMING** of engine is checked with indicator held on cylinder head as here

6 **CRANKCASE** pressure shows up on this gage hooked directly to crankcase

7 **AIR-BOX** pressure is read from low-pressure gage that registers from 0 to 15 psi. Recheck if engine is adjusted

8 **OIL PRESSURE** must be known to avoid serious breakdown. Gage is hooked to parts of engine needed for information

8
Air compressors, air systems

Piping Tips for Air

INTAKE AIR LOCATION

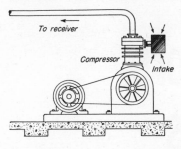

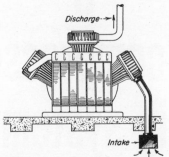

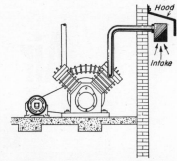

INTAKE at compressor in machinery room is common practice in many small units. This is not best hookup because air is usually warm close to the compressors

BASEMENT intake doesn't help much because air is usually as warm there as in machine room, especially if taken from basement ceiling. Basement air is damp

OUTSIDE intake is usually best because cooler air takes up less volume and machine compresses greater weight of air on each stroke. Machine stays cooler

INSTALLING INTAKE LINES

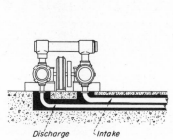

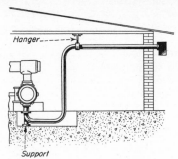

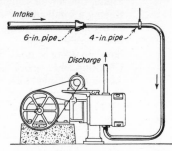

DISCHARGE and intake lines should never be placed in same duct. Intake-air temperature is raised. Then there's loss of about 1 percent for each 5-F intake rise

INTAKE AIR lines on large compressors need supports. Otherwise weight is on cylinder, and compressor is forced out of alignment. Expensive repairs follow

LONG INTAKE lines must be increased 2 in. in diameter for every 15 ft of piping. Always make pipe at least same size or larger than cylinder intake opening

OUTSIDE INTAKE POINTERS

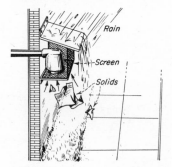

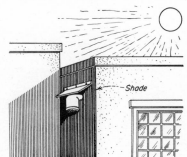

INTAKE FILTER on outside is protected from rain by hood and from solids by screen cage. This prevents paper and other large particles plugging the filter

OUTSIDE INTAKE close to drain or exhaust lines is bad. Vapors are naturally drawn into intake line. It's best to locate intake where there's only clean, dry air

WHILE PLACING air intake on outside, give some thought to selecting shady side of building, and overhanging roof. This is usually the north side of the building

Compressor Systems

DISCHARGE LINE HOOKUPS

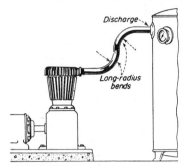

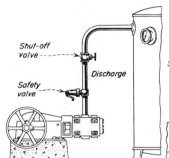

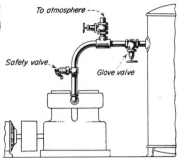

DISCHARGE line to receiver must be same size or larger than opening on machine. Make long radius bends to reduce friction or compressor will be working too hard

WITH SHUTOFF valve in discharge line, install safety valve between it and compressor. Set safety 5 to 10 psi above discharge pressure. This is important

DISCHARGE LINES shown in this notebook enter receivers near top or bottom, as in most plants. There's only one right way. You write us—we'll print answers

RECEIVER TANK LOCATION

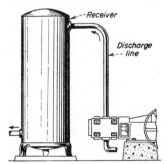

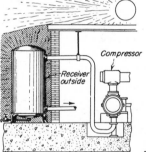

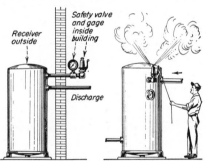

RECEIVER or tank stores air and acts as an aftercooler. Be sure to place it as close to compressor as possible. Also use large radius bends and large piping

OUTSIDE receiver on building's shady side radiates more heat. Then cooler outside air helps knock out moisture, furnishing drier air at same cost

TANK outside must have safety valve and gage INSIDE to keep from freezing. Lift safety valve every week by hand or by raising pressure to check if it works

AFTERCOOLER PIPING

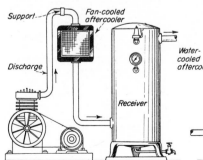

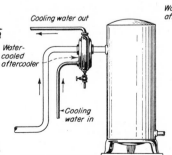

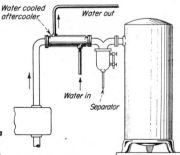

FACE THIS fan-cooled aftercooler to blow away from receiver. Cooled compressed air and moisture go to tank. Aftercoolers assure drier air needed for any service

WATER-COOLED aftercooler condenses moisture in hot, moist air from compressor. It prevents water hammer, freezing, leaky joints. Hookup to discharge

PIPE-TYPE aftercooler is usually used with separator on large jobs. Aftercooler condenses moisture and separator removes it from air. These two make good team

TWO-STAGE air compressor supplies air for all instruments in a candy factory. It was selected because of longer life and higher efficiency than single-stage machine

Buying a Small Compressor?

What size must it be? Should it be one- or two-stage? Is it for intermittent, continuous or mixed use? What size must receiver be?

► BEFORE YOU BUY a small compressor, better look through this article and do a little headwork. The tables will help you buy the right machine for your service. One thing to consider is whether single- or two-stage job is best for your work. Here are a few facts that hold true for most compressors:

SINGLE-STAGE

1. Initial cost is less.
2. Low pressure (under 150 psi).
3. OK for intermittent service.
4. Not quite as efficient (under 70 percent).
5. Operating cost a little higher.
6. Maintenance cost slightly more.

TWO-STAGE

1. Built to last longer.
2. Efficiency usually higher (above 75 percent).
3. Better for continuous service.
4. For high press. (above 150 psi).

5. Usually less maintenance.
6. Up to 25 percent power saving.

It's hard to give a fast rule for selecting the right compressor. There are too many variations in operating conditions. Also there's much argument about equipment used intermittently and continuously. In many cases, an increased storage volume overcomes periodic compressed-air loading. That allows the shop to use, for short periods, stored air at a faster rate than the compressor's actual capacity.

For example, the compressor output may be 20 cfm. For periods of two or three minutes (if there's enough stored air piped through a reducing valve from higher than usable pressure) the shop could consume 30 cfm. This only works where peak use is only three, four, or five times per hour.

What's Intermittent Service? A cleaning or blow nozzle is on intermittent service because a valve's opened for about 30 seconds every quarter to half hour. However, if a job requires this blast of air, say, once a minute for 15 seconds, it approaches continuous service.

Excluding unusual conditions, here are three examples that give you a fairly accurate way of finding compressor needs.

First—your shop has all intermittent-service equipment. Assume you have seven blow guns (cleaning nozzles), one hoist, two pneumatic door openers and one 8000-lb hydraulic lift.

First survey it for the highest pressure in use. In this case, the hydraulic lift takes 145-175 psi. That calls for a two-stage compressor. Now let's add the air requirements:

7 blowguns @ 2.5 cfm	17.5 cfm
1 hoist	1.0 cfm
2 door openers @ 2.0 cfm	4.0 cfm
1 hydraulic lift	5.25 cfm
	27.75 cfm

Since these are all intermittent service devices, look at chart No. 2 column two, opposite 145-175 psi line. It shows maximum of 24.3 to 36.4 cfm. In horsepower column we find a 3 hp, two-stage compressor is needed. Chart No. 3 shows that the minimum storage volume for a 3-hp unit is an 80-gal tank. If the blowguns or pneumatic doors are used often, it might be best to buy an extra 60-gal tank (see note on chart No. 3) and reducing valve. Pipe the 60-gal tank to all the low-pressure equipment. Connect hydraulic lift (high pressure) to 80-gal tank.

Second—all continuous-service equipment. Your shop has two production-

1—CFM USED BY PNEUMATIC DEVICES

Air Pressure (psi)	Type of Device	Average Free-Air Consumption (cfm)
70–100	Small sanders*	5.0
70–100	Blow guns (cleaning, etc.)	2.5
120–150	Grease gun*	3.0
70–100	Air hammer*	16.5
70–100	One ton hoist	1.0
120–150	Door opener	2.0
70–100	Paint spray* (production)	2.25
70–100	Paint syray* (touch-up)	8.5
70–100	Polisher	2.0
145–175	Hydraulic lift (for 8000-lb. cap. Add 0.65 cfm for each extra 1000 lb.)	5.25

* Considered continuously operating devices when in normal use. All other devices are intermittent.

Consider air tools such as torque wrenches, drills, ejection systems, automatic clutch devices, as continuous. Find their rated consumption prior to compressor selection.

2—SIZE AND TYPE MACHINE FOR BEST SERVICE

Compressor cut-in and cut-out (psi)	Intermittent (a) air used (total cfm)	Continuous (b) air used (total cfm)	Compressor hp required Two Stage	One Stage
70–100	Up to 6.6	Up to 1.9		½
	6.7– 10.5	2.0– 3.0		¾
	10.6– 13.6	3.1– 3.9		1
	Up to 14.7	Up to 4.2	1	
70–100	13.7– 20.3	4.0– 5.8		1½
	14.8– 22.4	4.3– 6.4	1½	
	20.4– 26.6	5.9– 7.4		2
	22.5– 30.4	6.5– 8.7	2	
70–100	30.5– 46.2	8.8–13.2	3	
	46.3– 60.0	13.3–20.0	5	
	60.1– 73.0	20.1–29.2	7½	
	73.1–100.0	29.3–40.0	10	
120–150	Up to 3.8	Up to 1.1		½
	3.9– 7.3	1.2– 2.1		¾
	7.4– 10.1	2.2– 2.9		1
	Up to 12.6	Up to 3.6	1	
120–150	10.2– 15.0	3.0– 4.3		1½
	12.7– 20.0	3.7– 5.7	1½	
	15.1– 20.0	4.4– 5.7		2
	20.1– 25.9	5.8– 7.4	2	
120–150	26.0– 39.2	7.5–11.2	3	
	39.3– 51.9	11.3–17.3	5	
	52.0– 67.5	17.4–27.0	7½	
	67.6– 92.5	27.1–37.0	10	
145–175	Up to 11.9	Up to 3.4	1	
	12.0– 18.5	3.5– 5.3	1½	
	18.6– 24.2	5.4– 6.9	2	
	24.3– 36.4	7.0–10.4	3	
145–175	36.5– 51.0	10.5–17.0	5	
	51.1– 66.0	17.1–26.4	7½	
	66.1– 88.2	26.5–35.3	10	

(a) These figures not to be regarded as actual capacity of the compressor in free-air output. A factor has been used to take into account intermittent operation.

(b) These figures to be used when nature of device is such that normal operation requires a continuous compressed-air supply. Figures represent actual free air delivered for compressors listed.

3—RECOMMENDED RECEIVER CAPACITY

Compressor hp	Tank size (gal)
½	30
¾	30
1	60
1½	60 or 80
2	80
3	80
5	80
7½	120
10	120

Note. Two tanks give increased efficiency and better moisture control. This is done best by doubling capacity up to 1½ hp and adding another 60-gal tank to all compressors up to and including 10 hp.

---

paint spray guns, three touch-up paint guns, two small sanders and perhaps one grease gun. The highest pressure used is 100 psi. Considering pressure alone, a single stage machine is OK. But the amount of air used also is important:

```
2 paint guns @ 8.5 cfm. .17.0 cfm
3 paint guns @ 2.0 cfm. . . 6.0 cfm
2 sanders @ 5.0 cfm. . . . .10.0 cfm
1 grease gun. . . . . . . . . . . 3.0 cfm
                                  _____
                                  36.0 cfm
```

In the third column of chart No. 2, we find two selections. First, the two-stage unit set at 120-150 shows a 10-hp machine. Since the shop needs fall between 27.1 and 37.0 cfm, we can use the 10-hp machines of 70-100 psi. Requirements cover a 29.3- to 40.0-cfm unit (at lower pressures, compressors can be speeded up to give greater delivery).

I would choose the 120-150-psi 10-hp unit because more free air is stored in a given volume at high than at low pressure. It allows the shop to use pressure-reduction system (reducing valve from tank to shop line), thus helping dry the air much more.

In Chart No. 3 we find that a 120-gal tank is minimum requirement. Here again, I would use an added 60-gal tank and the reducing-valve system piped between the tanks to give a storage at low pressure.

Third—your shop has mixed equipment; 8 blowguns, one air hammer, one touch-up paint gun, 1 pneumatic door, and one polisher. The top pressure used is for pneumatic door opener, 120 to 150 psi:

```
8 blowguns @ 2.5 cfm. . 20.0 cfm
1 pneumatic door . . . . . . . 2.0 cfm
1 polisher  . . . . . . . . . . . . 2.0 cfm
                                  _____
                                  24.0 cfm
```

These are the intermittent service devices. According to Chart No. 2, a 2-hp unit is needed. Column 2 shows it between 20.1 and 25.9 cfm.

```
1 air hammer . . . . . . . .16.50 cfm
1 paint gun . . . . . . . . . . 2.25 cfm
                                  _____
                                  18.75 cfm
```

These are continuous service devices and require, by chart No. 2, 7½ hp. Third column of chart No. 2 shows needs fall between 17.4-27.0 cfm. The total horsepower comes to 9½, or a standard 10-hp unit.

This method of deciding compressor requirements should stimulate your imagination and help in buying the right compressor. Where requirements exceed 10 hp, divide uses into less than 10-hp quantities so charts show approximate hp needed.

For example, if intermittent service devices total 30 cfm and continuous service devices total 45 cfm, at top pressure of 120-150 psi, proceed as follows: You need 10 hp for intermittent equipment alone. Chart No. 2, column 3, shows 27.1 - 37.0 cfm. 45 cfm on continuous equipment exceeds 10 hp. We can safely assume that 10 hp will take care of a figure between 27.1 - 37.0 cfm shown, or about 32 cfm. Assume that 10 hp will easily handle 35 cfm. This leaves 10 cfm or, by Chart No. 2, a need for 3 more hp.

Total hp requirements add up to 23, so you need a standard 25-hp two-stage, 120-150-psi unit. The storage requirements show 120 gal for each 10 hp, plus 80 gal for the 5, or a total of 320 gal minimum capacity for this job.

These intermeshing helical rotors supply oil-free air for

LYSHOLM-TYPE screw compressors are rotary positive-displacement machines. The male rotor has four lobes, the female six (sketch, left). Thus the male rotor revolves 50% faster. The female serves mainly as a rotating sealing member. Air is admitted at one end of the casing through a port and is carried within flute cavities of the rotors around the casing in an axial direction. Near the other end, and 180 degrees from the inlet, the male rolls into the female at the pitch line in a cycloidal action that forces the trapped air out.

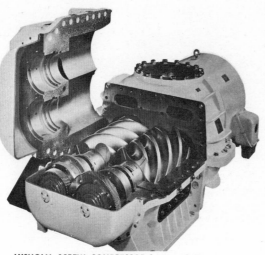

LYSHOLM SCREW COMPRESSOR has two hollow, intermeshing screws (for cooling fluid), a water-cooled housing and mechanical seals. Timing gears prevent metal-to-metal contact. Sealing strip runs along screw tips and down ends

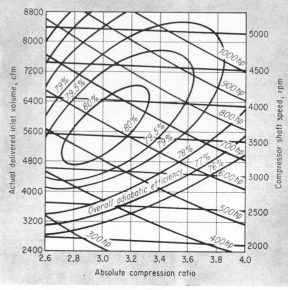

CHARACTERISTIC CURVES for the true Lysholm screw indicates hp, efficiency for given compression ratios

New rotary screw designs take on more jobs

Compressor designers have long dreamed of a machine combining the best characteristics of the positive-displacement reciprocating unit and the continuous-flow rotary centrifugal machine without the shortcomings of either. For general service or heavy-duty use, the rotary-screw compressor, above, comes pretty close to matching those wants. Today the screw machine is challenging the reciprocating unit.

Design details. Combining the aerodynamic stability of reciprocating units with simplicity and compactness of rotary compressors, the screw design uses two meshing rotors in a double-cylinder casing. Timing gears position the rotors so they intermesh, yet do not touch. To understand how a screw rotor compresses air, picture a bowling ball rolling through a pipe with 0.004-in. clearance. The ball advances like a piston, compressing the trapped air as it moves ahead.

During a portion of the rotation cycle, each groove-end is sealed by the compressor end-plate. As the reduction in volume and increase in pressure reaches the designed compression ratio, the sealed groove-ends arrive at the discharge port. Where moisture will pose no problem, treated water is sprayed into the suction for cooling and help in sealing.

Oil-free air. Since there is no metal-to-metal contact between the moving parts of the dry-screw compressor, no lubrication is required. Thus air output is oil-free. Cooling oil, to help stabilize rotor expansion, is sometimes circulated through the hollow screws. Casings are cored for water cooling of the unit.

Performance. With the compressor operating at optimum speed, maximum efficiency occurs when discharge pressure is compatible with the built-in pressure ratio. But since the efficiency curve is almost flat (study the performance curve, above), a single compressor can operate over a relatively wide pressure range. These units may also be operated over a wide speed range at near constant efficiency. The almost-flat efficiency curve comes about because the variations in aerodynamic and leakage loss tend to counteract each other as

general plant service; they need no valves, piston rings

SEVERAL VARIATIONS of the Lysholm screw are made for compressing air and pulling a vacuum. The helical, four-grooved female rotor and a mating two-lobed male rotor (sketch, right) is an example. In this unit the intake connection is on one side of the housing at the timing gear. The discharge opening is on the opposite side. Hence, flow of air is both rotary and axial.

Another spiral type has two screw-like rotors of identical design, in which the amount of internal compression is controlled by the discharge port.

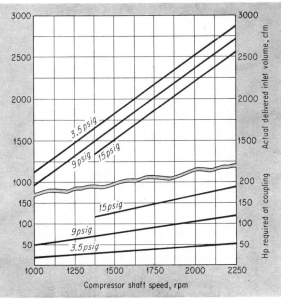

PERFORMANCE CURVES for screw, right, show how hp varies with pressure, speed and inlet air volume

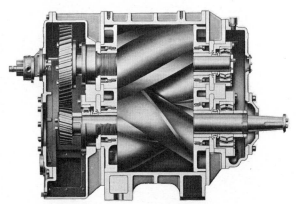

DESIGN VARIATION of Lysholm type differs mainly in contour of the screws, their lobes and grooves. Absence of metal-to-metal contact in compression chamber means there is some internal bypassing of air through these close clearances

speed changes during operation.

The dry-screw units run at best efficiencies in the capacity range between 5300 and 20,000 cfm, 6000 to 8000 rpm. Two-stage units compress air to 250 psi with stages in tandem or twin-pinion arrangement.

Wet-screw compressor designs (wet meaning lubricant) generally operate at 1800 rpm and boast of the need for little maintenance. One wet-screw design operates without timing gears. The machine is controlled from zero-to-full capacity by a throttling valve at the inlet. With an increased pressure ratio, temperature climbs somewhat, with the circulating oil transferring the heat created to an outside heat exchanger.

Oil in a wet-screw unit circulates over the rotors as a coolant and, incidentally, as a lubricant. The flood of cool oil picks up the heat of compression generated in the screw-meshing process. It also cools the compressed air, helps to seal rotor clearances, lubricates rotor surfaces. Temperature rise across a wet-screw machine in single-stage compressors is about 100 F at 100 psig.

Oil carryover is no problem in the wet-screw units. Claim is made that less oil is carried into the compressed air system by a wet-screw than from a lubricated piston machine using a few drops of oil per minute from a mechanical lubricator. One reason is the low temperature of the oil and air at discharge; thus oil is not vaporized or cracked into gas components. Oil remains as oil. What does not settle out in the receiver is filtered out with a demister as it leaves the receiver for plant service.

Variations of the Lysholm design are many. One such unit, above, has a four-groove rotor and mating two-lobe rotor. Typical performance curve, based on dry operation, is shown above. Treated water may be used with this unit for injection at the air intake to absorb heat and help seal clearances. Capacity may be varied with a bypass from discharge to inlet. Capacities are from 100 to 12,000 cfm at pressures to 15 psig.

Still another spiral design, not shown, has screw-type rotors with timing gears to prevent rotor contact. Here again, air is oil-free with the amount of internal compression controlled by the design of the discharge port. When used in single-stage, capacities range to 12,000 cfm, pressures to 30 psig. With double-staging, pressures of 100 psig are reached.

Let's take a look at

TWO-STAGE COMPRESSOR runs continuously, supplies 125 psi air to automatic extrusion machine in sewer pipe plant. Intercooler is in the base of this machine

SLIDING VANES move in slot; are held against the cylinder by centrifugal force

How compressor works

▶ POPULAR 2-STAGE AIR COMPRESSOR is the rotary, sliding-vane type. Air is trapped between the vanes as the eccentrically mounted rotor passes the inlet opening, making it a positive displacement machine. As rotor turns toward discharge port, volume of cell between any two vanes decreases and air pressure rises to rated discharge value. Centrifugal force holds vanes against bore, preventing trapped gases from slipping by. Thus vanes act as pistons and valves.

Machine is designed for common electric motor and internal-combustion engine speeds—needs no speed reducer. Since it's a rotary unit, air flow is free from pulsations.

Each unit or stage has a cylinder with heads and is water jacketed. Steel rotor, with integral shaft, is arranged eccentrically within the cylinder on same vertical center line. Shaft is supported on radial roller bearings in each head.

Blades, made of a thermosetting plastic composition or metal, slide freely in slots. Eccentric mounting of rotor inside cylinder forms a free space, crescent-shaped in cross section, divided into compartments by the blades. Volumes of compartments vary from minimum to maximum on intake or suction half of revolution; then from maximum to minimum on compression half.

Lube oil is force-fed to cylinder and to bearings, forming a thin film of oil on the running surface of cylinder wall. This lubricates blades and cylinder.

Single-stage machines are OK up to 40 to 50 psi. Use these units in zones requiring pressures lower than carried in central system. Zoning saves network of distribution system and need for reducing pressure from a central plant.

Pressure regulator. Pressure in system moves spool valve, Fig. 4, from bottom to top seat, admits air behind small piston. Piston acts against main valve, moves it to its seat. At same time, relief valve opens, venting air trapped in cylinder to atmosphere. Thus machine runs continuously, as regulator is controlled by pressure rise and fall in air receiver or discharge lines. Operation is automatic, needs no attention once it's set. Two simple adjustments allow variations in pressure or range setting if needed.

Multiple unit hookup. Have one unit on start-and-stop, while additional units run continuously. Two-stage machines are for pressures over 50 psi, as they are more economical at higher pressures than single-stage unit. Final air temperatures are also much lower. For capacities through 229 cfm (actual free-air delivery) at 100 psi, machines have intercooler tube nest in base. Larger units have separate intercooler. Two-stage machines are normally direct-line coupled to prime mover; no speed reducers are needed.

Maintenance. Open machine about once a year. Examine rubbing surfaces for wear. New vanes are same width and length as slots. Keep suction filter clean to keep contamination out of cylinder. Adjust mechanical lubricator feeds properly and see that reservoir always has enough oil so it is visible in the lubricator gage glass. Keep enough cooling water flowing through the cylinder jackets and intercooler to hold temperature where needed (usually about body temperature at final water outlet funnel).

Courtesy, Fuller Company, Catasauqua, Pennsylvania

the sliding-vane compressor

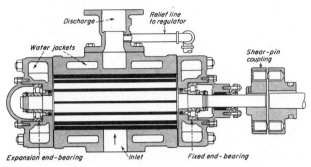

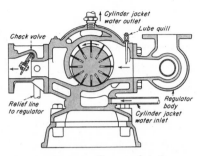

1 Rotor has fixed roller bearing at one end, floating bearing at other to take care of the longitudinal expansion and contraction differences

2 Blades automatically compensate for wear; check degree of wear when opening unit

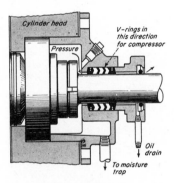

A   Compressor packing set

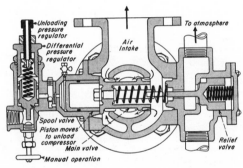

4 Regulator is controlled by pressure rise and fall in air receiver or discharge line as compressor runs continuously

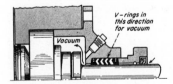

B   Vacuum packing set

3 Typical packing arrangements for air service only. As you can see, V rings are not installed in the same direction in the vacuum and compressor machines. Experience shows that these packing sets run for long periods, with little or no gland adjustment required

Used as a vacuum pump

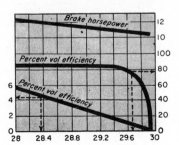

Vacuum pumps (sliding-vane type) are very similar to the compressor just described. Single-stage machines create vacuums up to 29.5 in.; compound machine up to 29.95 in. (referred to 30-in. barometer). Above 28 in. vacuum, 2-stage vacuum pumps are recommended because of their high volumetric efficiencies. Curve at left shows typical characteristics of these compound vacuum pumps between 28 in. and 29.95 in. mercury. Absence of vibration and air pulsations are often an important factor for application of these pumps.

Money burns when you let pressure fall below that for which air tools were designed. This article, based on new "Compressed Air Handbook", of Compressed Air & Gas Institute, shows how much you're losing and what to do about it

Stop Air-Dollar Waste in . . .

▶ COMPRESSED-AIR POWER is a wonderful thing, rightly used. It's too bad many users miss out by letting air pressure get low. By low air pressure, we mean low at the point of use when the tool is running.

Most air tools are designed to run at 90 psig (pounds per square inch, gage). When pressure falls below this, the tool can't do the work for which it was designed.

Causes of Low Pressure—Main causes of low air pressure are: (1) Compressor is too small (2) Piping is inadequate (3) Too much air leaks out of system.

Don't jump to the conclusion that low pressure shows the need for more compressor capacity. Maybe so, maybe not. To find out, first see whether the compressors are running full blast when the pressure is low at tools. Even if they are, don't overlook two other possible causes of low air pressure—leakage and inadequate piping.

If pressure drop from air receiver to point of use is more than 10% average, or 15% at the very worst point in system, correct it before you buy more compressors. Chart on next page will show how much air your present or planned piping can handle properly.

Air Leakage. It is just common sense that every cubic foot of leakage eliminated is another cubic foot available for useful work. Eliminating leakage cuts the power bill and restores the system capacity.

Low Pressure. With portable pneumatic tools you can boost production 37% for about 30% increase in air consumption if you run up the air pressure (at tool) from 70 psi to 90 psi.

What to Do. The cures for low air pressure are just as definite as the causes and effects. Faulty air-power conditions usually result from poor planning or increased air uses without corresponding expansion of the system. Make a survey to find out the extent of low-pressure conditions and the present air-flow requirements. Then try to estimate your future demands.

Pressure drop varies roughly as the square of the velocity of the air going through the pipes. So if the flow is doubled, the pressure drop is $2 \times 2 = 4$ times as great. This works the other way also. Enlarging the line reduces the velocity and very much reduces the pressure drop. For example, a 3-in. line, 1,000 ft long, will handle 500 cfm with a 2.5 psi pressure loss, while a 4-in line of the same length will pass about 1,000 cfm with the same drop. Remember, too, that the installed cost of a 4-in. line won't be much more than that of the 3-in. line because most of the cost is labor.

But what do you do if your line is already in and the pressure drop is excessive? One remedy is to run a new line parallel to the original with frequent interconnections. Or you may install a loop system with some outlets taken off the new line to relieve the old. Sometimes it will pay to put in a complete new air-distribution system.

Since everything in this country has a tendency to grow bigger, it's always smart to install oversize lines for branches and manifolds for tool attachments.

And don't forget the hose. Losses of 15%—even 25%—in the air pressure are frequent in the hose alone. What a pity to waste all this in the last few feet after you've spent good money to produce enough air and deliver it to the hose at the proper pressure. For one thing, don't use a hose longer than you need to make the tool reach the work and to give proper freedom of movement to the operator. Above all, make hose big enough (see chart).

Every cubic foot of leakage salvaged is pure gain. Some systems leak as much as 10 to 20% of the total air compressed. Individual air leaks may be small, but a lot of them add up to real money. Run an air-leak test on the whole plant and inspect every section of hose. Remember that a single $\frac{1}{4}$-in. hole wastes 182,000 cu ft of air per month, at a cost of $24.32.

The most likely locations of small leaks is around valve stems, in hose connections, unions, drains, home-made guns and lines leading to tools that are non-operating. When you have eliminated leakage, inspect system regularly to keep it tight.

Only after you've done all these things, do you have time to consider new compressor capacity if you can't get along with what you have.

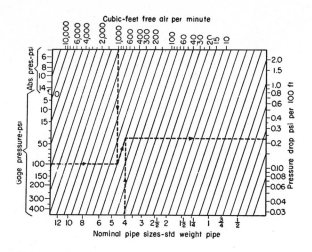

Cubic-feet free air per minute

Nominal pipe sizes-std weight pipe

▶ THIS CHART HELPS you figure pressure loss in any length pipe. In example (dotted line) initial pressure is 100 psi gage, quantity is 1,000 cu ft of free air per min and piping is 4-in. standard.

From 100 on left scale run *horizontally* until *under* 1000 cfm on top scale. Then move parallel to *slanting lines* until *over* pipe size on bottom scale. Now move *horizontally* to right scale and read pressure drop per 100 ft of pipe. Chart redrawn by permission, from original by Walworth Co.

.. 1-PIPE FRICTION

EQUIVALENT FRICTION OF FITTINGS

With factor below, to figure out how much pipe would have the same friction as the given fitting, multiply factor by nominal pipe size in inches and get equivalent pipe length in feet. For example, a 3-in. open angle valve would be equivalent to 3 × 14 = 42 ft of straight pipe:

Fitting	Factor	Fitting	Factor
Globe valve (open)....	26	Close return band......	6
Angle Valve (open)....	14	Standard elbow........	2.5
Swing check (open)....	7	Medium-sweep ell.....	2.3
Gate valve (open)......	0.5	Long-sweep ell.......	1.8
Standard tee, through side		45-degree ell..........	1.2
outlet.............	6	square ell.............	5.5

▶ THE FOLLOWING example shows convenient way to work out and total the pipe equivalents of all fittings. Add their total to actual piping length to get total equivalent piping. Then, in this case, multiply 3.18 by pressure drop per 100 pipe ft found from above chart.

Fittings (for 3-in. pipe)	Factor	Equivalent feet	No. of fittings	Total Equiv Feet
Standard Ell......	2.5	7.5	4	30
Tee, Side Outlet..	6	18	1	18
Angle Valve.....	14	42	1	42
Globe Valve.....	26	78	1	78
		All Fittings		168 ft
		Actual Piping		150 ft
		Total Equivalent Piping		318 ft

.. 2-FITTING FRICTION

PRESSURE LOSS IN 50 FT OF HOSE

Hose Size Inches	Line Pressure psi gage	Air flow in cfm (cubic feet free air per min)										
		50	60	70	80	90	100	110	120	130	140	150
½	50	18										
½	100	8	13	19	27	37						
¾	50	2	4	4	6	9	11	14				
¾	100	1	1	2	3	4	4	5	7	8	9	11
1	50	x	x	x	2	2	3	4	5	7		
1	100	x	x	x	x	x	1	1	2	2	2	2

▶ TABLE AT LEFT lists pressure drop (in pounds per square inch) to nearest pound. All drops of less than 1 psi are indicated by X. Empty spaces indicate hose size is too small for given air flow. Air flow (cfm) is in cubic feet of free air per min. Note how much the pressure drop is lessened in each case by jumping to the next larger size hose.

.. 3-HOSE FRICTION

Checking & Starting

CHECK BEFORE STARTING

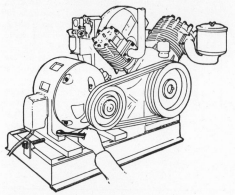

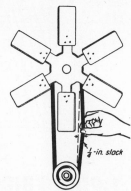

1 First hours of new compressor are important. Go over all bolts and screws that might have loosened during shipping and check for any loose wiring

2 Belt tension is ok if 2 lb force pushes belt down distance equal to the thickness of a belt

3 Fan belt tension is best with about one-quarter-in. slack from a straight-line path shown

START AND OBSERVE

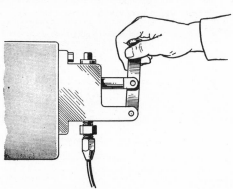

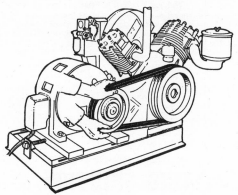

7 Unload the compressor by hand before starting. On some machines this is done by small hand unloader. On other types you unload machine by turning a switch to **off** position

8 Good practice calls for turning the compressor over by hand a few times before starting. This tells you if the machine is free to turn, also lubricates the internal parts well

LOW OIL-PRESSURE CAUSES

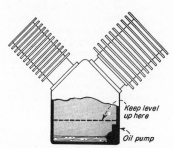

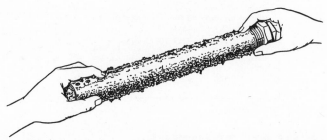

11 Low oil level in crankcase causes low oil pressure. This is serious. Oil drops below pump; air is pumped

12 Lube-oil-strainer plugging prevents oil from entering pump. Then oil pressure drops gradually in machine. Remove strainer and clean it at set periods. Never fill crankcase above normal or rings carbonize, stick, and the machine loses power

New Air Compressor

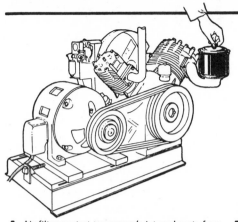

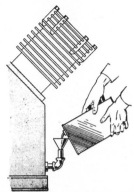

4 Air filters protect compressor's internal parts from being worn much by dirt. Fill with right oil and tighten the wing nut before starting the machine

5 Never start compressor without filling crankcase with lube oil of recommended type first off

6 Oil cups on compressor need oil before machine is started. Be sure to find all of the cups

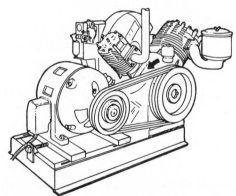

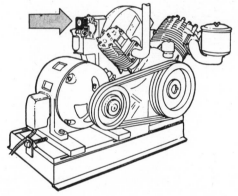

9 Arrow cast on most air compressors shows right direction of rotation. This is important. If machine rotates in reverse direction, oil pump fails and machine burns up very fast

10 Oil-pressure gage is most important instrument on any compressor. Watch it closely. Pressure above 35 psi at first means oil is cold. If it falls under 15 psi, stop compressor

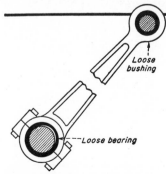

Loose bushing

Loose bearing

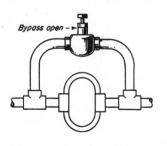

Bypass open

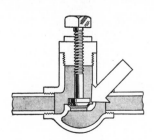

13 Low oil pressure is also from loose bearings and loose wrist-pin bushings. Only cure is to take up this wear

14 Check bypass for being cracked open if oil pressure doesn't build up in a hurry. Valve regulates pressure

15 If pressure still doesn't build up, remove bypass valve bonnet and check for foreign matter stuck to seat

SAFE TANK is clean inside and out, installed and piped right with safety devices and caged off from busy factory floor

Check Air Tanks NOW!

1. COMMON EXPLOSION CAUSES
2. HOW TO INSPECT TANKS INSIDE AND OUT
3. HOW TO SQUEEZE OUT MOISTURE
4. HOW TO HOOKUP TANK
5. HOW TO STOP LUBE-OIL HEADACHES
6. HOW FOUL AIR CAN FOUL UP PROCESS
7. WHAT ATTENTION COOLING WATER NEEDS
8. WHAT SAFETY RULES TO FOLLOW
9. WHY HOMEMADE TANKS ARE DANGEROUS
10. TANKS THAT COME UNDER ASME CODE
11. HOW PIPING GETS DAMAGED

National Safety Council

► AIR TANKS and receivers, the most common unfired pressure vessels, are relatively safe. However, they do explode unless they're properly designed for pressure and service used. They also explode when not installed right or not equipped with safety devices. So inspect them periodically and keep them in good condition.

Explosion Causes. Main causes for explosion are: (1) Vessel's metal corrodes. (2) Spontaneous combustion and explosion of lube-oil vapors in tank. (3) Vibration of tank or its connected piping. Vibration sets up fatigue stresses, eventually causing cracks. (4) Faulty or improperly installed safety devices. (5) Improper design or installation. (6) Water hammer in tank or piping.

There's no excuse for explosion caused by corrosion. If tank or receiver is inspected inside and out at set periods, corrosion can be spotted when it starts. Then arrest corrosion by cleaning and painting.

Inspection Opening. Air tanks should be equipped with inspection opening. According to ASME Code, a tank of over 36-in. diameter must have a manhole. Provide pipe plug openings on tanks of less than five cu ft. In small air tanks, remove intake and exhaust piping for inspecting interior.

Inspect small tanks inside and out for corrosion once every three months. When you first see corrosion spots, clean and dry thoroughly, coat with rust preventive paint. Spray paint on interior surfaces of smaller tanks.

Moisture. Prevent external moisture on air tanks by locating tank in a well-ventilated, cool, dry location. Internal moisture can be prevented to some extent with separators on air line from compressor. Place a drain valve or cock at lowest point of air tank. Open valve daily, or oftener to remove water. This prevents a bad water-hammer hazard.

Concave-Convex Heads. Many smaller air tanks are made with a concave head at one end and a convex head at the other. It's poor practice to install these tanks in a vertical position with tank resting on the concave head. Reason is space between concave head and shell cannot be drained. If you can't set up such a tank any other way, fill concave head with tar or cement. Then place a drain line to blow all moisture from bottom, Fig. 1.

Mount this type tank in a horizontal plane with drain line on bottom side of shell, Fig. 2. If tank must be installed vertically, place convex head on bottom. Then use supporting stand. Tap drain into convex head, Fig. 3.

Lube-Oil Vapors. Explosions in air tanks and receivers are caused by lube-oil vapors igniting. Prevent such explosions by installing a separator. Also by good maintenance of air system.

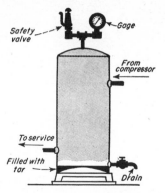

1 If tank must be set on concave end, tar the bottom to prevent corrosion

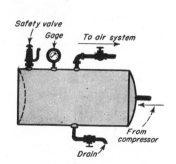

2 The right way to install a concave- or convex-head tank is horizontally

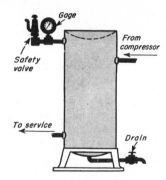

3 If tank must be vertical, give it a supporting stand on convex bottom

Air taken into compressed-air system through compressor's suction contains dirt and other impurities. That wears the compressor valves. Worn and faulty discharge valves tend to raise air temperature rapidly and progressively. That can cause spontaneous ignition of oil vapors.

Foul Air. Foul air is bad for most processes using compressed air. Locate compressors suction line, if possible, to take in clean, dry air. Also place a filter over suction line and activated carbon filters after the separator. Dismantle air compressor periodically and inspect all valves and piston rings. Replace faulty or badly worn valves and rings at once.

Cooling Water. Check cooling water several times daily for proper flowing. Compare water temperature at suction side with discharge side of cooling system. A sharp temperature rise shows that compressor jacket, or inter- or after-cooler is plugged up.

Clean scale and sludge from compressor water-cooling surfaces, intercooler and aftercooler. Best way is flush out at every inspection period. For more safety, install a fuse plug to relieve pressure caused by dangerously high temperature.

Lube Oil. Lube oil from compressor may carry into receiver or tank—often in the form of vapor. **Danger!** A spark, or a temperature rise of oil vapor above oil's flash point, will cause explosion.

To prevent explosion use best grade lube oil. Use only amount of oil recommended in manufacturer's instructions. Moisture and oil separator on tank's discharge line will greatly reduce amount of oil vapor entering tank.

Excessive oil and dirt, carried through air compressor, forms carbon scale in discharge pipe. As this carbon scale becomes dislodged and carried into tank, it may cause sparks. Then sparks ignite oil vapors in tank. Clean discharge pipe thoroughly at every inspection.

Cleaning Tank. When cleaning air tank's inside surfaces with caustic solution, also clean connecting piping at every inspection period.

Safety. Take usual precautions to prevent accidents in cleaning air tank and its piping. Open all manholes and handholes. Ventilate tanks thoroughly by removing all pipe plugs. If your tank is large enough for a man to enter, first test for volatile gases, carbon monoxide, and lack of oxygen. One way is with a flame safety lamp.

Have man entering tank wear goggles, rubberized gloves, and rubber boots to protect against caustic solution.

Testing Tank. Hammer test your air receiver at every annual internal inspection. If practical, hydrostatically test to one and one half times the operating pressure. Then shell plate or head cracks not visible to eye show up by leaking.

Improper Design. Many accidents are caused by improper design of non-code air tanks and receivers. Low 15- to 25-psi air is used in many plants. If receivers for this low pressure are made from water tanks or other containers, look out. Chances are your tank will explode. **Remember!** All air tanks should be designed and made in accordance with the ASME Code for unfired pressure vessels.

Piping Damage. Damage to compressed-air piping from vibration and pulsation occurs in two places. On air-compressor discharge piping and in the air line from portable air tank. A portable air tank like those used for construction work may have a manifold with openings for several air lines, connected directly to tank. Occasionally, a hose reel for a long length of air hose may be mounted directly on the air tank.

Pulsation set up by air tools causes vibration stresses at connection between air hose or manifold and tank. Install a U bend or vibration-dampener coupling at these points.

Stay Alive. Check all safety devices, such as safety valves and compressor unloaders, daily on a compressed-air system. Try vent and drains daily. If a safety device doesn't work, repair it at once.

Is this too much work? Accidents happen for first time in plants every day. Follow my advice and you'll live a lot longer. And you'll be $$$ ahead.

Did you know . . .

. . . that rubber gaskets don't work well on air compressors and air lines? Reason is that oil and heat soften the rubber in time. Use any asbestos-body sheet gasket between air cylinders and heads.

. . . that clearance between piston and compressor cylinder head will change unless you use same thickness gasket that the manufacturer used in new compressor? Install new gaskets, then pull bolts up tight after compressor runs a few days. Gasket will stay tight and keep from blowing out.

With help of this handy check chart you'll find that . . .

Compressor trouble shooting is easy

Worthington Corporation

SYMPTOM	POSSIBLE CAUSES OF TROUBLE

Water-cooled compressors

WON'T DELIVER AIR
1 Restricted suction line
2 Dirty air filter
3 Valves missing
4 Worn or broken valve strip
5 Defective capacity control

CAPACITY TOO LOW
1 Clogged suction line
2 Dirty air filter
3 Loose valve
4 Worn valves
5 Defective unloaders
6 Wrong speed
7 Worn piston rings
8 Leaky head gaskets
9 Poor capacity control

PRESSURE TOO LOW
1 Valves missing
2 Strips missing from valves
3 Worn valve strip
4 Loose valve
5 Worn valves
6 Unloaders defective
7 Excessive system leakage
8 Wrong speed
9 Worn piston rings
10 Leaky head gaskets
11 System demand greater than compressor capacity
12 Defective capacity control

COMPRESSOR OVERHEATS
1 Worn valves
2 Wrong speed
3 Poor capacity control
4 Not enough cooling water
5 Cooling-water temperature too high
6 Discharge pressure too high
7 Inadequate cylinder lubrication
8 Motor too small
9 Belt too tight

SYMPTOM	POSSIBLE CAUSES OF TROUBLE

COMPRESSOR KNOCKS
1 Loose valve
2 Loose unloader
3 Discharge pressure too high
4 Inadequate cylinder lubrication
5 Inadequate running gear lubrication
6 Loose flywheel or pulley
7 Excessive bearing clearances
8 Misalignment
9 Loose piston-rod nut
10 Loose motor rotor on shaft
11 Loose crosshead shoes

COMPRESSOR VIBRATES
1 Defective unloaders
2 Excessive system leakage
3 Discharge pressure too high
4 Loose flywheel or pulley
5 Loose motor rotor on shaft
6 Bad foundation
7 Poor grouting
8 Wedges left under compressor
9 Misalignment
10 Piping improperly supported

INTERCOOLER PRESSURE TOO HIGH
1 Strips missing from valves in h-p cyl
2 Worn or broken valve strip
3 Loose valve in h-p cyl
4 Worn valve in h-p cyl
5 Defective unloaders
6 Worn piston rings in h-p cyl
7 Poor capacity control on h-p cyl

INTERCOOLER PRESSURE TOO LOW
1 Valves missing
2 Strips missing from valves in l-p cyl
3 Worn or broken valve strip
4 Loose valve in l-p cyl
5 Worn valve in l-p cyl
6 Defective unloaders in l-p cyl
7 Worn piston rings
8 Leaky head gaskets
9 Defective capacity control in l-p cyl

SYMPTOM	POSSIBLE CAUSES OF TROUBLE
RECEIVER PRESSURE TOO HIGH	1 Defective unloaders 2 Poor capacity control 3 Poor capacity-control adjustment
DISCHARGE AIR TEMPERATURE TOO HIGH	1 Worn valves 2 Defective unloaders 3 Defective capacity control 4 Poor capacity-control adjustment 5 Not enough cooling water 6 Cooling-water temperature too high 7 Discharge pressure too high 8 Poor cylinder lubrication 9 Intercooler pressure too high 10 Dirty intercooler 11 Dirty cylinder jackets
COOLING-WATER DISCHARGE TEMPERATURE TOO HIGH	1 Worn valves 2 Defective unloaders 3 Not enough cooling water 4 Cooling-water temperature too high 5 Discharge pressure too high 6 Poor cylinder lubrication 7 Intercooler pressure too high 8 Dirty intercooler 9 Dirty cylinder jackets
MOTOR FAILS TO START	1 Defective unloaders 2 Poor capacity control 3 Wrong electrical characteristics 4 Motor too small 5 Belt too tight 6 Voltage too low 7 Excitation incorrect 8 Motor overload relay tripped
MOTOR OVERHEATS	1 Wrong speed 2 Poor capacity control 3 Discharge pressure too high 4 Poor cylinder lubrication 5 Poor running gear lubrication 6 Belt too tight 7 Voltage too low 8 Excitation incorrect
SYNCHRONOUS MOTOR FAILS TO START	1 Defective unloaders 2 Poor capacity control 3 Wrong electrical characteristics 4 Voltage too low 5 Excitation incorrect 6 Motor overload relay tripped

Air-cooled compressors

SYMPTOM	POSSIBLE CAUSES OF TROUBLE
PRESSURE TOO LOW	1 Excessive leakage in air lines, fittings, valves 2 Worn piston and rings 3 Demand greater than unit capacity 4 Wrong speed
COMPRESSOR OVERLOADS MOTOR	1 Restricted discharge line 2 Insufficient lube oil 3 Control line leak 4 Defective switch, or trigger valve set for too high a cut out 5 Belt too tight 6 Compressor or motor binding 7 Electrical-power characteristics incorrect 8 Voltage too low 9 Discharge pressure too high 10 Wrong speed

SYMPTOM	POSSIBLE CAUSES OF TROUBLE
INSUFFICIENT CAPACITY	1 Suction line blocked, dirty filter 2 Suction valve unloaders stuck in unloaded position 3 Excessive leakage in air lines, fittings, valves 4 Leaking intercooler 5 Valves not installed right 6 Strips missing from valves, or broken 7 Worn piston and rings 8 Blown cylinder-head gasket 9 Belt slipping 10 Demand greater than unit capacity 11 Discharge pressure higher than rating 12 Wrong speed
COMPRESSOR OVERHEATS	1 Suction line blocked, dirty filter 2 Insufficient lube oil 3 While running loaded: Broken or leaking h-p suction-valve strip, h-p unloader stuck in unloaded position, blown h-p suction valve seat 4 Control line leak 5 Valves not installed right 6 Strips missing from valves, or broken 7 Wrong speed 8 Wrong direction of rotation
COMPRESSOR KNOCKS	1 Loose valve 2 Not enough lube oil 3 Valves not installed right 4 Strips missing from valves, or broken 5 Motor rotor shunting back and forth, due to belt misalignment or mounting not level 6 Wrong direction of rotation 7 Loose flywheel or pulley 8 Too much wrist pin and bushing clearance 9 Main bearings need adjusting 10 Too much crank-pin bearing clearance
COMPRESSOR VIBRATES	1 Piping not supported right 2 Motor rotor shunting back and forth, due to belt misalignment or mounting not level 3 Motor rotor out of balance 4 Unit not properly secured to foundation 5 Defective foundation 6 Shipping block not removed under base
INTERCOOLER VALVE BLOWS	1 While running unloaded: Broken or leaking h-p discharge-valve strip, h-p unloader leaking air, blown h-p valve gasket 2 While running loaded: Broken or leaking h-p suction-valve strip, h-p unloader stuck in unloaded position, blown h-p suction valve seat 3 Defective or improperly set safety valve
RECEIVER SAFETY VALVE BLOWS	1 Leak in control line 2 Defective or improperly set safety valve 3 Pressure switch differential too narrow
UNIT BLOWS FUSES	1 Discharge line restricted 2 Not enough lube oil 3 Pressure switch differential too narrow 4 Compressor or motor binding 5 Defective motor 6 Fuses too small 7 Electrical-power characteristics incorrect 8 Unit starting against full load
EXCESSIVE OIL CONSUMPTION	1 Suction line blocked, dirty filter 2 Oil level too high 3 Oil viscosity too light 4 Oil pressure too high (if force-feed lubrication) 5 Worn piston and rings

What Do You Take Me For, a Fathead?

THIS HIPPOPOTAMUS IS A DEAD RINGER for a chief engineer I worked for many years ago. It was an industrial plant and I had the day shift.

Shortly after starting, I got interested in the old dc switchboard. The fuses and knife switches were pretty dirty, so I started checking all of them.

Some fuse clips were badly burned. And I found double links in many cartridges. It was the old "penny in the fuse" practice, but I was surprised to find that in a plant.

"If this board's so sloppy," I thought to myself, "there must be plenty wrong with it." First off, I checked the generator's field rheostat settings. They were pretty high. Then I noticed the main-circuit-breaker trip arm was tied down. I untied it in a hurry. Then I set to work polishing the fuse clips, cartridge ends and knife switches with fine sand-paper. Next I got rid of all the double fuse links, and there were plenty.

Two days later, on my morning watch, the breaker kicked out. "What's going on here?" thundered the chief, tearing into the engine room.

I explained about untieing the breaker because I didn't want the generator to burn up.

"I tied that down because I don't want it kicking out," he yelled. Then he warned, "When I tie down a breaker, I want it left that way. WHAT DO YOU TAKE ME FOR, A FATHEAD?"

I've learned there are plenty of his kind around even today. They're heavyweights, all right. About as heavy as a hippo, and about as fast . . . mentally. They don't know why protective devices are built, and they don't take the trouble to find out—until it's too late.

9
Pumps

Overhauling Duplex

(To set valves, see "Operators' Notebook I" Jan, 1948)

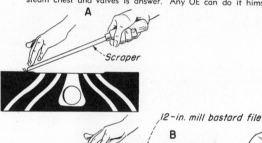

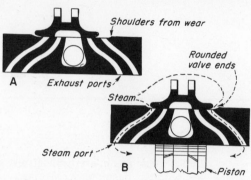

Shoulders from wear

A *Exhaust ports*

Rounded valve ends

Steam

Steam port

B

Piston

1 This steam duplex pump is so reliable it's often given no attention. Then it stops, refusing to budge. Overhauling steam chest and valves is answer. Any OE can do it himself

2 Shoulders wear at ends of valve travel on valve chest, **A.** Slide-valve face wears thin and ends round. Steam enters cylinder under rounded ends, stopping piston on midstroke, **B**

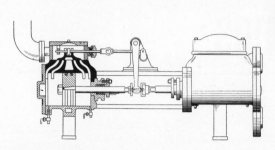

Scraper

A

12-in. mill bastard file

B

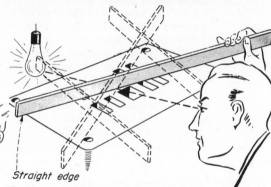

Straight edge

5 Scrape away shoulders if they are light. If heavy, remove chest studs and use 12-in. mill bastard file. First chalk file. Never rub fingers on file's cutting edge or it slides

6 Filing is an art. A good mechanic can file a valve-chest face level by checking with straightedge like this. Keep checking and file only the high spots until no light is seen

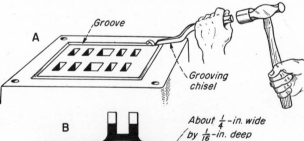

Groove

A

Grooving chisel

B

About $\frac{1}{4}$-in. wide by $\frac{1}{16}$-in. deep

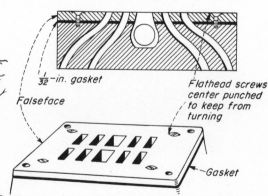

$\frac{1}{32}$-in. gasket

Falseface

Flathead screws center punched to keep from turning

Gasket

9 After valve chest is level, grooves cut with chisel give raised-valve-surface effect. This avoids shoulders, saves valves from hitting shoulders, prevents short pump strokes

10 When old pumps are faced often, false face may be needed to make up for lowered face. Cast iron, ½ in. thick, fastened with flathead screws on gasket like this is answer

Pump Valve Chest

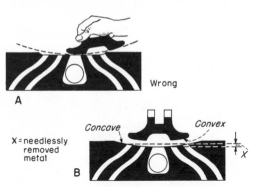

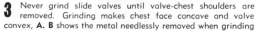

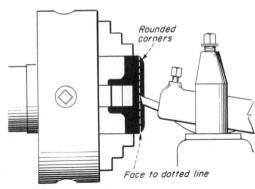

3 Never grind slide valves until valve-chest shoulders are removed. Grinding makes chest face concave and valve convex, **A**. **B** shows the metal needlessly removed when grinding

4 Right way to overhaul valve is to place in lathe chuck and face it down until rounded corners are machined away. Use pointed tool, then blunt tool for smooth finish

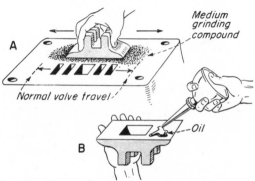

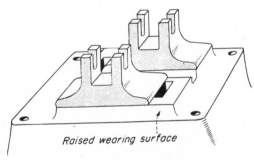

7 After face is finished level with single-cut smooth file and valves are faced in lathe, grind with medium grinding compound, **A**. Remove compound, give "oil rub" for finish

8 Some steam pumps with slide valves have a raised valve face like this. Since valves travel to ends of raised surface, no shoulders form. Light grinding is all that's needed

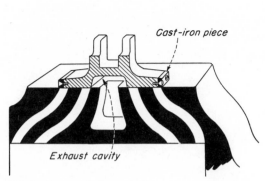

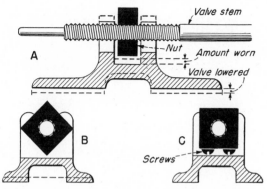

11 D-valves must be long enough to cover outside steam port edges. If right-length valves aren't in stock, attach cast-iron pieces. Make valve cavity cover ports as shown

12 Worn valve lowers and valve moves away from the valve-stem nut, **A**. **B** shows how lowered valve allows nut to turn. To hold nuts, drill and tap for roundhead screws, **C**

Don't Forget to Renew

TYPES

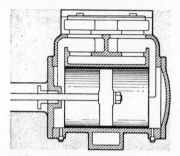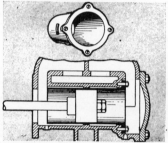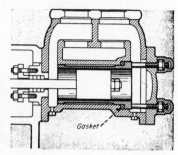

PRESSED liner is thin sleeve that is a pressed fit into cylinder bore. Material is bronze, monel, cast iron or others

REMOVABLE liner is held by four bolts and is centered by four lugs. Easy to renew because it is not a pressed fit

REMOVABLE cylinder with bronze liner is held by binder bolts through cylinder head against gasket to prevent leakage

REMOVING PRESSED LINER

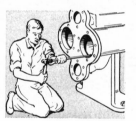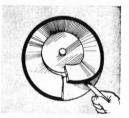

WEAR is on bottom of horizontal liner. Unless renewed, the piston rides on casting

SPLIT liner with chisel bar along bottom, where the metal is already thin from wear

CAUTION! It's easy to damage cylinder casting under the liner. Split only the liner

COLLAPSE the split liner with chisel bar. This isn't easy, also depends on the material

PRESSING IN NEW LINER

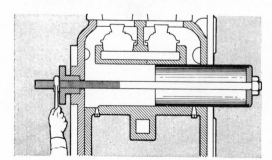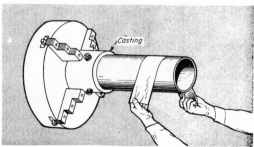

PRESS new liner into cylinder gradually like this. First clean cylinder and oil liners outside. If liner is tight after first few inches, remove it before it gets stuck tightly, for good

POLISH liner outside with coarse grinding cloth by holding it in lathe like this. If inside and outside micrometers are handy, polish liner so it is only about 0.0005 in. larger than cylinder

Your Reciprocating Pump Liners

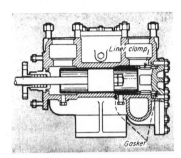

REMOVABLE liner with clamp (ported distance piece) between cylinder head and liner. Cylinder bolts support the liner

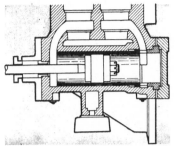

REMOVABLE liner with double shoulders, held by distance piece against cylinder head. Small bearing surface on cylinder

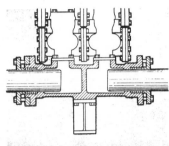

NO LINER. Whether outside-packed (like this opposed plunger pump) or inside-packed type, plunger takes all the wear

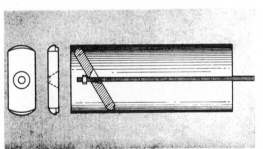

PULL worn liner in one piece. It's much easier but you must have strongback shown to hook in back of liner. Turn groove in lathe to fit liner; then grind back edges round as shown

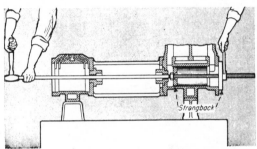

STRONGBACK against open cylinder keeps steady strain on liner while small sledge pounding against bar bumps it out. This method is much quicker and easier than splitting liner

SHRINK liner with Dry Ice. Place in wooden box and pack with sawdust. Break up Dry Ice and fill inside of the liner

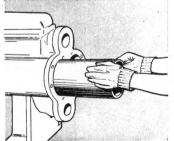

WEAR GLOVES, slide liner in fast. You may have to tap liner with mallet. Bag of Dry Ice inside liner keeps it shrunk

WHAT IS DRY ICE?

This is solid carbon dioxide, (CO_2), and is 109 F below zero. It's solid in 10-in. cubes, weighing 50 lb. Count on about 0.001-in. shrinkage per inch of liner diameter. To keep liner expanding while fitting into cylinder, place bag of broken-up Dry Ice inside liner. *Caution!* CO_2 displaces air, so ventilate working spaces well. Also, avoid frostbite by covering bare skin.

Piston Centering and Valve

FIRST CENTER PISTONS

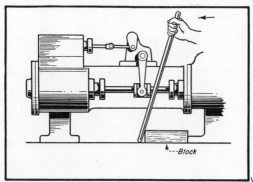

1 Open pump steam drains. Relieve liquid pressure. Bar crosshead knuckle until piston bumps cylinder head

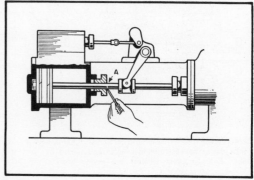

2 Make mark A on piston rod against the gland. Use scriber, pencil or soapstone as convenient

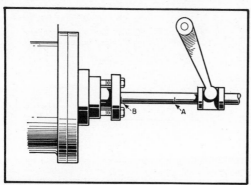

3 Bar in opposite direction until piston bumps cylinder head. Make mark B against steam gland as before

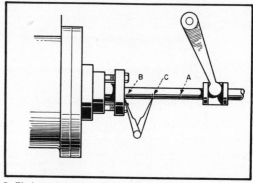

4 Find center mark C between marks A and B with dividers or scale. Avoid burring at B or C rod

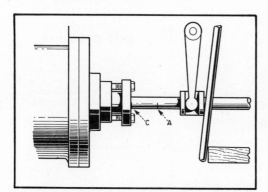

5 Final operation is barring piston rod until center mark C is brought back to the steam-end packing gland

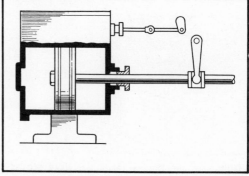

6 Steam piston is now in exact center of cylinder regardless of rocker arm angle. Center the other piston

Setting for Duplex Pump

THEN SET VALVES

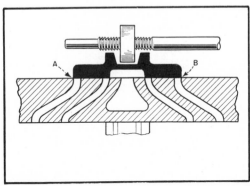

7 With piston in center of cylinder, valve ends should just cover steam ports at A and B as shown

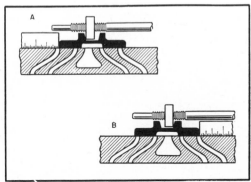

8 A, Move valve against nut and measure steam port opening. B, Move valve in opposite direction and measure

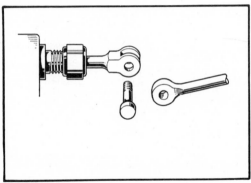

9 If port openings at both ends are equal, valve is set. If unequal, remove pin from valve link for adjustment

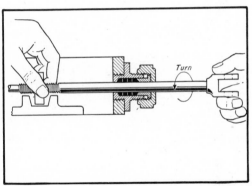

10 Hold nut and turn valve stem to correct the inequality, screwing stem into or out of nut as shown

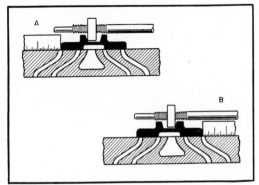

11 Measure port openings at A and B at ends of valves as before. If they are equal, valve is properly set

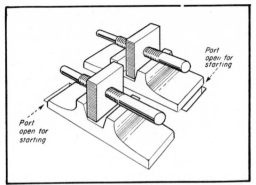

12 Before closing valve chest, move valves in opposite directions to uncover ports, or pump won't start

Do You Have Inside Dope

ROTARY feed pump in this tannery plant pinch-hits for old reciprocating duplex

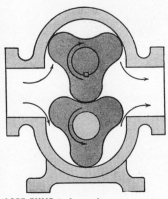

LOBE PUMP is form of gear type, must have gears to drive each shaft, or lobes would slip. Liquid travels around lobes and casing. No valves needed in pump

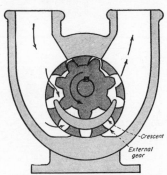

INTERNAL GEAR has stationary crescent to prevent losing liquid back into low side. Liquid from suction fills spaces between the gear teeth as they unmesh

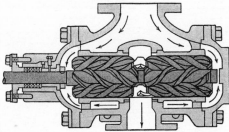

SCREW PUMP has one power rotor and two idlers. Liquid from suction enters at rotor's ends, is trapped by meshing screws and carried along between threads to discharge. Pump works against pressure of 1000 psi while handling oil at 7000-rpm

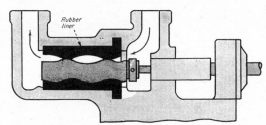

SINGLE-SCREW pump has one screw turning in double-threaded helix. Rotor is metal while helix is hard or soft rubber, depending on service. Softer helix bears against rotor, doing away with clearance. Use either end of the pump for suction

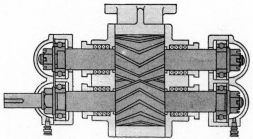

SEPARATE-BEARINGS pump is for abrasive liquids. Helical gears don't build up pressure at teeth ends as spur gear, so holes needn't be drilled. Outside bearings prevent abrasive wear by liquid. Watch shaft deflection by not overloading pump

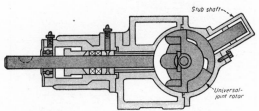

UNIVERSAL-JOINT pump has two shafts connected in casing at angle through universal joint. Joint is spherical and, while revolving, pumps because of closing and opening action of joint. Spherical parts fit snugly, preventing liquid slipping back

on Your Rotary Pumps?

GEAR TYPE shown is simplest rotary pump. Liquid is moved between gear teeth and casing. Meshing teeth act as valve, preventing liquid sliding back to low side. Trouble is high pressure builds up from liquid trapped between ends of the meshing teeth. That forces gears apart against casing, second sketch. Solution is drilling holes in idler gear, third sketch. Then machine slot in the pin so pressure can escape to high side. Helical gears do not need this change as oil is not trapped

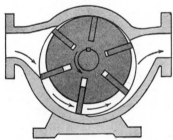

SLIDING-VANE—throws vanes against casing bore. Vanes move out from off-center rotor, sealing casing bore, and prevent slippage while pumping liquid

SWINGING VANE is much like sliding vane but vanes are hinged, and offer less friction. Vane is sturdier; so chance of damage is less if the discharge is closed

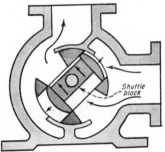

SHUTTLE-BLOCK pump has sliding block in rotor slot. Hole in block fits over off-center stationary pin fastened to pump cover, which slides block in the rotor

Rotary and Reciprocating Pumps Are Positive Displacement

► MOST PLANTS have rotary pumps of one kind or another. These sturdy work horses move oil, water, chemicals and many other liquids. Because they are compact and do their work so well, tendency is to forget all about them until they start giving real trouble. A quick look at the insides of the more popular rotary-pump types will help you understand them.

Principle. Because outside of these pumps resembles the centrifugal type, that's what many people take them for. But they are next in line to reciprocating pumps because both are positive-displacement pumps. The water is not "thrown," as by a centrifugal pump, but actually trapped and pushed around in a closed space as in a plunger-type pump. Rotary type combines constant-discharge characteristic of centrifugal with positive-discharge feature of reciprocating. Only difference from reciprocating is that discharge pulsates

with piston, while rotary's discharge is constant. And, of course, one rotates while the other reciprocates.

Construction. Casing of all rotary pumps is fixed, or stationary. But insides have tightly fitted gears, vanes, pistons, cams, segments or screws, depending on design. Working parts of most common types shown have two external spur gears; two spiral gears; two herringbone gears; external-internal gears; screws, swinging-vane, 2-lobe, 3-lobe or cam types; special lobe type; guided-vane types, cam and piston; rotary piston; and universal joint.

Maintenance. Packing trouble is cut down to minimum where only low pressure or suction end of pump has packing gland.

For best results, don't wait until these pumps lose much of their capacity from slippage. Inspect insides at set intervals and check up on clearances that allow liquid to slip.

End-suction centrifugal

Some handy pump applications

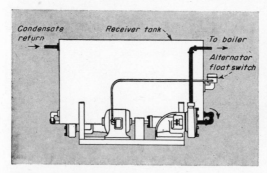

DUPLEX CONDENSATE UNIT has an extra pump for occasional heavy demands; reliability. Condensate returns to boiler from the steam coils, radiators or steam-operated equipment

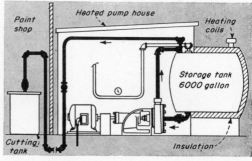

PAINT STORAGE TANK is outside the plant's many buildings. Paint is pumped as far as a mile to various buildings. Temperature is maintained at the right level for easy mixing

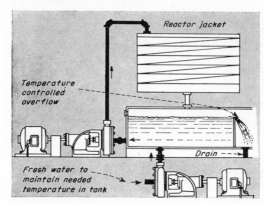

REACTOR JACKET needs continuous cooling to a constant temperature. Water is conserved by returning it to supply tank. Self-acting temperature-flow valve maintains temperature

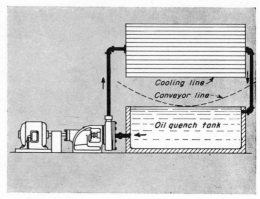

QUENCHING BATH must be agitated to keep its temperature uniform in all parts of tank. At same time, oil is recirculated at constant temperature for each piece of metal quenched

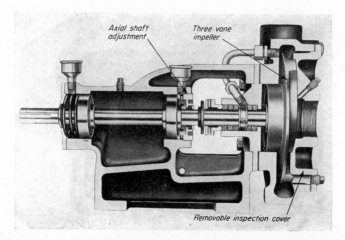

What you should know

▶ END-SUCTION centrifugal pumps are reliable old workhorses of industry. As name implies, fluid is drawn into end or "eye" of pump at impeller's center, then forced out at impeller's rim.

Semi-open, three-vane, nonclogging type impeller shown has only one side-wall. Being semi-open, it allows dirty water and many kinds of foreign materials to pass through impeller vanes.

You can adjust running clearance between open side of impeller and cas-

pumps do many tough jobs

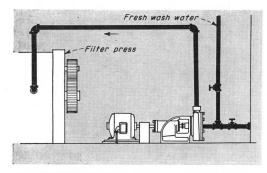

ACID-RESISTANT PUMP transfers slurry to a press where the liquid is removed. Then the pigment is dried and prepared for making ink. This is one of the many process applications

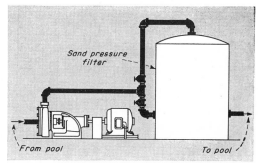

SAND FILTERS are serviced by 5-hp pump that recirculates 122,000 gallons of water through filters in about 8 hours. It also backwashes each filter at rate of 330 gallons per minute

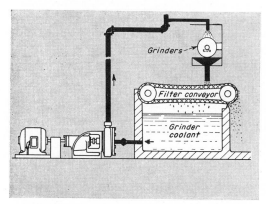

METAL-CUTTING OPERATIONS bring up several problems. Fluid is expensive and must be saved. Hookup, shown, conserves the cutting fluid for filtering and also for immediate reuse

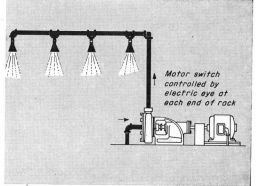

LIMESTONE SCREENINGS carried by truck must be wet down. Electric eye at spray track's end operates pump. Truck is rinsed and limestone wet down without the driver dismounting

about this type of pump

ing while pump is running. Move shaft axially with screw, drawing, lower left, p 116, then lock it in place. Axial movement also affects pump's capacity and head, reduces impeller wear.

Suction cover can be removed to inspect the impeller. Some impellers have wiping vanes on back. You need these vanes for handling liquids with lime, chips and solids that tend to coat metal surfaces or clog space between casing and impeller. These conditions cause

needless wear and also eat up costly power.

Another advantage of these pumps is that liquid end can be supplied in various metals without changing the power end. Where standard cast-iron pumps can't be used, liquid-end parts may be of special alloys, while the support head can remain cast iron. These features can save money.

Pumps shown, range from 1- to 5-in. discharge size with capacities up to

1000 gpm at 250-ft head. Either standard packing or mechanical seals are used, depending on service. Flushing-type stuffing box can be used for shaft cooling or high-vacuum sealing on pumps operating at 200 F or over. Some of the many uses are shown. Pumps are designed so when pumping liquids of viscosities and specific gravities not greater than water, motors won't load above 10% of their standard hp ratings. *Courtesy, The Deming Co, Salem, Ohio*

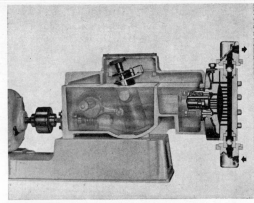

OIL IMMERSION lubricates all working parts. Integral vent and fill valves assure constant oil supply to reciprocating meter pump

Is your chemical feed toxic? hazardous? expensive? Then your feed pumps must meet two needs: controlled volume, no leaks. Find out how . . .

Piston-diaphragm pumps cut

Recent water-treatment developments have introduced many new chemicals. They can be toxic, hazardous and sometimes quite expensive—characteristics which can lead to severe pumping problems. And some old standbys—acid and caustic—are also hard to handle. So from the angles of safety and expense, stuffing-box leakage cannot be tolerated. The usual metering pump is a reciprocating type with extremely high volumetric efficiency. Stroke count serves to meter flow—but these pumps must have a stuffing box. Leakage here may affect metering ability. Where dangerous chemicals are involved, personnel and other hazards are high, housekeeping problems aggravated.

Diaphragm pumps, very much like the ordinary gasoline pump in your automobile, have been used to solve this leakage problem. But this can bring up an out-of-the-frying-pan-into-the-fire situation—the severe hazard of diaphragm breakage. It is virtually impossible to mechanically load the flexible diaphragm so nonuniform loads and diaphragm stretching never happen. In continuous-duty operation, these can lead to diaphragm failure.

Piston-diaphragm pumps get around both of these shortcomings.

Here's how these unique pumps work.

Basically, a piston-diaphragm pump is two pumps built together. First a simple reciprocating metering pump handles a nontoxic hydraulic fluid. Slight stuffing-box leakage here presents no problem—no danger and wastage only of relatively inexpensive hydraulic fluid. But this reciprocating pump doesn't really "pump", it "pulses" the hydraulic fluid in a fixed time-volume relationship. Pulsing fluid acts evenly and uniformly against a flexing diaphragm. Then the chemical-handling side of the diaphragm is fitted with suction and discharge valves in the regular way. Exaggerated drawings, above right, show suction and discharge strokes, diaphragm position in each case.

To insure accurate metering the hydraulic-fluid supply *must* be kept constant. It cannot be allowed to accumulate air or vapor—compressibility of these would take up part of the metering pulse and destroy metering accuracy. In some cases, constant supply depends on immersing the entire pumping action in the hydraulic fluid. Then two spring-loaded valves—one at the top and one at the bottom—vent trapped air and vapor and assure a constant supply of hy-

draulic fluid to the pumping element.

In other designs, a separate hydraulic system mounted on top of the pumping unit vents air and keeps the pumping element full of hydraulic fluid. This system operates mechanically directly from the crankshaft and functions on each stroke of the pump.

Metering capacity can usually be varied in infinite increments from zero to full flow by one or both of two means. First, the driving motor may be a variable-speed type. Second, the stroke of the reciprocating pump may be varied. This is done in some cases by shortening the crank throw. Or an intermediate wobble type crosshead can be adjusted to alter plunger stroke without varying the crank throw. In some cases the pump plunger is not directly attached to the connecting rod. Rather it bears against a spring-loaded plate whose lateral position can be externally adjusted. Then extreme position of the plunger is limited; effect is similar to short stroking and regulates capacity per stroke.

Latter two flow-changing methods can be adjusted while pump is running; adjustment can be manual or controlled by an ordinary pilot device actuated by process flow, pH or temp-

EXTERNAL HYDRAULIC SYSTEM operates mechanically on each revolution to vent hydraulic chamber and assure constant supply

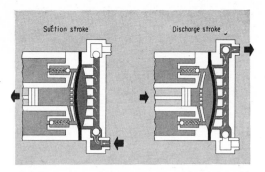

LIQUID END of piston-diaphragm pump is exaggerated to show diaphragm and piston positions during suction and discharge strokes

metering headaches

erature. Similarly, automatic control is possible from a remote location.

In another variation, air pressure is used to pulse the hydraulic fluid and flex the diaphragm. Pneumatic pump, right, uses a timer to operate a 3-way valve alternately pressurizing and dumping the hydropneumatic cylinder. The timer can, of course, be located at any convenient place to control from zero to full flow.

Or flow can be controlled by a stroke-changing device. An electric probe in the hydropneumatic cylinder makes contact with the rising surface of the conducting hydraulic fluid. This then controls the 3-way valve, admitting air and forcing the liquid level down to pulse the diaphragm. After a short delay, air pressure is dumped, liquid level raised by chemical suction pressure which reflexes the diaphragm until surface again touches the probe. Operating cycle repeats for each metering stroke of the pump.

Materials of construction are one big advantage of the piston-diaphragm pump. Leakage from stuffing box is noncorrosive hydraulic fluid, so most of the pump can be built of cast iron or steel. There's no need to take into account corrosion character-

istics of chemical or liquid handled.

Diaphragm, pump casing and valves are only components actually contacting the pumpage. For these parts, materials can be selected to handle almost any chemical. Highly corrosive service commonly uses AISI-300 series steels, with diaphragm made of an inert synthetic. For very-high temperatures the diaphragm can be a thin stainless-steel sheet with high ductility and low metal fatigue characteristics.

Diaphragm-head designs are available to meet many other special services. Example: head at right is built to handle slurries. Note that outboard ported head is omitted to prevent possible clogging. Extreme flex of diaphragm is controlled by another, conventionally confined diaphragm in series with it. Liquid sealed between the two keeps the unconfined diaphragm exactly in phase.

Double-diaphragm construction has other uses, too. For safety's sake in extremely hazardous services, double diaphragm prevents contamination of major portion of hydraulic fluid. If primary diaphragm fails, a gage glass between the two diaphragms acts as a telltale, permitting repair before damage goes too far.

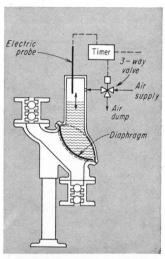

AIR-OPERATED PUMP is controlled by timer or electric probe which limits stroke

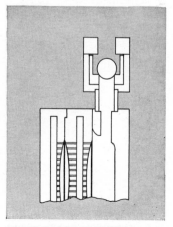

DOUBLE-DIAPHRAGM DESIGN for slurry service keeps unconfined diaphragm in phase

For Long Life, Check

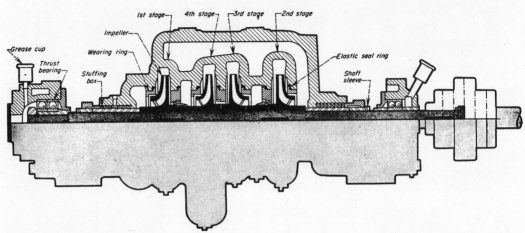

Grease cup — Thrust bearing — Stuffing box — Wearing ring — Impeller — 1st stage — 4th stage — 3rd stage — 2nd stage — Elastic seal ring — Shaft sleeve

1 Centrifugal pumps take a beating from continuous operation. Open boiler feed—and other vital pumps—once a year for a good look-see inside. Check for wear, corrosion, fractures and cavitation. Poor circulation during the off-pumping periods in high-pressure pumps causes plenty wear. Boiler feed pump's arch enemy is low head of water on suction side from sudden head drop, causing water to flash. Then the pump overheats, wearing sealing and channel rings, or bending shaft

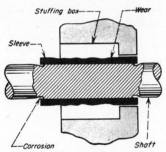

Stuffing box — Wear — Sleeve — Corrosion — Shaft

4 Check shaft sleeve through stuffing box for wear from packing. Turn down worn sleeve or replace with new. Look for shaft corrosion at ends of the sleeve

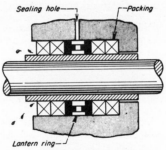

Sealing hole — Packing — Lantern ring

5 Renew shaft packing, align lantern ring with sealing hole. When running, lubricate shaft with slight gland leakage. This keeps air out if under vacuum

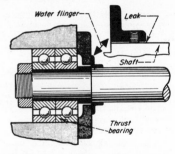

Water flinger — Leak — Shaft — Thrust bearing

6 Water-flinger ring protects thrust bearing from water leakage. See that ring is snug fit on shaft. If not, seal between shaft and ring with sealing paste

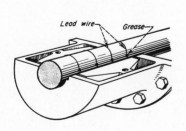

Lead wire — Grease

9 For shell-type bearing, check clearance by placing lead wires on shaft and screwing down bearing cap. Mike wires. Remove shims for right clearance

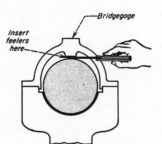

Bridgegage — Insert feelers here

10 On shell-type bearings, take bridgegage readings to find wear on lower half of bearing. Insert feelers between gage and shaft. Rebabbitt if low bearing

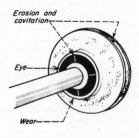

Erosion and cavitation — Eye — Wear

11 Examine impellers carefully for wear at wearing rings. Check vanes for erosion, cavitation and corrosion. Vanes at eye and discharge are trouble spots

Centrifugal Pumps Annually

In these times of peak loads, high prices and slow deliveries,

it's cheaper to look inside the pumps now than be sorry later

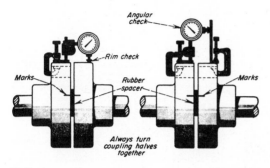

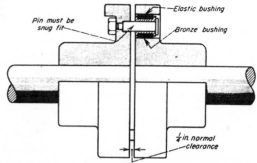

2 Break coupling and align pump and motor shaft while pump and lines are hot, as hot piping tends to throw pump out of line. Align pump again after it's cold. Mark halves, place rubber, rotate halves together and make rim and angular checks

3 If coupling is flexible pin-type, remove all pins and bushings. Clean wearing surfaces of bushings and pins, check for roundness. Lubricate pin and replace elastic bushings that are shot. Check wearing surfaces of all type couplings

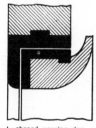

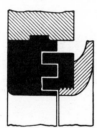

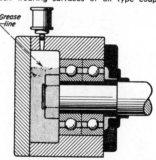

7 Keep clearance between impeller and casing wearing rings close to design recommendations. This opening affects pump's efficiency. Usual clearance should be no more than 0.003 in. for each inch wearing-ring diameter. Never let this opening double the original clearance. Labyrinth type leaks less than L-shaped shown here

8 Clean oil or grease out of bearings. Check for water leakage that knocks bearings out. Clean out all the old lubricant and inspect bearing for wear

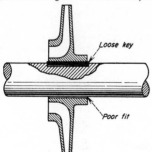

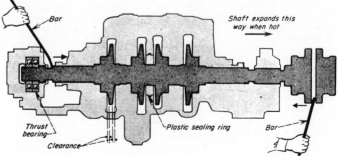

12 Tap key in impeller for being tight. Shaft twisting while under load, expansion, corrosion, or original poor fit makes impeller progressively loose

13 Check thrust-bearing clearance by breaking shaft coupling and barring shaft both ways against thrust bearing. Measure shaft movement from each side of impeller, near tips, housing. Consider expansion of shaft from cold to hot running condition. Before closing pump, renew plastic sealing rings as old ones harden from heat

Casualty experience gives centrifugal-pump users a good guide to preventing failures—shows areas of design, installation and operation that are most apt to cause trouble

EMERGENCY SERVICE-WATER PUMPS should be protected by suction strainers. Strainers keep foreign material from getting into the impeller and damaging it

To guard against centrifugal-pump failures, know their primary causes

Probable failure, the sword of Damocles in any centrifugal-pump installation, depends on factors ranging from application and pump type to degree of preventive maintenance. But a statistical review of past failures can guide users by showing them the sensitive areas. Forewarned is forearmed; emphasis on the weak links will go a long way toward catching potentially serious troubles before they get a chance to become actual.

A recent statistical survey of centrifugal-pump failures, made by The Hartford Steam Boiler Inspection and Insurance Company, provides a good basis for users to predict trouble areas and guard against them.

Table, right, breaks down initial failures of 132 centrifugal pumps by general service area and part that initially failed. Further analysis shows *why* the majority of these failures happened.

Pump-shaft failures predominate—they represent

Based on information supplied by The Hartford Steam Boiler Inspection and Insurance Company, Hartford, Conn.

50%—so let's look at them first. Many feed-system centrifugal pumps (including boiler-feed) operate at relatively high speeds. So the wear and tear of misalignment is magnified. Failure percentages reflect this.

On the other hand, the possibility of foreign objects entering pump suction is far greater with other energy-system pumps (which include service-water pumps). But since a service-water pump usually runs at lower speed for higher capacities, it has generally larger internal passages. Many solids will pass through without causing enough internal stress to actually break the shaft, so here, the percent of failures tracing to foreign objects is relatively small.

Progressive cracking because of corrosion or vibration is a disease common to all pumps; this area needs maintenance emphasis regardless of pump type. At the design stage, proper material selection can minimize corrosion failures. First step here is to know all properties of the liquid pumped; then you're on reasonably safe ground as you go ahead with material selection.

Vibration can come from hydraulic cavitation or mechanical sources. Former should be a function of correct original application—but if you have a pump in operation which is cavitating badly, take immediate steps to relieve it. The process may not be cheap, but in the long run it will be far less expensive than letting the pump run to destruction. There are two ways to attack cavitation: First decide whether your system can be altered to increase suction head, by reducing pipe friction, increasing static head or perhaps cutting pumping temperature. Second, if these steps are not practical, perhaps the pump manufacturer can give you a different impeller design with lower npsh requirements.

Progressive shaft cracking due to mechanical vibration is inexcusable. All rotating equipment should be inspected frequently to be sure that vibration is well within allowable maximum set by the manufacturer. If it exceeds these limits, the pump should be taken off the line immediately and the cause of vibration determined and corrected.

Bearing failures are next on the list. The vast majority stem from lubricant difficulties, regardless of pump service. Loss of lube occurs when oil level falls too low or, in pressure systems, when the oil pump fails. Former is an inspection fault; latter can be overcome by providing a backup pump to cut in on low oil pressure.

Contamination of lubricant takes many forms. It may just be dirt. Even if an oil system is not exposed to the air, vents are provided to equalize air pressure with changing lubricant temperatures—letting the system breathe. These vents alternately draw in and expel air which may be loaded with dust or other plant impurities. Then, over a period of time, the impurities collect and form sludge. Answer here: inspect the lubricant as frequently as necessary to assure yourself that it remains satisfactorily clean. In critical equipment where bearings are pressure lubricated, a filter with bypass arrangement may be a good investment to continuously clean system oil.

Water is a contaminant too. It can enter a lubrication system through cracks in cooling jackets and oil coolers. But another water source is less obvious than these—condensation. As we have said, any lubrication system breathes—and to some degree it inhales water vapor. Temperature within the bearing housing dictates how much of the intake water vapor condenses out. When bearings are overcooled—to a temperature below the ambient wet-bulb—water vapor will condense out within the bearing housing. Since this water vapor has no other place to go, it will collect in the lubrication system and build up. Water contamination then promotes corrosion and ultimate bearing failure.

Loss of main-pump suction is another source of bearing troubles, especially in relatively high-speed feed-system pumps. Vibration, caused by loss of suction, overloads the bearings, and they fail prematurely. Defective bearings show up more in feed-system pumps, also because of higher operating speeds which accelerate failure and will not long tolerate a marginal bearing.

Internal pump damage again breaks sharply between feed-system pumps and others. Condensate in feed systems is normally clean from a physical viewpoint—at least there are no sticks and stones in it. But these pumps

Pump-failure study shows you what to watch out for

	Feed-system pumps	Other energy-system pumps	Total
Pump-shaft failures	14	52	66
Reasons for shaft failures			
Misalignment or poor shaft installation	50%	26%	
Progressive cracking (vibration, corrosion)	42%	54%	
Foreign objects	—	12%	
Other causes	8%	8%	
Bearing failures	23	32	55
Reasons for bearing failures			
Loss or contamination of lubricant	45%	71%	
Loss of main-pump suction	14%	3%	
Misalignment	9%	10%	
Defective bearings	14%	6%	
Other causes	18%	10%	
Damaged pump internals	4	7	11
Causes of internal pump damage			
Loss of pump suction	72%	8%	
Foreign objects	14%	80%	
Other causes	14%	12%	
Total failures	41	91	132

are more prone to suction-side disturbances. Their high speeds accelerate the effects of suction troubles so that internal damage to impellers, wearing rings and interstage bushings can be extensive.

On the other hand, other energy-system pumps are frequently not blessed with such physically pure water. They're the ones that do catch the sticks and stones, so their greatest internal damage stems from foreign objects. The cure for feed-system pumps lies in proper operation and design—plenty of suction pressure margin. In other energy-systems pumps, suction screens will keep out trash that might cause internal damage. Be careful of suction strainers, though. Unattended, they can be worse than none at all. If suction strainers are permitted to load up, pressure drop across them increases with attendant loss of suction pressure. And so, out of the frying pan into the fire—you have kept out the trash only to ruin your pump because of loss of suction pressure.

Of course, the list of things that can go wrong with a centrifugal pump is endless, but these failures top the list. Odds are that special attention to these potential trouble areas will pay off in reduced maintenance and pump-replacement costs. And downtime will be cut too— a real money saver.

1 End thrust of impeller shafts is balanced hydraulically in many centrifugal pumps but they also have thrust bearings

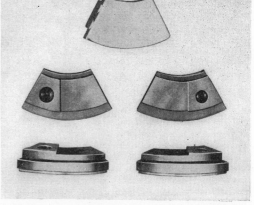

2 Shoes or pads in thrust bearing, Fig. 1, are adjusted to take equally unbalanced end-thrust load on impeller shaft

Watch Wear and Tear on

It's not so much how you correct trouble that counts, but what you do to keep it from happening. Here are ounces of prevention that are worth pounds of cure for your pumps

▶ For MAINTENANCE purposes rotating pumps may be classified as centrifugal, turbine, propeller and rotary. When used right, and given adequate inspection and maintenance, these pumps give long service with little trouble. Knowing the kinds of troubles that may happen and what to do to prevent them helps lengthen pump's life.

Bearings. Treating first the wear of oil or grease lubricated parts, we usually think of the different types of bearings. While propeller pumps are generally vertical-shaft design, others, outside of deepwell turbine types, usually have a horizontal shaft.

Sleeve bearings are commonly used on these units and are the simplest design. They are lowest in first cost, but unless given proper attention they may be the most expensive to operate.

For trouble-free operation, sleeve bearings must (1) have an adequate supply of proper grade oil, (2) be kept clean, (3) not run too hot.

Oil Rings. Inspect oil rings frequently to make sure they do not "hang up." These rings must turn freely with the shaft to supply the bearing with oil from the oil well, Fig. 3.

Dirt in the oil will cause oil rings to stick. Then the bearing, not getting enough oil, heats and may burn out. So be sure the oil is clean and there is plenty of oil in the wells.

Keep Oil Clean. If you have any doubt about the quality of the oil, shut the pump down at the first chance. Drain the oil and flush the bearing and oil wells first with kerosene and then with light lube oil.

Refill the bearings with new oil of proper grade. Keep oil at proper level in oil wells, usually indicated by a line on the bearing side or on overflow connection or gage.

A low level may prevent enough oil getting to the bearings. Oil too high causes waste and oil may work along the shaft and get onto motor windings to destroy their insulation.

Don't let babbitted bearings get hotter than 160 F. If you can't hold your hand on a bearing cap for more than a few seconds without discomfort, the bearing is getting too hot.

Anti-Friction Bearings. Small and medium-size pumps frequently have ball or roller bearings. Lubricant in

these is sealed in and requires little attention, except when the bearings are cleaned at scheduled intervals (every two or three years) and the oil or grease replaced. If anti-friction bearings get noisy it shows they have gone bad and should be replaced.

Balls, rollers or races sometimes crack and become noisy. When this happens shut the pump down and replace the bearing, because it is probably seriously damaged and may injure the shaft.

Thrust Bearings. Rotating pumps usually have a thrust bearing, anti-friction or Kingsbury, to take the end thrust of the impeller shaft. Many pumps are designed to have their end thrust hydraulically balanced. But these also have a thrust bearing to take any unbalanced end thrust caused by imperfections in the pump or from wear that may develop during operation under normal conditions.

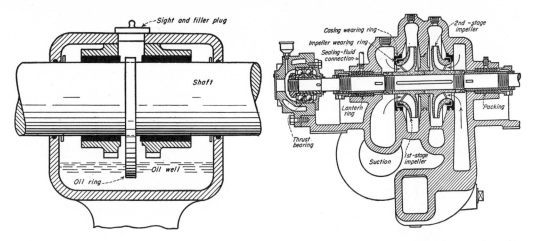

3 Inspect oil rings often to make sure they rotate freely with shaft to supply bearing with oil from the well

4 Wearing rings, have close running clearance to prevent fluid leaking from discharge to suction side of impeller

Your Centrifugal Pumps

When anti-friction bearings are used on horizontal-shaft pumps, the bearing that takes end thrust may also carry the radial load. Even on some sleeve-bearing pumps anti-friction thrust bearings are used. But in many pumps a Kingsbury bearing takes the thrust. These bearings are usually installed on the outboard bearing pedestal, Fig. 1.

The Kingsbury thrust bearing, shoes or pads, Fig. 2, are adjusted to share the load equally. They are usually oil lubricated by a pump gear driven from end of the impeller shaft. On large high-speed pumps the lube oil is water cooled.

Check thrust-bearing temperature often to be sure that it doesn't exceed 160 F. If bearing temperature rises above normal: (1) check oil supply, (2) see if it is clean, (3) check cooling-water flow and its temperature, (4) see if oil-cooler tubes are fouled.

Sludge on the oil surfaces, or scale on the water surfaces of the tubes, can cause high bearing temperatures. It is usually necessary to stop the main pump to take the cooler apart for cleaning. Sometimes soaking the cooling coils in an inhibited acid simplifies cleaning.

Assembly of a Kingsbury thrust bearing is usually a job for a machinist. Careful adjustment is needed to hold running clearances in the pump.

Lubricating-Oil Pumps. For most bearings, these are usually of the rotary type. They need little attention as long as they supply plenty of clean cool oil to the bearings.

At regular intervals, depending on the application, drain the oiling system, including the oil pump and piping, and flush with kerosene to remove sludge.

Following the cleaning, flush with light oil before putting new oil into the system. Consult the manufacturer's instruction book on how to make oil-pump adjustments.

Wear on Parts. Wasting away of metal in pumps may be caused by (1) corrosion, which is an electrochemical action, (2) erosion or scouring action of solids in the fluids, (3) cavitation, the forming of bubbles in a low-pressure area and their sudden collapse later in a high-pressure region. Collapsing of these bubbles frequently causes erosion and pitting of the metal part in their region. Cavitation also may cause noise or a drop in head, capacity and efficiency, and the pump may vibrate.

Pump Stands Idle. If a pump is left full of water when idle for long periods, the inside of the casing and other internal parts may corrode in certain waters. It is advisable to drain the pump, unless it must be held in standby service. Pumps may corrode less in some waters if left full than if left dry.

If water does not seriously corrode the internal parts when your pump is running it should not do so when the pump is idle. Under such conditions it should be safe to leave the pump full of water when idle for a long time.

If water attacks parts during normal operation, then the pump probably should be drained, if it is to remain idle for a long period. Under such conditions its a good idea to open the pump and put it in shape to go into service again when needed.

Unless you may need the pump on short notice, you'd better leave out its packing. Some packings have corroded impeller shafts when left in idle contact with them for a long period. Also some packings tend to dry out and harden more rapidly when idle.

Wearing Rings. Sealing or wearing rings between casing and impeller, Fig. 4, wear from several causes. If the clearance is too small the rings may wipe. Abrasive material in the fluid pumped may wear these rings rapidly. Sealing rings may also wear rapidly if the pump runs dry or becomes vapor bound and loses its prime. When sealing-rings wear and cause a pump to lose capacity and efficiency, you should renew them immediately.

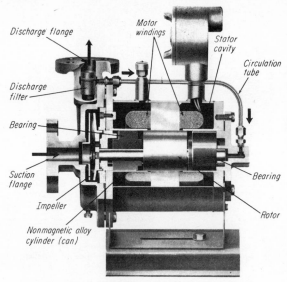

INTEGRAL MOTOR PUMP has small portion of pumped fluid recirculated through motor section (can) and oil-filled stator cavity

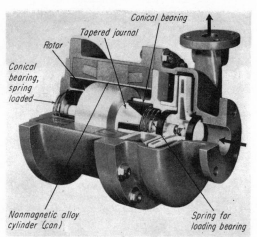

ANOTHER ZERO LEAKAGE DESIGN has conical journal-actuated bearings to maintain minimum bearing clearance at all times

Integral motor pumps or separate motor pumps, how do they compare?

One hears more and more about the canned pump because it has made some impressive inroads into several areas long dominated by the separate motor centrifugal pump. One firm alone claims to have sold over 100,000 canned pumps since this design first made its bow a few years ago. Thus, a number of manufacturers are jumping on the bandwagon and producing the unique design.

Because of the confusing and often conflicting statements one hears when canned pumps are mentioned, we asked pump manufacturers to give us their views on eight areas of controversy that keep coming up. Most of these manufacturers make both integral and separate motor-type pumps. They are primarily interested in seeing that you apply the right pump. We hope that you put the following summary of their comments to constructive use.

1—The motors of canned pumps are low in efficency because of the can material on the rotor and stator. Conversely, separate motor pumps are "as efficient as modern technology can make them." Also, no material is required for covering the rotor and stator.

It is true that electromagnetic efficiency of a canned motor pump is partially reduced because of increased air gaps and the presence of nonmagnetic liners. But these losses

are somewhat compensated for by the elimination of stuffing boxes and mechanical seals. Don't forget that seals also contribute to the inefficiency of conventional pumps through friction losses.

The efficiency of a separate motor pump is normally based on the brake horsepower input to the pump versus the work done by the pump. No consideration of electric motor efficiency is made. Since it is impossible to separate pump efficiency and motor efficiency in a canned pump, the normal method of evaluation is to look at *power input* versus *work done* by the pump, or *wire-to-water* efficiency.

The same figure is obtained for the separate motor pump by multiplying motor efficiency by pump efficiency. When all of these factors are considered, the difference in ratings is frequently negligible.

Also, because a zero leakage pump is liquid-cooled, heat is removed more efficiently from the motor area. This cooling obviously permits the motor to work at higher amperage levels than would be possible in a standard air-cooled application. In other words, more work can be actually obtained from an electric motor of a given size when applied in a canned pump.

Amperage draw in canned pumps is usually insignificant in the smaller sizes up to about 7 hp. But, for the larger rating units, amperage draw can be important. While most zero leakage pump motors may be less ef-

Canned pump has zero leakage, does a specific job

Originally, the integral motor pump (also known as *sealless, zero leakage* and *canned*) pump, was applied only on highly specialized applications, where absolutely NO leakage could be tolerated. An example is the pumping of liquids subjected to radioactive contamination in nuclear energy electric generating stations. This unit has now invaded the general process field, and some other industrial areas as well.

Basically, the canned pump differs from other centrifugal pumps in that the pump and motor are combined in a single unit, and completely enclosed within a corrosion-resistant, nonmagnetic alloy cylinder (can). Part of the pumped liquid circulates through the motor section to cool the motor and also lubricate the bearings. One design has been simplified so that now it has only 14 basic parts. This compares to 50 basic parts required for a standard cen-

trifugal pump of separate motor design.

However, all canned pump motors suffer losses because of the larger air gap required to accommodate the stator and rotor cans. But most manufacturers measure these losses and take them into account when writing amperage draw ratings for their canned pumps.

Bearings are the single, most critical parts of a canned pump, primarily because they are in the pumped fluid. The choice of bearing material often presents a complex problem. Why? Because the choice must be made from hundreds of available materials.

And when it comes to bearing life, it may depend on the clearance between shaft and bearing. This was pointed out in a study of submerged bearings conducted by Battelle Memorial Institute some years ago.

An automatic thrust balance feature by one manufacturer has completely

eliminated the axial bearing wear that was a serious source of bearing failure in earlier designs. Another improvement is the use of an oil-filled stator which extends motor life by keeping the windings cooler while protecting them from condensation damage problems.

Since the pump and motor are built as a single unit, without an external rotating shaft, no coupling, seals, or stuffing boxes are needed. Elimination of mechanical seal or packing maintenance and lubrication, and housekeeping problems from fluid leakage, is dramatically extending the range of potential applications for this pump.

For most applications, canned pumps are lighter and smaller than separate motor pumps. This makes their foundation cost considerably less. Smaller sizes can even be mounted directly in the piping without additional support.

---

ficient, there are many applications, such as those for hot oil, when the *can* barrier is not necessary. The motor efficiency of the zero leakage pump then approaches that of the conventional motor.

Pump manufacturers will agree that all industrial-type pumps can be built with efficiency as high as modern technology can make them. The only question is: Will the customer pay the price?

2—Standard canned pumps are designed for 3500 rpm only. This is critical because 3500 rpm requres more NPSH (net positive suction head) than 1750 rpm, for example. Also, 1750-rpm canned motors are special, which means more money and longer delivery. In contrast, separate motor pumps have motor and pump combinations available with a variety of speeds to suit pumping requirements.

Nothing could be further from the truth. Canned pumps are available in both 1750- and 1150-rpm speeds (from one firm), as a completely standard pump with price and delivery on the same basis as a standard line of 3450-rpm zero leakage pumps (from the same firm). And there are many lower-speed pumps furnished as a complete line.

While the 1750-rpm canned pump (for example) is not special, it is only more expensive if a larger pump is required to do the job at 1750 rpm than at 3500 rpm. But deliveries in either case are the same. Also, there is no good reason why these pumps cannot be made for various other speeds. It depends on the definition of *standard*.

3—The direction of shaft rotation is difficult to establish in the integral motor pump because no moving parts are visible. On the other hand, the rotation is easily determined in the separate motor pump by observing the shaft or coupling.

While it is true that the direction of shaft rotation cannot be observed in canned pumps, the pressure gage on the discharge side of any centrifugal pump will indicate correct rotation. The reason is that a centrifugal pump develops only about 60% of its head when running in reverse. Thus, it is a simple matter to run the pump a few moments against a closed discharge valve as soon as it is wired into the system. Then reverse the electrical connections and again run the pump for a few moments. The connections which result in the higher discharge pressure are obviously the correct ones.

While observing pressure is easy enough to do, it may not always be practical, especially during the installation period of new, larger plants. The wiring is usually completed long before the process equipment is ready for initial startup. By the time the piping is flushed and the plant is ready for its trial run, the operators may naturally assume that all equipment is properly hooked up.

To make sure pump rotation is correct, a pressure check would obviously have to be scheduled for all pumps as soon as flushing or process fluid is available in the plant. Conversely, the wiring in an existing plant is usually done by the plant electrician, who can check rotation as soon as the pump is hooked into the system.

Another method is to use a relatively inexpensive phase-sequence meter that checks actual connections against the phase sequence indicated on the pump nameplate.

Also, a shaft rotation indicator is now available which quickly gives the answer. It can be used on any type of centrifugal pump or other rotating machinery to insure initial operation in the proper direction, instead of the usual "hook up and try" approach.

Don't forget that while there are no external rotating parts in an integral motor pump, this feature is basic to its main advantages. Under some conditions, pumps can be checked by looking into the suction opening, but no

Separate motor pump is the workhorse of industry

In contrast to the integral motor pump, the separate motor type comes in numerous designs. Regardless of the design, they all have a separate motor and a separate housing for the pump (see illustrations on the opposite page). All separate motor pumps require a seal around the shaft to prevent the pumped fluid from leaking out of the casing. It may be a mechanical seal (face type), or jam-type packing in a stuffing box around the pump shaft. The stuffing box may have a connection for pumping a clean protective fluid into the lantern ring to keep abrasive or corrosive liquids away from the packing.

Although leakage from a mechanical seal is usually so slight that it may not be apparent, it still cannot be depended upon as zero (see *Mechanical seals*, 24 pages, *Power* magazine). But where zero leakage is not the main consideration, the separate motor pump with seals has been widely used for years.

For one thing "standard" centrifugal (off the shelf) pumps cost less initially, and there is usually a standard size, design and speed for almost every application. As one pump manufacturer told us, "The only reason some commercial centrifugal pumps may not be as efficient as modern technology can make them is because they have to be sold for prices which are not very much higher than the cost of the raw materials they are made from." Thus, they are popular.

Just as the canned pump (because of its more compact design and lighter weight) may be installed in the line in smaller sizes without a foundation, so may a new design of separate motor pump. This is known as the in-line pump, which is mounted in the line like a valve. Thus, the ability to locate an in-line pump where it is needed, instead of in a fixed bay, avoids long piping runs and cuts installation expense.

For example, one recently introduced in-line design lets you take out the entire pump cover and rotating assembly for service without disturbing the pump casing, piping connections, or driver. The mechanical seal in this design is also serviced, or replaced, by removing the rigid coupling and gland, again without disturbing the rotating assembly or driver. Features of this type keep the separate motor pump in competition with most special designs introduced from time to time.

In a way, it is unfair to compare the integral motor pump with the separate motor design, since the canned feature was originally developed only to prevent leakage. But it is a fact that the canned pump is invading the general pump market in some areas because it has proven to be highly reliable in respects other than zero leakage. Our aim in this report is to help bring the outstanding features of both types of pumps into proper focus.

---

manufacturer would like to see this information in his installation instructions for obvious reasons.

4—Clearances in the plain bearings of the canned pump must be maintained to assure close running clearance between the rotor and stator. But wear, destroying the bearing clearance, will result in short pump life due to contact of the rotor and stator. On the other hand, the motor bearings of the separate motor pump are isolated from the pumpage, or any liquid carrying abrasive particles. Thus there is no problem from abrasive wear.

It is true that bearings and journals are two of the most critical components in a canned motor pump. In addition, proper materials must be selected for each application to insure long service life. But today, manufacturers of canned pumps select the proper bearings and journal combinations for a broad spectrum of applications.

Although bearing tolerances must be maintained, there is ample allowance for reasonable wear. Therefore, there is no reason to state that "wear will destroy the bearing clearance, resulting in short pump life." As a single, isolated example, one paper plant has had over 35,000 hours of trouble-free operation from a canned pump. In fact, this pump has required much less attention and maintenance than could normally be expected from their older-design pumps with stuffing boxes or mechanical seals. That is why the canned pump was initially installed since, in this case, zero leakage was not the main problem.

As for bearing life, it can be lengthened in abrasive service by inserting a filter in the external piping for circulation from pump cavity to motor cavity. And the filter can be cleaned periodically without opening the pump.

One new type of rotating bearing, for instance, claims

to reduce bearing wear over stationary bearings and increase its life by a theoretical factor of four. In actual practice, bearing life has been increased as much as 10-to-1. Wear resulting from thrust loads is also claimed to be minimized by the use of a thrust control design.

Another new design of bearing for canned pumps is a conical bearing (photo, right side of p 171) spring loaded against a tapered journal. This feature is claimed to assure bearing journal contact at all times and compensates for wear on the bearing faces.

Again, it is generally true that bearing wear is critical on canned pumps. But remember that close-coupled, separate motor pumps may develop leaks around the shaft which can damage motors. And, in some cases, special enclosures may be required to prevent this damage. So, if bearings are considered the Achilles' heel of canned pumps, you might say that seals are just as critical in separate motor pumps.

5—Repair of the zero leakage pump is expensive because it requires replacement of the rotor and stator *can*, the bearings and the shaft. By comparison seal component replacement for the separate motor pump, when necessary, is inexpensive.

There is no good reason why a proper inspection and preventive maintenance program will not eliminate expensive canned pump repairs, just as with other types of pumps. Bearings are usually the only items required when the pump is serviced intelligently. Canned pumps used by the AEC, for example, have run for years without repairs in critical nuclear energy electric generating applications. These large canned pumps have chalked up an enviable record. Not only that, but replacement of bearings in canned pumps usually takes less time than the replacement

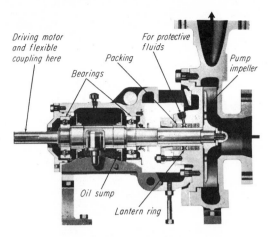

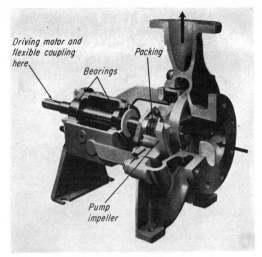

SEPARATE MOTOR PUMP is sealed with jam-type packing and is connected to driver through flexible coupling on end of shaft

ANOTHER SEPARATE MOTOR PUMP DESIGN is sealed with packing backflushed with a clean fluid to prevent excessive wear

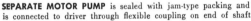

of a mechanical seal. And if a mechanical seal fails, pumped liquid may be sprayed over the surrounding area. This is a potential hazard which usually cannot be tolerated when zero leakage pumps are installed.

Replacement frequency of bearings and gaskets in canned pumps depends on the application. The best way to determine frequency is to monitor bearing wear over the first few months of operation. Wear can then be extrapolated in terms of time based on bearing tolerances, and tabulated in instruction books.

Of course, there have been failures. But the misuse or misapplication of any complex equipment can cause catastrophic failure. One big saving with a zero leakage pump is in installation cost. By comparison, separate motor pumps require special technical care in leveling, alignment, packing break-in, and mechanical seal adjustment.

Complete failure of the canned unit requires expensive replacement of bearings, journals and cans. However, there are records of maintenance-free design life-ratings in excess of 75,000 hours.

6—If the _can_ material thickness on the rotor and stator must be increased to combat the effects of erosion, corrosion, or pressure, motor efficiency will drop as the gap between the windings increases. In contrast, the separate motor pump has no problems with the effects of erosion, corrosion, or pressure on the motor since the motor is not in contact with the pumped fluid.

Basically this is correct. But can thickness and air gap are standard and fixed throughout in some manufacturers' lines of canned pumps. To combat corrosion, a variety of materials is available. Also, special backup plates around the can are used to increase the pressure rating of the

pumps as required. Thus, the academic truth of this statement has no bearing on pump's practical applications.

7—Compared to canned pumps, the separate motor pump costs much less.

There is a differential in original cost, although one must look at the complete picture. When compared on a quoted price basis, canned pumps are almost invariably more expensive. However, one must not overlook their compact size, lighter weight, and elimination of the need for heavy bases, complicated alignment, and leveling procedures. Thus the overall cost of an installed canned pump may be less than that for other types.

Even when the initial cost figure is higher, don't forget that the zero leakage feature commands a premium price. Then again, as the canned pump gains widespread acceptance and production runs increase in quantity, this price differential will probably be reduced.

8—The auxiliary equipment that must be included to protect the motor from abrasives, temperature or corrosion, must be added to the cost.

Once again, the integral motor pump has never been offered as an overall solution to all centrifugal pumping problems. As a result, it is rarely offered for abrasive applications. The only exceptions are when a toxic or explosive fluid cannot be trusted to mechanical seals—no matter how elaborate they might be.

Heating or cooling jackets and similar accessories are available for special applications, but such applications normally require similar auxiliary equipment for the separate motor-type pump. An example is the use of steam or cooling water for mechanical seals. ■

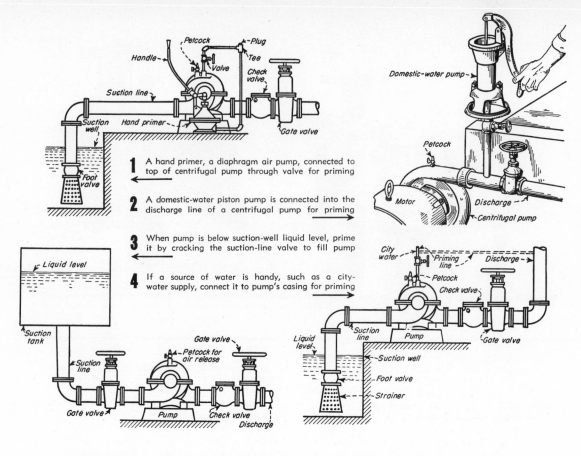

1 A hand primer, a diaphragm air pump, connected to top of centrifugal pump through valve for priming

2 A domestic-water piston pump is connected into the discharge line of a centrifugal pump for priming

3 When pump is below suction-well liquid level, prime it by cracking the suction-line valve to fill pump

4 If a source of water is handy, such as a city-water supply, connect it to pump's casing for priming

How to Prime That Pump

Before a centrifugal pump can operate it must be primed. Ways to do this range from pouring liquid into the pump to automatic pumping. Here you are told how a pump is primed by hand

▶ Positive-displacement pumps, such as reciprocating and rotary types, are self-priming for total suction lifts up to 28 ft, if they are in good condition. But there are installations where these pumps must be primed to insure good operation because of long suction lines, high suction lifts and other abnormal conditions.

Centrifugal pumps are not self-priming, and so when operated on a suction lift, they must be primed. Sometimes it is necessary to install equipment to keep them primed.

Priming Methods. Priming of pumps may be divided into three general classes of installations:

1. Liquid is admitted to the suction pipe until the air is displaced and the suction pipe and the pump are filled.

2. Air is exhausted from the suction pipe and pump by a vacuum-producing device.

3. The pump is made self-priming by recirculating liquid in the pump to exhaust the air, or by special devices in the suction.

Manual Priming. Priming may be done by hand or automatic means. There are several kinds of hand primers. When the pump is below the suction-well liquid level, Fig. 3, it can be primed by cracking the suction-line valve to fill the pump with liquid.

Let the liquid flow slowly into the pump and open the petcock on the casing to release the air. Open the petcock even though the discharge line may be full open at the same time.

If the petcock is not opened, air may be held in the top of the casings and cause trouble after the pump is running. Leave the petcock open until a clear stream of water flows from it. Close the petcock, open the suction and discharge valves, and start the pump.

On installations like Fig. 3, after the pump is once properly primed it usually remains that way and can be put on automatic control, without further attention. But be sure the suction tank will remain full so the pump will not run dry.

Suction Well Below Pump. When the suction well is below the pump, priming must be done in either of two ways: (1) Fill the suction line and

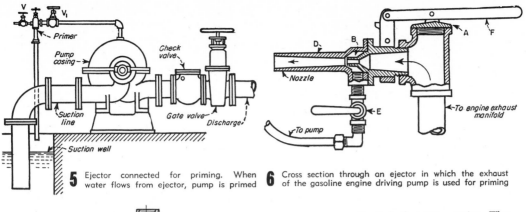

5 Ejector connected for priming. When water flows from ejector, pump is primed

6 Cross section through an ejector in which the exhaust of the gasoline engine driving pump is used for priming

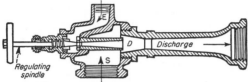

7 Cross section through an ejector used for priming centrifugal pumps if air, steam or water is available

pump casing with liquid from an outside source. (2) Exhaust the air from the suction line and pump.

The former requires a foot valve on the intake end of the suction line, Fig. 4. If a source of water is available, such as a city supply, it may be connected into the pump casing as shown. To prime the pump, crack the valve in the priming line and open the petcock to release the air.

Slowly fill the suction line and pump until a solid stream runs from the petcock. The pump is now primed, so close the petcock and the valve in the priming line. Start the pump and after it is running for a few minutes open the petcock to release any air that might collect in the casing and to make sure the pump is primed. If the pump is properly primed, a solid stream of water should flow out of the petcock.

When the pump discharges into an overhead tank, it may be primed through a small connection with a valve from the discharge line into the pump's casing as indicated by dotted lines, Fig. 4. The pump is primed as previously explained. In either hookup, after the pump is primed make sure the priming valve is closed.

Priming by Vacuum. Because of the trouble with foot valves and the screens on them, these valves are often omitted and the pump primed by a vacuum-producing device. These may vary from power-driven vacuum pumps on automatic pumping system down to a hand-operated piston pump. Fig. 2.

Here a domestic-water piston pump connects into the discharge line of a centrifugal. Before starting the centrifugal, it is primed by operating the piston pump. Because the centrifugal's discharge is at the top of the casing, when water flows from the hand pump the centrifugal is primed and can be started.

Hand Primers. Specially designed hand primers are available for use where an automatic primer may not be justified.

Fig. 1 shows a hand primer, a diaphragm air pump, connected to a centrifugal pump through a valve at the top of the casing. Before the first priming operation it may be necessary to remove the plug from the tee and fill the primer with water to seal its valve.

To operate the primer, move its handle back and forth until air is pulled from the pump's casing and water comes over into the primer. Its action then becomes stiffer, which indicates the pump is primed and can be started.

Priming With Ejectors. When steam is readily available, compressed air or pressure-water ejectors may be used for priming pumps. Fig. 7 is a cross section through an ejector. The ejecting medium comes in at E and passes out through D. When doing so it draws air in at S, which is connected to the pump's casing as in Fig. 5.

To prime pump, Fig. 5, open valve V in steam, air or water line, then open valve V_1 in the ejector suction. When water flows from the discharge pipe, the pump is primed and valve V_1 and then V may be closed. If the suction line is long, a foot valve should be used, to hold the line at least partly full of water. This reduces the priming time.

Where a vacuum source is always available, such as a condenser in a power station, it may be connected to the top of the pump's casing and controlled by a valve for priming.

Engine-Exhaust Priming. On contractor's pumps and other portable units driven by gasoline engines it is common practice to use the engine's exhaust in an ejector for priming. Fig. 6 shows a cross section through such a primer. The ejector connects to the engine exhaust manifold as indicated, and its induction chamber B connects to the pump's casing through valve E.

To prime the pump, open priming valve E. Exhaust valve A is closed by lever F, to pass the exhaust gas through the ejector. The rapidly moving gas passing through the ejector nozzle D draws air from chamber B. This action rapidly removes air from the pump's casing. A check valve in the discharge line prevents air entering the pump.

As soon as the pump is primed, water vapors discharge from the ejector and the priming valve is closed. Then open exhaust valve A, by throwing lever F up and over 180 deg.

Ejector priming is quite rapid on suction lifts up to 15 ft. Greater suction lift can be primed, but priming time increases quite rapidly above a 15-ft lift.

This method has the advantage of having no moving parts, and of using an ejector medium that is readily available. It also eliminates pouring the casing full of water, and a foot valve is not needed.

273

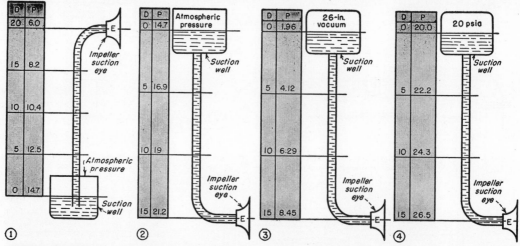

SUCTION SYSTEMS. Four typical pump-suction systems, together with the pressures at different points inside the suction line. 1—Impeller eye 20 ft above suction-well level. 2— Impeller eye 15 ft below suction-well level. 3—Impeller eye 15 feet below suction-well level which is under 26-in. vacuum. 4—Impeller eye 15 ft below suction-well level is under 20 psia

Shoot Your Pump Troubles

▶ UNEXPECTED SHUTDOWNS of centrifugal pumps may be prevented by a study of how they operate. When trouble is approaching, the pump usually indicates if the source is mechanical or hydraulic. Centrifugal pumps are simple machines and have few mechanical troubles. Most of them are due to faulty hydraulic conditions in suction system.

Centrifugal pump troubles may be easily found and remedied, but why certain conditions cause faulty operation may not be readily understood. If so a brief study of what takes place in a suction system will be helpful.

Suction Conditions. Fig. 1 to 4 show four typical suction conditions and the pressures at certain points inside the suction line. Column D shows the distance in feet from the water level in the suction well to centerline of the impeller's eye E.

Column P gives the absolute pressure at the impeller's eye. For example, in Fig. 1, the centerline of the impeller is 20 ft above the liquid level in the suction well. Atmospheric pressure, 14.7 psi absolute, is applied to the liquid's surface in the suction well. This is all the pressure available to cause the water flow into the pump. But this pressure is opposed by a column of water 20 ft high, which exerts a downward pressure of 20x0.434 = 8.68 psi.

To get the absolute pressure at the

Centrifugal pumps are pretty reliable but if something goes wrong you might have to get going fast. So here are 80 time-tested remedies that will tell you what to do in a jiffy

impeller's eye we must subtract pressure of the water column from atmospheric pressure, or 14.7—8.68 = 6.02 psia (lb. per sq in. absolute) as shown in column P. This assumes no friction losses in the suction line.

If the pump is set so that its suction is only 15 ft above suction-well level, pressure at the impeller's eye will increase to 8.2 psia. The lower the pump is set the higher its absolute suction pressure.

In the other figures, pressure at the pump's suction is obtained as in Fig. 1. In Fig. 2, it is 15 ft from suction-well level down to the pump's suction eye. Pressure due to this water column is 15x0.434 = 6.51 psi which added to that of the atmosphere is 6.51 + 14.7 = 21.2 psia.

Rated Capacity. For a centrifugal pump to deliver its rated capacity, a given pressure must be applied at the

suction eye E of the rotating impeller. If the pressure at E drops below that needed for the pump to operate at rated capacity, it will cause trouble.

For example, in Fig. 1 and 3, pressures in the suction line are below atmospheric and a defective joint would allow air to leak into the line. In Fig. 2 and 4, pressures in the suction line are above atmospheric and liquid would leak out through a defective joint.

Air or Gas. Either causes trouble in suction line because both increase in volume when pressure on them decreases. When air or gas expands in a suction line, liquid is displaced and pump capacity is reduced.

The amount of air or gas released and volume of liquid displaced cannot be found exactly. However, experience shows that air or gas set free in a short suction line causes less trouble than in a long line. For this reason, short lines

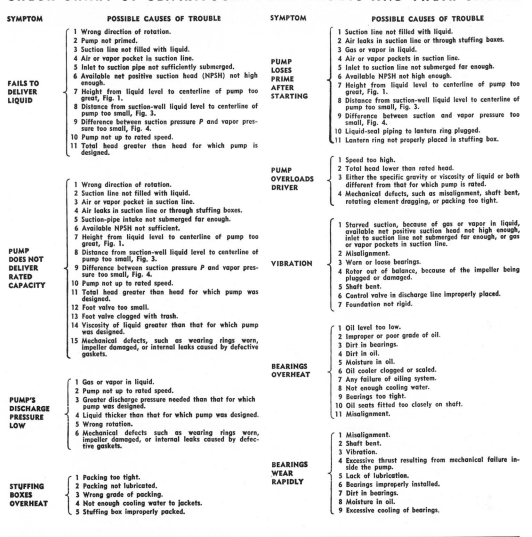

SYMPTOM	POSSIBLE CAUSES OF TROUBLE
FAILS TO DELIVER LIQUID	1 Wrong direction of rotation. 2 Pump not primed. 3 Suction line not filled with liquid. 4 Air or vapor pocket in suction line. 5 Inlet to suction pipe not sufficiently submerged. 6 Available net positive suction head (NPSH) not high enough. 7 Height from liquid level to centerline of pump too great, Fig. 1. 8 Distance from suction-well liquid level to centerline of pump too small, Fig. 3. 9 Difference between suction pressure P and vapor pressure too small, Fig. 4. 10 Pump not up to rated speed. 11 Total head greater than head for which pump is designed.
PUMP DOES NOT DELIVER RATED CAPACITY	1 Wrong direction of rotation. 2 Suction line not filled with liquid. 3 Air or vapor pocket in suction line. 4 Air leaks in suction line or through stuffing boxes. 5 Suction-pipe intake not submerged far enough. 6 Available NPSH not sufficient. 7 Height from liquid level to centerline of pump too great, Fig. 1. 8 Distance from suction-well liquid level to centerline of pump too small, Fig. 3. 9 Difference between suction pressure P and vapor pressure too small, Fig. 4. 10 Pump not up to rated speed. 11 Total head greater than head for which pump was designed. 12 Foot valve too small. 13 Foot valve clogged with trash. 14 Viscosity of liquid greater than that for which pump was designed. 15 Mechanical defects, such as wearing rings worn, impeller damaged, or internal leaks caused by defective gaskets.
PUMP'S DISCHARGE PRESSURE LOW	1 Gas or vapor in liquid. 2 Pump not up to rated speed. 3 Greater discharge pressure needed than that for which pump was designed. 4 Liquid thicker than that for which pump was designed. 5 Wrong rotation. 6 Mechanical defects such as wearing rings worn, impeller damaged, or internal leaks caused by defective gaskets.
STUFFING BOXES OVERHEAT	1 Packing too tight. 2 Packing not lubricated. 3 Wrong grade of packing. 4 Not enough cooling water to jackets. 5 Stuffing box improperly packed.

SYMPTOM	POSSIBLE CAUSES OF TROUBLE
PUMP LOSES PRIME AFTER STARTING	1 Suction line not filled with liquid. 2 Air leaks in suction line or through stuffing boxes. 3 Gas or vapor in liquid. 4 Air or vapor pockets in suction line. 5 Inlet to suction line not submerged far enough. 6 Available NPSH not high enough. 7 Height from liquid level to centerline of pump too great, Fig. 1. 8 Distance from suction-well liquid level to centerline of pump too small, Fig. 3. 9 Difference between suction and vapor pressure too small, Fig. 4. 10 Liquid-seal piping to lantern ring plugged. 11 Lantern ring not properly placed in stuffing box.
PUMP OVERLOADS DRIVER	1 Speed too high. 2 Total head lower than rated head. 3 Either the specific gravity or viscosity of liquid or both different from that for which pump is rated. 4 Mechanical defects, such as misalignment, shaft bent, rotating element dragging, or packing too tight.
VIBRATION	1 Starved suction, because of gas or vapor in liquid, available net positive suction head not high enough, inlet to suction line not submerged far enough, or gas or vapor pockets in suction line. 2 Misalignment. 3 Worn or loose bearings. 4 Rotor out of balance, because of the impeller being plugged or damaged. 5 Shaft bent. 6 Control valve in discharge line improperly placed. 7 Foundation not rigid.
BEARINGS OVERHEAT	1 Oil level too low. 2 Improper or poor grade of oil. 3 Dirt in bearings. 4 Dirt in oil. 5 Moisture in oil. 6 Oil cooler clogged or scaled. 7 Any failure of oiling system. 8 Not enough cooling water. 9 Bearings too tight. 10 Oil seats fitted too closely on shaft. 11 Misalignment.
BEARINGS WEAR RAPIDLY	1 Misalignment. 2 Shaft bent. 3 Vibration. 4 Excessive thrust resulting from mechanical failure inside the pump. 5 Lack of lubrication. 6 Bearings improperly installed. 7 Dirt in bearings. 8 Moisture in oil. 9 Excessive cooling of bearings.

should be used whenever possible.

If air leaks into the suction line or if the liquid contains air or gas it will cause serious trouble or even complete failure of a pump located as in Fig. 1. In theory, air or gas when expanding from 14.7 to 6.1 psia increases in volume $14.7 \div 6.1 = 2.4$ times and displaces 2.4 volumes of liquid.

If the liquid pumped contains 2% of air or gas in solution at 14.7 psia and the suction line is air tight, the air or gas in solution expands and the mixture increases $14.7 \div 6.1$ x 2 = 4.9% in volume at E. Even though the expanded volume at E, in either case, is less than in theory, it is enough to cause lots of trouble.

In suction system, Fig. 3, pressure in the suction line is higher at E than in the suction well, but lower than atmospheric pressure at all points in the line. Air leaking through a poor joint rises to the space above the liquid and lowers the vacuum. Air leaking through the packing on the suction side has the same effect.

Troubles that may occur on a centrifugal pump and their causes are given in the table shown above. Keep this table handy and you'll always have your answer ready when pump troubles pop up.

Corrosion in centrifugal pumps

ELECTROLYTIC CORROSION often attacks centrifugal pumps because dissimilar metals are in contact with the water handled, which then acts as an electrolyte. This forms local currents. These may not be sufficient to cause serious corrosion but they do cause pitting, especially in bearings.

Remedy is to ground the pipe work of pump, and mount a slip ring on impeller shaft. Then connect carbon brush bearing to a ground. Ground must be perfect to handle any local currents.

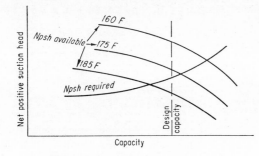

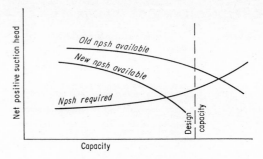

1 **TEMPERATURE CHANGE** can alter npsh available—you may not even realize that it's happening. System changes to trim operation are often the culprit here. In curve above, temperature climbed from original design of 165 to 185 F. Severe cavitation resulted. Later, calculations showed that 175 F was about maximum if a little margin was to remain. System was adjusted to meet this need by controlling temperature to 175F.

2 **LOCATION CHANGE** to suit plant modernization may save money in new equipment but add operating grief if you don't watch npsh carefully. In this job, suction lift didn't go any higher—elevations were unchanged —but length of suction piping increased. So available npsh dropped far below pump needs. Solution: enlarge suction piping, eliminate unnecessary fittings to cut losses and restore close-to-original available npsh.

4 net-positive-suction-head troubles—

Just what is npsh?

Net positive suction head—npsh—is a measure of pump-suction conditions common to manufacturer and user alike. Any pump manufacturer can give you a curve plotting the npsh-vs-capacity his pump requires; with the formula below, you can easily figure a curve of available npsh-vs-capacity for your system. Then all you have to do is be sure that, at all flows, available system npsh (npsh$_a$) exceeds npsh required by the pump (npsh$_r$).

$$npsh_a = \frac{(p - p_v) \times 2.31}{sp\ gr} - h_d \pm Z$$

where:

p = pressure at liquid surface, psia (atmospheric pressure in open tanks)

p_v = vapor pressure of liquid at pumping temperature

$sp\ gr$ = specific gravity of liquid pumped at pumping temperature, expressed as a decimal

h_d = sum of all dynamic losses in the suction system, ft. These include pipe friction and entrance loss

Z = static elevation difference, ft, between liquid level and center line of horizontal pumps or first-stage-impeller eye of vertical pumps

Net positive suction head lies at the bottom of many hydraulic troubles you may meet in applying centrifugal pumps. At first glance npsh may seem complex, but don't let it scare you—it's just a number, and formula at left will show you how to figure it for your system.

Npsh is absolute energy that produces flow. At any operating capacity and speed, two npsh numbers must be considered: npsh *available* (npsh$_a$) in your suction system and npsh *required* (npsh$_r$) by your pump. Both of these numbers are in terms of ft of liquid pumped. For successful operation, npsh$_a$ must always exceed npsh$_r$.

Since a centrifugal pump cannot use vapor energy, npsh is measured over and above vapor pressure. It's figured to the center line of the first-stage impeller for static elevation. Friction, on the other hand, is included only to the pump suction flange. Beyond that point the manufacturer figures it into pump needs.

Npsh$_r$ data can be obtained as a curve from the manufacturer—npsh$_r$ plotted as a function of capacity at operating speed. But you must figure npsh$_a$ from conditions of your own suction system, using formula at left. Solve this formula for several capacities in your operating range and plot a curve of npsh$_a$-vs-capacity. Then compare these two curves on the same paper. Npsh$_a$ curve must exceed npsh$_r$ curve at all flows within your operating range. The npsh$_r$ curve supplied by the manufacturer will rise with capacity,

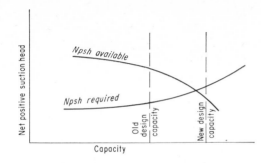

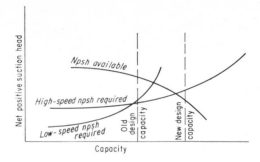

3 INCREASED CAPACITY may be required by plant expansion, and your pump may be perfectly capable of meeting this need. But will available npsh be adequate? Here it wasn't. Required npsh increases roughly as the square of capacity, while available npsh usually drops as capacity increases. With these two factors working against you, odds are that blindly increasing capacity without making other changes will lead to trouble.

4 SPEED INCREASE is usually a good way to raise pump head and capacity capability. But required npsh goes up too—and fast. Before proceeding with a speed change, go back to your pump manufacturer and get a new required-npsh curve from him for your proposed new speed. Check this against the npsh you will have available at your new high-speed capacity to be sure you don't have to cut suction lift to increase your npsh.

and how to avoid them

varying about as the square of capacity change. The $npsh_a$ curve you figure will slope down, since friction losses increase with capacity. So they're bound to cross somewhere, and this point is the limit of pump operation. It's no good to try for more capacity by opening control valves; operation near this point for any length of time will produce cavitation that damages the pump.

How much leeway should there be between $npsh_a$ and $npsh_r$? Of course there should be *some* in any application. 25 to 40% is a good general target. Where suction conditions swing widely, provide more margin for insurance. But condensate and h-p boiler-feed pumps are important exceptions here.

How can npsh get you into trouble? We'll assume your pump was properly applied when you originally bought it. But now there are subtle signs of cavitation: the pump sounds as if it's pumping marbles; the impeller seems to have had a bad case of smallpox. What has happened? There are four major possibilities (shown in curves above); let's look at them one by one.

1—Pumping temperature often changes with process optimization. If npsh was tight to begin with, a slight temperature increase can easily eliminate original margin. Answer here is to review actual pumping-temperature needs and see if some in-between value won't do. If not, perhaps the pump manufacturer can provide a new im-

peller capable of satisfactorily taking your reduced $npsh_a$.

2—Pump location may have changed as a part of plant modernization. While the pump may meet new service conditions—head and capacity—it's suffering from npsh starvation. Even though static suction lift is the same, $npsh_a$ in the new system may well be far less than originally set, especially if suction lines are longer. Solution would be an increase in suction-pipe size to minimize friction losses and restore close-to-original $npsh_a$.

3—Increased pump capacity goes along with plant expansion. This gain may have been automatic, brought about by your control system, or perhaps you actually opened up the discharge throttle valve. In any event, operation beyond original design capacity cuts into npsh margin, may eliminate it altogether. See if you can get along with an intermediate capacity on a slightly longer pumping cycle to preserve some npsh margin.

4—Speeding up the pump is usually a good way to get more capacity out of it. But $npsh_r$ goes up fast. Check with manufacturer to be sure that high-speed $npsh_r$ will stay within system $npsh_a$.

These are just four bad actors—there are many more. If operation of pumps you're responsible for seems to be taking a turn for the worse, check out possible system changes that may have altered the original npsh figures. If you find any, act at once to prevent damage.

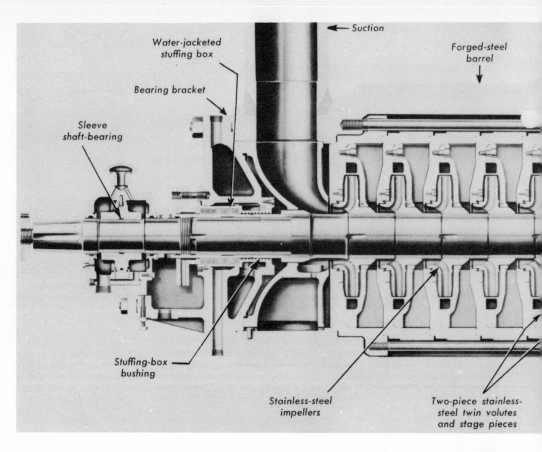

Suction

Water-jacketed
stuffing box

Bearing bracket

Forged-steel
barrel

Sleeve
shaft-bearing

Stuffing-box
bushing

Stainless-steel
impellers

Two-piece stainless-
steel twin volutes
and stage pieces

Now let's open, inspect and overhaul

▶ IT WILL HELP YOU understand all pumps so much better if you know construction details of a high-pressure boiler-feed pump, and how to inspect and assemble it, even though you don't have this type of unit yourself. Modern double-case barrel-type centrifugal pump is today's answer for pressures up to 6000 psi, to 600 F. In the common 2600-psi range, this type runs at about 3600 rpm, has up to 12 stages.

Long service without major overhaul is common for these pumps. But when you do open one, do your job thoroughly because this unit is vital to plant operation. Photos show every important step. They should guide you to do the job right, even if you have never opened your pump before.

Pump overhauled here is a Worthington, 9-stage unit, with discharge diameter 8 in., design capacity 657,000 lb per hr, suction pressure 102 psi, discharge pressure 2582 psi, suction temperature 320 F, speed 3500 rpm, driven by electric motor through hydraulic coupling.

We start right out by overhauling pump parts shown in cross section at top of page. Of course, oil lines and other piping connections are disconnected. Here we show only major pump parts that you will have to give special attention.

1 Here's how to open

1 Remove coupling, gland, packing, then bearing cover on inboard end. Next, take weight off lower bearing bushing as shown, roll out the bushing by sliding it up over the shaft. With load off bearing housing, you can unbolt and remove it

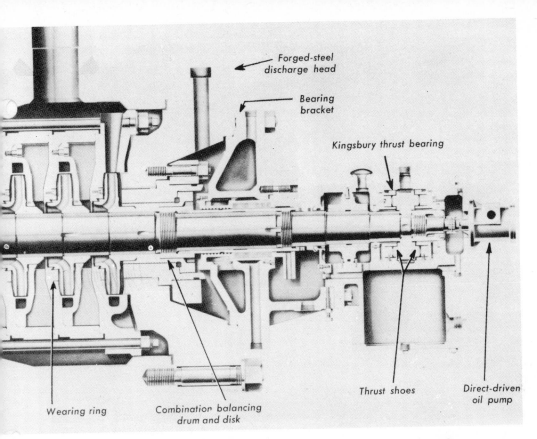

Forged-steel
discharge head

Bearing
bracket

Kingsbury thrust bearing

Thrust shoes

Direct-driven
oil pump

Wearing ring

Combination balancing
drum and disk

this barrel-type feed pump in 3 steps

2 Now let's switch to the outboard end. Remove direct-driven oil pump and bearing cover. Remove thrust shoes, roll out leveling plate and the base ring

3 Screw off the thrust collar-nut by first loosening with special wrench. Now slide thrust collar off shaft. Next, remove the bearing so shaft is clear

4 Remove nuts from outboard bearing bracket. Fasten wire sling to eyebolt and swing bearing bracket clear of shaft. Guide so it does not touch shaft

5 Special tools come with all barrel-type feed pumps. Set tool up on shaft to remove the shaft sleeve as here. Tighten each nut a little until the sleeve starts to move on shaft

6 Remove shaft-sleeve packing with packing hook. Aim hook at stuffing-box bore—not at shaft as hook will scratch it. Always remove all the old packing from any stuffing box

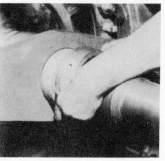

10 Move rotor carriage, raise or lower with hand wheel to align with casing studs. Lock it in place with two nuts

11 Place the carriage crosshead over the shaft at discharge end of pump. Screw crosshead onto the locking nut

12 Next, slide the special inboard shaft support-head and shaft protecting sleeves over the shaft, bolt in place

2 And here's how to inspect_____

15 Heat each impeller to 300 F to expand over shaft. Keep moving torch. Use special clamp to lift the impeller

YOU MUST WATCH THESE DETAILS WELL
A: wearing rings for wear, clearance
B: metal-to-metal joints for mating
C: impeller for a good shrink fit
D: 2-piece twin volutes and stage pieces' condition
Compare clearances with those of manufacturer

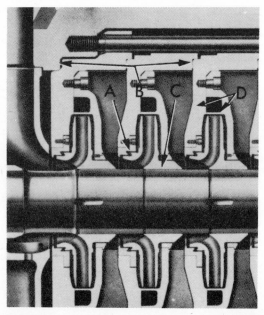

7 Remove balancing disk from supported shaft with special wrench. *Never* turn rotor if bearings are out

8 Next, remove nuts, swing discharge head clear. Remove stay-bolt nuts (small nuts in cylinder), washers and keys

9 Remove nuts from suction head at other end, take weight with chain fall. Guide carefully, swing clear of shaft

13 With carriage bolted to two cylinder studs, slide out pump's inner assembly. Two mechanics can easily do this job by turning pinions over the pinion rack on each side of carriage. Heavy rotor is easily moved thus to any part of your plant

14 Lift rotor assembly off the carriage with crane and place vertically on wooden blocks. Remove the stay-bolt nuts

1 Examine thrust shoes carefully. Wear should be very little as relocating balancing device takes the thrust load

2 Mike thrust collar for wear. There is usually little or no wear here as softer babbitt on shoes takes wear

3 With rotor assembly removed, inspect the inside of the barrel thoroughly. Look for signs of erosion, dirt, scale

281

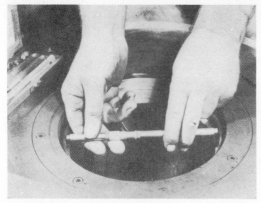

4 Measure ID of interstage bushing, balancing-disk, wearing rings and OD of mating parts. Renew parts if clearance is about double that in the manufacturer's instruction book

5 Place impeller in stage piece, move laterally against a dial indicator to see diametral clearance. Compare this reading with data given in your manufacturer's instructions

6 Now install twin-volute diffuser and repeat the process to determine diametral clearance in interstage bushing. Compare this with the data given by the maker of your pump

7 Inspect impeller bore carefully. Since the impeller is shrunk on the shaft and keyways do not cut through, there should be no sign of leakage here. But check carefully anyway

8 Check the tongue of the volute carefully for any signs of erosion. Volutes are of stainless and usually this is not a problem. But don't overlook anything while the pump is open

9 Check metal-to-metal joints between stage pieces for signs of erosion. Since stage pieces are stainless, this is usually not a problem either. But, again, don't miss anything

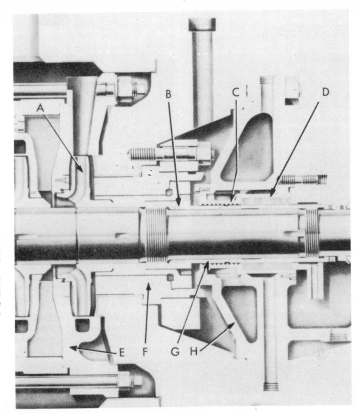

HERE'S DISCHARGE END

A: 9th-stage impeller   **B:** sleeve seal
C: water-jacketed stuffing box bushing
D: packing set   **E:** last-stage diffuser
F: balancing-disk head
G: shaft sleeve   **H:** bearing bracket

Please note: This is the high-pressure end of the pump. Make sure you know these details. Give each part your careful attention for long troublefree service

3 Now let's close it up right _____

1 To expand over shaft, heat impeller with flame if lacking an oil bath. Here hot plate distributes heat evenly

2 Slide heated first-stage impeller in place. Make sure that bearing surface is coated with molybdenum disulfide

3 Blow compressed air through hot impeller to cool. Parts are numbered. Fit into the right position on the shaft

4 With first-stage impeller shrunk on shaft, mount assembly vertically in first-stage stage-piece, as shown here

5 Next, install first-stage twin-volute diffuser in stage piece. Shaft is machined slightly smaller for each successive impeller. Joints are metal to metal, so be sure to get them perfectly clean before mating or you might run into some trouble

9 Thoroughly clean metal-to-metal joint, between discharge head and casing barrel, with crocus cloth. Tight joint here is important, depends on perfect contact of mating surfaces

10 As Frank Anderson, *right*, installs balancing disk with special wrench, Editor Steve Elonka and Ted Edwards of Worthington Corp, *left*, look on. Shaft is on hydraulic jack

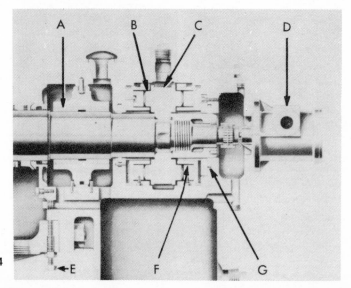

BEARINGS ARE IMPORTANT
A: bearing bushing   B: thrust-bearing shoe
C: thrust collar   D: oil pump on shaft
E: jacking screw for centering the shaft
F: thrust-collar nut   G: thrust sealing ring

6 When replacing wearing ring in stage piece, always lock Allen setscrews by upsetting with center punch as here

7 Put molybdenum disulfide on bearing surface before shrinking impellers in place to prevent galling next overhaul

8 Install subsequent stage pieces, impellers and the twin-volute diffusers until the complete rotor is assembled

11 With rotor assembly in place and pump casing ready to close up, install shaft sleeve and tighten shaft sleeve-nut with spanner wrench. Sleeve takes all the packing wear

12 Check the vertical movement of shaft in wearing rings and diffuser bushings by jacking up on the shaft against dial indicator. Refer to recommendations for the maximum clearance

13 Bolt thrust-bearing housing in place, install lower journal bearing-bushing. Adjust jackscrews on housing sides for total vertical clearance, split in half to center shaft

14 Pack pump, assemble oil and cooling-water lines, drains. *Caution:* Before starting overhauled pump, always roll pump by hand while cold, also afterward warm with suction open

FOUR REASONS AGAINST <u>PERIODIC OVERHAULS</u>

1 High cost

2 Pump unavailable for emergencies

3 There's chance of faulty reassembly

4 It is uneconomical with this type pump

1 **CALIBRATED ORIFICE** on this barrel-type centrifugal boiler-feed pump is used to measure leakage past balancing device, indicates degree of internal leakage

How often should you overhaul your centrifugal boiler-feed pumps?

► WHILE PUMP OPERATORS have long argued over the advisability of opening feed pumps at regular intervals for inspection and overhaul, pump designers are almost unanimously united in believing there is no advantage gained until performance shows a maintenance need.

Modern designs. Table, above left, shows four good reasons against periodic overhauls—cost, availability, damage and poor economy. Modern designs and materials-selection practices are primarily directed at lengthening the time between shutdowns for overhaul and maintenance.

Higher investment for equipment, keynoted by greater use of stainless-steel alloys, would make it extremely uneconomical to subject these units to unnecessary overhauls, if they were chosen to run for extended periods.

Schedules. The cardinal principle in scheduling complete overhauls is that a centrifugal boiler-feed pump need not be opened for inspection unless (1) factual or (2) circumstantial evidence indicates it is necessary. There is but one exception to this rule.

Factual evidence constitutes (1) a fall-off in pump performance justifying the expense of an overhaul (2) pump noise (3) driver overload indicating trouble (4) excessive pump vibration. In extreme cases, the pump may have seized at the internal clearances as a result of foreign matter, overheating or a mechanical defect.

Circumstantial evidence is accumulated from past experience with a particular pump or from similar equipment on the same service. For example, if a group of chromium stainless-steel alloy pumps run 60,000 hr continuously without needing a complete overhaul, an additional duplicate unit can be run for the same period.

Exception. This rule does not hold for corrosion-erosion troubles. These are rare, however, because much attention has been devoted in recent years to proper materials for high-pressure high-temperature boiler feed pumps. Today, 5% chrome stainless casings and 13%, or higher, chrome stainless fittings are established practice for all feed pumps serving 850-psi and higher-pressure plants. The practice is beginning to spread to plants operating at lower pressures.

But since ordinary materials (cast iron for casings, bronze for fittings) are still used in many low- and medium-pressure plants, corrosion-erosion troubles are still a potential danger. So when pumps of these less-resistant materials are used with feedwater that may cause severe and rapid deterioration, it is advisable to open the unit for inspection soon after initial operation. It should also be opened at intervals thereafter to establish an experience pattern that proves the equipment is satisfactory for its job.

Causes of wear. Erosive action of the water flowing past wearing rings and interstage bushings causes part of the wear at all internal clearance joints. Another portion is due to infrequent momentary contacts, which occur on relatively rare occasions during pump operation.

No matter how rigidly a pump is designed, certain conditions arise during transient loads (partial cavitation or similar sudden hydraulic impacts) that impose a temporary vibration on the rotating element sufficient to cause a slight momentary contact at the running joints.

This contact may not be enough to cause permanent damage or seizing, but it does affect the running clearances. A similar condition may exist if the shaft deflection under stationary condi-

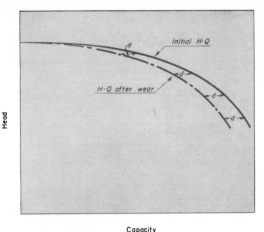

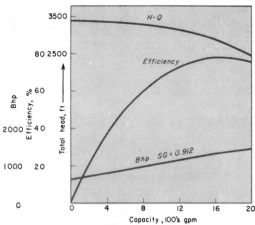

2 EFFECT OF WEAR on head-capacity curve of a centrifugal boiler-feed pump. Leakage is nearly constant at all heads

3 CHARACTERISTIC CURVES of a typical centrifugal boiler-feed pump, which is rated at 1645 gpm with a 2840-ft head

tions exceeds the existing clearances. This can occur even though the deflection is reduced to less than the running clearance by the supportive effect of internal clearance joints acting as additional steadying bearings. In these cases a slight amount of wear occurs every time the pump starts.

It is impossible to state the exact amount of wear attributable to erosion or to momentary contact. But the more rigid the pump construction, the less the effect of transient conditions on wear of the running joints.

Varying wear in exactly similar pumps operating under apparently similar conditions indicates existence of a varying frequency of transient conditions and vibrations.

Effect of wear. As running clearances increase with wear, a greater portion of the gross capacity of the pump is short-circuited through the clearances to a lower-pressure stage. The effective or net capacity delivered by the pump against a given head is reduced by an amount equal to the leakage increase.

While in theory the leakage varies with the square root of the total head, it is sufficiently accurate to assume the leakage increase remains the same at all heads. Fig. 2 shows the effect of increased leakage on the shape of the head-capacity curve. Subtracting the additional internal leakage d from the initial capacity at each head in the pump range gives a new head-capacity curve, as shown in Fig. 2.

Performance. If the pump whose characteristics are shown in Fig. 3 delivers, when new, 1645 gpm against a 2840-ft head, while absorbing 1370 hp when handling 314-F water at an efficiency of 78%, let's see what an in-

crease in internal leakage of 2% of the initial capacity will do to performance.

With increased leakage the pump delivers 1610 gpm against the same head and requires the same power input. Efficiency is reduced by the ratio of the initial capacity to the present capacity, or to 76.5%. At the original capacity, 1645 gpm, the pump now develops a 2825-ft head at 77% efficiency.

Since a capacity margin is always provided in feed-pump design, the unit in Fig. 3 may normally be required to discharge only 1480 gpm at 2840-ft head, with the excess pressure throttled by the feedwater regulator if the pump operates at constant speed. So the additional leakage does not impair normal performance. The decision to overhaul the pump at this point depends solely upon evaluating the reduced economy against the cost of the overhaul.

In high-pressure feed-pump operation it is good practice to renew internal clearances when the leakage doubles. At this point the added power to meet station loads becomes appreciable and usually justifies the overhaul cost.

Performance tests. An accurate test of the pump head-capacity curve gives a reasonable indication of the internal condition. Proper instrumentation for determining pump capacity and pressure is a must, both for routine operation and for tests.

A rigid schedule should be set up for pump head-capacity tests so the performance can be compared with the original values when the unit was new. This comparison, not the mere passage of a certain time interval, should establish the degree of internal wear and the need for an overhaul. Tests, which are cheaper than overhauls, can be made

without taking the unit out of service.

Power-consumption increase for a given flow rate, change in pressure drop across the feed regulator and speed measurements on variable-speed units are approximate means of determining performance when full head-capacity tests cannot be made or are not accurate enough.

Measuring leakage. A calibrated orifice, Fig. 1, fitted with a differential pressure gage or manometer can be installed in the balancing-device leakoff line if the leakage is returned through external piping. By inference, the orifice readings can give an accurate indication of the condition of the internal running joints. They give a direct indication of the condition of the balancing device.

If the pump design is such that the pressure drop per inch of wearing-ring and balancing-device length is about the same, the wear rate of the balancing device can be taken as indicative of that of other parts.

For example, if after 50,000 hr operation, balancing-device leakoff doubles, it is likely that wearing-ring clearance has about doubled.

Spare parts. To insure restoring a pump to service promptly after an unexpected overhaul, the following spare parts should be kept on hand: one set of casing wearing rings, one set of impeller rings (if used), one set of interstage or diffuser bushings, one spare balancing device made up of rotating and stationary parts (if used), two sets of shaft sleeves, several sets of packing, one set of bearing bushings, and parts for the thrust bearing. A completely assembled spare inner assembly should also be kept on hand.

Pump Suction Do's & Don't's

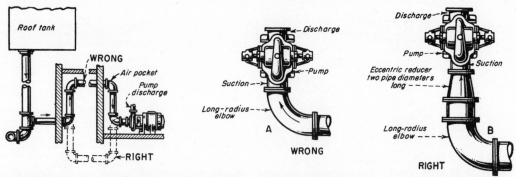

UPWARD LOOP. Never put an upward loop in the suction line of a centrifugal pump. It may form an air pocket and reduce capacity. Full lines show how a pump was installed. Even with pressure tank, pump would not deliver expected flow. Swinging the loop downward (shown dotted) corrected the trouble

HORIZONTAL ELBOW. Never place elbow in horizontal plane at suction nozzle, **A.** Elbow throws flow too much to one side. Connect straight pipe, 2 to 7 pipe diameters long, between elbow and pump suction to straighten out flow. You can use reducer, 2 pipe diameters long, between elbow and intake, **B**

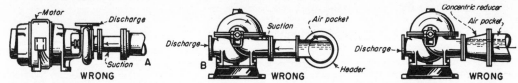

PIPE SIZE CHANGES. Avoid sudden changes in suction-pipe sizes. They produce air pockets, **A** and **B**, that reduce pump's capacity or cause it to lose prime. To change the pipe size, use tapered increasers or reducers instead of flanges shown

PIPE REDUCERS. Don't use a concentric reducer between suction pipe and pump's inlet as it forms an air pocket

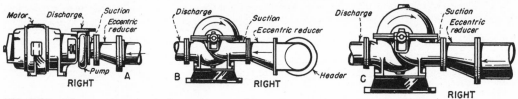

PIPE SIZES. Where suction pipe is larger than pump's intake, use eccentric reducers as in **A, B** and **C.** Where suction pipe branches from a header make connection same size as the header and use an eccentric reducer as at **B**

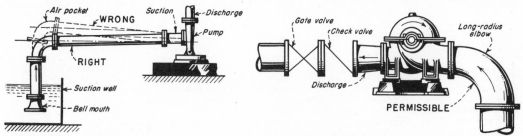

LONG SUCTION LINE. When suction line is long and water level in suction well is below centerline of pump, have line slope up to the pump as shown in full lines. If the pipe slopes downwards, as indicated by the dotted lines, an air pocket forms at the high spot and reduces pump capacity

VERTICAL SUCTION LINE. When suction line rises up to pump, a long-radius elbow connected to suction is permissible, but it's better to insert a piece of straight pipe between the elbow and pump intake. On discharge side, a concentric reducer or increaser is satisfactory for most installations

10

Electrical equipment

Circuit

▶ ELECTRICAL CIRCUIT TESTING is generally a simple job. The hokus-pokus and trade-secrets connected with trouble-shooting are stripped of their mystery when a few basic principles are understood. Boiled down to simple facts, circuit testing used by the operating engineer involves only two checks: (1) measuring voltage drop and (2) tracing through the circuit. Electrical troubleshooting is merely a simple application of these two tests.

Voltage Drop. Measuring voltage drop is nothing more than connecting a voltmeter across two points and taking a reading. This may be further simplified in some cases by touching the leads of a test lamp across two points. If the lamp lights with normal brilliance, you have, in effect, determined that the voltage drop across the two leads is equal to the voltage rating of the bulb.

In checking 220-v circuits, connect two 110-v bulbs of equal wattage in series. When the lamps give normal light, the voltage drop across this test set is 220 volts.

Continuous Circuit. Testing through a circuit shows whether there is a complete electrical path of low resistance between two points. The hookup in *A*, Fig. 3, would show a complete circuit between points *a* and *b*, since there is a continuous path through the motor armature and its field. The circuit in *B* shows an open when tested between points *a* and *b* because of the blown fuse. This is called an "open circuit." It is only one example of the many troubles that interrupt a circuit.

The terms "short" and "ground" sound a familiar note to even those whose only association with electricity has been removing a light bulb or replacing a fuse. These terms are almost household words and as such have been used rather loosely. As most operating engineers know, a "short" is entirely different from a "ground". The tests for checking an open, short or ground and the location of the trouble are applications of the circuit test.

Open Circuits. Opens are usually caused by blown fuses, open switches,

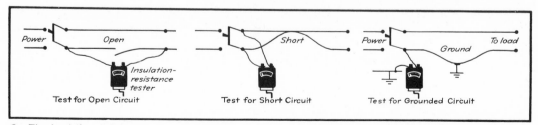

2 The insulation resistance tester reads infinity on a line that has an open circuit. It reads zero if the line is shortcircuited, and when connected between ground and the line shows zero if the line is grounded

Testing Made Easy

loose terminal connections or defective control devices. Rarely the cause is an actual break in a wire. One of the simplest ways to check for a continuous circuit is with an insulation-resistance tester. This instrument is handy when no source of electricity is available for testing. It is also useful on high voltage equipment where trouble shooting on live circuits would be dangerous.

The insulation tester has its own source of power—a hand-cranked generator. It develops either 500 or 1000 volts on open circuit. Fig. 1 shows a test for grounds as performed on a dc generator. Remember that this instrument can give the operator an unpleasant shock, although it can do him no real harm. Readings are indicated in ohms and megohms (1,000,000 ohms).

Connect the instrument as shown in Fig. 2 to the line to be tested. If the line is open it will read infinity, which is maximum resistance. A low resistance break often causes trouble which the instrument may fail to indicate as such because it is not a low-resistance reading meter. Values under 100,000 ohms cannot be read accurately. A bell testing set or low-reading voltmeter is most effective in locating breaks of this kind.

Short Circuits. To check a line for short-circuit conditions, connect the instrument between both lines as shown in Fig. 2. Continuity between the two lines is maintained by the short. It will read zero in such a case. When making this check it is necessary to disconnect the line from any equipment to which it is ordinarily connected. For example, a light bulb, a motor or a transformer connected across the line will falsely indicate short circuit conditions on the instrument.

Grounds. Fig. 2 also shows how to test for a grounded line. With one lead

connected to a good ground (such as a water pipe or motor frame) and the other lead connected to the line to be tested, continuity will be established through any ground that may exist in the line, so the instrument will read zero.

If in doubt as to whether the lead is connected to a good ground, connect the other lead to another point thought to be a good ground. If the instrument reads zero, either ground connection may be used for testing purposes.

When using the insulation testing instrument for testing as in Fig. 2, make certain that power is off the line, and that other equipment or connections to the circuit are disconnected. This second precaution eliminates the possibility of misleading effects from other parts of the circuit.

Testing With Lamps. A test light can be used to check for grounds. First, open the line switch and ground one lamp lead. Then check each of the two power wires. If one is grounded, the lamp will light when the ungrounded wire is touched. The line that does not light the lamp is the grounded circuit.

Where no ground exists on the power source, but one is suspected on the load side, the test lamp will reveal it. With the line switch open, ground one side of the power line temporarily. Then check across the open switch with the test lamp to see if the circuit is complete.

The three checks described above were all tests of continuity. There are many variations in testing depending on the means available. Voltmeters, ohmmeters, test-lamps, bell sets and insulation resistance testers all can be used. Of these the test lamp is the simplest and may also be used effectively in checking a circuit like the one in Fig. 3.

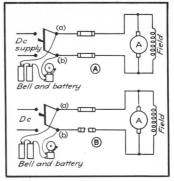

3 The bell and battery shows a complete circuit at A and open at B

With the switch closed, the motor will not run because of the blown fuse between *b* and *d*. Most fuses for industrial application are not as obviously blown as the one in the diagram. Usually they appear much the same whether good or "shot". Sometimes you can tell which fuse is blown by feeling each and noting which one is warmer. Do not be misled by the high temperature of fuses carrying full load.

Fig 4 shows a more certain way. Close the switch and connect a test lamp across the fuses at one end, as at *a* and *b*; then connect the lamp across the other end, *c* and *d*. If the test lamp lights when connected across *a-b* but does not light across *c-d*, it identifies *a* and *b* as the supply side of the fuses and also proves that one or both fuses are blown.

If the lamp lights both times it proves the fuses are good. Connecting the test lamp across *b* and *c* tests the condition of fuse *ac*. If the lamp lights, as shown, the fuse is good; if it does not light the fuse is blown.

If the test lamp is connected across *a* and *d*, the condition of fuse *cd* may be tested in like manner. In Fig. 4 it is clear that the lamp can't light because of the blown fuse.

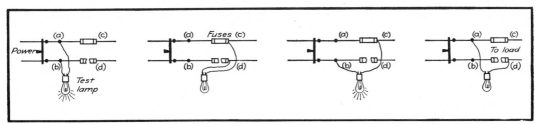

4 A positive method of testing for a blown fuse is shown above. Close the line switch and check the power source, then the back side of the fuses. Cross test each fuse separately to find the one that is burned open

1 After making all connections, check insulation resistance by connecting megohmmeter between leads and frame. Keep line switch open when making test

Connecting DC Motors

Puzzled about motor hookup? Reconnecting untagged motors and starters is a cinch

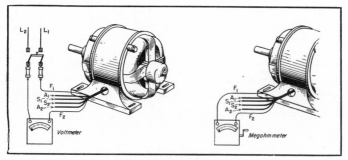

2 A test lamp may be used if a voltmeter is not handy. The lamp will burn dimly when connected in series with the shunt field. Use a low-wattage test lamp. Armature circuit will test open if one set of brushes is lifted off commutator. Remaining pair of leads must connect to series field

▶ LET's SAY you have the job of reconnecting a dc motor that has been disconnected without tagging the leads. Manufacturers use standard markings for the leads. Reconnecting the motor is only a matter of knowing the meaning of these markings. If the leads are not marked or tagged, you can make the hookup after a few simple tests.

Compound Motor. Every dc motor has an armature and one or more field windings. The compound motor has a series and shunt field. Commutating field (not shown) is connected internally, in series with the armature. Manufacturers mark leads as in Fig. 2.

First let's consider the motor whose leads carry manufacturer's marks. Usual direction of motor rotation is counter-clockwise, looking at the commutator end. To hook up a motor for this rotation, connect A_1 and F_1 to one side of the line. Connect S_2 and F_2 to the other line wire. Leads A_2 and S_1 are spliced together at the motor's terminal block.

Starting box resistor is connected in series between A_1 and the line. To change rotation, reverse leads A_1 and A_2 at the terminal block. Do not change the shunt- or series-field connections. If current is reversed through armature and both fields, by reversing the leads at the line switch, motor rotation will not change.

Fig. 2 shows the shunt field F_1 and F_2 connects across the line. The series field S_1 and S_2 connects in series with the armature, and leads A_1 and A_2 are from the armature. In motors having a commutating field, the terminal A_2 connects directly to this field while terminal A_1 goes directly to the brush holder.

Motor Hookup. Assume your compound motor is to operate at constant speed in one direction only. A 3-point motor starter is generally used. Fig. 3 shows schematic diagram and Fig. 4 external connections. The switch has two terminals L_1 and L_2. The starter has three terminals L, A and F, and the motor has six leads A_1, A_2, S_1, S_2, F_1 and F_2. Fig. 4.

There are two electrical paths for current to follow. One is from L_1 through the starter resistance, armature and series field back to L_2. The other is from L_1 through the shunt field to L_2.

If you want normal counter-clockwise rotation, current must flow from L_1 through the starter to A_1 out of A_2 into S_1, and out S_2 back to L_2. Current in the other path must flow from L_1 through the starter to F_1, and out F_2

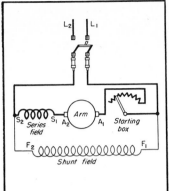

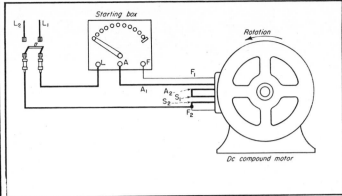

3 A compound motor connected as shown will rotate counter-clockwise

4 When making connections at the motor's terminal box use lugs and bolt together. It's a simple job in the future to disconnect such leads when shooting trouble

back to L_2. Rotation is the same whether L_1 is positive or negative.

Leads Not Marked. Here's what to do when you want to reconnect a motor whose leads are not marked. A dc motor rarely has more than six leads. We know that one pair is for the armature, another for the series field and the third the shunt field. A continuity test with a megohmmeter, voltmeter or test lamp, as in Fig. 2, quickly identifies the three pairs.

The shunt field has considerably more resistance than either the series field or armature. For this reason, a voltmeter connected in series with the shunt field across the line will read lower than when connected in the armature or series-field circuit.

To pick out the armature leads, lift one set of brushes from the commutator and apply the continuity test again. The armature leads will now test open and the remaining pair of leads must be the series field.

Checking Rotation. Connect any shunt-field lead with any series-field lead and run a line from the connection to the L_2 terminal of the line switch. Let us call these motor leads F_2 and S_2. The remaining shunt-field lead, armature lead and line-switch terminal are connected to the starter as shown in Fig. 4. With the line switch closed, move the starter handle to the first button to check the direction the motor turns.

Mark the direction of rotation on the motor frame with chalk. Now disconnect the shunt field from the connection nect the shunt field from the F terminal of the starter. The machine is now connected as a series motor. Again give the motor a quick spin to find the direction of rotation. This must be done

quickly or the motor will run away and wreck itself if not loaded.

If the motor rotates in the same direction under both tests you have connected the field windings correctly. Furthermore, if rotation was counter-clockwise we have made good guesses on all the leads and each may be marked as in Fig. 4.

If the motor does not rotate in the same directions on both tests, the fields

are bucking. Reverse the field that gave a clockwise rotation. Repeat both tests. When both field tests give a counter-clockwise rotation the motor leads may then be marked as in Fig. 4.

Because the compound motor includes both series and shunt fields, the problem of connecting shunt or series motors is a cinch. The same rules hold true, there are fewer connections, and there is no difficulty from fields bucking.

Locating Cracks In Porcelain

▶ A CRACKED PORCELAIN BUSHING on an oil circuit breaker is bad but one on an oil-insulated transformer is worse. There is always the hazard of it flash-

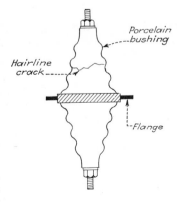

ing over and starting an oil fire. A power-factor test will show up a cracked bushing but the test requires expensive instruments and an elaborate setup. On solid porcelain supporting insulators neither a power-factor nor a high-potential test will show a hairline crack oftener than 50% of the time.

A good way to find cracks in both new and old porcelain insulators is by using a solution of chalk and carbon tetrachloride. Wipe the outer glazed surface with a cloth dampened with the solution. After the carbon tetrachloride evaporates, the bushing will be covered with dried chalk. Wipe the surface with a dry cloth and if a hairline crack is present it will stand out as a white line. The carbon tetrachloride carries chalk into the crack.

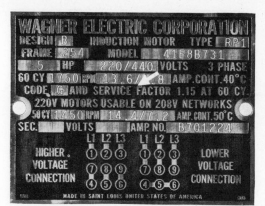

1 Multiply the hp rating by service factor to find maximum load motor can handle continuously. Use 50-C temperature rise rather than 40 C when using the service factor rating

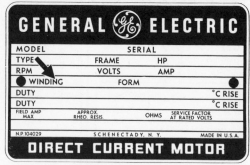

2 Dc-motor nameplates tell whether winding is shunt, series or compound. Approximate resistance of rheostat used in the field circuit and max field current may also be included

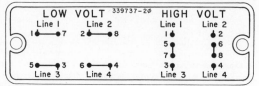

5 Some manufacturers use a separate tag like above to show connections for a two-voltage job. Actual voltages along with other standard info are listed on the regular nameplate

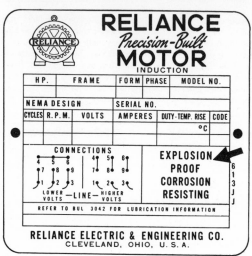

6 Case of an explosion-proof motor is designed to withstand internal explosion and at the same time prevent ignition by explosion within casing of gas or vapor surrounding motor

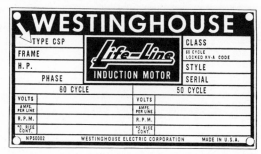

7 Type letters pin the motor down as far as the manufacturer is concerned, but no type markings used by two companies are alike. Design letters now give dope on motor performance

What Your Motor Nameplate

Your electric motor nameplate is chock full of info. Here's a quickie telling you how to get the most from it along with dope on recent changes

▶ MOST MOTORS you will come across have nameplate markings that follow a standard established by the National Electrical Manufacturers Association. To agree with this standard, most motor nameplates are stamped with the manufacturer's type and frame number, motor rating in hp, time rating, temperature rise, rpm at full load, voltage and full-load amperes.

In addition to this list you'll find the nameplates on dc machines also tell whether the winding is shunt, series or compound. On top of that ac motors also include frequency and number of phases the motor is designed for in addition to design and code letters. The list is pretty much rounded out by mentioning power-factor and dc excitation information on synchronous-motor nameplates.

Types. For many years motor manufacturers have included type markings on the nameplate. It is a mystical

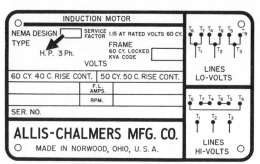

INDUCTION MOTOR			
NEMA DESIGN	SERVICE FACTOR 1.15 AT RATED VOLTS 60 CY.		
TYPE			
H.P. 3 Ph.	FRAME 60 CY. LOCKED KVA CODE		
VOLTS			
60 CY. 40 C. RISE CONT.	50 CY. 50 C. RISE CONT.		
	F.L. AMPS.		
	RPM.		
SER. NO.			

ALLIS-CHALMERS MFG. CO.

MADE IN NORWOOD, OHIO, U.S.A.

LINES LO-VOLTS

LINES HI-VOLTS

CAP.			FRAME		
VOLTS 220/440			NO.		
AMPS.			TYPE S.C.		
R.P.M. AT FULL LOAD			ROTOR V. A.		
PH. 3			CODE DESIGN		
CYC. 60			DUTY CONTINUOUS		
RISE 40°C					

LOW VOLT STAR

HIGH VOLT STAR

CROCKER-WHEELER ELECTRIC MFG. CO.

DIVISION OF JOSHUA HENDY CORPORATION

AMPERE, N.J. MADE IN U.S.A.

N P 397049

3 Hp rating is the full-load output of motor keeping within rated temperature rise shown. When handling this load, motor draws current marked after amp listing on the nameplate

4 The **slip** is difference between speed listed on nameplate and synchronous speed of rotating magnetic field. Slip is usually expressed as percent of the motor's synchronous speed

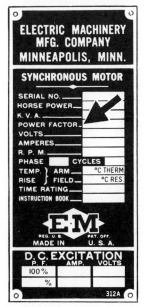

8 Power-factor marking on synchronous motor nameplate may be either unity or leading depending on specific design

Means

combination of numbers or letters that help each motor manufacturer keep his records in order but mean little to the repairman or operating engineer. Rather than change this established system, the NEMA added a series of *design* letters for polyphase squirrel-cage motors to give the motor user additional specific information about the motor's performance characteristics.

The design letters used are *A, B, C, D* and *F*. We won't go into all the specifications covered by each design letter in this short article. The following data will probably be sufficient for most operating men. For the lads who want it, all the dope is available in tabular form from the NEMA at 155 East 44th St, New York City. The point we want to get across is that motors with design letter *A* are normal-torque, normal-starting-current motors while design *B* refers to normal-torque, low-starting-current jobs.

When you meet a *C* design, it's a high-torque, low-starting-current motor while design *D* is a high-slip machine. The picture is rounded off with the *F* design which is a low starting-torque, low starting-current motor. Design *B* is by far the most popular of all types.

Code Letter. The code letter on an ac-motor nameplate (ac motors rated 1/20 hp and larger except polyphase wound rotor) is simply an indication of the kilovolt amperes per hp the motor will draw if the rotor is locked to keep it from rotating. There are a string of code letters running from *A* to *V* with a few left out in between. This series boils down to a graduated scale with code *A* motors pulling between zero and 3.15 kva per hp and code *V* motors at the other end drawing 22.4 kva per hp and up. This little nugget of knowledge comes in handy in determining the setting of motor branch-circuit protective devices.

Horsepower and Service Factor. The hp marking on any motor is its mechanical output at full load. Of course the input will be greater because of losses in the motor. Incidentally, it's smart to select motors to operate at or near their hp rating. Your motor will not only be running at a higher efficiency but at better power factor to boot.

Service factor ties in with the hp marking since it tells you the permissible overload the motor can carry continuously. For example if a motor has an allowable service factor of 1.15 (most 40 C rated motors have), that motor can carry 1.15 times its rated hp continuously when operated at rated voltage and frequency. Keep in mind that a motor marked for a 40-C rise above room temperature at full load will not exceed a 50-C rise when carrying its service-factor load.

Another marking to keep in mind is the *time rating*. This gem of information tells the length of time the motor will carry it's full load. Most motors you'll run across will be stamped "Continuous," meaning they will carry rated load continuously without overheating. Still others will be marked 15 min or ½ hr which is the maximum time the motor will supply full load output without overheating. After that period, the motor must be allowed to cool to room temperature.

By the way, the figure 40 C stamped on most nameplates means the manufacturer guarantees temperature rise in the motor will not exceed 40 C at rated load. This holds true provided there is nothing to block the free flow of air around the motor and the ambient (surrounding air) temperature is not more than 40 C. The word "Open" on your nameplate refers to the motor housing. It means the motor frame is open, permitting free passage of cooling air through the motor. Enclosed motors have a higher temperature rating.

That figure following the ampere marking is the current the motor draws when the baby is pulling it's full load, hence delivering rated hp output. In a two-voltage job there will be two values of current listed, Fig. 1. The larger of these will be the current drawn at the lower voltage; this is usually twice the current pulled at the higher voltage.

1 Remove heavy surface dust with vacuum hose. Finish off the job with compressed air, keeping the pressure to 50 psi

2 Make sure the metal protecting hood is either large enough or properly ventilated to prevent the motor overheating

Meaty Tips on Maintaining

Bearings and insulation are two most common reasons for motor failure that can be licked by smart maintenance. Here are facts that should be woven into any good maintenance plan*

▶ IT'S REALLY A TOUGH JOB to sit down and set up rules about how often electrical equipment should be inspected and maintained. Every operating man you meet has his own ideas and propably some pretty good reasons to back up those ideas. We're going out on a limb with the following suggestions, based on conditions you are liable to run into in the average plant. Our guess is you can take most of the following time schedules and adopt them in your own plant. As the years roll by, you may find the schedule should be changed. If so, go right ahead and custom-tailor it.

Weekly. Make a weekly check on bearing-oil level, and see that the oil rings turn freely. At the same time, use your hand to check temperature of motor bearings and motor frame. Sniff the warm air coming from the motor; you can't miss the smell of overheated insulation.

Monthly. Make a thorough check on brush holders, pigtails and brushes. While you're at it, blow out the motor with compressed air. After the blowing-out job, check that the brushes ride free in their holders. See that the brush springs are doing the job they were designed for. Last but not least, check the insulation resistance with a megohmmeter. And keep a record of that reading, so you can compare with previous and future readings.

Yearly. Just about once a year check the motor air gap with a feeler gage. After that, go ahead and check line voltage with a voltmeter, and motor load with an ammeter. During this checkup, replace grease in ball and roller bearings.

Every Two Years. Figure on a general overhaul about every two years. Take the motor apart and do a class A-1

*Based on information supplied by the Westinghouse Electric Corp

cleaning and inspection job. See that all windings are tight. Replace any loose wedges and bands before dipping in varnish and baking. Inspect commutator and risers.

In addition to the many things that common sense will tell you to take care of during the general overhaul, wash out the bearing housing while you have the end bell off. A good way is with hot kerosene followed by compressed air.

Bearings. After sounding off about setting up the time schedule, let's spend a little time talking about bearings. You know that bearings rank right up near the top when it comes to causes of shutdowns, delay and expense. This isn't surprising when we remember that they are the victims of poor foundations, misalignment, vibration, dirt mixed in with lubrication and frequently the wrong lubricant. All this is in addition to their real job of supporting a rapidly revolving part.

Lubrication is the first requirement for successful bearing operation. And this means more that just squirting in oil. That all-essential oil film between bearing and shaft prevents metal-to-metal contact. Properly designed bearings have adequate area for the load and proper oil grooving. In a sleeve bearing, oil sticks to the shaft and is dragged around by rotation of the shaft. This action forms an oil film between shaft and bearing. It is this film that carries the load and prevents metal-to-metal contact.

There are two things to keep in mind for proper maintenance of sleeve bearings. First, use the right oil to be sure you'll have an oil film once the shaft has started rotating. The second is to reduce damage of metal-to-metal contact by using the right babbitt. Remember, you'll have metal-to-metal contact when the oil film is lost either by accident or during the starting period.

Oiling Sleeve Bearings. Keep your oil well filled to the proper level, and see that the oil rings turn freely. Add new oil only when the motor is at rest, to prevent flooding. When you run into a sleeve bearing that needs too frequent refilling, it is probably leaking oil into the stator windings. A smart thing to do is replace it with a sealed sleeve bearing.

Now for a few tips about that felt seal at the bearing end:

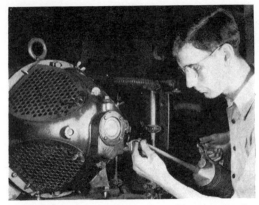

3 That fellow with the oil can plays a big part in bearing maintenance. Overlubrication is just as bad as not enough

4 Thanks to intelligent maintenance, this bearing wore less than 0.0015 in. after eleven years of continuous service

Your Motors

The real reason for using this felt seal is to keep out air and dirt. It does little in preventing oil leakage, once it becomes oil soaked. Hence, whenever you replace a sleeve bearing, put in a new felt seal. Your best bet is to order these seals from the motor manufacturer. If you have to make them in an emergency, use a high-grade felt not less than a ¼-in. thick before compression. Make the seal's inside diameter equal to the shaft diameter or slightly less.

Under normal operating conditions, the bearing temperature shouldn't rise more than 40 C above that of the surrounding air. A bearing will feel comfortably warm to the hand at this temperature. Look into the matter if your bearing rise is more than 40 C. Keep in mind that oil leakage is generally much worse at high temperatures.

Anti-Friction Bearings. When the shaft is moving in a sleeve bearing, shaft and bearing are separated by an oil film. But in a ball bearing, the balls themselves do the separating. This is true both when the shaft is turning and standing still.

Most ball bearings used in horizontal motors are grease lubricated. Follow the advice of the motor manufacturer when it comes to selecting a suitable grease. Soda-base greases are usually preferred because of their high melting point. When a ball bearing is in trouble. you can usually spot it, since it will overheat or give off an unusual noise. Remember that broken or nicked balls will spell a quick finish to any bearing. You can usually pick out these cases by the *clicks* as the shaft turns.

Check for overgreasing when the bearing temperature runs high. The first thing that will show up after overgreasing is excess heating caused by churning of the grease. A general rule to follow is that the bearing should not be over half full. Leave the drain plug out for a few minutes after regreasing. This will let the excess grease drain off as the motor spins.

Insulation. Poor insulation runs hand in hand with bearings as a main cause of motor failure. In addition to the regular insulation around the individual wires, you have insulation between commutator bars and individual coils.

5 To seat new brush, fit a strip of No. 00 sandpaper around commutator. Hug commutator, pull in direction of rotation

The first rule for maintaining good insulation is: Keep it clean; keep it dry.

Dust and dirt not only contribute to insulation breakdown, but they tend to increase motor temperature by preventing proper ventilation. The best way to remove dust and dirt is with compressed air at about 50 psi. Don't blow compressed air against the insulation until you are sure the compressed air is free of any moisture that may have gathered in the air line. And remember that too great a pressure may loosen the binding tape or injure the insulation by sand-blasting with the abrasive dirt that is nearly always present.

Where you find that compressed air doesn't remove all the dirt, try using a solvent such as cleaner's naphtha or carbon tetrachloride, or a mixture of the two. The cleaner's naphtha is not the safest thing in the world, so take precautions to prevent a fire or explosion. A mixture of 50% carbon tetrachloride and 50% cleaner's naphtha won't burn. But their vapors, mixed with the right proportion of air, are explosive. In extreme cases it may be necessary to use straight carbon tetrachloride, but remember that the vapors are poisonous. Apply your varnish treatment after the motor has been thoroughly cleaned. Two dips and bakes are enough for most normal jobs.

Here Is a Review of Maintenance Practices In Step With Growing Use of Dc in Industry

FINISHING STANDS in Pittsburgh Steel Co strip mill use 4000-hp 600-v dc motors

Shooting Trouble on Dc Machines

SYMPTOM   POSSIBE CAUSE OF TROUBLE: CURE

General Overheating of Frame, Field Coils

1 Machine overloaded. Proper protective gear; fuses, relays, temperature detectors generally keep this symptom in check
2 High ambient temperatures. May be simple matter to improve ventilation with ducts, fans
3 Restricted ventilation through machine because of dirt in vent ducts or screens
4 If shunt field coils appear as heat source, check for shorted turns, too-high voltage, part of winding shorted-out
5 Where series field is overheating look for short-circuited turns; current density too high indicating wire size is inadequate; machine overloaded

Overheating of Armature Coils, Commutator

1 First make sure machine is not overloaded
2 Megohmmeter insulation test spots grounds; try for minimum of one megohm for units to 600 v
3 Individual coils may be shorted, open, incorrectly connected to commutator segments (adjacent leads to segments may be accidentally swapped in rewind)
4 Commutator overheating may indicate too great brush friction, contact drop. Here are some points to check: tarnishing of commutator (may stem from presence of gases from production equipment), rough surface, excessive brush pressure, wrong brush grade

Excessive Sparking at Brushes

1 Once again, make a close check on machine loading to rule out possibility of overloading

SYMPTOM   POSSIBE CAUSE OF TROUBLE: CURE

2 If brushes appear to vibrate when machine is running, check for eccentric commutator, low bars or flats, rough or grooved commutator. In each case the cure lies in turning, polishing commutator, preferably in its own bearings
3 If commutator appears distorted with some bars high, the answer is to "re-season." This means heating commutator uniformly to about 100 C, running full speed, then shutting down and tightening commutator bolts. Finish off with a fine stoning to take off any high spots
4 Commutator may be out of dynamic balance or vibrating at frequency of some external unit: namely, transmission gears
5 Do brushes slide easily in their holders?
6 Brushes may be off neutral; shift brush boxes to improve commutation
7 Brushes may not be equally spaced around commutator. Check this by actually measuring distances between all adjacent brushes; unequal spacing leads to circulating currents being set up
8 Incorrect brush pressure, brush type, unequal air gaps can cause sparking. Unequal gaps can stem from worn bearings
9 Open, short or loose connection in armature circuit. Open or shorted armature coils can usually be spotted at the commutator bar or bars connecting to the faulty coil; copper or mica will be burned. There may be signs of solder throwing out of the commutator joints where a loose connection is the fault
10 Brush arms and interconnections may have unequal

298

TIPS TO CUT DOWN THE NUMBER OF TROUBLE CALLS

GENERATOR OPERATION. Where a compound generator feeds a motor load, disconnect both electrically before bringing the units to rest. If both are left connected when prime mover is shut down, generator may come to rest before the motor, leaving the latter to act as a generator. The motor would then pass current through generator series field reversing its polarity. You'd then run into trouble on start-up if fields were not flashed first.

Where a shunt generator is one of several connected to the same bus, the field can be energized from bus by closing main machine switch *after* raising the armature brushes. But hang a resistor across field terminals to prevent voltage build-up when field is opened. Series field of compound generator will be energized in correct direction if equalizer and main switch are closed with armature brushes raised.

GENERATOR LOADING. If you run into trouble balancing load between generators where field rheostat is hand operated, try slight shifting of brush arms; forward movement increases, backward decreases. But you're limited by the effect brush-shifting will have on commutation.

Remember, for stable operation of compound generators in parallel an equalizer connection is necessary between inner terminals of the series field. For a true indication of load being carried, make sure the machine ammeter is in the generator lead that's not connected to the series winding. Otherwise, with current flowing in the equalizer, the ammeter would give a false reading of load distribution between machines.

MACHINE MAINTENANCE. Don't overlook the importance of staggering brushes on commutator to prevent grooving. By staggering brush arms in pairs you'll also reduce the tendency toward "copper picking" at brush faces.

I've found the following brush pressures work out best: carbon and graphite brushes, generally between 1.5 and 2 psi. Where motors must face up to heavy overloads it may be necessary to go to 2.5 or 3 psi. Still higher pressures are required for metal-graphite brushes; minimum about 3 psi.

INSULATION. There are three practical methods of measuring resistance of electrical insulation with a megohmmeter: (1) *short-time* or spot reading (2) time-resistance or *dielectric absorption* (3) *2-voltage* The short-time test is the most popular. But in the past some confusion has existed about when to stop the test since the value of insulation resistance would ordinarily keep climbing with time. So for purposes of standardization, 500-v dc applied for 60 sec is recommended where short-time single readings are to be made.

SYMPTOM   POSSIBE CAUSE OF TROUBLE: CURE

resistances, resulting in unequal current distribution in the arms. Thorough maintenance on all joints and terminals generally rules out this as a factor in brush sparking

11 Remember, a partially shorted or grounded main or interpole field winding upsets the magnetic balance much as would uneven air gaps

12 Are interpoles reversed? Check this with compass. Commutating pole for motor should have same polarity as the main pole *following*, traveling in direction of rotation. For a generator it should be same polarity as main pole *preceding* it, still considering the normal direction of rotation. At same time, check equalizer connections to assure that circulating currents are not at root of sparking

Motor Running at Other Than Normal Speed

1 Overspeeding of series motor generally indicates load was accidentally removed. Where a compound motor speed is unstable, check that the series winding is not reversed

2 A shunt motor overspeeds when field is too weak or a portion of the winding is shorted, grounded. Supply voltage may be too high. If it's running too slow, check for overloading, low supply voltage, short in field rheostat

3 Sudden speed and current spurts may mean a change in load, defective transmission, such as slipping belts or clutches. Where a shunt motor "hunts" it may point to commutator brushes that have shifted too far backward

SYMPTOM   POSSIBE CAUSE OF TROUBLE: CURE

4 Where a motor is to be reversed: Reverse either the field or armature leads for a series motor, reverse field leads only in a shunt motor, reverse both fields in a compound motor

Generator Fails to Build Up Voltage

1 No excitation may mean an open in field circuit, high resistance at joints or terminals—or the residual field may be lost. In the latter case, disconnect the field leads and separately excite from a dc source

2 Individual field coils may be reversed

3 Dirty commutator, low brush pressure, can build up contact resistance at commutator preventing voltage build-up

4 Is generator turning in proper direction?

Generator Voltage Too High or Low

1 Check speed and excitation amperes

2 Where voltage is low and field current OK, check for shorted portions of field coils. Also look into brush setting

3 If voltage is too low after the machine has been repaired, suspect wrong reconnection of fields, too large an air gap

4 If generators fail to share load evenly, machines may have different voltage characteristic curves. Also check through the excitation circuit

5 Where compound generators are unstable when paralleled, check equalizer connections.

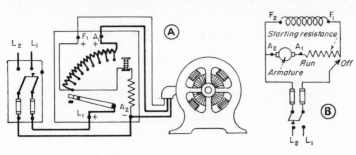

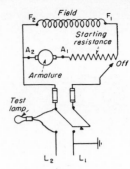

1 Motor circuit drawings should be studied first before shooting trouble. Working knowledge of both wiring and schematic drawings shown above, is half the job

2 A circuit ground is quickly located using the simple series lamp test

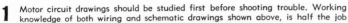

Seven Steps For Testing

▶ You can talk all you want to about circuit testing, but putting to practical use proves the pudding. Let us say we have a 120-v dc shunt motor with a manual starter and the motor won't run. Fig. 1A shows the wiring diagram of the equipment as it appears in the plant. Fig. 1B shows the schematic or simplified layout.

Trouble Shooting. In shooting trouble take the following steps:

1. See that the motor armature turns freely. Try turning the motor by hand because drag or overload can cause trouble. The high load demand causes fuses, overload trips and such devices to open the circuit.

2. See that the line switch is closed. You might consider this too elementary but, fortunately, the most common causes of trouble are usually simple.

3. Place a test lamp across the switch terminals L_1 and L_2 to see that power **is available** from the supply. If it isn't, **check** the voltage at the next switch ahead of the supply. Continue testing back to the source until the supply interruption is located.

4. When the test lamp shows voltage at the motor switch, check the motor fuses.

Fuses may be blown by trying to start the motor too rapidly or by overloading the motor; by a ground or short; by a short circuit in the armature, field, starter or leads and by open circuit in the field.

There are other causes of motor failure that do not blow fuses.

Finding An Open. 5. To check for an open circuit in the motor note whether there are two paths for current to flow from L_2 to L_1, Fig 1B; interrupting current through either path affects operation of the motor. When you move the starter handle from the off to the first position, one path is from L_2 through the fuse, motor lead, A_2 brushes, armature windings, A_1 brushes, starter resistance, starter handle, motor lead back to the fuse and L_1.

All parts of this circuit are of low resistance, so a test lamp may be used to check the circuit without being dimmed to any extent. Place one lead of the lamp on starter terminal L_1. The points marked L_1, A_1, and F_1 in Fig. 1A are identical points in Fig. 1B. The test lamp should light if the line switch is closed and the other test lead is placed on A_2.

The most convenient place to get at A_2 is on the motor brushes. If the lamp lights at both A_2 and A_1, you can be certain that the circuit is continuous from L_2 to A_1.

If the lamp doesn't light at the brush terminals A_1 and A_2, open the motor

switch and inspect the brushes and wiring between L_2 and the point where the lamp fails to light. The fault may be a loose wire or terminal connection, or the brushes **may not be touching** the commutator.

After correcting any faults, close the switch and try to run the motor. If it won't start keep on checking the armature circuit by placing one lead of the test lamp on L_1 and the other on the resistor button for the run position. As long as the lamp lights with each check, the circuit between L_2 and the test-lamp lead is complete.

By trying each button of the starter resistance you may find that the lamp won't light beyond a certain button. This means that the armature circuit is open in one of the resistor coils or grids. Spot the break by tracing the lead from the button that won't light the test lamp, back to the resistor.

Testing the Field. Having checked and corrected any breaks in the armature circuit, check the path through **the** motor field. One reliable check for open circuit in the field is moving **the** starter handle from the first position **to** the off position with brushes removed. If there is no arc the field circuit is open at some point.

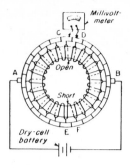

3 Armature faults, whether an open or shorted coil, are easily found

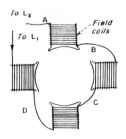

4 A defective coil in the field usually is replaced rather than repaired

Motor Circuits

Some motors are wired so that the field current is not interrupted by the starter handle. Instead current is maintained on the field as long as the line switch is closed, so that the heat will prevent condensation in the motor windings. With this arrangement opening the line switch will cause an arc if the field wiring has no open circuit.

The field circuit may be checked with a test lamp in the same manner as the armature circuit. However, because of the higher resistance of the field circuit, the lamp won't last as brightly as in other tests. Try connecting the lamp across L_1 and F_1, Fig. 1B. If the lamp lights, it proves the open is in the starter and at the same time proves the motor fields are not open.

If the open is in the motor fields, each field coil will have to be checked separately by untaping the leads connecting the individual field coils. Refer to Fig. 4 and note the method of checking. The field-coil leads are untaped at points A, B, C and D. The line switch is closed and the starter handle placed on the first position. Apply the test lamp leads to AB, BC, CD and DL₂. If the field coil is open at BC the lamp will light when connected across it, but will not light across the other three coils. It is advisable to replace the open coil with a new one, rather than attempt to repair the break.

Locating Grounds. 6. Before checking for grounds, make the following test. With the line switch open, ground

one of the test-lamp leads. Connect the other lead first to L_1 and then to L_2. Assume that the test lamp lights when connected to L_2. This shows that L_1 is grounded and a circuit established from L_2 through the lamp, the grounded lead and back to L_1 through the ground in that line.

The motor and starter may be checked for ground by connecting the test lamp as in Fig. 2. Leave the line switch open and if one side of the supply is not grounded, it should be temporarily grounded through another lamp. If a ground exists in either the motor or starter the test lamp will light, except under one condition. This condition is where the ground exists in the starter handle or the wiring from that point to the switch. However, if the starter handle is moved to one of the starting buttons, this part of the circuit will also be included in the test.

If a ground is indicated by the test lamp the next step is to determine whether it is in the starter or motor. Disconnect the motor leads from the switch and starter terminals. Then by the same test as shown in Fig. 2, touch the test lamp lead to the starter terminals and then to one of the motor leads. Lamp lights on the grounded circuit.

If the ground is in the starter, the most probable fault location is in the resistor at some point where insulation has failed. The quickest way to find such a ground is by close inspection. If necessary, the location may be tracked

down by disconnecting a suspected part and repeating the ground test. When the ground appears to be in the motor, lift the brushes and check the armature by touching the test lamp lead to any commutator segment. To test the field for ground, touch the test lead to any field wire.

Testing the Armature. A ground in either the armature or the field is invariably a result of insulation failure. A grounded field coil may be found by opening the wires between field coils and checking each coil separately. Usually it is advisable to replace, rather than repair, a grounded armature or field coil.

7. Short circuited armature coils are easily located by means of a low-reading voltmeter (millivoltmeter) and a couple of dry cells. To test for a short in an armature refer to Fig. 3. The diagram shows that armature coils are connected in series, and that the ends of each armature coil are connected to the commutator bars. A low voltage from two dry cells is applied to the commutator at points 180 deg. apart. This may be done through two motor brushes provided all other brushes are lifted and any connecting wires to the brush wiring are disconnected. If the voltage is checked between adjacent bars on the commutator, it will be noted that when the millivoltmeter is across bars C and D which lead to an open armature coil, the meter will give a higher reading than when across any other adjacent bars.

On the lower half of the armature the readings between adjacent bars are about the same, except across EF. When the millivoltmeter is across the adjacent bars E-F which connect to a shorted coil, it reads zero. Between the points A-C-B-D the voltage across adjacent bars reads zero because that path is open circuited.

A shorted field coil is checked in a similar manner. Low voltage dc is applied to the fields in series similar to Fig. 4 and a voltmeter with a suitable scale is connected across the bared leads of each coil. The voltage drop as measured by the voltmeter will be less across a shorted coil than across the others. All coils that are satisfactory will give readings nearly alike.

An open-circuited coil in a field having a normally high resistance, may be located quickly using a voltmeter rather than a test lamp. There will be no voltage across the coils which are good. The open coil will give a voltage indication, the value of which depends on the resistance of the field coils.

Finding motor faults fast saves you cash and care.

...Synchronous Motor Faults

SYMPTOM	POSSIBLE CAUSE OF TROUBLE: CURE

MOTOR WON'T START OR COME TO SPEED

1 Low starting voltage; a 15% supply voltage drop can prevent starting. Higher voltage tap, mechanical auxiliary starting may do trick. If reactors are used for starting, they may be too big—cutting starting current too much.
2 Rotor may be tied in to direct-connected exciter during starting. This increases required starting torque. Best bet: leave rotor field circuit open and in parallel with discharge resistor. Close in when motor is near synchronous speed
3 Open in stator circuit or field starting winding
4 Starting load too big. Raise power, cut load

LOUD AND SOFT HUMMING: HUNTING

1 Alternator speed may be changing. Improve speed regulation of prime mover running alternator that feeds motor
2 Rapid changes in load. If frequent, keep motor over-excited (limited by heat-up factor)
3 Voltage and frequency variations. Could be due to: (a) short circuit in supply (b) wrong excitation (c) mechanical overloading
4 You may be weakening excitation at no-load to improve power factor. Salient-pole-type motors can take lowered excitation up to about 20% of unity pf without pulling out. Avoid large excitation reduction

MOTOR CIRCUIT-BREAKER TRIPS

1 Line surge. Restart when line is cleared
2 Low voltage because of supply fault

MOTOR CIRCUIT-BREAKER TRIPS *(cont)*

3 Excitation stops because of open circuit in motor field, rheostat or wiring between exciter and motor field. Seen by exciter ammeter reading
4 Exciter not working (seen by lack of voltmeter reading). Check exciter brushes, field coils, armature, rheostat to trace and repair fault

MOTOR WON'T SYNCHRONIZE

1 Too much load torque during speed-up. With a pump, a partially closed valve could cause this ill
2 Starting voltage too low. Higher voltage tap is the answer
3 High resistance in damper (amortisseur) winding
4 Faulty exciter brushes
5 Reversed field polarity; stator current doesn't drop to minimum at unity power factor after exciter is connected. Repeat starting cycle with correct polarity
6 Current surges when full supply voltage is applied to field-excited motor. Remedy by selecting suitable excitation value to give *over-excitation* during start. Then switch from START to RUN without delay

OVERHEATING OF MOTOR FRAME

1 Over-excitation. To cure, reduce excitation until stator current comes down and tries to build up again. This betters motor pf, but reduces system pf
2 High resistance fault to ground in winding
3 Short circuit in turns or coils of winding

When a motor goes sour your best leads often come from looking, listening, smelling, feeling.

PROBABLE CAUSE of motor troubles can often be found *before* taking the unit apart

...Motor Troubles You'll Meet

SYMPTOM	POSSIBLE CAUSES OF TROUBLE: CURES

MOTOR WON'T START

1 One phase open; may be blown fuse or poor contact
2 Rotor hitting stator; bearings may be worn
3 Open in stator winding, switch, starter or wiring
4 Low starting torque; check voltage and motor size
5 Excessive friction in driven machine or line-shaft bearings; overhaul, lubricate, check belt tightness
6 One phase of stator winding reversed. In this case unloaded motor will come up to speed but hum loudly
7 Windings connected wye rather than delta, or in series rather than parallel. Result is low phase voltage

MOTOR-FRAME OVERHEATING

1 Overloading caused by increased load, fault in driven machine, tight bearings
2 High ambient or too near heat source; improve ventilation, move motor or design heat barriers
3 Poor motor ventilation because of dirt in vent holes; clean thoroughly, install motor vent ducts or replace with totally enclosed motor
4 Low supply voltage; check with local utility. May call for booster transformer or new feeder
5 Winding connected wye rather then delta; motor-starter handle in wrong position or windings connected wrong
6 Poor connection in squirrel-cage rotor
7 Single phasing; blown fuse, poor contact or open circuit in winding, starter, switch or wiring
8 High-resistance ground in winding
9 Few turns or coils in winding shorted out

MOTOR VIBRATES

1 Poor foundation
2 Insecure mounting
3 Rotor out of balance; see *Power,* Nov '50, p 84
4 Pulley, coupling, pinion or fan out of balance
5 Winding partly shorted; check for current balance
6 Coil or coil groups reversed in stator winding. With reversed coils, motor will be noisy too
7 Loose squirrel-cage rotor connections (lights connected in same line will flicker if rotor connections are poor)
8 Poor end connections in wound rotor or bad contact at short-circuiting ring

SYMPTOM	POSSIBLE CAUSES OF TROUBLE: CURES

HOT RING-OILED BEARING

1 Not enough oil in bearing well; leak in drain plug or gage glass, overflow pipe bent or plugged
2 Oil-ring sluggish or not moving; ring tight in slot, ring damaged, dirty oil, wrong-viscosity oil
3 Ring jumps out of slot; fit retaining strip on slot
4 Bearing grooves clogged
5 Oil grooves cut wrong in bearing (See *Power's Handbook of Bearings and Lubrication,* p 87)
6 Bearing too tight
7 End thrust on sliding bearing designed for radial load only; faulty drive, incorrect assembly, or job may call for thrust bearing

HOT BALL AND ROLLER BEARING

1 Not enough clearance in bearing; check whether bearing number and type agree with manufacturer's specs
2 Inner or outer race turning in shaft or in housing
3 Too much grease; pack bearing only about 1/3 full; leave relief plug out for 1 minute after the machine starts
4 Poor grade of grease. Soap and oil separate

HOT BEARING ... ALL TYPES

1 Poor alignment of coupling or gear drive
2 Driving belt too tight (flat belt or V-belt)
3 Belt too far from bearing; reduce pulley overhang; or add another bearing to handle belt load
4 Middle or outer bearings of 3-bearing units overheating. Hot bearing takes more than its share of load; uneven bearing wear, poor alignment

NOISY BALL OR ROLLER BEARING

1 Bearing worn; replace bearing
2 Flattened or chipped ball or roller; replace bearing
3 Fractured cage or race; replace bearing
4 Inner race loose on shaft; build up shaft by metal-spraying, then machine to fit the new bearing
5 Outer race turns in housing; bore and bush end bracket. Or build up to fit outer race of new bearing

Moisture Damages

SPLASH from hose while washing floors grounds out motors in plants where floors are washed around electric motors

DRAINS running to floor from stock chests in paper mills. Motors are stuck under tanks in overcrowded room

CURES

CHECK COILS for dampness with megohmmeter. Caution! Even 100v can puncture wet insulation—so crank slowly. Have only qualified electrician make test

WASH INDUCTION MOTOR that's been submerged with warm water to remove oily film. Use 25-psi pressure (not more) and not over 194 F. Oil shaft, bearings

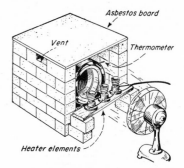

DRY OUT WET MOTOR in temporary oven. If no oven, make one of bricks, asbestos-board and heating elements. Blow heated air through motor like this

PREVENTIONS

FIRST STEP to stop needless damage to electrical equipment is to train personnel. Tack this OPERATOR'S NOTEBOOK on wall and have all persons working around equipment read it

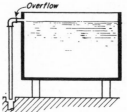
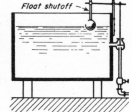

STOP OVERFLOWING TANKS by placing overflow pipe at highest safe level. Another way is float shut-off to cut off incoming water. If not practical, move electric motors

Electric Motors

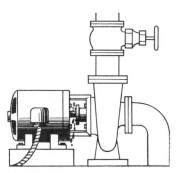

GLANDS LEAKING from motor-driven pump shoot water into motors in numerous plants. Then motor burns out fast

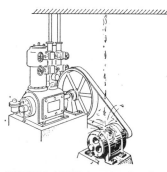

OVERHEAD LEAKS through ceiling from floor above. Having floors above that are continually wet is common cause

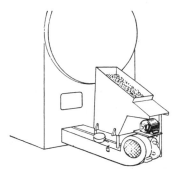

DAMP AIR during idle periods coats expensive motors with moisture. This cuts down the insulation resistance

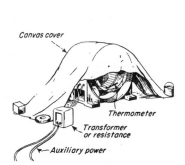

PASS LOW-VOLTAGE (about half-load value) through windings if no oven. Cover with canvass partly for ventilation. Don't let temperature go over 175 F

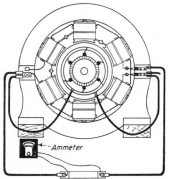

TO DRY dc generator, short the armature, reverse series field, weaken shunt field. Use ammeter like this, then check with megohmmeter after a few hours

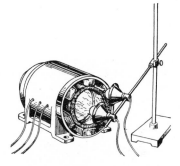

INFRARED LAMPS speed drying-out process. Position lamps carefully and check winding temperature often; otherwise the insulation may be damaged badly

DRIP-PROOF   SPLASH-PROOF   TOTALLY-ENCLOSED   FAN-COOLED

REPLACE MOTORS that are in trouble spots (if not practical to move them) with one of these four types. Money saved from no outage and repair bills will easily pay for change

NEVER, NEVER start motor after long shut-down, without having electrician check for moisture. That holds true after wet cleaning, long shut-down or suspected moisture

To Clean Motors Safely...

... *use common sense, compressed air at proper pressure, avoid all highly flammable solvents while inside plant**

Don't let air pressure exceed 40 psi

Goggles

Respirator

1 Compressed air does good cleaning job on assembled motors. But first use a vacuum rig to pick up light surface dirt before it's scattered by air pressure

▶ SAY WHAT YOU MAY, one of the most important jobs in maintaining electric motors is keeping them clean. And generally speaking there are only two ways to clean motors. One is with compressed air as sketched up in Fig. 1 and the other is by use of solvents, Fig. 2.

Hazards. Before we say much more about either method of cleaning, let's consider the safety angle. First off, no cleaning job should be attempted unless the motor is shut down and the safety switch open. On top of that, it's wise to padlock this switch in the open position.

When using compressed air for cleaning, protect yourself with safety goggles *and* a respirator. With both flamable and non flamable solvents, there's the everpresent danger of breathing the vapors. Then we have the fire and explosion hazard.

Compressed Air. Wherever possible, a brush or vacuum system should be used instead of compressed air for cleaning light dust, chips and oily dirt. On the other hand, compressed air may be the only practical method unless motor is taken apart.

Where you use compressed air, make sure there's a separator in the line to remove all impurities and moisture. Pressure should be regulated, and *not* over 40 psi. Where you go above that pressure, there's danger of damaging motor's insulation.

In the case of grain elevators, candy factories and cotton mills, dust can become an explosive mixture if blown into suspension by an air hose. Hence, never use an air hose for cleaning motors in these locations.

Now here are a few tips when using an air hose: Before turning on pressure, check all air line connections for tightness. Then while opening or closing valve hold the air hose nozzle. Otherwise the hose may whip about and hurt somebody.

Solvents. Unfortunately, its a common but unsafe practice to use solvents with low *flashpoints*. Flashpoint is lowest temperature at which solvent will vaporize into an inflamable mixture. And low means anything below 100 F. This includes gasoline, −45 F, naphtha, 20 F, and benzol 12 F. Such solvents create a serious fire hazard and may produce an explosive mixture with air at room temperature. *They should never be used for cleaning purposes inside a building.*

In a small enclosed space such as an engine room, only a small amount of vapor from these solvents is necessary to set up an explosive mixture. In the case of gasoline vapors only 1.3 to 6 percent by volume in air is needed. A spark or exposed flame coming in contact with this mixture could cause a violent explosion. Then again, small amount of naphtha or benzol left on a motor's windings could form an explosive mixture that would explode when motor is started.

On top of that, some low flashpoint solvents are highly toxic. You're running the chance of being overcome by fumes when used in an enclosed place.

For instance, benzol in concentrations greater than 100 parts per million in air is unsafe for an 8-hr exposure.

High flashpoint solvents—those having a flashpoint of 100 F or above—are often used for cleaning motors. Kerosene and the Stoddard solvents are in this family. Although the explosion hazards of these solvents are not as great as the low flashpoint babies, they do present a fire hazard when not used safely. Keep a carbon-dioxide fire extinguisher handy when using these solvents.

If high flashpoint solvents are used where the temperature may reach 100 F, they will vaporize rapidly and may form explosive vapors. And at high temperatures these vapors may have the same toxic effect as vapors of low flash point solvents.

Other Solvents. Vapors given off by petroleum solvents (both low and high flash points) are heavier than air. They travel rapidly and in confined spaces tend to settle and remain in low areas. They will enter pits and flow through drains and sewer lines. A spark may easily ignite them and cause a serious explosion.

Carbon tetrachloride and trichloroethylene won't burn. They do vaporize rapidly, and their vapors are poisonous. Carbon tetrachloride is unsafe for continuous exposure if it is present in quantities greater than 50 parts per

KEEP WORK AREA WELL VENTILATED . . .

Keep carbon dioxide fire extinguisher handy

Use approved one-gallon safety can

Elbow grease helps too!

2 Solvents do trick when cleaning frames. Good mixture is 25% methylene chloride, 70% Stoddard solution and 5% perchloroethylene (use it, don't spell it!)

SAFE STORAGE . . .

Fireproof cabinet

Dont store more than 10 cans in one cabinet

1-gal safety can

3 Storage of all solvents should agree with Fire Underwriters and local laws

HERE'S HOW TO GET RID OF USED FLAMMABLE SOLVENTS . . .

DON'T....

Flammable solvent

Drain

4 Disposal in sewer drains is taboo. Solvent may give off flammable vapors that can be ignited at some distant point

DO...

Firefighting equipment

Wind in this direction

5 Disposal of used solvents is best handled outside plant. Dump flammable solvents in small quantities, then burn

---

million in air. So where carbon tet has to be used in confined areas, provide either forced ventilation or have the job done under an exhaust hood with a downdraft flow of air. Make sure the exhaust fan is big enough to hold a negative pressure under the hood at all times. Another out is using an air-supplied gas mask.

If carbon tet is mixed with low flashpoint solvents, will the explosion hazard be eliminated completely? Nope, the fire and explosion hazard may still be present if one of the fluids evaporates

faster than the other fluid. Then when the mixture is being used in a confined space the hazard of toxic vapors still remains. A reasonably safe cleaning mixture contains 25% methylene chloride, 70% Stoddard solution and 5% perchloroethylene.

Storage. When it comes to storing any of these solvents, your first step is checking with the Board of Fire Underwriters and your local fire-ordinance body. Then follow whichever regulations are most strict.

It's poor practice to store any type

cleaning solvents in the work area. If you must keep them in the building, they should be stored in one-gallon safety cans in a fireproof cabinet, Fig. 3. If you find it necessary to store more than ten such cans, use a spearate fireproof storage vault away from the building.

As practical men, we know there are many times when such solvents must be stored in the plant. But in such cases make provision for adequate ventilation. In confined areas use down-draft forced ventilation.

Where to look for trouble

Shading coil

Usually copper or brass stamping. Loop may have split open because of vibration. Causes chattering of contacts, or at least, noisy operation. Remedy: braze loop closed or replace with copper wire using long pigtail splice. Fasten tightly.

Magnet-pole face

Must be kept clean, rust free. Gummy dirt can prevent magnet from dropping out; rust or heavy dirt can cause chattering. Remedy: clean with fine sandpaper. If badly worn, file flat. Note: center pole of E-type magnet should not touch or contactor may not drop out.

Coil

Overheating coil can be caused by overloading (too much contact pressure; worn or misaligned mechanical parts). Also check: pole-faces not seating correctly; wrong coil; shorted coil; too high voltage. Shorted coil or with intermittent open can cause chattering, unnecessary tripping. Remedy: remove coil, check rating, squeeze (coil should be firm, insulation unimpaired).

Mechanical parts

Worn bearings and linkages can cause chattering, unnecessary tripping, poor alignment of contacts and magnet. Remedy: clean, sandpaper, lubricate, realign. If badly worn, best bet is to replace.

Auxiliary contacts

Poor contact caused by dirty contact surfaces, weakened spring, misalignment, can cause chattering, unnecessary tripping, failure to hold after start. Remedy: clean, realign, check tension.

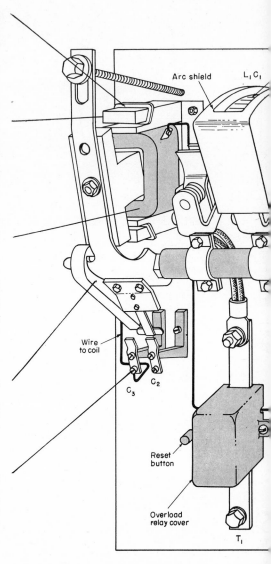

Arc shield $L_1 C_1$

Wire to coil

C_3

C_2

Reset button

Overload relay cover

T_1

in ac magnetic starters

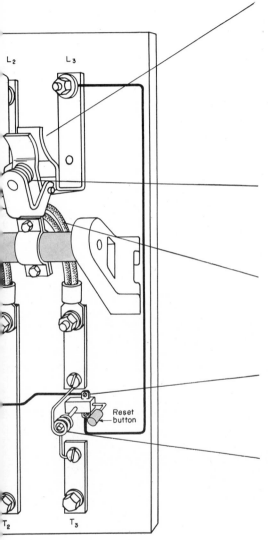

Contacts

Short tip life can be caused by weak contact pressure, interrupting heavy currents (frequent jogging; opening on short-circuit current); chattering; wrong contacts; misalignment; excessive filing. Overheating comes from weak pressure, inductive loads, copper oxide on tips. Remedy: install new contacts of proper rating; check contact pressure with spring scale. Look into operating requirements for possible misapplication of starter. Replacing copper with silver-plated or alloy contacts may help. (Don't file silver or alloy contacts.) Check arc chutes for proper alignment.

Contact springs

Weakened springs cause poor contact pressure. Springs that are too stiff overload magnet, result in chattering. Remedy: best bet is to replace with manufacturer's spring.

Shunts

Burned or broken shunts are due to misapplication (too short, too stiff, etc); corrosive atmosphere. Remedy: check regularly for signs of deterioration; replace only with manufacturer's shunt.

Overload trip mechanism

Damaged by frequent overload or by resetting too soon after tripping. Causes unnecessary tripping, sometimes chattering. Remedy: replace.

Overload thermal coil

If too small, unnecessary tripping results. If too large, main contacts will carry overload current. Remedy: replace with recommended element, check peformance if test equipment is available. Ambient temperature has effect; never remove cover or other associated parts from relay. Leaving cover off controller may expose relay to drafts.

● Clapper-type starter is shown because it's easier to identify parts. Plunger type has same essential parts, except shunts. Troubles follow same pattern although improved designs minimize them.

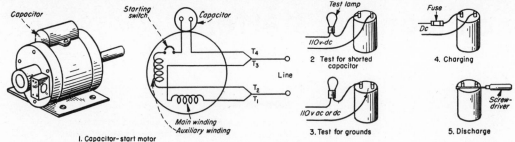

1. Capacitor-start motor

CAPACITOR-START MOTORS just won't start if capacitor is bad. Note how condenser connects with auxiliary winding

SIMPLE TESTS to help you spot a faulty condenser shown above. If ohmmeter is handy, use it to check for shorts and grounds

Spotting Faulty Capacitors

You can save a repair bill next time one of your capacitor motors gives trouble

▶ SMALL MOTORS seldom get the care and attention that large ones do. As a result many small motors are unnecessarily sent to the shop for repair or replacement, when the plant operator with a little "know-how" could put the motor back in service.

Most small motors are single-phase, alternating current, using either capacitor or repulsion starting. Unless the service requires high starting torque, capacitor-start motors are satisfactory.

The capacitor-start motor may be readily recognized by the capacitor unit mounted on top or side of the motor frame, Fig. 1. Most troubles with capacitor-start motors involve the capacitor, therefore, it is mounted so that testing and renewal is simple.

If your capacitor motor blows fuses, has poor starting torque or hums without turning over, the trouble is usually in the capacitor. To check it, remove its cover, disconnect it and make the following tests:

Shorted Capacitor. A test for short circuit in a capacitor must be made with direct current since even a capacitor in perfect condition will pass alternating current. If a 110-v dc source is handy a test light connected as in Fig. 2 will light when the capacitor is shorted. Otherwise, use a bell and battery test set. If the capacitor is shorted, you probably have located the cause of

blown fuses. Replace the shorted capacitor with one of equal voltage and current rating.

If the capacitor is not shorted, it should be checked for grounds as in Fig. 3. Either ac or dc may be used to make this test. Be sure to make the test on each of the capacitor terminals. If the lamp lights, it indicates a grounded capacitor and a possible cause of blown fuses.

Loss of Capacity. A simple though not accurate test for possible deterioration of a capacitor is checking its spark. The capacitor should be charged by connecting it across a dc supply (properly fused) for a few seconds, as in Fig. 4. The capacitor is now "charged" and the terminals should not be touched, due to the possibility of a severe shock.

With the capacitor disconnected from the line, the terminals are shorted with a screwdriver having an insulated handle, Fig. 5. If the spark is snappy, the capacitor is probably in good condition; if the spark is weak or cannot be seen, the capacitor has either lost capacity or is open. In such a case, the motor starting torque will be affected. This may prevent the motor from coming up to speed and cause overheating or damage to the starting winding.

Checking Capacity. A capacitor of the *electrolytic* type may deteriorate with age causing a gradual decrease in

capacity. An electrolytic condenser is formed by winding two sheets of aluminum foil into a cylindrical shape. The dielectric is usually gauze treated with an electrolyte. If such a condenser loses more than $\frac{1}{4}$ its rated capacity, it should be replaced. The previous test of charging and noting the spark of a capacitor on discharge is simple but not accurate.

Capacity may be accurately measured by connecting the capacitor in an alternating-current circuit with a voltmeter and ammeter. Connect capacitor across the line with an ammeter in series. The voltage rating of capacitor should equal that of line. Close the switch and take voltmeter and ammeter readings. If no current flows, you know the capacitor is open circuited.

If you do get a flow of current, take the current reading and divide it by the voltage across the condenser; then multiply by 2650. The result will be the capacity in micro-farads. The value 2650 holds true only when the frequency is 60 cycles. As mentioned above, if the capacitor has lost more than $\frac{1}{4}$ its rated capacity, a new one should be installed.

Although the capacitor is not always responsible for troubles in capacitor motors, it is a likely suspect. For this reason the above tests will solve the majority of your capacitor-motor faults.

Put Your
Ammeter To Work

▶ CONSIDER THREE THINGS before using an ammeter: (1) Check whether the current you plan to measure is direct or alternating. (2) Determine about how much current is flowing. (3)

Select the proper ammeter and connect it in the circuit as shown in the sketches below. Always connect an ammeter in series with the load. Select your meter so the reading falls about

mid-scale. Close short-circuiting switch before starting a motor. This prevents damage to the instrument by heavy starting current. Keep shorting switch closed except when reading meter.

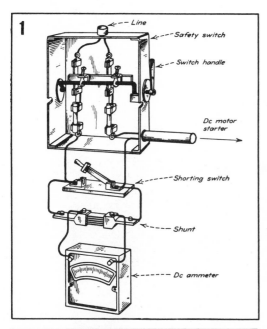

1

— Line

— Safety switch

— Switch handle

Dc motor
starter

— — Shorting switch

— Shunt

— — Dc ammeter

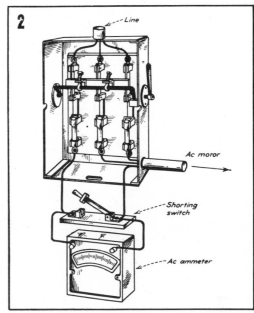

2

— Line

Ac motor

— — Shorting
switch

— Ac ammeter

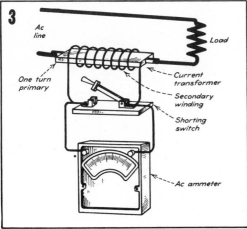

3

Ac
line

Load

One turn
primary

— Current
transformer

— Secondary
winding

— Shorting
switch

— Ac ammeter

1 The dc ammeter measures voltage drop across a shunt. Select the shunt whose current rating equals or exceeds the maximum current to be measured. Each meter should have a set of calibrated leads. Millivolt rating of the shunt and meter must be equal for the needle to read full scale

2 Some ac ammeters measure to 200 amps directly. Don't use shunts with ac meters. Many ac instruments can be used for dc measurement. When so used take the average of normal and reversed meter readings. Use heavy meter leads. A split-core ammeter is handy for large current measurement

3 Current transformers are used with ammeters on high voltage systems. They are also used when measuring large currents. Normal secondary current is 5 amperes with full load in the primary. Never open-circuit the secondary of a current transformer while current is flowing in the primary

Need Another Outlet?

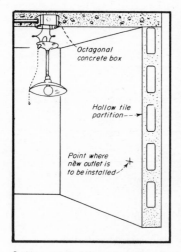

1 You can tap the new outlet from nearby lighting circuit—if new load is small (extension light, clock etc). First step is to kill circuit and drop fixture

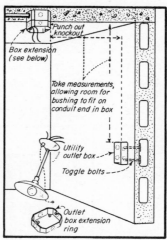

2 Box extension saves chopping concrete. Fasten it with 8-32 machine screws to ears of ceiling box. Set new wall box and take measurements shown

3 To bend **offset** lay conduit on floor and slip bender in place. Can use a 1-in. pipe tee instead of regular conduit hickey. Bend offset before making L bend

▶ FROM TIME TO TIME every small-plant operating engineer juggles the question of installing an occasional electrical outlet in the plant. Many engineers farm out all electrical wiring. But where local laws do not require all electrical wiring be done by a licensed electrician, the operator is often the boy called on to do it.

Keep the National Electrical Code as your bible on all wiring jobs. You can get a copy of the Code by dropping a card to the National Board of Fire Underwriters, 85 John Street in New York City. Many localities have in addition to the NEC, local laws which must be lived up to.

Wiring Methods. Your best bet in running in an extra outlet is to follow the present type of wiring used in your plant provided it is in accord with the code. In other words, if branch circuits are wired with No. 14 wire in ½-in. conduit, follow through using the same materials. Many plants use ¾-in. conduit and No. 12 wire for branch circuits. Also your plant may be wired in electrical metallic tubing which cuts out threading.

Separate Circuits. Tapping your new outlet into the existing lighting circuits should be done only where the new

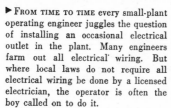

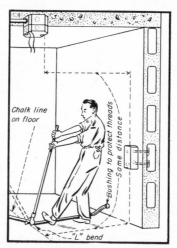

4 Chalk line guides you in making L bend. Don't make bend in one step; move the hickey forward and take small bites till bend checks with the chalk line

outlet is to be used for small loads such as a portable extension cord. If load is to be heavy (grinding wheel, heavy soldering outfit, etc), use a separate circuit. You can use the same wiring method for new circuit by looping the

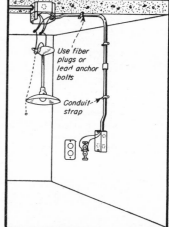

5 Put conduit in place, using locknut and bushing at each box. Use **snake** to pull in wires. Splice black wire to **black** and white to **white;** solder and tape

wires through each ceiling outlet to the panelboard.

Where ceiling lighting fixtures are controlled from a distant panelboard, you will have to run your wiring back to the panel to pick up a "hot" leg.

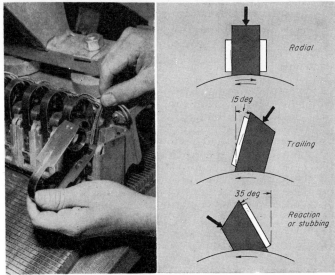

PRESSURE AND ANGLE of brush on commutator or slip rings must be correct for particular kind of service. Constant-pressure brush holder is designed to be self-adjusting

Brush holders: key to good commutation

For trouble-free performance of commutators and slip rings it is not only important to select the right brushes and maintain a proper rotating surface, but brush holders must be correct for the job and kept in proper adjustment.

There are three basic brush holders: *radial, trailing* and *reaction*. Radial brush is primarily for a reversing surface, but it's mechanically unstable and requires relatively high brush pressure to maintain contact. Many times, during reversal of machines, the brush develops a double fit and fails to maintain adequate contact, which is detrimental to performance.

Trailing brush holder is most popular type. This holder, in order to be stable, should make a 15- to 20-degree angle with the vertical, and have a bevel on top of the brush. The reaction brush holder should be operated between 35 and 40 degrees from the vertical, and also be beveled at top.

Each of the three types can be used on various types of machines when properly applied. For instance, the radial holder is primarily for reversing machines, but on larger machines the radial brush does not function too well because of mechanical instability and double fits. On heavy load applications requiring numerous brushes, loss of brush face and overlap prevents successful commutation. In such applications you can often use a holder which contains both leading and trailing brushes.

The trailing holder is used primarily for one direction of rotation. Reaction holder is used mainly on high-speed applications. It has the disadvantage of losing much of usable brush because of steep angle from the vertical.

Brush pressures of optimum values to minimize harmful arcing have been established by tests. It is essential to maintain uniform pressure on all brushes in the machine. Generally speaking, brush pressure that is too light is more harmful than that which is too heavy, for it will speed up not only brush wear but also commutator wear. It's not practical to specify a brush pressure that would cover all brushes; brush pressure must be set according to type of material and application requirements.

Electrographitic brushes should be operated between 3.5 and 4 psi. This value is increased for many special applications. Some slip-ring applications using low-density graphite carbons require pressure as low as 2 psi. Usually the instruction book provided with a new machine specifies correct brush pressure.

Brush pressure should be kept constant throughout the life of the brush. The usual type of brush holder doesn't maintain constant pressure, but must be adjusted as the brush wears. There are now available constant-pressure brush holders for both ac and dc machines which keep pressure uniform as well.

Maintain brush holders in good condition; set them at the spacings recommended at the factory. Keep hubs and pin mechanisms free and easy to operate. Always keep the reaction surface clean, and blow out the holder itself occasionally. Replace all springs subjected to current which have lost their temper. Make sure the brush holder finger is riding on the brush clip. These occasional checks will enable you to get the most out of your carbon-brush applications and help avoid trouble.

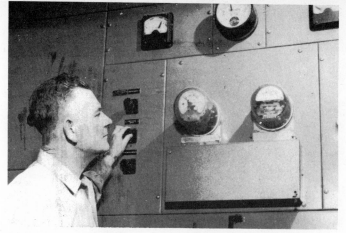

1 Typical departmental meter board in an industrial plant. This installation facilitates keeping accurate records of the electrical consumption in key areas

2 Polyphase watt-hour meter with case removed to show the motive element

First Aid for Plant W-Hour Meters

Here is handy information for the man metering his plant's electrical use: fundamentals of watt-hour meters and practical trouble shooting for five common causes of inaccuracy

▶ BEFORE DIGGING into this matter of watt-hour-meter maintenance, let us get a few fundamentals under our belts. First of all, the polyphase watt-hour meter is a combination of the single-phase meter's elements, driving a shaft at speed proportional to total power. Meter consists of a multi-element motor, means for balancing the torques of all elements, a magnetic braking system, a register, and the necessary compensating rigs. All these components are put together in a meter frame.

Wattmeter measures basic unit of electrical power, the *watt*. Electrical power is the rate at which electrical energy is used. And this unit of measurement, the watt, implies a time factor. Hence, the term watt includes the implied meaning "per hour."

But when we think of watt-hour meters we are dealing with electrical *energy*. We can define that as the total use of electricity over a period of time. It is the product of rate of use (power or watts), multiplied by the number of

time units, usually measured in hours.

Meter Operation. The instantaneous speed of the watt-hour meter's motor is proportional to the power passing through it. On top of that, total revolutions in a given time are proportional to total energy, or watt-hours, drawn during that time. Besides the motor, this meter has a magnetic braking system and a register.

Calibration. Now for a thumbnail sketch of calibration and adjustment. You have seen men from the local utility run checks on watt-hour meters in your plant. They use a standard indicating wattmeter to measure the average power over a stated period. And, at the same time, they count revolutions of the watt-hour-meter disk with the aid of a stop watch.

Then they check the average meter watts by taking the watt-hour-meter constant K, and multiplying it by the number of disk revolutions. This quantity is multiplied by 3600, and the product is divided by the time in seconds, ticked off on the stop watch. Value of watts they get through these mental gymnastics is compared with the watts read on the standard wattmeter. That constant K, mentioned above, is the number of watt-hours represented by a single revolution of the aluminum disk.

Meter is adjusted to read properly at full load and unity power factor simply by varying the position of retarding magnets. It is adjusted at light load by the light-load adjustment lever

that you can readily spot once the glass disk is removed.

Keep in mind that the above facts are set down in a very general way. The actual job requires more specific information and a better knowledge of the meters. But plant men should give careful thought to the trouble-shooting side of the picture, since watt-hour meters are widely used to measure electrical energy generated or consumed in various parts of a plant.

Trouble Shooting. The modern watt-hour meter is an inherently accurate and a stable device. When properly calibrated it retains its accuracy for a long time unless subject to influences outside normal operating conditions. So, in brief, the purpose of your maintenance of a trouble-shooting program is to (1) identify meters that have become inaccurate (2) find out what has caused the error (3) restore the meters to normal accuracy.

A watt-hour meter may be found inaccurate because of improper calibration or adjustment, incorrect installation, or the adverse effects of foreign influence.

Actually, when you boil it down there are few meters that are found inaccurate as a result of the first two causes. Possibility of mistakes from incorrect installation should not be ruled out when a maintenance check is being made, but outside influence is a much more common cause.

For instance, when the meter runs slow, the things to look for are: pos-

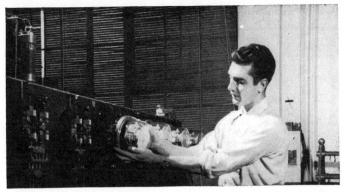

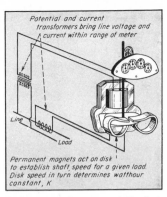

3 Testing boards like the one shown here are sometimes used to carry out meter trouble shooting. This method centralizes all checking and repairs in one area

4 Meter may be connected directly in a low-voltage, low-current circuit

sible dirt; a magnetic short; bearings that are dry, worn or out of line; shaft bent; gears too tight; shorted current coil; open potential coil; meter not level; incorrect adjustment.

On the other hand, if the meter runs fast, think in terms of a possible weakened braking magnet or incorrect adjustment. If meter runs backward, you have an incorrect connection.

A creeping meter may result from vibration, stray fields, excessive voltage, shorted current coils, incorrect adjustment, voltage leakage or ground in wiring.

Dust or Dirt. Dust, dirt or other foreign particles, like tiny pieces of metal, are the most common cause of incorrect operation after watt-hour meter has been in service. They may tend to close the air gap between the disk and the electromagnetic polepieces, and thus cause friction. This is especially likely if the dirt includes particles that cling magnetically to the magnet, and align themselves with the flux so they rub on the disk. Dirt in any of the gearing will increase gear loading, and may cause the meter to slow down. The answer is simple. Just give it a thorough cleaning.

Bearing Troubles. Worn or poorly aligned jewel-and-pivot or ball-type bearings increase value of the friction load, and thus make the meter run too slowly. Wear from this source is greater with light loads than with heavy, and becomes progressively greater as long as the worn bearings are left in service. Bearing friction may also be increased to an objectionable degree if the meter is out of level. Lack of lubrication of the pivot bearing is another source of increased bearing friction.

Misalignment or improper adjustment

of bearings is generally an indication that the meter has been subjected to severe shock, either mechanical or electrical. Best solution is simply to replace the bearings. Of course, there may be cases where it is best to oil the pivot bearings or level a tilted meter. On the other hand, it may be best to have a repairman completely overhaul the meter.

Bent Shaft. When you come across a bent shaft or a warped disk, you invariably meet up with measurable errors in watt-hour-meter registration. A bent shaft or a warped disk generally indicates that the meter has been roughly handled or subjected to excessive surges. Your best bet is to return the meter to the shop for a thorough and complete overhaul.

Defective Coils. A shorted turn in a current coil of a watt-hour meter lowers the torque near or at full load. At light load, however, meter may run fast or slow depending on which of the two coils is shorted. A short circuit in the potential coils changes the meter's torque and also its lag adjustment. Shorted turns may indicate that surges have been applied to the meter. Once again, your best bet is to return the meter to a competent shop for an overhaul job.

Weak Magnet. A braking magnet, whose strength has been reduced by exposure to excessive stray fields or surges, allows the meter to run faster than it should. In this case, the answer is simply to replace the weakened magnet with a new one.

Creeping can mean a headache. By this term we mean a continuous, slow rotation of the disk when there is no load on the meter. The disk may creep either forward or backward. Although

the creeping has to be rather fast to have any effect on meter registration, it is standard policy among meter-test men to permit no creeping of meters in service.

Watt-hour meters are intended for use where they are not subjected to vibration. And any unusual amount of vibration may cause creeping.

But if stray fields are strong enough to produce eddy currents in the meter disk, these currents may react with the potential coil flux to develop a torque that causes creeping. Once again, the only answer is to relocate the meter, or shield it magnetically so the stray fields will not affect it.

Or if the line voltage is considerably above normal, the compensating torque due to a light-load adjustment is proportionately higher, with the result that the meter may creep. The solution is simply to recalibrate the meter to operate at the higher voltage.

In short-circuited turns in the current coil, flux of the potential coils induces voltage. This causes eddy currents in the disk, and results in creeping. You will find it best to return the meter to the shop for complete overhaul and replacement of the defective coil.

Creeping may also be traced to improper connection. For example, when the meter is installed with potential coil connected on load side of the meter, creeping results.

Sometimes creeping is caused by leakage or grounds in the metered circuit. This false creep disappears when meter is tested with load wiring disconnected from meter. You can make a quick check on this by disconnecting meter and running a megohmmeter test on the load wiring.

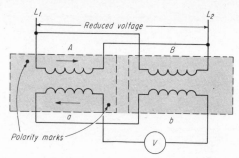

1 Polarity test with standard transformer A of known polarity matches voltage with unknown transformer joined

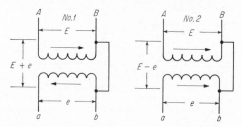

2 Polarity test by differential ac method has directly opposite leads of primary and secondary connected together

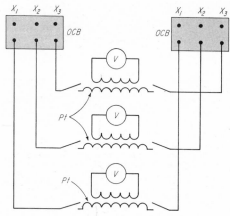

3 Paralleling and phasing of 3-phase transformer banks fed from same source can be done with 3 pt's and voltmeter

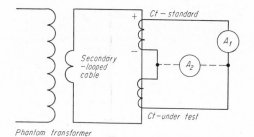

4 Polarity check is simply made with phantom transformer while making ratio check using known ct and ammeter

Make practical transformer tests in the field

Three methods are in common use for testing polarity, which means the instantaneous voltage relationships: (1) comparison with a standard transformer; (2) using a dc inductive "kick"; (3) differential ac method.

When a standard transformer of known polarity, having the same ratio as the transformer under test, is available, connect the high-voltage side of both units in parallel, Fig. 1. Connect secondaries in series in the way shown, and attach to a voltmeter. If zero volts is read,

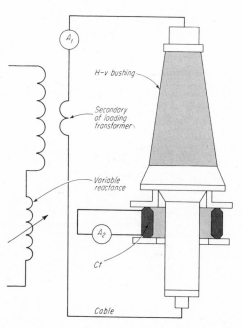

5 Ratio of ct can be checked in place using phantom load. It saves labor and hazard of lifting heavy bushings

polarities are identical; if voltage is double, transformer B is subtractive while A is additive.

Inductive kick method uses same primary hookup as Fig. 1, but secondary leads are open. With high-voltage dc voltmeter connected across left-hand secondary, induce a small kick in positive direction by opening dc excitation current of primary. Next, transfer voltmeter to right-hand secondary and repeat procedure. If pointer swings in the same direction, polarity is additive. A negative swing indicates subtractive polarity.

For differential test, connect primary and secondary together as shown in Fig. 2. Apply a convenient value of ac voltage to h-v terminals A and B. Note applied voltage, then read voltage between free h-v lead, A, and l-v lead, a. If this voltage is $E + e$, transformer is of additive polarity; if $E - e$ is read, polarity is subtractive. In fact, dividing applied voltage by the difference in voltage in both cases gives the ratio.

Phasing test is essential when a transformer bank is newly installed or re-installed where it must operate in parallel with another transformer or system. *Take care when making these tests that connections are correct and that clearance is sufficient, phase-to-phase and phase-to-ground.* If the transformers to be phased are supplied from the same system and lack of synchronism is impossible, phasing may be checked using 1 or 3 pt's as shown in Fig. 3. If only one transformer is used, each phase must be checked separately. Pt primary voltage should be twice

system voltage since 180 deg out-of-phase may exist.

If transformers are energized from different sources, lack of synchronism may exist. Phasing is then done with 4 pt's, a pair connected in open delta to each source. Pt secondaries are connected in open delta, and are joined together through lamps, one per phase. One set of pt's is purposely connected 180 deg out-of-phase with the other. If in-phase condition exists, all three lamps will be bright; if not, one lamp will be bright and two dark. Where phasing is correct, but systems are not in synchronism, all lamps are alternately bright and dark. If out of phase and out of synchronism, alternate lamps become bright, then dark, and out of step with one another. Flicker of lamps shows systems are on widely different frequencies.

Loading transformer, frequently called a *phantom,* can be used for a number of tests. While it may be purchased, we found it advantageous to make our own (box below). Our core had a 7-7/16 in. ID, a 12 in. OD and was 7½ in. long. The variable reactance shown is a valuable adjunct for controlling current output.

Polarity may be checked as shown in Fig. 4. With ammeter in position A_1, it will read 5 amp if ct's have same polarity, and zero amp if opposite. Double check with ammeter in position A_2 where it will read zero if polarities are the same, double amp if opposite.

Ratio of ct's may be checked in place, after lowering the oil, as shown in Fig. 5. Ammeter A_1 should use ct of known ratio. A_2 should be in 5-10 amp range.

Phantom load transformer can be used to simplify field testing

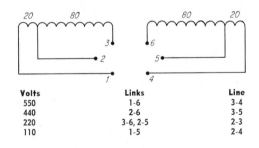

Volts	Links	Line
550	1-6	3-4
440	2-6	3-5
220	3-6, 2-5	2-3
110	1-5	2-4

How winding was built up

2 layers of 1-in. cotton tape half-lapped and air-drying varnished
1 layer 1/32-in. molded mica or 20-mil fish paper
1 layer 1½-in. cambric tape
2 layers 1½-in. webbing tape
1 coat varnish (orange shellac or synthetic varnishes may be used)
1 layer 100 turns No. 10 magnet wire
2 layers 1½-in. cambric tape
1 layer 1½-in. webbing tape
1 coat varnish
1 layer 100 turns No. 10 magnet wire (wound in same direction)
2 layers 1½-in. cambric tape
1 layer 20-mil fish paper
2 layers 1½-in. asbestos tape
2 layers 1½-in. webbing tape
2 coats varnish, as above

PHANTOM LOAD transformer is demonstrated using two turns of 500-Mcm cable. The adjustable reactor holds current to 1000 amp

Let this handy check chart be your guide when . . .

Tracking down transformer troubles

TROUBLE SIGN	POSSIBLE CAUSE AND WHAT TO DO
General overheating	1 Overload is possible. *Reduce load or improve power factor* 2 Improper division of load between transformers in parallel. *Make sure transformers have same turns ratio, percentage impedance, and ratio of resistance to reactance* 3 Unbalanced single-phase loading of polyphase transformer. *Redistribute load between phases* 4 High ambient temperature due to inadequate ventilation of transformer area. *Improve air circulation or install forced ventilation* 5 Heat absorbed from adjacent equipment. *Relocate transformer or source of heat, or build insulating barrier between them.* 6 Insufficient cooling liquid. *Check for leaks or cracks in tank and fill to proper level* 7 Sludge in the cooling ducts or on the windings. *Clean thoroughly by spraying with coolant liquid under pressure* 8 Defective pump where transformer liquid is force circulated 9 Air ducts of forced air-cooled transformer plugged by dust deposits 10 Defective watercooling system in watercooled transformers. *Check that water supply is adequate and reliable, that circulating pumps are functioning properly, and that tubes in transformer and radiator are not obstructed by deposits* 11 Power losses in core increased by deterioration of lamination insulation. *Requires dismantling core to reinsulate it, which usually means rewinding transformer*

TROUBLE SIGN	POSSIBLE CAUSE AND WHAT TO DO
General overheating and breakdown indications (confirmed by voltage radio test)	1 Short circuit between turns of winding caused by sharp edges of copper conductor piercing insulation as a result of vibration, or of shock from excessive currents of external short circuits or switching in on heavy loads 2 Short circuit resulting from dislodging of one or more turns, caused by heavy external short circuit, followed by abrasion of the insulation between turns. 3 Insulation breakdown due to moisture. *After repair of damage, dry out transformer and test insulation. Test the oil for presence of moisture. Check for place of moisture entrance, through gaskets or breathers, including effectiveness of breather drying agent.* 4 Deterioration and breakdown of insulation by corona in very high-voltage transformers. Corona may occur around sharp edges or corners of conductors, especially if air is trapped in part of the winding. *After repair and test, refill with oil, being careful to prevent air pockets forming* 5 Pounding of insulation as result of rapidly fluctuating heavy load with consequent expansion and contraction of conductors and varying mechanical pressure on the insulation. *Reinsulate damaged winding, securing coils to prevent vibration. If possible, change load to reduce fluctuations* 6 Insulation breakdown caused by line surges and lightning. *Better protective equipment, such as reactors and lightning arresters, should be considered*

TROUBLE SIGN	POSSIBLE CAUSE AND WHAT TO DO	TROUBLE SIGN	POSSIBLE CAUSE AND WHAT TO DO
General overheating without load on Y-Y transformers	1 Increased core loss caused by distortion of transformer's magnetic circuit as result of grounding only secondary neutral. *Condition can be corrected by employing delta-connected tertiary winding on transformer. Use of underground secondary may reduce core loss, but may lead to unbalanced phase voltages. Best practice is to avoid Y-Y connections where possible*	Localized overheating	1 Eddy current losses in the core caused by breakdown of bolt insulation, probably due to vibration. *Reinsulate the bolts and secure insulation to prevent vibration* 2 Deterioration of insulation between laminations. *Reinsulate part of core affected. General condition of core, as well as extent of dismantling necessary, usually makes rebuilding of entire transformer advisable* 3 Breakdown of insulation between yoke and yoke clamping plates. *Reinsulate and secure to prevent vibration* 4 Eddy current losses caused by short circuits between burred-over edges of laminations. *Carefully remove burrs across laminations*
Unequal heating in Y-Y connected bank with grounded neutral	1 Ground fault exists on one phase as result of insulation failure, either inside transformer or somewhere on external circuit. *Disconnect bank and find where ground fault is located. After repairing ground fault, give insulation test to each transformer in bank*		
Unequal heating in Y-Δ connected bank	1 Ground fault on one of the primary phases due to insulation failure, either internally or externally. *Disconnect the bank, locate and repair ground fault*	Protective alarm indicates gas in tank	1 Test of gas sample shows gas is flammable and when passed through silver-nitrate solution white precipitate is formed, showing that acetylene is present. Acetylene is given off by breakdown of oil or other insulation caused by development of internal fault. *Take the transformer out of service for detailed internal examination and tests to locate the fault before it causes serious damage* 2 Test of gas sample indicates it is air. *If air is present regularly after unit is first put in service, it is probably air trapped in windings or circulating system that is being driven off as transformer warms up. To avoid such false alarms it is advisable to excite the transformer for a period before switching on the load. Another cause of air in tank may be from ingress into circulatory system caused by excessive suction. Reduce the suction head by using a larger-diameter pipe, or by increasing the height of the conservator*
Low insulation resistance	1 General overheating, caused as described on facing page, results in deterioration of insulation 2 Moisture. *Drain oil and dry out windings, and test again. Check oil for moisture content; if moisture content is high, recondition oil or replace it. Examine tank, conservator and radiators for possible place of entrance of moisture* 3 Insulation weakened by line surges or lightning. *Repair insulation if damage is localized, or rewind or replace transformer. Consider installing better protective equipment to prevent recurrence* 4 Shifting of conductors caused by mechanical shock resulting from excessive currents of external short circuits or switching of heavy loads. *Replace coils in proper location, reinsulating where necessary, and secure against further movement* 5 End-turn insulation not adequately reinforced. *Reinforce insulation on all end turns and test windings again*		
		Protective alarm indicates vacuum in tank	1 Oil level in conservator tank has fallen below conservator pipe. *Test sample of oil and refill conservator to proper level. Check tank for leaks that may be responsible for loss of oil to prevent recurrence*
Output voltage too high or too low	1 Leads connected to the wrong ratio taps, or tapping links incorrectly positioned. *Reconnect the leads or reposition the tapping links to make correct connections*	Unequal phase voltages in 3-phase bank	1 Ground fault (either internal or external) on one phase of wye-to-delta connected 3-phase bank. *Locate and repair ground fault, and test insulation resistance of transformer to avoid recurrence* 2 One transformer connected with polarity opposite to that of others in bank. *Determine correct polarity connections of transformers (additive or subtractive) and reconnect them*

Can transformer oil dielectric tests be improved?

Dielectric test is the most important check on condition of oil in your transformers. For years, standard gap test prescribed by the American Society for Testing Materials has been used. Recently, however, it has come under close scrutiny because of exacting demands of modern transformer design. Standard test places oil in cup in which are two disk terminals, one inch in diameter and spaced exactly 1/10 inch apart: 60-cycle voltage is applied across gap; voltage is then increased in steps until oil breaks down in gap. An average breakdown voltage is determined from several tests. Oil in service should have a breakdown voltage of at least 22,000 volts; new oil may test as high as 30,000 volts.

Recent research into effects of shape and area of electrodes indicates that much more accurate results can be obtained than with standard gap. Modified hemispheric electrodes designed to produce uniform electric field in gap have been developed by the General Electric Company.

More reliable checks on your transformer oil are provided by new gaps. Since transformer oil needs good care to prevent the formation of sludge, acids and explosive gases, and to keep down moisture absorption, regular tests for dielectric strength are a must.

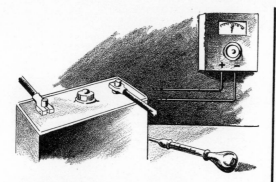

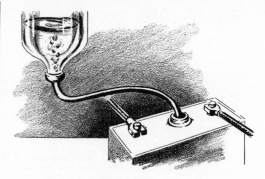

1 Charging Methods

Float method of charging is used with most glass-jar batteries. It means maintaining an average charging voltage of 2.15 v per cell at battery terminals.

Periodic method calls for normal "finishing rate" (as specified by battery manufacturer) as charge rate. But use charge rate equal to 8-hr discharge rate if battery is 25% discharged. Then when cell voltage exceeds 2.3, reduce charging rate to normal finishing rate. Stop charge when specific gravity is 10 points below maximum.

Equalizing charge should be given every 3 months. Use a total charging voltage of 2.33 times number of cells in battery. For example, total charge voltage would be 60 × 2.33 or 140 v for 60-cell battery. Continue equalizing charge till there is no increase in specific gravity of pilot cell for 24-hr period *after current to battery has tapered to constant value.* Record voltage and specific gravity of all cells at end of charge.

Emergency charge should follow any emergency discharge. Charge battery at any current value in amperes not greater than 1½ times the 8-hr discharge rate. When gassing starts, reduce current to finishing rate and continue till voltage and specific gravity remain constant.

2 Adding Water

As a rule stationary batteries don't need water more than twice a year . . . may even extend to a year before electrolyte level drops to low-level mark. Maintaining level of electrolyte is simplest phase of battery maintenance. Don't overlook it. Here are a few tips.

Select one cell as a pilot cell. Any cell except charge-indicator cell will do. Add water so solution is just at level line. Best practice calls for adding water just before an equalizing charge. It will then mix with the electrolyte during charge. Never allow electrolyte-level to drop below low-level mark. And, on the other hand, never fill cells above the high-level mark.

Use approved water in all batteries. Distilled water is approved but local water is often OK. To be sure send water samples to battery manufacturer for analysis. Water loss varies with amount of charge and discharge.

Store water in a glass, plastic or rubber container. If possible, have water container elevated and movable. Fit container with hose and nozzle to make watering fast, easy. Keep record of date, amount of water added.

Never add so-called special battery solutions, powders, jellies. Most of these harm the battery.

5 Sound Tips For Stretching Battery Life

Information From Gould-National Batteries, Inc, Trenton, N. J.

CHECK-LIST FOR STATIONARY BATTERIES IN NORMAL SERVICE				
OPERATION	EVERY DAY	EVERY MONTH	EVERY 3 MONTHS	EVERY 6 MONTHS
INSPECT BATTERY TERMINALS AND CONNECTORS				✔
CHECK ELECTROLYTE LEVEL		✔		
CHECK AND RECORD FLOATING VOLTAGE OF BATTERY	✔			
CHECK AND RECORD FLOATING VOLTAGE AND SPECIFIC GRAVITY	PILOT CELL		ALL CELLS	
GIVE EQUALIZING CHARGE			✔	
CLEAN BATTERY				✔
CHECK CHARGING EQUIPMENT		✔		
GIVE IDLE BATTERIES FRESHENING CHARGE			✔ or when the specific gravity drops to 1,200	

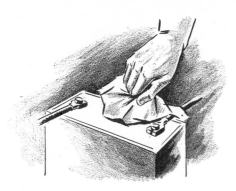

3 Cleaning Cells

Clean batteries prevent current leakage across terminals and possibly to ground. Wet or dirty battery tops offer good current-leakage paths. Cleaning is a simple job when you follow the plan outlined below.

First step is checking that all vent plugs are in place and tight. If battery tops are wet from gassing or spilled electrolyte, neutralize with a concentrated solution of soda and water.

Cell terminals and inter-cell connectors may be cleaned with a wire brush. Clean lead-plated copper connectors with No. 00 sandpaper or a suede brush.

Never scrape connectors. Scraping can damage the lead plating of copper connectors. Removing the protective lead covering exposes the copper beneath to corrosion.

Wash top of battery with plain water. Then wipe dry with a clean cloth. After thorough drying, coat terminal surfaces with battery grease or vaseline.

Modern glass-jar cells have a vent plug with a built-in condensing chamber. This chamber traps any acid spray and funnels it back into the cell while at same time letting the gas trickle out through the vent openings. Only hydrogen and oxygen escape from the battery.

4 Keeping Records

Accurate records tell you (1) whether batteries are fully charged when placed in service (2) number of months each battery has been in service (3) which cells often need equalizing charges (4) whether batteries are being properly charged (5) when battery needs replacing. Here are the highlights of a simple record system.

Number each cell to speed identifying. When the battery is installed, record specific gravity and temperature on record chart. Also note the date each battery is put in service. Use prepared numerals fastened to battery rack to keep track of cells.

Specific gravity, temperature and charging rate should be recorded along with the date, whenever an equalizing charge is put on. Note the battery floating voltage at least twice daily. Record the floating voltage and specific gravity of pilot cell every day. The floating voltage and the specific gravity of all cells should be checked and recorded monthly.

Water added should be logged, both amount and date. Review all the readings from time to time and compare them with the previous data to spot any irregularities. Excess water consumption is one sign of overcharging.

5 Proper Installation of Batteries

There are a few simple precautions in locating, installing and connecting batteries for full-float service that will go a far way in making your job easier.

Hot locations speed up self-discharge of battery, which means shorter battery life. So keep batteries out of direct sunlight, and away from steam pipes, radiators and similar heat-generating equipment. Ample ventilation in battery room is a must to carry off battery gases. Louvers at top and bottom of door panels usually do the trick.

Install batteries on racks at suitable height. This will simplify adding water and taking hydrometer readings. Coat racks with acid-resistant paint. Make them plumb and level and insulated from the floor. When two or more racks are used they should be insulated from one another.

Clean contact surfaces when installing batteries. Then apply a thin coat of vaseline. Use a cord for a straightedge in lining up cells. When they are right-in-line, put on your inter-cell connectors. Start this cell-connecting job at center of battery. Connect positive to negative tightly. Check polarity of all cells. See that electrolyte level is right.

1 Special tool drills an annular hole, permits removal of connectors and covers without disturbing terminal post

2 Alternate method of disassembling storage battery is to drill out the posts with an ordinary 15/16-in.-dia drill

3 Cover and element may be pulled simultaneously if cover is undamaged. Clips and chains hold down the battery jar

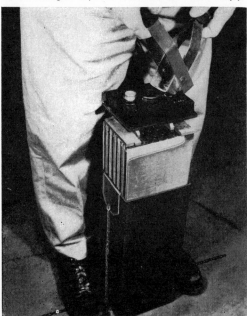

Follow These Tips...

For Your Storage

▶ HERE ARE THE STEPS to follow when replacing a cracked storage-battery cover or taking out the cell for general maintenance. The technique is simple—just a straightforward job of cutting, sealing and rebuilding.

Cutting. The hard-rubber battery covers have post openings lined with lead inserts. These inserts are welded to the lead terminal posts to prevent leaks. In a nutshell, the procedure for taking out a damaged cover is to cut out the metal between the post and its opening.

One way is to use a special tool, Fig. 1, that cuts an annular hole around the post. With this tool, the lead posts remain solid, upright and of normal height. You merely lift off the cell connector and battery cover when your hole is deep enough.

The other way is to cut out the metal in the posts with a 15/16-in. drill, Fig. 2. When you get down to ⅜ in., the intercell connector will be free. Drilling down through the cover frees it in turn.

Sometimes the battery is designed so there is clearance between the intercell connector and the top of the cover. In this case you can eliminate the first drilling by making a horizontal saw cut under the connector.

If cover is undamaged, but you want to remove the battery elements from the jar for basic cell repair, remove the intercell connector and lift out both cover and elements as a unit, Fig. 3.

By making a saw cut through the connector above junction-line between battery cells, you do not even have to remove the connector.

The lifting tool can be a post cutter with dull cutting edges or any other suitable instrument. An easy way to hold down the battery jar is with clips and chains, as shown in this photo.

Sealing. To assure a proper seal between cover and jar when reassembling the cell unit, use a good sealing compound and apply it properly, Fig. 4. Battery manufacturers ordinarily supply compound in packaged form, ready for use.

Melt sealing compound over a burner or blowtorch in any metal saucepan or pot with a good pouring lip. Keep temperature at the lowest point that will permit easy pouring, as too rapid heating and too high temperatures cause it to catch on fire. Burning consumes oil and impairs effectiveness. The melting pot should be about one-quart size. Do not use too large a pot because more time will be required to melt the compound completely.

If some compound remains after pouring a seal, it may be cooled, stored and reused. But when it is being reheated, do not puncture the unmelted top layer with a screwdriver or other pointed tool because pressure in the lower melted section may cause hot, melted compound to squirt through the hole. The hands, face or body of persons nearby may be severely burned.

Information for this article, and photographs of the procedure described, were supplied by the Service Department of Guild-National Batteries, Incorporated.

Battery Repairs

Clean the surfaces to be sealed, neutralize with ammonia or baking soda, and dry carefully. Be sure cover is properly seated on top of cell and level with other covers in battery. Pour the sealing compound into space between cover and jar, being careful to avoid spilling. Space must be filled to same level as the other cells so there are no recesses where acid spray can collect. Use a sharp knife to smooth over rough spots, and remove excess after compound has solidified.

Rebuilding. Next step in reassembling the battery is to rebuild the drilled posts. This is a simple "puddling" operation, Fig. 6. Use a carbon electrode to apply heat within the cavity of a simple post mold that can be made in the shop or purchased outside. Take care that the carbon rod does not touch side of mold.

Carbon electrode can also be used for burning (welding) the connections. A carbon burner consists of an electrode holder and a carbon rod. It is designed for 6-v power supply (three cells of a battery), and can use power from the battery being repaired. To insure a good weld, use exactly three cells, as less is inadequate for a good weld, and more actually endangers battery, operator and joint. If battery under repair is entirely discharged, connect the carbon burner to three cells of an adjacent unit.

Rod should be pointed to a $\frac{1}{8}$-in. tip and extended about $1\frac{1}{2}$ in. from the holder. If possible, connect it to the power source so the carbon rod has negative polarity. This makes welding easier as it tends to keep joint and working surface clean.

Lead surfaces to be welded or built up should be neutralized and cleaned with a wire brush. Clean inside hole of connector with a knife, and wire-brush tip of the carbon.

To make a joint, thoroughly heat center of the part being worked until you have a puddle of molten metal. Then add additional lead, Fig. 5, working rapidly to complete the operation before entire part melts. When possible, surround the part worked on with a damp rag. Use molds made for the purpose when building up posts or splicing a connector.

To add new lead, touch it to the carbon rod and let it flow into the puddle rather than plunging cold lead into the puddle itself.

When burning a connector to a post, one slow trip around the post should do the job. This is also true of burning a post to the lead-insert of a new cover. But joint is so deep that considerable lead will be removed from the post in the circuit around the joint. To build it up, return carbon point to center of the post and add new lead to raise post height to level of the top of post mold.

After making the burn, let joint cool undisturbed. Moving it during cooling may cause lead crystallization and result in a poor weld. After cooling, you can determine whether weld has been made properly by testing with a pair of pliers. Joint should strongly resist separation.

If separation occurs, examine the two surfaces to determine the points and nature of weakness so the situation can be corrected on the second try.

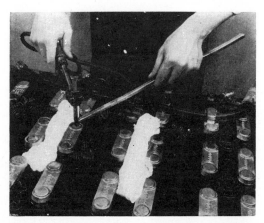

4 Pour sealing compound into space between cover and jar. Space must be filled to same level as the adjacent cells

5 Burn joints with a carbon rod connected to three battery cells. Apply fresh lead so it flows into melted puddle

6 Use a mold when building up battery posts, but be careful that the carbon rod does not touch side of the mold

Good Maintenance Cuts Fluorescent Lamp Costs

These pointers on fluorescent lamp maintenance will help you reduce lighting troubles and cut costs. Good service means good light which is an important factor in any plant

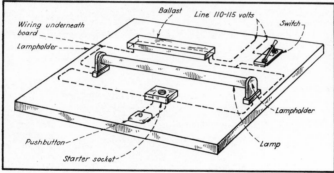

1 Board for testing lamps and starters before putting them in the fixture, should also be used to test them before discarding

► IN PRACTICALLY EVERY TYPE of power plant and industry today, fluorescent lights are used more and more. Although most fluorescent lights give little trouble, a certain amount of attention will pay dividends in low-maintenance costs.

Here are some practical reasons why prompt and careful attention should be given to the small amount of maintenance required by fluorescent lamps. In most cases, the fixture and its installation represent a considerable investment of time and money. Prompt attention, and proper care guard this investment, insure better light, and reduce the number of replacement parts required.

Whenever something goes wrong with the light, auxiliaries or fixture, turn the light off immediately. This requirement should be thoroughly advertised and preached throughout the plant so that maintenance can be started before troubles get out of hand. Some defects put abnormal loads on the lamps themselves or the ballasts and starters. Turning the light off when trouble starts quite often saves burned-out starters, lamps, ballasts, and other ruined equipment.

Trouble Symptoms. Common symptoms of trouble include blinking or flickering lamps and smoke. For other troubles see check list opposite. The fixtures quite often hum and smoke, and sometimes even smell.

Some of these conditions are normal developments, while others result from one or more simple causes. First, let us look at certain indications, plus a few experimental tests that tell where the trouble lies.

When testing defective or damaged equipment practice maintenance economy by trying the simple, least expensive remedies first. For example, if a lamp flickers for some time, it may be the fault of the voltage, ambient temperature, tube, starter or ballast. So check voltage and ambient temperature and look for a cold draft striking the lamp.

Sometimes a lamp fails to light because the starter has worked loose from its socket. Making such a checkup first quite often saves the cost of a lamp or starter.

If at all possible, keep a few spare lamps, ballasts and starters on hand for emergencies. They often help to make quick checks right in the fixture and

thus determine which element is at fault.

Tests. Here are some trouble-shooting tests that can be used to advantage on most fluorescent lighting fixtures:

Sometimes the cathodes are broken or air leaks into the lamp. In this case test each end of the lamp separately by connecting the two base pins in series with a 60-watt light bulb on a 115-v circuit. If cathodes are in good shape, the lamp filament will light and the fluorescent bulb will glow. If the fluorescent lamp does not glow, it indicates a burned-out cathode or that air has leaked into the tube.

A 60-watt light bulb is all right for testing fluorescent lamps from 14 to 40 watts. For larger fluorescent tubes of from 65 to 100 watts, use a 200-watt, 115-v test lamp. Commercial cathode testers are also available, and can be purchased for this test if desired.

Test Board. A simple test board, as shown in Fig. 1, will also help in checking starters and lamps right on the job. This test board, of course, checks only the tubes and starter and does not show wiring defects or ballast troubles in the fixture itself. When testing the various parts, be sure that test conditions such as correct voltage and ambient temperature are ideal. Be sure to test each lamp and starter.

When using the test board, place the fluorescent tubes in the holder so that they make good contact and then close switch A. Press pushbutton B and hold it down several seconds. If the lamp is good, both ends glow and it lights immediately upon releasing button B.

To test a starter, place a good lamp in the holder of the test board. With switch open, insert the starter in the

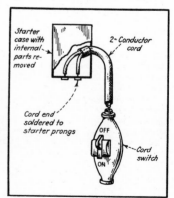

2 Made from an old starter switch, this manual starter helps in testing lamps, starters and auxiliaries without taking the lamp fixture apart

FLUORESCENT LAMP RECORD

LAMPS						STARTERS						BALLASTS					
Originally Installed			Replaced			Originally Installed			Replaced			Originally Installed			Replaced		
Date	Number	Cost	Date	Number	Cost	Date	Number	Cost	Date	Number	Cost	Date	Number	Cost	Date	Number	Cost
Total																	

Use this record sheet to keep track of number and cost of equipment replacements. It also serves as a check on lamp life

Fluorescent lamp troubles listed with possible causes and suggested remedies. List serves as a maintenance check chart

Chart of TROUBLES vs. POSSIBLE CAUSES and SUGGESTED REMEDIES

Possible Causes: Normal lamp failure; Defective lamp; Defective starter; Defective ballast; Loose, broken contacts; Exhausted electrodes; Defective wiring; Insecure lampholder; Improper connections; Wrong voltage; Wrong auxiliaries; Open circuit; Air leak in lamp; Wrong temperature; Dirty fixtures; Line, bulb radiation; Too-frequent starts; Cold drafts; Compensator omitted; Mercury deposit

Suggested Remedies: Replace lamp; Replace starter; Replace ballast; Secure lampholders; Check voltage; Check wiring; Check connections; Check temperature; Clean fixture; Rotate lamp 180°; Move, shield aerial; Install compensator; Insulate fixture; Enclose lamp; Check for grounds; Ventilate fixture; Apply line filter; Turn lamp on—off; Operate lamp; Remove and shake; Check frequency

Troubles: Old lamp blinks; New lamp blinks; Lamp won't start; Lamp starts slowly; Lamp goes out early; Lamp spirals, snakes; Ends stay lighted; Ends blacken early; Rings appear; Spots appear; Light output drops; Fixture hums; Cathodes burn out; Static on radio; Fixture overheats; Fixture smells; Flicker occurs suddenly; Lamp leaks air; Stroboscopic flicker

---

socket and close the switch. If the starter is in good condition, the lamp will light and burn steadily.

If lamp ends glow continually, but the lamp does not light, it shows that starter contacts are stuck. If the lamp does not glow, this shows that starter contacts have not closed. This defect generally indicates that breakdown voltage is too high. For satisfactory testing results, be sure that voltage at the test board is never less than 110 v.

Portable Starter. The manual starter shown in Fig. 2 checks starters and lamps while they are mounted in their own fixture. When trouble symptoms appear, remove the starter from the lamp fixture and insert the manual starter. Move the switch on the end of the cord to the "on" position. The ends of the bulb should glow if the lamp is good and the wiring correct. When no glow appears, insert a new test tube in the fixture. If the ends of the new unit do not glow with switch in the "on" position, it indicates that lamp holders, wiring or ballast is defective.

On a good lamp the ends glow with the switch "on," and it starts immediately when the switch is moved to the off position. If the lamp does not light, snap the switch several times from on to off before condemning the tube. When the light comes on properly with the manual switch, but does not operate with the old starter, this indicates that the trouble is in the starter.

Clean Lights. In addition to testing and replacing lamp parts when they are defective, it is also a good idea to keep reflectors and glass light bulbs clean. This alone can increase light output as much as 30 to 50 percent.

To clean porcelain or synthetic enameled reflectors, remove lamps, then wash reflectors with soap and water and rinse with clean water and dry with a cloth. Dry-wiping leaves much of the dirt on the reflector and does very little good. Wash the lamp tubes thoroughly, especially if the fixtures are mounted in a dirty, smoky atmosphere. Caution: Don't use either a strong alkaline or acid solution on aluminum reflectors.

Hum produced by the magnetic forces within the transformer or ballast may be objectionable in certain office or shop locations. If the lamps are in a room with noisy machinery, obviously this small hum does not matter.

Manufacturers design and assemble ballasts to minimize hum, but sometimes the fixture is noisy in spite of this. If the fixture is noisy, first check ballast mounting, and, if necessary, attach it to a rigid part of the fixture. Second, fasten all metallic parts securely because loose screws or bolts amplify the noise. If excessive hum persists, and cannot be traced to a defective ballast, replace it with a new ballast unless the noise can be eliminated upon disassembling.

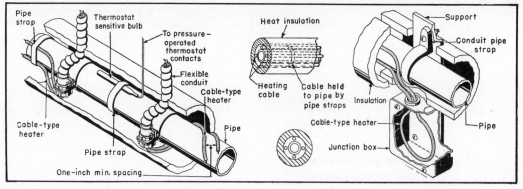

1 Worthwhile heating-cable tricks. In most cases you'll find the above setup (left) OK. Remember you can run the cables parallel to the pipe (center). When running a cable on pipe, 1-in. or less dia, use the loop box (right)

Heating Pipelines Electrically

When wind is blowing icicles down your neck a few months from now, you'll appreciate the author's warm advice on toasting pipelines.

Typical Applications For Pipeline Heating

1. Cold water tap off main line.

2. Fuel-oil line from storage tank to point of use.

3. Pipelines carrying heavy (viscous) chemicals in cold weather.

4. Where chemicals with high melting points are flowing through pipelines.

Here Are The 4 Ways to Heat Pipelines

1. Steam

2. Low voltage dc or ac

3. Heating Cable

4. Tubular heaters

▶ YOUR PLANT and mine have pipelines that we just can't let freeze. In fact we can't even allow them to drop below a set temperature.

This whole question of heating pipelines breaks down into two general problems: (1) how to keep liquids fluid (so even molasses will flow in January), and (2) how to prevent a water solution from freezing in pipelines during cold weather.

How It's Done. Just about any heat source along a pipeline would answer both these problems. But simple economics and overall efficiency limits the practical methods to steam and electrical applications.

More engineers are going to electrical heating of pipelines. And there are several practical ways to heat pipelines with electricity.

First, there's the old standby of heating with dc from a welding machine (or any dc generator that's available). Then there's the simple lead-sheathed heating cable with a resistance wire core. Heating cable is simply wrapped around or run along the pipe as shown in Fig. 3. This is an easy installation since the cable is laid along the pipe and held in place with straps. Outer insulation and covering then applied.

Heating by sending ac through the pipe itself is gaining wide acceptance. We've used this method in one job as shown at right, Fig. 4. It's essentially the same thing as using a dc welding generator. General advantage of sending current through the pipe proper is uniform heating of pipe's contents.

In the case of ac, keep in mind that the effective resistance of the pipeline is more than the pure dc resistance. For instance, take a length of 2-in. iron pipe to be heated by passing ac through the pipe. Effective resistance of this pipe at 32 F is about $4\frac{1}{2}$ times the dc resistance. With this in mind the current flow can be calculated easily.

Then there's the tubular-type electric heater. This is made by centering a resistance wire in a seamless metal tube. The space between the wire and tube is then filled with magnesium oxide insulation. These heaters can be bent or formed into practically any shape. Sheath materials are available in steel, stainless steel, chrome steel, copper etc. Temperatures up to 1500 F can be handled.

Heating Cable. Fig. 1, 2 above give installation know-how at a glance. Both pipe and heater should be insulated for maximum efficiency and

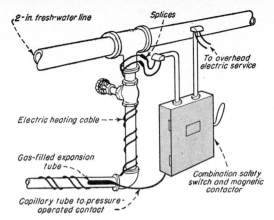

2 Cable on water line tap. Magnetic contactor may be eliminated by putting thermostat contacts in cable line

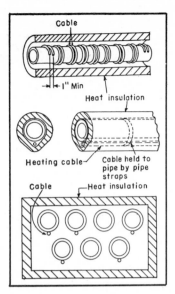

3 Typical ways to install lead heating cable. Note 1-in. minimum spacing

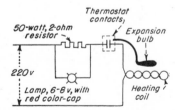

4 Here's lowdown on sending ac right through pipeline. Note primary setup

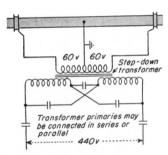

5 Reliable on-off indication is given to a remote point with this hookup

economy. However, in applying the insulating material make sure you don't pack it firmly around the heating cable. Tightly packed insulation will blanket the heat, making the cable overheat and burn out.

Rule of Thumb. *Cable lengths should never be less than 60 ft on 110 v or less than 120 ft on 220 v.* This is plain common sense. Shorter lengths would allow more than the maximum-allowable current to flow.

Here's a Typical Case. Say your fuel-oil storage tank is above ground. It's located so the heavy No. 6 fuel oil is gravity fed to the boiler-room burners through two 60-ft lengths of 2-in. pipe. About 40 ft of each pipe is above ground. That means heat must be supplied to the pipeline to keep the heavy oil from getting thick in the lines during the winter months. Otherwise you'll have to shut down the heating plant every so often to heat the pipe lines.

Here's a way to solve the problem. Take two 120-ft lengths of the lead-covered heating cable. That 120 ft means 220-v feeds. Total distance is about 40 ft from the tank to 1 ft underground. Form each cable like a hairpin and wrap it around the pipe. Then cover with insulation and a heavy coating of asphalt paint.

Your automatic thermostatic controls, see Fig. 5, hold proper fuel-oil temperatures. These units operate by expansion of the gas within bulb and capillary tube. They automatically flip heaters on and off holding an almost constant temperature. Remember the maximum-allowable sheath temperature is 165 F. And minimum bending radius of cable is one inch.

Then there's the question of how much heat must be supplied. As a rough guide install about 0.24 watts per square ft of pipe surface (pipes with 1-in. insulation) for each deg F difference between the pipe and ambient temperature. Then for uninsulated pipes install about 0.8 watts per sq ft of pipe surface per deg F temperature difference.

Figurin'. Now for this practical matter of getting needed wattage per sq ft of pipe surface. First take a string and wrap it around the pipe to find the circumference. Lay the string out and measure it in feet. Then take this figure and multiply it by the length of pipe in feet. You now have the total outside pipe area.

Next step is multiplying this area by 0.24 or 0.8 watts as the case may be. Then multiply this figure by the temperature difference you want to hold between the pipe and outside world (ambient temperature). Now you have your total wattage needed.

But that's not the whole story. You then have to figure how to run the heating cable to get the required watts per sq ft. Remember that most heating cables run 400 watts for each 60 ft length. And a 60 ft run taps across 110 v, 120 ft on 220 v, 240 ft on 440 v.

Installing. There are five ways to run the cable. You can shape it like a hairpin (with both ends at same end of pipe) and start wrapping it around the pipe allowing so many watts per sq ft, Fig. 3. This method will give the highest temperature.

If you don't need that much heat keep the hairpin arrangement but run the cable lengthwise without wrapping. That means the cable will run alongside the pipe in both directions, Fig. 1. You can also use two cables, hairpin fashion, and run them parallel to each other on either side of the pipeline.

The other method is stretching out the hairpin like a single cable. Then you can either wrap the cable around the pipe or strap it right alongside depending on how much heat you need. The least amount of heat is given off when the cable is running straight without doubling.

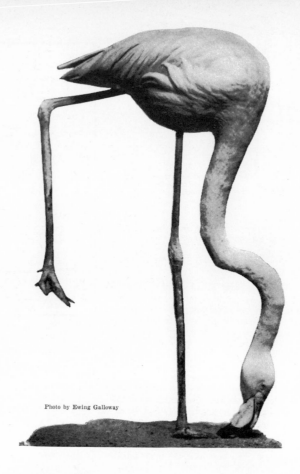

Photo by Ewing Galloway

I'm Using My Head to Balance My Load

SURE OSCAR, WE'RE LOOKING—and we think that's a real cute trick, but why not try keeping both feet on the ground instead? Then you could raise your head high enough and open your eyes wide enough to see what's going on in the world. You'll then discover many more practical ways to balance your load.

This flamingo's trick of solving a purely mechanical problem might look silly to most power engineers, but I'm not laughing. I've been in this game too long and operated too many plants to laugh at his antics.

Too often a plant operator keeps his head so close to his feet that he can't balance his load. Like this silly flamingo, he thinks he's doing a real job, but that's only because he hasn't had his head far enough from home plate to compare his wasteful methods with neighbors' more advanced practices.

And characters like this guy don't even bury their

heads in a technical magazine to see what's going on in their field. That's too bad, because as Fred Low, the grand old chief editor of POWER magazine said many years ago, "You don't have to be a college graduate to run a smooth plant. Just go through each POWER issue thoroughly, then put the information to work in your plant."

So raise your head and see what's going on, Oscar. You'll find that money spent on balancing your load the right way ups efficiency and cuts labor. And it's the ONLY money that pays dividends in a power plant.

Think that over, my fine-feathered friend.

11
Well water
for industry

DIFFERENT TYPES OF WELLS ARE SUITABLE FOR INDUSTRIAL PLANTS

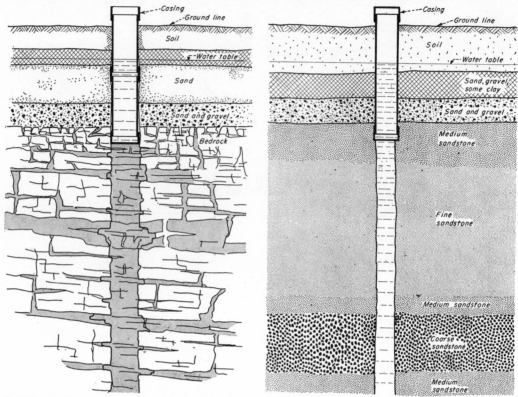

1 **ROCK HOLE** in broken limestone is cased with wrought-iron or steel pipe driven through soil, clay, sand layers

2 **SANDSTONE WELLS** often produce usable quantities of water for industrial plants. Yield depends on stone porosity

For Low-Cost, Dependable Water . . .

Choose Industrial Wells Carefully

American Water Works Service Co, Inc

More and more industrial plants are becoming major users of underground water. Unwise practices, drought and pollution create area shortages, lead to regulation, restriction. Here are basic facts to help select the right well for your plant

WELLS DO RUN DRY; RIGHT SELECTION, CONSTRUCTION ARE IMPORTANT

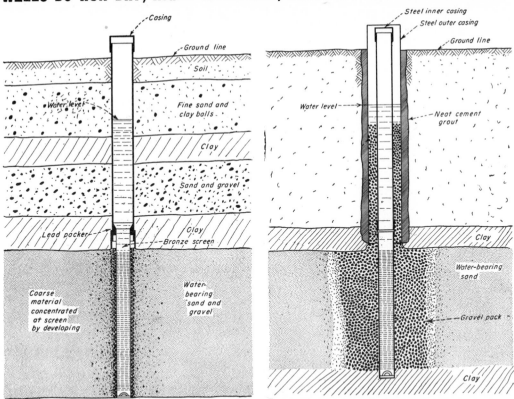

3 **DEVELOPED WELL** has coarse material concentrated at screen to permit free water flow without excessive fines

4 **GRADED GRAVEL** is used in this under-reamed gravel-packed well to reduce flow velocity around screen at entrance

►WATER-BEARING FORMATIONS, suitable for construction of wells of greater capacity than household types, fall into three rough classes: (1) rock having many cracks or fissures, usually limestone (2) sandstone (3) sand or gravel deposits, or both. The last may have several subdivisions like glacial deposits, prehistoric river beds or ancient sea beds. Glacial and river beds may be related.

Limestone Formations. Most wells in Indiana, Ohio and Pennsylvania are "rock holes" in broken limestone. They are cased with wrought-iron or steel pipe driven through upper layers of soil, clay and sand. Pipe is driven one to six feet into bedrock. From there on well is an open hole, Fig. 1.

Yield of a rock hole depends on many variables. Basic one is amount of water fed entire well-field area. Others are general permeability of limestone aquifer, well diameter, and number of water-bearing cracks and fissures intercepted by hole. Cracks and fissures may be irregular in size and location. They have much influence on value of the well. Of two wells with the same diameter and depth, separated by a distance of only a few hundred feet, one may yield 200 gpm, the other 50 gpm.

Water from rock holes may be hard, with many dissolved minerals and gases like iron, manganese, CO_2 and hydrogen sulfide in solution. It may be highly corrosive and objectionable for use. Gases are removed by aeration (spraying the water in air or allowing it to splash over coke trays). Iron oxidizes and can be removed by sand filters. Iron and manganese are fixed in solution by lime and ferric chloride, ferric sulfate or zeolites. Since this is a complex job you get best results by consulting water-treatment specialists.

Removal of gas from water is important because it may vapor-bind centrifugal or turbine well pumps. Unit then acts like a centrifuge, separating water and gas. Water passages are restricted by gas, flow is reduced, or stopped intermittently, or altogether. Take a look at any turbine-well-pump manufacturer's guarantee. It says, in effect, "performance is based on pump handling clean water free of entrained gas or air."

Sandstone Wells. Sandstone is sand grains cemented together by some mineral. It may be porous and hold much water. If sand grains are uniform in size and shape, and coarse enough so spaces between are large, it may yield water in usable quantities. With fine grains plenty of water may be present but capillary action or surface tension can prevent flow of usable amounts.

There are, however, many satisfactory sandstone wells, Fig. 2. In Iowa we "shot" a 2190-ft-deep sandstone well with nitroglycerine at 1400 ft and at the bottom. Static water level is six feet below ground, yielding about 1200 gpm

A GOOD WATER SUPPLY IS AN ASSET FOR EVERY INDUSTRIAL PLANT

with a pumping level of about 150 ft. Shooting shatters the sandstone, creating large cavities that expose more area to yield water.

Wells in Sand, Gravel. Sand and gravel formations are very satisfactory water producers. They present problems in formation hydraulics similar to sandstone. The perfect situation compares to a bed of marbles all the same size, large or small. Size doesn't affect amount of water the bed can contain, but does influence rate of flow through bed and its ability to yield water.

Going back to our marble illustration, the larger the spheres, the less total surface exposed to cause friction and the less tortuous the flow path. Now if we mix different sizes of marbles, small ones fill some of the spaces between larger ones, reducing room available for water, rate of flow and yield. So uniformity of size is desirable in formation material.

Construction of wells in sand and gravel differs widely from rock holes. Hole must be cased to top of water-bearing formation. A screen to prevent entry of sand but allow passage of water is placed in formation. Formation material, Fig. 3, is a mixture of different sizes. If screen selected prevents passage of fines it will also restrict water flow. Properly chosen screen allows a percentage of fines to pass but holds back coarser material. Fines enter well until they are exhausted from a large enough radius around screen. Flow velocity is then so low that water won't move finest material present. Such a well is called a "developed well."

Non-movement of fines is the best argument against not pumping at highest possible rate. Many a good well is ruined by over-pumping that creates velocities sufficient to continually move fines in and plug screen. Redevelopment may or may not be possible. A developed well is built by selecting and installing a proper-sized screen and surging well by one of several means so flow reversals and high velocities are created. This washes fines into well where they are periodically removed.

Another method of reducing flow velocity around screen uses graded gravel packed at bottom of well, Fig. 4. Gravel size selected gives low velocity at pack edge without heaving and surging, and with little disturbance of natural arrangement of formation.

Who Builds Wells. Most important point to remember is that well building is an art, not a science. If possible select a drilling firm with an established reputation and experience in your im-

mediate locality. This may be a nationally known firm or a home-town driller with a good record. Beware of the unknown and untried; you're betting a large sum of money on the rig operator's skill, knowledge, experience and integrity.

Ask for a performance bond equal to the value of the contract. Premium for this runs about 1% and is borne by purchaser. If the bonding company won't issue a bond for your contract, you must be doubly careful. Don't allow a driller to move in on your site unless his equipment is sound and well kept. Even if you collect on the bond it may not pay for a well site that is irretrievably lost.

Don't be "oversold" on a well. Some firms specialize in high-capacity gravel-packed wells, refusing to build anything else. Worth of this type well depends entirely on your needs and nature of formation in the locality. Under certain conditions and in given localities, we would consider no other type. But this is not always true.

Drilling Problem. Here's a practical example of how you may be oversold on a well. Recently we had test holes drilled at a New England site where glacial deposits prevail. Water demands permitted use of 500 gpm, and no more. The drilling company recommended a gravel-packed well giving a high specific capacity (gallons of water delivered per foot of lowering of the pumping water level in well). Price was $18,000. This seemed rather high and a bid on a developed well was requested. Price was $10,800, but we were assured we couldn't obtain a specific capacity equal to that of a gravel-packed well.

Test samples and data showed that a developed well was best for our gpm requirements. It was built and gave 800 gpm with a 13-foot drawdown, or a specific capacity of $800/13 = 61.5$. At 500 gpm, drawdown would be only 7 to 8 feet.

Suppose the specific capacity could have been doubled by a gravel-packed well. Drawdown would be halved but pumping head for 500 gpm would be only about 1.5 psi less. This is negligible compared to $7200 difference in cost.

Well Pointers. Wells should be reasonably plumb, free from kinks and "dog legs." Many are surprised to find that a well is almost never perfectly straight and plumb. Drilling tools tend

to offset or slope away from formation irregularities. This may be overcome to some extent, depending on conditions.

Don't be stingy with well diameter. Manufacturers size turbine well pumps 6, 8, 10, 12 in., and larger. This is not the exact pump size; instead it is the particular manufacturer's size for installation in a well casing of that nominal *inside* diameter. With a 12-in. casing and 10-in. pump there is enough clearance to move pump to compensate for drift or other irregularities in borehole. And an air line for checking static and running water levels is easily installed. In an open rock hole, some water may enter well above pump bowl. If annular space between pump and well bore is too small, water flow will be restricted, causing loss in capacity. You will also have false pumping-water-level readings. Our experience shows that money spent for a large enough hole is well repaid by better performance during life of the well.

Instruments for checking well and pump performance are extremely important. You are completely in the dark unless you have accurate means of finding water level in well, measuring pump output in gallons per minute and pressure or head against which pump works. Measurement of these items and no others, is the minimum to assure satisfactory operation.

It's also desirable to check pump input horsepower, amperage at motor terminals and motor rpm. With these additional items you can conduct complete field tests of both pump and well. Made periodically and recorded, such tests indicate pump wear in advance. You can plot a continuous record of water levels in the well to obtain a clue to the behavior of the water-bearing formation in general and also the well itself. Specific capacity of a well doesn't depend on nor is it affected by pump performance. It is only a measure of well performance. A well good for 400 gpm that is tested at 300 gpm will have about the same specific capacity at both rates.

If the static water level drops off materially from a general lowering of the water table in your area, you can expect well specific capacity to fall off, too. Without appreciable change in static levels and a lowering of specific capacity you have well trouble and must cure it fast.

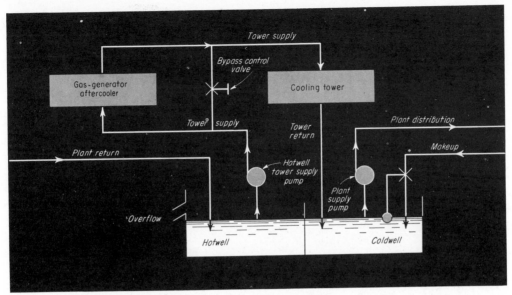

MODEST CHANGE in plant piping cuts costs by taking full advantage of unused cooling-tower capacity

Conserving utilities can mean big savings. This
scheme allowed a plant to . . .

Cut water cost by $37,500

Rochester Prod Div, General Motors Corp

We generate inert atmospheres for our metalworking furnaces by both exothermic and endothermic processes. Gas from the former contains water vapor, an oxidizing agent that does metal-treating processes no good. Generally, vapor is reduced by depressing gas mixture's dew point using a shell-and-tube heat exchanger. We were buying about 100-million gal of 70-F water per year for this dew-point control. Even at 70 F enough water vapor remained so we had to overcome oxidizing characteristics by adding richer endothermic gas. This gas is less productive volume for volume, and, since heat must be added to complete its generation, it costs about three times more to produce. Thus if we could use reclaimed water for cooling—releasing heat at the tower—and remove more water vapor, we would save both in water purchase and expensive endothermic gas. Truly a goal to work toward—but with several obstacles in the way.

Applying reclaimed water in the exchanger for dew-point control and releasing heat in our cooling towers dictated an even higher dew point since the tempera-

ture limit was 85 F. Too, because the major part of our tower system was being used for hydraulic-oil cooling, water returning from the oil coolers didn't exceed 95 F. Since our towers were designed for an inlet temperature of 115 F, this meant that much tower capacity was stagnating.

We solved the dilemma by designing and installing a direct-expansion aftercooler to take the furnace atmosphere gas as it left the shell-and-tube exchanger. Gas went from 2000 to 120 F in the exchanger, then to 40 F in the aftercooler. With this dew point we reclaim enough water to eliminate expensive endo-gas use.

We wanted to avoid added pumping capacity so we revamped our piping. We used hotwell tower-supply pumps to push water through the gas-generator aftercooler, raising temperature from 95 F leaving the oil coolers to 115 F to take up the tower's full capacity. Our scheme sharply reduced amount of endothermic gas we need and is saving 100-million gal, or about $37,500 in water costs per year.

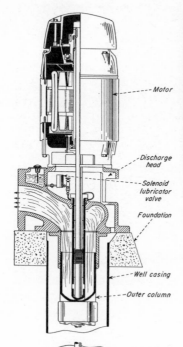

- Motor
- Discharge head
- Solenoid lubricator valve
- Foundation
- Well casing
- Outer column

- Inner column
- Shaft bearing
- Shaft coupling
- Shaft
- Bleed-off port
- Diffuser vanes
- Impeller
- Pump bowl
- Guide vanes
- Grease-lubricated bearing
- Suction screen

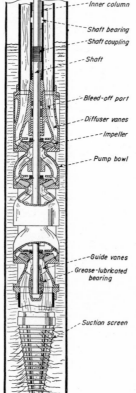

WELL PUMP, vertical-turbine type, is installed indoors on steel framework

OUTER COLUMN of 300-ft-deep well was neglected; corroded and parted at bowl

Smart Maintenance Keeps Wells on the Job

Here's what gives you that all-important water for plant operation.

► YOU MAY RUN INTO two major troubles with a well—loss of capacity or a pump that misbehaves in one way or another. Either trouble can be serious enough to put you behind the eight ball where water is concerned. To make things easy we'll stick to well troubles on this page, pump troubles on next three.

Spot Well Trouble. If static water level in well stays about the same but pumping level falls off, you can be fairly certain the well is to blame. Easiest way to cure the trouble is to find it and go to work pronto. If you can't do the job yourself, call in someone who can. Don't waste time—you may regret it.

Clogged or corroded intake screen is most common cause of capacity falloff. Minerals in water may deposit on screen or there may be an organic growth that plugs screen openings. Worse yet, pumping at too high rates may have pulled a large amount of sand into the screen, developed area or gravel pack. In rock wells mineral deposits or organic growths can give trouble. Or well may partly fill with sediment that closes off some water passages. In certain sections of the country your casing rusts fast, pushing capacity down.

Cure Trouble. You can often remove mineral deposits from screen by chemical treatment. Acid or Calgon, alone or together, may cure balky wells. Screens aren't damaged by inhibited acid, but if they are badly corroded it's best to replace them with new ones.

Never try to salvage a well without expert advice. You can spend a lot of money on the job and not get much in return unless you do it right and use proper materials. So you'll know what to expect when salvaging a well, we'll quickly run over some methods.

In acid cleaning, inhibited muriatic acid is poured into the well to fill the space around the screen. Well is then capped for 8 to 12 hours. Water in well

FOUR-STAGE motor-driven vertical turbine pump of oil-lubricated enclosed-shaft type. Solenoid-controlled oiler feeds lubricant to the bearings in inner column

TURBINE WELL PUMPS ACT UP? USE THIS CHECK CHART TO FIND TROUBLE CAUSE, CURE

SYMPTOM	CAUSE OF TROUBLE; REMEDY
	1 IMPELLERS LOCKED. Sand causes many locks. Try raising and lowering impellers by adjusting nut. This may free them. If it doesn't, backwash with clear water. Or try turning shaft at top with small pipe wrench. Be careful—you may damage shaft. If you can't free impellers, pull pump and tear down bowl assembly to get at rotating parts
	2 TRASH IN CASING. Rags, wood or metal jammed in pump may prevent turning. Tear down pump and remove. Fit with a suction strainer to keep trash out
	3 CORROSION OR GROWTHS. Pumps that are out of service for long periods may be locked tight. Use acid or Calgon to remove corroded matter or growths
	4 PACKING TIGHT. Adjust so there's enough leakage for shaft cooling and lubrication
PUMP WON'T START	5 TOO MUCH BEARING FRICTION. Use right oil; consult builder for viscosity range. Oil bearings before starting. Check tube tension nut for tightness. See if shaft is bent; replace if needed. Check anchoring of pump head to see it hasn't caused bending and distortion of pump. Return bent shafts and columns to factory for new ones. See that water-lubricated rubber bearings are wet and sand-free. Wrong tension on shaft-enclosing tube of oil-lube pumps may throw bearings out of line. If well is so crooked that it causes misalignment, have it reamed to larger diameter or put in smaller pump
	6 MOTOR OR WIRING FAULTY. Check circuit breaker or fuses for open line. If starter overload relays have tripped, reset. Disconnect motor from pump and see if it starts. If it doesn't, have a manufacturer's engineer look it over. Check motor wiring against wiring diagram for pump
	7 IMPELLERS NOT ADJUSTED PROPERLY. Set them high enough so there's room for shaft stretch caused by hydraulic thrust. Adjustment should allow shaft to turn freely; then stretch caused by rotor and shaft weight won't bind pump
	8 WELL CAVE-IN. You need outside help to repair a collapsed well
	1 WRONG ROTATION. Change rotation of motor. With a 3-phase motor, just switch any two power leads
	2 SPEED TOO LOW. Check voltage and frequency of power supply. See if excessive bearing friction, corrosion or obstruction of impeller slows pump. Check gear ratio and motor speed if pump is being operated for first time. Look over belt-driven units for slippage or wrong pulley size
	3 PUMP NOT PRIMED. Vent well to atmosphere so there isn't a vacuum at pump suction. All impellers of vertical turbine pumps must be under water because these units won't start discharging against a suction lift. A 4 to 10 ft net positive suction head is needed for good operation. Have enough head on pump to allow it to discharge at rated capacity
PUMP DOESN'T DELIVER WATER	4 FAILURE OF PUMP PARTS. Look for broken shaft, broken bowl assembly and loose column-pipe joints. Tighten all impellers
	5 PUMPING HEAD TOO GREAT. See that discharge valves are open and that check valves don't stick. If water table has fallen, suction lift may be too large. Or discharge resistance may have increased to too high a value. Increase size of discharge line or reduce discharge pressure. If this doesn't help, you may have to install a new pump with a greater head
	6 CLOGGED SUCTION. Clear clogged suction pipe, strainer or impeller by backwashing. If well screen is plugged, you probably need help from an experienced well driller
	7 WELL OVERPUMPED. With excessive drawdown the pump may break suction, fail to deliver water. Reduce pump capacity by throttling discharge

SYMPTOM	CAUSE OF TROUBLE; REMEDY
	1 PUMP NOT RIGHT FOR JOB. Study performance curves. If power demand can't be reduced a larger motor may have to be installed. You can reduce impeller diameter to cut input hp; head and capacity go down too. Reducing speed of belt- or gear-driven pumps has same effect. If pump has been incorrectly selected for job it may have to be changed for one with right head, capacity
	2 OVERSPEEDING. Check for high frequency, voltage. See that pulley sizes and gear ratios are right
PUMP USES TOO MUCH POWER	3 WRONG LUBRICANT. See that you have enough of right oil where needed. Check motor bearings for oil or grease quantity and type. Be careful with water-lubricated bearings. You may need an air-relief valve in column to allow water to enter bearings
	4 TIGHT PACKING. Adjust so there's enough leakage for shaft lubrication
	5 IMPELLER RUB. See item 7 under "Pump Won't Start"
	6 WRONG ROTATION. Change two power leads of 3-phase motors
	7 OTHER CAUSES. Look for misalignment, bearings that are too tight or vibration in pump or piping. Check for excessive discharge pressure
	1 LOW PUMPING WATER LEVEL IN WELL. Vent well. Check pump inlet for excessive turbulence, vortexing or eddies. Liquid velocity entering pump must be that recommended by manufacturer and submergence sufficient. See item 3 under "Pump Doesn't Deliver Water." Check bowls, well screen for sand, rust or bacterial blocking
	2 IMPELLER WEAR. Metal loss from outer tips of impeller vanes pushes capacity down. Loss at inner or suction end hasn't much effect. If fully enclosed impellers have usual wearing rings, trouble may be in them. Look for excessive clearance. With semi-open impellers that don't have bottom shroud or wearing rings, you need close-running clearance at bottom of vanes
PUMP CAPACITY LOW	3 FAULTY INSTRUMENTS. Make sure that your water-level reading is right. See that flowmeters for measuring pump capacity read correctly. Check pump pressure gage
	4 HEAD ON PUMP TOO GREAT. As you increase head on a pump its output falls off. Look for plugged pipes, closed valves or other obstructions that may increase head
	5 LEAKS. Openings in pipe at flanged or threaded joints can cause loss of water. Same is true for gaskets and packing. See that there aren't any holes in bowl, column pipe or pump head
	6 SPEED LOW. Check as given above
	1 BEARING TROUBLES. Check your lubricating oil or grease for grade and quantity. Look for too much sand in water. See that pump is aligned properly
TOO MUCH VIBRATION	2 ROUGH OPERATION. Check to see that impeller and bowl passages are free of wood, rags, sand and other material that might throw pump out of balance. Also check driving motor by disconnecting from pump and running alone. Look for too much wear in rotating parts
	3 PUMP TAKING AIR. Check on water velocity at pump inlet. See if there are leaks in well vent. Find if suction head on pump is sufficient. Overpumping a well so pumping water level is intermittently drawn down may cause pump to "grab air" and is a common source of severe vibration

is surged gently with a plunger, by pouring more water into it, by using solid carbon dioxide (dry ice), or by capping and forcing compressed air into well at a moderate pressure. After gentle surging well is again capped for 2 to 4 hours. This is followed by violent surging. Sulfuric acid may be used with muriatic to dissolve minerals that have collected on screen.

With Calgon, well is taken out of service for 24 to 48 hours. Charge of about 16 lb of it for 1000 gal of water in well is put into casing, in solution. If there's bacterial growth, enough calcium or sodium hypochlorite is added to disinfect well. The pump is started and stopped to surge well about 12 times every four hours for 24 to 48 hr. When finished, well is flushed to waste until water clears up. On many jobs this is enough to restore original well

capacity, increase pressure and reduce drawdown.

No matter what salvage method you use, be sure to get competent advice before spending any money. You can waste plenty if you don't.

Pumps. Table, above, gives low-down on well-pump troubles, causes and cures. On the next two pages are sketches showing how to do many parts of job.

For more on wells and pumps turn page

Sketches, *this and next page*, give you a quick look at typical jobs you meet in well and pump maintenance. They're handy to have for study before or during every new job. Here are some extra pointers:

Tearing down a well pump is heavy work. So equipment you select to lift parts must be strong enough to safely support them. Wooden "A" frame, column clamp and sling, and other rigs shown are safe for all usual jobs. But if ever in doubt see an engineer who knows well and pump maintenance.

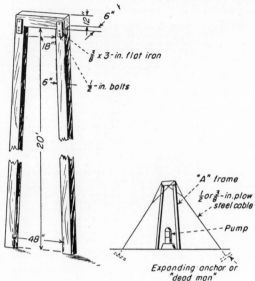

1 **"A" FRAME** safely handles 5 tons when guyed or blocked properly. Use when lifting pump head, column sections

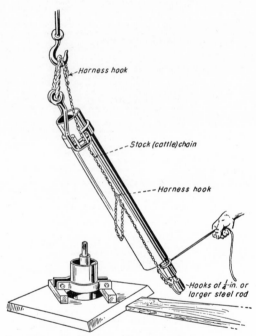

3 **TAILING OUT** a column section using the column clamp and sling shown in Fig. 5. Have sling strong enough for load

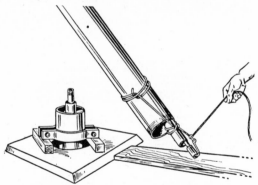

4 **CLOVE HITCHES** in ⅜-in. manila rope can be used when hooks and chains aren't on hand for tailing out or in

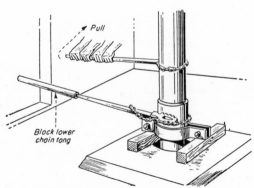

7 **USE CHAIN TONG** to make up outer column. Have two men pull up as tight as possible on 4-6 ft leverage to assemble

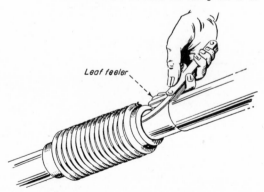

8 **MEASURE** clearance between worn shaft and bearing. Slide bearing to unworn section; measure. Difference is wear

TOOLS TO MAKE WELL AND PUMP MAINTENANCE EASIER, SAFER, SURER

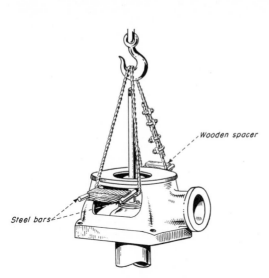

Wooden spacer

Steel bars

2 **USE SLING** to lift pump head after motor removal. This is the heaviest lift you make, so be careful with job

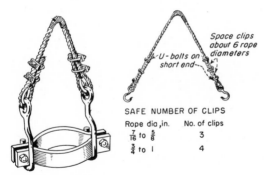

Space clips about 6 rope diameters

U-bolts on short end

SAFE NUMBER OF CLIPS

Rope dia., in.	No. of clips
$\frac{7}{16}$ to $\frac{5}{8}$	3
$\frac{3}{4}$ to 1	4

5 **COLUMN CLAMP** and sling. Many drillers use patented "elevators" for this job. *Right:* safe number of Crosby chips

6 **ASSEMBLE** shaft by butting ends firmly but lightly in the coupling, using small wrenches. Don't pull hard on them

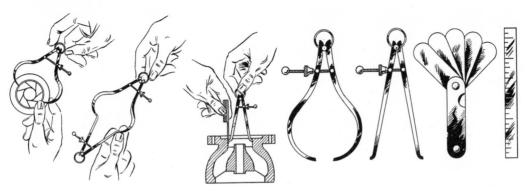

9 **WEARING-RING** wear measurements are easy to make when you have the right tools and know how to use them. Inside and outside calipers are used on impeller wearing rings, inside calipers and leaf feelers for bowl. Use light touch on work

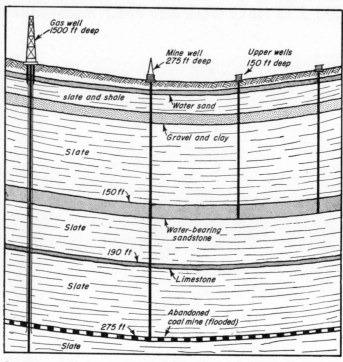

1 John Murphy lends a hand to save 17 wells draining into old mine

2 Upper wells supply water the year round. Mine well is good for three months, then must rest to replenish water. The new gas well casing wasn't sealed right

Watch Your Neighbor's Well

EDITORS NOTE. *With well water levels sinking lower each year, plant owners and operating engineers must know more about this vital water source. Knowing how wells are "lost" and what to do about it, may save your water supply some day. Read carefully.*

Your ground well may be a good producer now, but will a new well nearby affect it?

▶ THE WELL LAYOUT at one of our eastern plants is unusual. There's an upper formation of water-bearing sandstone about 150 ft down, Fig. 2. We have 17 wells in this field, equipped with turbine pumps or airlifts.

Further down, at 275 ft, there is an abandoned coal mine which holds 600 million gal of water. Tests showed mine water to be usable with treatment, so we drilled into it five years ago.

Neighbor's Well. This summer a driller bought a gas lease and started a wildcat well one quarter mile from our nearest water well. We warned him about draining *our* water down his hole, but he didn't understand the problem, and merely took steps to prevent

the mine draining into *his* gas well.

When sinking our well to the mine, we did the job right. Well has cast-iron casing, cemented full length in 18-in. bore hole. The cement charge was placed through a 1½-in. pipe from a 500-gal tank. We used compressed air to push it through a plug in the casing's bottom and up into the annular space between casing and earth.

After cement had set, the plug was drilled out and the well drilled on through the mine's roof. Cementing was to prevent water from the 150-ft field following the casing's outside and draining to the mine.

This mine water is used only three months a year. The other nine months

are needed for water to replenish itself.

Driller Begins. The driller bored a 10-in. hole through the mine, setting a temporary 8-in. steel casing through it, Fig. 3. He then started a 6-in. hole, following with a 6-in. casing driven to 900 ft. His total depth was 1500 ft. After feeding powdered coal down around the 6-in. casing to pack it at the mine floor, he removed the 8-in. casing.

Everything was okay until the day he drilled through the mine roof. The water level in our 150 ft sandstone wells dropped 45 ft. Reason was that the 10-in. gas-well hole was open around the 6-in. casing. The water drained freely around the casing down to the mine from the water-bearing sandstone.

338

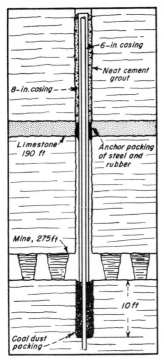

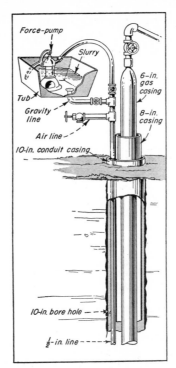

3 Anchor packing and cement stopped water from wells draining into mine

4 Cementing hookup for gas well used to pour cement with ½-in. pipe

The coal packing kept water out of the gas well, which turned out to be a good producer. But losing that water from our 17 wells sure gave us a headache.

Stopping Drainage. My problem was to stop the loss without interfering with his gas production.

The log records of his hole indicated there was only one point between the 150-ft level and the mine for setting a mechanical packer. This was a 5-ft limestone bed. The rest was slate or shale, and wouldn't drill with a smooth wall for packing.

We reinstalled the 8-in. casing with an anchor packer. This is a patented steel and rubber section that expands when the holding pins shear. The pins shear when bottom end of casing comes to rest and pressure is applied to upper end. Packer was set in 8-in. casing at limestone section, Fig. 3. We waited two weeks before cementing to make sure packing was holding.

Hot Dope. In cementing, we mixed one bag of portland cement with 7.5 gal of water. Don't use sand, it separates. Drillers won't put cement through pipe smaller than 1½-in. They use large mixers and sludge pumps. Lacking these, they suggest pouring it in

annular space at top—hoping it settles to bottom where it's needed. We couldn't take that risk as failure meant losing our entire field.

Cementing. The largest pipe we could drive into annular space was ½ in. We had plenty of trouble working pipe past the slate shelves, but finally got it down 185 ft to packer's top.

We mixed cement grout in three barrels, with board paddles. The slurry was poured into the tub with buckets, Fig. 4, so flow wasn't stopped. A portable air compressor stood by to blow into ½-in. line if it got plugged. The hand force-pump was to keep slurry moving if it got sluggish. One hundred bags of cement were placed in 4 hr. The first day we removed the ½-in. pipe and flushed it out. The next day we drove it in until the end plugged with partially set cement. This showed there was 40 ft of cement on top of the packer.

We cleaned cement out of ½-in. pipe, reinserted it 120 ft., and started pouring again. In 3½ hr. we placed 107 bags, filling the annular space to ground level. This settled for a few hours with makeup being poured in at the surface. The well accepted a

total of 219 bags of cement, mixed with about 1600 gal of water.

Success. The repair worked fine. We have a casing completely enclosed in cement and the upper field is gradually recovering. At no time was cement flow interrupted, or the air used. We did use the pump at times to determine, by feel, if grout was rising above the ½-in. pipe's end. Gas-well production was stopped only about a half day while lowering the 8-in. casing.

Here's hoping our experience saves you trouble from your neighbor's well.

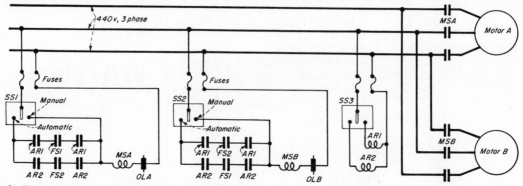

1 This is the control arrangement. Either pump may operate from the high-level float switch **FS1**. In case water level still drops, low-level float switch **FS2** starts the standby pump. Selector **SS3** determines which pump operates normally

Relays Give Four-Way

Here's a simple design of well-pump controls to meet most water-supply conditions

▶ WHEN WE DESIGNED our water-supply system, we selected two 500-gpm vertical turbine pumps to provide 480 gpm for our power house and mill. This arrangement permitted one pump to be out of service while the other took care of the demand. In case of fire we have 1000 gpm available in the two pumps for emergency.

Float Switches. The pumps discharge to a concrete tank that has two float switches. These automatically start and stop the pumps according to water-height, at the 80 and 60% full level. Since normal water consumption is less than 500 gpm, one pump can keep the tank supplied.

This could mean one pump does all the work while the other remains idle most of the time. But we wanted both

pumps to wear about equally, so we installed the simple control system shown in the diagrams.

Control System. In these circuits, Fig. 1, we use two manual-automatic selector stations SS_1 and SS_2, two auxiliary relays AR_1 and AR_2 and two float switches FS_1 and FS_2 to actuate the motor magnetic contactors for across-the-line starting. From selector station SS_3, the coil on either auxiliary relay can be energized.

Float-switch contacts, FS_1 and FS_2

are connected between auxiliary relay contacts. These connections are so made that when AR_1 is energized, high-level float switch FS_1 controls motor A and low-level float switch FS_2 operates motor B. When relay AR_2 is energized, high-level float switch FS_1 controls motor B and float switch FS_2 starts and stops motor A.

Pump Operation. Let's study how these circuits give the desired pump control. To simplify the diagrams, coils are not placed at the contacts they op-

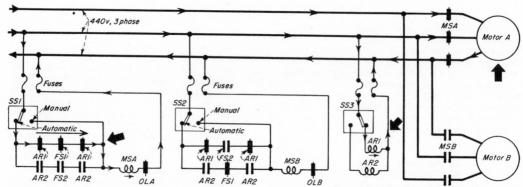

2 Pump **A** normally operating. Selectors **SS1** and **SS2** are set to automatic. Throwing switch **SS3** to right energizes coil **AR1** to close contacts **AR1**. When level drops below 80%, high-level float **FS1** closes to energize coil **MSA** to start motor **A**

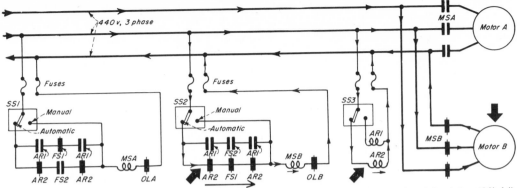

3 Pump **B** normally operating. Closing **SS3** to left energizes coil **AR2** to close contacts **AR2.** When **FS1** closes, **MSB** is energized to start motor **B.** Should level drop below 60% full, **FS2** will close, energizing coil **MSA** to start motor **A** running

Control to Well Pumps

4 All controls are mounted in a compact group right at each motor. Pumps discharge into header **H.** From here line **L** runs underground and up the side of the tank

erate. Coil *MSA* closes contacts *MSA* to start motor *A*, and coil *MSB* closes contacts *MSB* to start motor *B*. Coil AR_1 closes auxiliary-relay contacts AR_1, and coil AR_2 closes contacts AR_2.

Assume we wish to use the pump driven by motor *A*. This means this pump must be operated from high-level switch FS_1. We then close selector switch SS_3 to the AR_1 position, Fig. 2. Selector switches SS_1 and SS_2 are closed to automatic.

Pump "A" Operates. The circuit completed by switch SS_3 is indicated by arrowheads, Fig. 2. Energizing coil AR_1 closes contacts AR_1 as shown in both

float-switch circuits. When water level has dropped in the tank to 80% full high-level float switch FS_1 closes.

This completes a circuit from the middle power line through selector switch SS_1, closed contacts AR_1 and FS_1, coil *MSA*, overload contact *OLA* back to the line, as indicated by the arrowheads. Energizing coil *MSA* closes contacts *MSA* to start motor *A*. This motor runs until tank is filled and float switch FS_1 opens.

Motor Protection. The motor is protected by a thermal overload relay. In case of excessive motor overload, relay *OLA* opens to break the circuit through

coil *MSA* and shut down the pump.

If it were not discovered that the relay had tripped off motor *A*, water level in the tank would slowly drop. At 60% tank level, float switch FS_2 closes and starts pump motor *B*. Note that AR_1 contacts are closed on selector station SS_2, so that current flows through *MSB* to start motor *B* any time that low-level float switch FS_2 closes.

Pump "B" Operates. If selector switch SS_3 is closed to the left, Fig. 3, relay coil AR_2 is energized to close AR_2 contacts in both float-switch circuits. But, high-level switch FS_1 is now in AR_2 circuit of motor *B*. When FS_1 switch closes as shown, the circuit for coil *MSB* is energized. This coil closes contacts *MSB* to start motor *B* and its pump. This pump would then supply the tank under high-level float switch FS_1 control until someone throws selector SS_3 to the right. Then, motor *A* would again take over.

This system has been working well for three years. Each week when the oiler checks the motors, he changes the position of selector SS_3. Therefore, each pump operates every other week.

This system also insures that both pumps will automatically operate in case of a fire or whenever the water demand is high. Assume pump *A* is holding tank level at 80% full on float switch FS_1. If water demand increases to where this pump cannot supply it, water level continues to drop. When the level falls to 60% full, float switch FS_2 closes and pump *B* comes into service. This pump makes available a total of 1000 gpm to meet the demand.

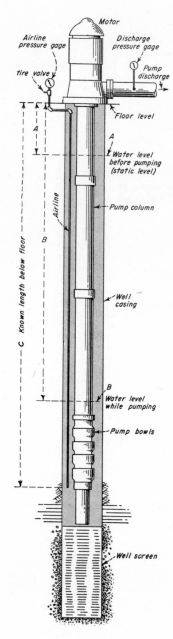

Labels on diagram:
Motor
Airline pressure gage
Discharge pressure gage
tire valve
Pump discharge
Floor level
Water level before pumping (static level)
Airline
Pump column
Well casing
Water level while pumping
Pump bowls
Well screen
A
B
C Known length below floor

Testing Wells and Well Pumps in the Field

A, water level before start of pumping = static level

B, water level while pumping = pumping level

B-A, amount water is lowered by pumping = drawdown

$$\text{Specific capacity} = \frac{\text{GPM}}{\text{Drawdown}}$$

This measures performance of well (not pump), gpm per ft.

C, air line of known length in ft, part of system for measuring water level. Air pumped into air line through a tire valve stem, with a hand pump, reaches its maximum stable pressure when all water is displaced through bottom, open end, of air line.

One psi of air pressure displaces 2.31 ft of water; then from known length of air line:

Distance from floor line to well-water level = $C - 2.31 \times$ air-line psig

FH, Field head, ft = discharge-pressure psig $\times$ 2.31 + elevation of discharge-pressure gage above floor, ft + distance below floor to well-water level, ft

$$\text{Water horsepower} = \frac{\text{FH} \times \text{GPM}}{3960}$$

Brake horsepower = Motor horsepower $\times$ motor efficiency

$$\text{Motor horsepower} = \frac{\text{Motor kw input}}{0.746}$$

$$\text{Pump field efficiency} = \frac{\text{Water horsepower}}{\text{Brake horsepower}}$$

$$\text{Wire-to-water efficiency} = \frac{\text{Water horsepower}}{\text{Motor horsepower}}$$

12

Gaskets, packing,
O-rings, mechanical seals

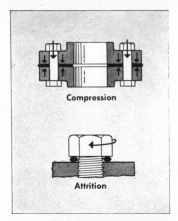

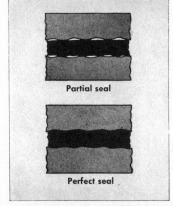

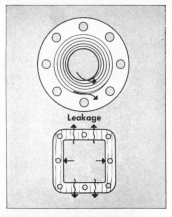

Gasket material must flow

To seal tightly, gasket must conform to joint's mating surfaces. Most gaskets are made relatively soft, so compressing them causes the material to squeeze into surface irregularities. With some materials, heat does the same thing. Another form of "flow" results from combined compression and dragging action, called attrition.

Bolt pressure, surface finish affect sealing action

For any given gasket material, a certain pressure produces just so much flow. That may be enough to fill irregularities of a fairly smooth surface, say an ordinary machined flange, but not those of a rough surface, like a serrated flange (*top sketch*). If joint can take more bolt pressure, tighter squeeze will fill both sides (*bottom*).

If flange serrations are "phonograph" type (continuous spiral) they must be completely filled or there's a path for fluid pressure to escape. With concentric serrations, partial filling of grooves may be enough. Parallel shaper or planer grooves make life tough for gasket—there are paths for pressure escape on two sides of the flange.

Gaskets supply the <u>give</u> that makes a tight

In piping, and all kinds of machinery, we've got the problem of making pressure-tight joints between two rigid elements —flanges, say. We can do it, but only if flanges are machined to mate perfectly, and can be counted on to stay that way. Such a joint is practical only in rare cases, so we turn to gaskets. Because they're designed to give, they make up for imperfections of the average joint—surface roughness, small misalignments, etc.

At relatively small cost, gaskets do away with need for precision machining. And it's easier to replace a gasket than to re-machine surfaces when something goes wrong.

How Gaskets Seal. To stop leakage, flat gasket in a flanged joint must be squeezed between joint faces, tight enough so it exerts more pressure against faces than fluid pressure tending to leak past it. Since most joint surfaces aren't perfectly smooth, gasket material must "flow" into hills and valleys of joint faces to get good mating and a tight seal with no leakage pathways.

Usually, gasket materials are made to flow by compressing them between flanges. In some cases, heat may do the job, as when a bell-and-spigot joint is sealed with molten lead. In other cases, copper ring on screw plugs, for example, flow is caused by combined compression and dragging action— called attrition.

Surface Condition. In ideal contact, gasket material fills every depression in joint surfaces completely. This depends on gasket material and condition of joint surfaces (*sketches above*). Whether partial filling of tool marks, etc, is good

enough depends on nature of grooves. Ordinary spiral tool marks (so-called phonograph finish) leave a continuous path for leakage unless filled. But concentric serrations leave no such path and complete filling isn't as important. Really tough problem comes on joints finished with planer or shaper tools—parallel grooves give paths for pressure leakage on two sides.

What Finish? Smooth surface is usually best. Serrations —V-grooves 1/32 in. apart—are OK with soft gaskets. Never use phonograph finish with metallic gaskets, or any gasket for high pressures.

If joint surfaces are smooth, problem of leakage can't be solved by merely refinishing with rougher tool marks. If you do that, bolts must be strong enough to stand extra stress needed to flow gasket into deeper grooves.

Forces on Joint. Pressure in pipe tends to push flanges apart. Bolts resist this. In addition, they must be loaded enough to squeeze gasket and keep pressure from leaking past. Since pressure pushing joint apart (*hydrostatic end force*) works against exposed portion of flange faces as well as internal diameter of pipe, always try to use gasket with about same ID as flange. This keeps hydrostatic end force to minimum, reduces needed bolt load.

Seating Force. The force needed to make gasket material flow into surface imperfections is known as *seating force*. It's usually expressed as unit stress in gasket, psi, and is considered independent of internal pressure.

For most gasket materials, it's important to apply a high

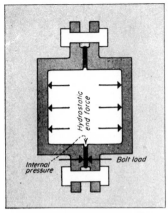

These forces act on joint

In a working joint, with pressure in the pipe, hydrostatic end force tries to push flange faces apart. Bolt load resists this action, and must be enough greater to put sufficient squeeze on gasket to maintain a tight seal against internal pressure.

Carbon print of pressure distribution shows most pressure, and hence most

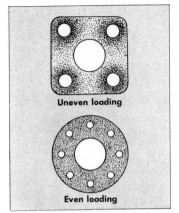

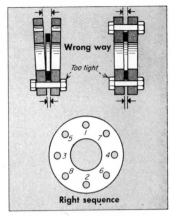

Bolt layout and tightening make a big difference

flow, is around bolts (*upper sketch*). Joint of this type will give trouble because sealing effect midway between bolts is poor. Solution lies in better bolting layout, with more uniform pressure distribution.

If you tighten one bolt full, flanges cock out of parallel. Taking up on another bolt won't bring them back even. Resulting poor pressure distribution encourages leaks, speeds failure. Right way is to take up all bolts hand tight first. Next tighten two bolts diametrically apart, then two as near 90 deg from first two as possible, and so on as indicated in sketch. Tighten bolts again after joint comes up to operating condition.

---

seal in a rigid joint

stress initially, to get full flow and good seating. Then load on gasket can be reduced without hurting seal. Such reduction of load on gasket usually occurs naturally as a result of bolt expansion and creep, gasket "settling."

Bolting Layout. Most gaskets are used in bolted joints. Carbon prints, *sketch above*, show pressure is greatest close to bolts, weakest midway between. This puts a premium on good bolting layout—adjusting number and spacing of bolts to distribute total load uniformly around gasket. It also means knowing the operating conditions to be met and the kind of gasket best suited for them. In other words, the gasket should be designed into the joint, not selected by cut-and-try after the joint gives trouble.

Tightening Bolts. Follow right sequence in bolt tightening (*see sketch above*). Equally important: Don't overstress bolts by whanging away at the end of a long hickey or oversize wrench. Overstressed bolts take a permanent set, won't keep the squeeze on gaskets.

Gaskets at Work. With joint at working conditions, hydrostatic end force tries to separate flanges. Bolts, gasket and flanges expand. Net effect is less contact pressure on gasket. That's why bolts are easy to take up after they've been in service. At high temperatures, don't overlook effect of bolt creep, slow flange deformation. And sudden temperature and pressure changes give joints a real wallop, hence often cause serious leakage problems.

To see how designers build resiliency into gaskets to meet operating conditions, turn to following pages.

Four types of gasketed joints

Confined

Unconfined

Partly confined

Self-energized

Gaskets may be broadly classified by the type of joint they seal:

Confined gaskets have no freedom to flow except for slight clearance between mating faces.

Unconfined gaskets are free to flow away from and toward fluid pressure.

Partly confined gaskets are free to move on either inside or outside rim.

Which Joint? In unconfined joint, gasket can be blown out if it isn't loaded enough. With confined or partly confined joint, worst that can happen is a slow pressure leak.

Because unconfined joint is most practical it's commonly used for piping and many other joints. Most metallic gaskets can be considered as self confining. Some gasket makers design semi-metallic gaskets for unconfined joints by putting a retaining ring, or similar device, outside.

Self-energized joints are used mostly on boiler handholes and manholes. These may have only one bolt; don't depend on bolt pressure to keep them tight. Fluid pressure working on entire plate area provides gasket load.

Critical period is while pressure is being brought up—initial bolt pressure drops to zero as fluid pressure climbs. So you have to tighten bolt from time to time until final pressure is reached. In this type of joint, make sure surfaces are even, gasket area is big enough for load.

Right Way to Remove, Lay Out,

First: We remove old gasket

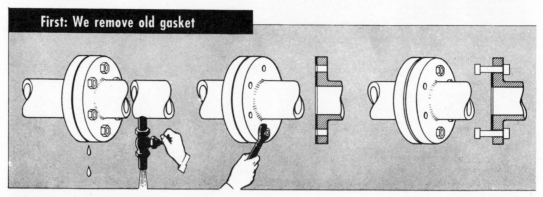

1 Open drain to relieve pressure before breaking any pressure-vessel joint

2 Remove all bolts if flanges are sealed with a full gasket (shown at right)

3 If gasket is ring type, remove only enough bolts to slip old out, new in

Second: We lay out new gasket . . .

1 Bolt circle is found from bottom (or top) of holes

2 Radius of bolt circle is equal to half of diameter

3 Draw line through circle's center for opposite holes

4 On 6-hole flange, radius is distance of bolt holes

Third: Now the new one is cut out

1 Gasket cutters turn out clean holes fast. Top unit needs only one hand

2 Tin snip is next best way. Never hammer gasket as that causes lumps

3 Punch holes after placing material on hard wood to protect the punch edge

Cut and Install Your Gaskets

4 Flange spreader: handy to open joint. You might require two on some jobs

5 Steel wedges, driven in at opposite edges, take place of flange spreader

6 Scrape out old gasket. Be sure every particle is removed. Don't harm face

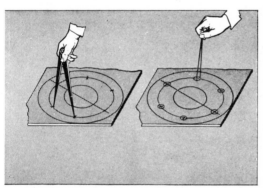

5 To check distance, walk dividers around the circle. Lay out holes a little larger than their actual size

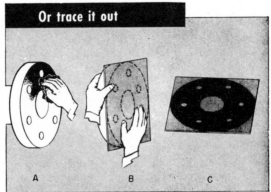

Or trace it out

TRACING a gasket's form (if you can get at flanges) is often faster. *A*, graphiting the flange; *B*, making impression; *C*, finished

And last: We install it

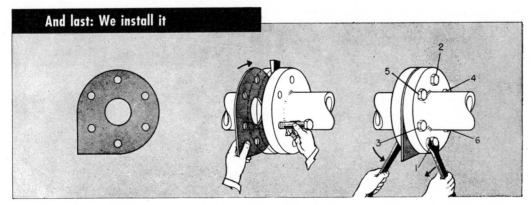

1 Leave one corner on gasket so you can hold it for positioning in flanges

2 Slip one bolt through flanges and gasket. Position gasket, place bolts

3 When line is under full pressure, tighten again in the sequence shown

PUMP GASKET MUST BE MADE RIGHT

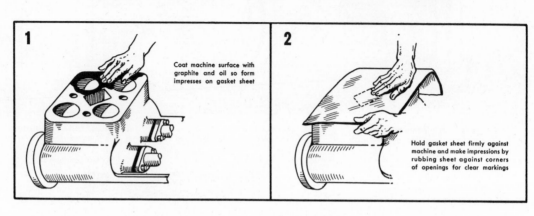

1 Coat machine surface with graphite and oil so form impresses on gasket sheet

2 Hold gasket sheet firmly against machine and make impressions by rubbing sheet against corners of openings for clear markings

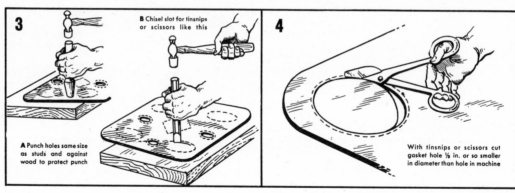

3 **B** Chisel slot for tinsnips or scissors like this

A Punch holes same size as studs and against wood to protect punch

4 With tinsnips or scissors cut gasket hole ⅛ in. or so smaller in diameter than hole in machine

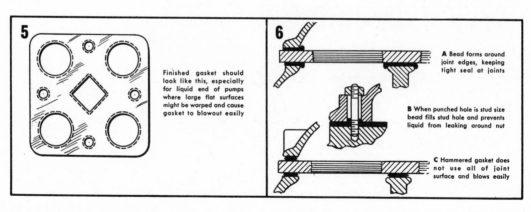

5 Finished gasket should look like this, especially for liquid end of pumps where large flat surfaces might be warped and cause gasket to blowout easily

6 **A** Bead forms around joint edges, keeping tight seal at joints

B When punched hole is stud size bead fills stud hole and prevents liquid from leaking around nut

C Hammered gasket does not use all of joint surface and blows easily

13 ways you can stretch the life of your O-rings

Hydraulic-seal application seems to generate plenty of questions. Of course, most of the time I'm solving specific problems. With so many variables in each situation there can be no one formula for success: as in any engineering equation, the solution is usually the compromise that works best.

A lot of design data has been published and is available to anyone. Manufacturers offer excellent manuals. But some basic answers aren't spelled out, and nevertheless we must cope with the problems. From my experience I've distilled 13 fairly general questions about O-ring design and application that crop up again and again. Now I'll try to answer them to help with your stubborn sealing problems.

1 **Some static O-ring seals show the same wear you'd expect to find in a dynamic seal. Why?**

A This happens in a pulsing-pressure system: the O-ring moves with the pressure pulses. As a field remedy, add a backup ring to prevent this motion. If you're tackling the problem in the design stage, change the groove width to equal O-ring's cross section plus an allowance for squeeze.

2 **What is the limit of pressure an O-ring can seal?**

A There is no limit—theoretically. If the O-ring doesn't get a chance to extrude into a clearance gap, it's unaffected by pressure. But from the practical viewpoint that's rarely attainable at very high pressures because of bolt stretch, diametral growth, etc.

A single backup ring with normal dimensional toler-ances seals up to 3000 psi. By putting two backup rings downstream you can extend this to about 4500 psi. But beyond that point you must resort to more drastic measures. Using an O-ring plus a fairly heavy Teflon ring, backed with a bronze ring, it's possible to seal 100,000 psi statically and 50,000 psi dynamically. In this design the O-ring expands the Teflon ring, which fits into lips of the bronze ring, in turn expanding them to produce the required zero clearance gaps.

3 **Why do O-rings sometimes extrude in situations where system pressure and downstream clearance gaps are within accepted limits?**

A There are two likely reasons. In systems where the pressure pulses rapidly, short-duration transient pressure peaks may shoot up to two or three times the indicated system pressure. Gages won't show those peaks, but suitable pressure pickups feeding an oscilloscope will tag them. So you see, your O-ring *isn't* working within accepted limits—it's subjected to shock loads beyond its ability to resist.

A less-frequent cause stems from industry's longtime reliance on Durometer hardness as a measure of an O-ring's limits. From the extrusion standpoint, modulus of elasticity is the value that really sets the limits of O-ring application, and the value may vary considerably from the Durometer reading.

In any given class of compounds, Durometer and elasticity values are close to parallel. But if we shift from, say, Buna N to silicone, we find that a 70-Durometer silicone will have about the same modulus of elasticity as a 50-Durometer Buna N. This is also true to a lesser degree of other types of compounds. Since it's actually the modulus of elasticity which resists stress caused by pressure, there can be varying degrees of extrusion resistance in compounds of the same Durometer value.

4 What can we do about spiral failure of O-rings?

A Spiral failure, which looks like a long-pitch screw thread cut around the O-ring, is caused by eccentricity between the groove root and the mating surface. That produces a varying squeeze. This in turn causes varying amounts of roll around the periphery of the O-ring on the pumping stroke. The return stroke usually will not have exactly the same cycle of eccentricity, so the ring may not completely recover. Repeated again and again, this cycle eventually overstresses the ring to the point of failure.

That's the cause. The cure isn't that simple. Since eccentricity may result from excessive clearances, high side loading, breathing cylinder tubes, too-short bearings, etc, you'll have to review the entire mechanical picture so you'll know what measures to take to correct it. In many cases it is better and more economical to select a different type of seal, such as a V-ring.

5 How coarse a surface finish can we use for static O-ring seals?

A This is always a good topic for a hot discussion between design engineers and the machine shop. Break the answer down into three categories: (1) In relatively steady-state pressure systems or where pulsations are not violent, O-rings in 3/16- and 1/4-in. cross sections can tolerate 125-micro-in. finishes. (2) If pressure pulsations are violent these rings should have a 63-micro-in. finish at the groove root and opposite face but can still live with 125 micro-in. for the groove walls. (3) Because of more limited sealing-contact area, the smaller 1/16-, 3/32- and 1/8-in. sections should have a finish at least as smooth as 63 micro-in. on all four sides. And if pressure pulsations are violent, they will be better off with 32 micro-in. at the groove root and opposite face.

6 What are the upper and lower temperature limits of present-day O-ring materials?

A That's a simple question, but it won't get a simple answer. It's impossible to be general in certain areas.

Fluid or gas environment, type of equipment, duty cycle, life requirement and other factors all enter the equation.

In normal design, where liquid or gaseous environments are compatible with use of silicones, −130 F is the accepted low limit. But in some situations, using squeeze in the 80% range, you can get good static O-ring seals in the cryogenic region of −350 F.

At the other end of the thermometer, in the 450-500 F range, silicones offer good life sealing gas environments, and fluoroelastomers like Viton and Fluorel can handle many fluids. As an extreme, successful tests of 1-minute duration have been run at 1000 F.

Here are two vital rules of thumb: (1) Don't exaggerate temperature problems. That will probably cost money and may result in inferior performance. (2) Do give all—and I mean all—details to your O-ring manufacturer or supplier. Take advantage of his know-how.

7 What about rotary O-ring seals?

A Rotary O-ring applications must be approached with caution! I've seen successful applications under conditions which seem impossible. For example, I've seen an O-ring working as a shaft seal in fairly high-pressure pumps. It's not really a full-fledged seal, but allows a controlled leak with leakage bleeding back to the intake side. What happens here is that the O-ring wears enough to leak a little; leakage then provides a heat-removal and lubricating film.

A rotary O-ring seal always contacts the same small area, so it retains its own frictional heat. Also the system fluid tends to squeeze out, so the seal must run dry. These two factors can become a vicious circle, each reinforcing the other. So unless rubbing speeds and pressures are low, say 250 rpm and 25 psi, or required life is short, better consult with your sales engineer before making a decision.

8 An O-ring rod seal is running leakfree and dry to the touch. How does it get lubricated?

A Looks like a paradox, doesn't it? But there's a fairly simple answer: the rod is *not* dry. Viscosity and film strength of the fluid actually carry a microscopically thin film of fluid past the O-ring. Since there is no accumulation of fluid on the rod, the same forces are carrying the film back past the O-ring against pressure on the retraction stroke.

This can be easily demonstrated in tests. When the rod is solvent-cleaned after extension and before retraction, wear rate increases very markedly.

9 What is the ideal surface finish for dynamic O-ring seals? By the way, can it be *too* good?

A Whether a surface finish can be too good for O-ring seals has been a moot point for years. I believe that your surface can be refined to the point where the fluid film's adhesion to the surface is reduced, so that friction and wear are actually increased! But this is only an opinion. I have never seen a surface this highly refined and I'm sure no one would want to pay for it.

Barring this questionable situation, we can say that the better the finish, the longer O-ring life will be. However, as pages in this chapter explain by illustrations and text, one should always keep in mind that it is the character of the surface profile rather than the RMS micro-in. value which is important. Cylinder tubing, with a maximum roughness of 35 or 40 micro-in., is an excellent surface from the seal's standpoint—its profile is like rolling waves. But the same RMS value produced on some materials by a hard grinding wheel could have sharply defined peaks and valleys, 100 micro-in. or more in total deviation. This would be like the same waves approaching the beach: with the same volumetric deviation, they've assumed the sharper, more violent surface of breakers.

To demonstrate the effect of surface profile on the seal there's a small experiment requiring only simple equipment that you *could* make, although I don't particularly recommend it. First, sit on a $\frac{1}{4}$-in. bearing ball, then on a $\frac{1}{4}$-in. thumbtack. One might be very uncomfortable, but the other is intolerable!

10 Does pressure affect life of dynamic O-ring seals?

A Of course it does, to an extent which depends on surface finish, fluid, O-ring material and clearance gap.

Better surface finishes, as we've said, mean longer life; in addition, pressure increases won't chop as big a percentage off the ring's life. But the O-ring material is usually determined by the seal's environment, so you'll seldom get the chance to improve life this way.

Since size of the clearance gap together with O-ring's modulus of elasticity affects extrusion, keeping the gap to the lowest practicable value will improve the ring's life under increased pressure. All else being equal, if you can prevent extrusion, lower-Durometer compounds will resist wear better. This indicates that if extrusion failures do occur, backup rings would be a better solution than increasing the O-ring's Durometer.

Take two hypothetical cases. In one case extrusion can occur, in the other it's prevented. Let's tabulate the relative effect of pressure on life, assuming both rings have a 1-million-cycle life at 1000 psi.

Pressure, psi	Cycles of life with extrusion allowed	Cycles of life with no extrusion
1000	1,000,000	1,000,000
2000	300,000	800,000
3000	150,000	600,000
4000	70,000	400,000
5000	20,000	200,000

These hypothetical figures show that the decrease in life is exponential when extrusion can occur, practically straight-line when it's prevented.

11 How can we increase friction in O-ring seals?

A Frictional resistance of an O-ring is made up of two components. One is a function of internal energy of the ring itself, generated by the squeeze. The other comes from energy generated by the applied pressure.

If the ring is not under pressure, the second component is zero. At about 250 to 300 psi the two components become equal; beyond this point the component caused by squeeze becomes relatively unimportant. Hence, reducing the squeeze helps in low-pressure applications.

In higher-pressure situations where the squeeze component is unimportant, we can resort to a mechanical device in addition to refining the surface finish. We can use an *undercut* or *dovetail* groove to hold the ring. In this type of groove, pressure will force part of the O-ring down the undercut sidewall, allowing less area of contact and therefore reducing friction. Best results come when the sidewall angle is adjusted so contact length equals about 25% of O-ring section diameter at operating pressure. It's also necessary to drill bleed passages to the bottom of the undercut wall, so fluid can't get trapped to prevent movement.

Sometimes friction can be cut by using a dry lubricant. A number of preparations on the market consist of Teflon or molybdenum disulfide dispersed in resin. In some types of applications they'll be very useful, but only trial will determine which ones.

12 Will it help the situation if I can identify the cause of an O-ring failure? If so, how should I go about it?

A Knowing the mode of failure can be extremely helpful in correcting the equipment. Barring faulty O-rings, which do sometimes get installed, the basic causes of failure are normal wear, extrusion, nibbling, spiraling, pulsation cutting, installation damage, damaged equipment such as scored bores or rods. Appearance of the ring helps you decide which of these is shortening its life:

Normal wear will appear as an more or less evenly worn flat area at the rubbing surface.

Pure extrusion first appears as an undercut groove around the periphery of the ring near the ID or OD on the side towards the pressure. By the time the ring fails an entire section of it may be sheared away.

Nibbling is a form of extrusion which occurs when side loading or some other cause results in shaky travel of the piston or rod. Chunks are bitten out around the periphery. Preventing nibbling, like preventing extrusion, is a question of clearances or anti-extrusion devices.

Spiral failure appears like a very long-pitch screw thread cut around all or a part of the O-ring. We talked about spiraling in answer 4.

Pulsation cutting looks somewhat like extrusion damage. But instead of being undercut, the ring is gouged toward the center of the section.

Installation damage usually appears as a single chunk sheared off the periphery of the ring. It's usually caused by lack of a suitable entrance chamfer or an attempt to install an oversized ring, or perhaps the ring was pushed into place across a threaded shaft. A second rubbing surface will produce fine grooves across the face of an otherwise usable ring. Grooves in a badly worn ring are caused by high-pressure jets of fluid through leakage areas.

13 How can I be sure that a new O-ring-seal design will work out?

A This question is the easiest of all to answer. Your O-ring manufacturer has had long experience dealing with the various combinations of pressure, temperature, fluids, pulsations, equipment design. When in doubt, consult him. Take advantage of his experience with other people's problems. But let me stress that optimum O-ring design, like any other, is usually the compromise that best suits a variety of factors. So give your expert all the details.

First: Let's see what motions we seal, what packing types we have for doing job

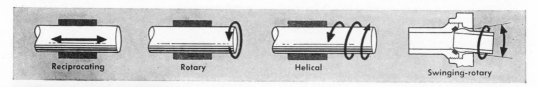

Reciprocating   Rotary   Helical   Swinging-rotary

Four types of motion

Sealing a static joint by clamping a gasket between surfaces isn't always easy, as we learned in *Gaskets*, March 1954. With thermal shock, pressure and chemical action all working on the sealing medium, gasket takes a beating.

Now packing must seal a *moving* part. If shaft is perfectly round with fine finish and running dead true, job is easier. But shafts usually run out at least a few thousandths. With grooves or shoulders on the shaft you have a really tough job, regardless of how good packing may be for licking problem.

Packing's job. When you try to seal by tightening packing to varying degrees with a wrench, you go against "good mechanics." Reason: Machine designer tries to reduce friction but packing acts as a brake on moving shaft.

Many operators don't know that packing in, say, a centrifugal or reciprocating pump must *throttle* leakage—not stop it altogether. That's because packing acts like a bearing and must be lubricated like one. Lubrication comes from slight leakage of fluid inside machine or, in emergencies, from a sat-

urant in the packing. Where these aren't practical, packing must be lubricated some other way. Remember, if packing is dry it runs hot, hardens and scores the shaft—like any other bearing that fails.

Four motions are sealed by packing: (1) *reciprocating* (2) *rotary* (3) *helical* (as in a valve stem), and (4) *swinging* (as in a ball joint). Usually, packing seals against outward leakage, but where inside pressure is below atmospheric it must seal against air leaking inward and destroying the vacuum.

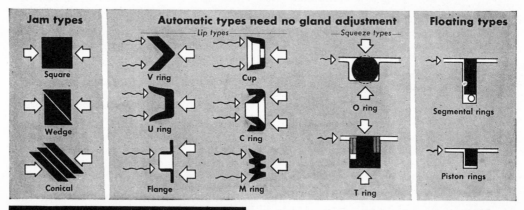

Jam types
- Square
- Wedge
- Conical

Automatic types need no gland adjustment
Lip types
- V ring
- U ring
- Flange
- Cup
- C ring
- M ring

Squeeze types
- O ring
- T ring

Floating types
- Segmental rings
- Piston rings

Here are packing classifications

Packings fall into three broad classes. First, there's the so-called "jam type," which includes any packing that is jammed into a stuffing box and adjusted from time to time by tightening nuts on a gland. These packings are braided, twisted, woven or laminated of rubber, leather, fiber, etc. As we'll see, other constructions and materials may be used.

Next come automatic types—so named because usually they need no gland adjustment. The fluid sealed supplies the needed pressure by forcing packing against wearing face. These packings can be further divided into lip and squeeze

types. Better known among lip types are the cup, flange, U, etc. Best known of the squeeze types is the O ring. Here, interference is built into ring; it's thus squeezed into the groove. Fluid pressure holds it against wearing surface.

Third are the floating types. These include segmental rings of carbon, metal, plastic, etc, held around shaft or against cylinder by springs. Piston rings, with built-in tension, also fall into this group.

Now that we have a general idea of packing classes, let's see how jam types seal, what part packing volume plays.

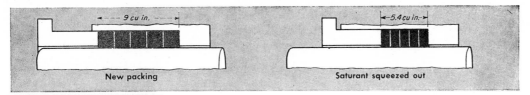

New packing | Saturant squeezed out

How packing seals

Volume taken in stuffing box by jam-type packing is key to understanding how this commonly used packing works. Manufacturers say packing is wasted, rods and shafts scored and packing needlessly burned up because few mechanics working daily with it understand this basic principle.

Left sketch shows a stuffing box filled with five rings of new packing. In this example, braided packing is saturated with lubricant, coated throughout with graphite and takes up about 9 cu in. of space in box. Here the gland nuts are

only hand tight because the new packing was just installed.

Right-hand sketch shows same packing completely compressed until it takes only 5.4 cu in. Volume is less because all saturant has been melted, squeezed or washed out. Because packing can't be compressed further, it has reached end of its useful life. You must renew it or it will burn up and score the shaft, causing more damage than the cost of the most expensive packing. Let's read on and learn why packing has saturant, what happens when volume is reduced.

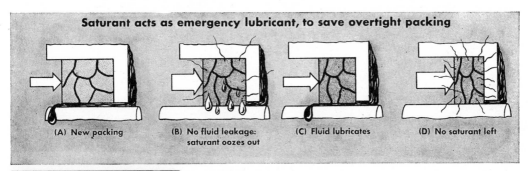

Saturant acts as emergency lubricant, to save overtight packing

(A) New packing | (B) No fluid leakage: saturant oozes out | (C) Fluid lubricates | (D) No saturant left

Packing fails like this

We told you, p 100, that packing acts like a bearing, needs lubrication as any other bearing does. Sketch *A* shows new packing, before gland has been tightened. As machine runs, some saturant in packing is washed or squeezed out. As saturant is lost, packing shrinks away from shaft because its volume has been reduced. Here operator usually tightens gland, stopping all fluid flow.

With no leakage, there is no lubrication from fluid flow. So rod heats up and temperature of packing goes up. That's when lubricant, placed inside packing material by manufacturer, goes to work, sketch *B*. High temperature starts to melt oil out of packing, which lubricates shaft for just such an emergency. Because more oil has oozed out of packing, space taken by it is again reduced. That starts flow of liquid from casing, which again lubricates rod and packing, also carrying away heat, sketch *C*.

But here is where trouble can start. If operator tightens gland enough to stop leakage again, liquid flow is lost. Again high temperature melts more packing lubricant, reducing volume and

starting flow from casing automatically as before. This can go on until no saturant is left in packing, until packing's volume cannot be further reduced. When gland is next tightened to stop leakage, packing burns up and rod scores, sketch *D*.

This simple principle, once understood, will get the service from packing that is built into it. Every man given authority to tighten a packing gland, regardless of how unimportant the job may seem, must understand this or he will score rods, waste packing and cause untold damage in outage and maintenance bills.

Now that you know packing volume taken by lubricant is important, you might ask how much saturant a packing should have. The Navy does not accept a packing that contains more than 30% of saturant by weight. Usually it's much less in most braided or woven packings.

If you buy packing that's over-lubricated, excess lubricant washes out or is squeezed out after being in operation

a short time. Then, because weight is taken up by excess saturant that does no good anyway, you have less fiber to seal joint.

How to tighten packing. Here's something to pound into the head of every operator. There is only *one* way to tighten a packing gland. If joint leaks too much, tighten each hexagon nut only *one* flat. This is one sixth of a turn. Then wait for about ten minutes. Reason: It may take that long for packing to adjust itself and reduce excessive leakage.

After ten minutes, packing has distributed itself in box and you can tighten nuts another flat if leakage is still too much.

Glands are usually tightened when machine is shut down overnight, to prevent leakage. But trouble is that operator starting machine next time does not loosen gland. This is common cause of packing failure and of grooved shafts. It will pay you to use a sign reading, "Loosen packing gland before starting this machine." Place sign on machine.

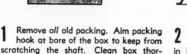
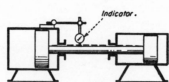
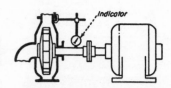

1 Remove *all* old packing. Aim packing hook at bore of the box to keep from scratching the shaft. Clean box thoroughly so the new packing won't hang up

2 Check for bent rod, grooves or shoulders. If the neck bushing clearance in bottom of box is great, use stiffer bottom ring or replace the neck bushing

3 Revolve rotary shaft. If the indicator runs out over 0.003-in., straighten shaft, or check bearings, or balance rotor. Gyrating shaft beats out packing

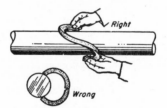

6 Cutting off rings while packing is wrapped around shaft will give you rings with parallel ends. This is very important if packing is to do job

7 If you cut packing while stretched out straight, the ends will be at an angle. With gap at angle, packing on either side squeezes into top of gap and ring, cannot close. This brings up the question about gap for expansion. Most packings need none. Channel-type packing with lead core may need slight gap for expansion

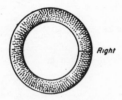
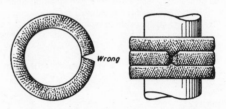

Now let's learn how to install packing correctly and cut down costly outage

11 Open ring joint sidewise, especially lead-filled and metallic types. This prevents distorting molded circumference—breaking the ring opposite gap

12 Use split wooden bushing. Install first turn of packing, then force into bottom of box by tightening gland against bushing. Seat each turn this way

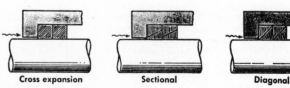

Cross expansion   Sectional   Diagonal

15 Always install cross-expansion packing so plies slope toward the fluid pressure from housing. Place sectional rings so slope between inside and outside ring is toward the pressure. Diagonal rings must also have slope toward the fluid pressure. Watch these details for best results when installing new packing in a box

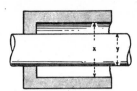

4 To find the right size of packing to install, measure stuffing-box bore and subtract rod diameter, divide by 2. Packing is too critical for guesswork.

5 Wind packing, needed for filling stuffing box, snugly around rod (for **same size** shaft held in vise) and cut through each turn while coiled, as shown. If the packing is slightly too large, never flatten with a hammer. Place each turn on a clean newspaper and then roll out with pipe as you would with a rolling pin

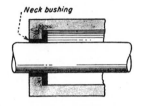

8 Install foil-wrapped packing so edges on inside will face direction of shaft rotation. This is a must; otherwise, thin edges flake off, reduce packing life

9 Neck bushing slides into stuffing box. Quick way to make it is to pour soft bearing metal into tin can, turn and bore for sliding fit into place

10 Swabbing new metallic packings with lubricant supplied by packing maker is OK. These include foil types, lead-core, etc. If the rod is oily, don't swab it

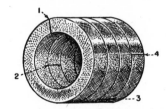

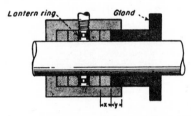

13 Stagger joints 180 degrees if only two rings are in stuffing box. Space at 120 degrees for three rings, or 90 degrees if four rings or more are in set

14 Install packing so lantern ring lines up with cooling-liquid opening. Also, remember that this ring moves back into box as packing is compressed. Leave space for gland to enter as shown. Tighten gland with wrench—back off finger-tight. Allow the packing to leak until it seats itself, then allow a slight operating leakage.

Hydraulic-packing pointers

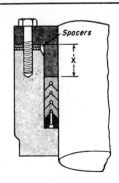

First, clean stuffing box, examine ram or rod. Next, measure stuffing-box depth and packing set—find difference. Place ⅛-in. washers over gland studs as shown. Lubricate ram and packing set (if for water). If you can use them, endless rings give about 17% more wear than cut rings. Place male adapter in bottom, then carefully slide each packing turn home—don't harm lips. Stagger joints for cut rings. Measure from top of packing to top of washers, then compare with gland. Never tighten down new packing set until all air has chance to work out. As packing wears, remove one set of washers, after more wear, remove other washer.

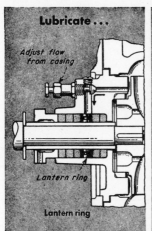

Lubricate...

Adjust flow from casing

Lantern ring

Lantern ring

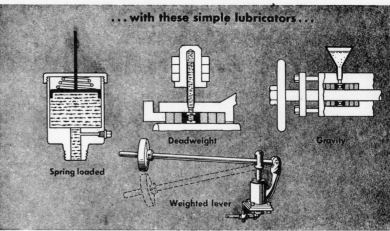

...with these simple lubricators...

Spring loaded

Deadweight

Gravity

Weighted lever

Lantern rings and lubricating systems

For some types of jobs,

We've learned that most types of packings must leak slightly to lubricate the shaft. But sometimes you can't count on such leakage. And if fluid sealed is abrasive or corrosive, you want to keep it out of the packing. One answer is to use a *lantern ring*.

Typical lantern ring, first sketch, is a channeled ring, with radial holes, fitting loosely into the stuffing box. Connection from the casing feeds fluid under pressure to outer channel, radial holes carry it to inner channel. It then flows along the shaft in both directions, insuring lubrication.

When you need to keep abrasive or corrosive fluids out of the packing, you can't feed the lantern ring from the casing, must turn to an outside source. Water, grease, oil, other fluids, used for

this purpose, are usually called "sealing liquid." Under pressure, the sealing liquid bucks fluid trying to leak out of the casing. That reverses flow through packing on the casing side of the lantern ring.

Ring position is important: With lantern ring nearer casing, flow into casing is increased as sealing liquid meets less resistance, flow to atmosphere is decreased. For maximum flushing effect into casing, put lantern ring at bottom of packing set, with no packing between it and casing. This "flushing connection" is used on equipment handling highly abrasive fluids. If you change ring position, be sure ring lines up with sealing connection.

Feeding schemes for sealing liquid are numerous (sketches). For some

services, a spring-loaded grease cup is a simple solution. Dead-weight and weighted-lever lubricators are easy to maintain and work without trouble. Just set weight for pressure needed and keep lubricator filled. Large valves are often sealed against vacuum with a funnel filled with water. This also keeps packing from drying out.

Sealing fluid can be injected continuously or periodically, as by a pulsating lubricator or grease gun. Trouble with periodical injection is that fluid from casing gets into stuffing box as soon as injection pressure drops. Continuous sealing flow, at pressure only slightly higher than casing fluid, is best, prevents contaminating casing fluid with sealing liquid.

Oil-bottle hookup, sketch, uses casing

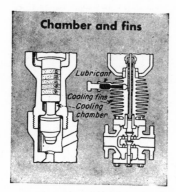

Chamber and fins

Lubricant

Cooling fins
Cooling chamber

Cooling the packing set

Heat developed in a packing is proportional to product of friction and surface speed. Packing temperature rises until rate of heat dissipation equals rate of heat generation. You can hold down heat by dissipating it with a water-cooled shaft, or by water-jacketing the stuffing box.

Valves often have a so-called cooling chamber between packing and valve body. But there is little temperature drop at packing because most of heat flows through metal, first sketch.

Regulator valve, second sketch, is for extremely high pressures and temperatures. Here cooling fins around stuffing box of long valve stem dissipate heat. Because valve stem moves constantly, must not stick, cartridge of molybdenum-disulfide lubricant is fitted into stuffing box. It won't carbonize or oxidize, helps keep stem free.

High packing temperature leads to excessive thermal expansion that increases load, causing seizure. Cooling and dry lubricant may be only answer.

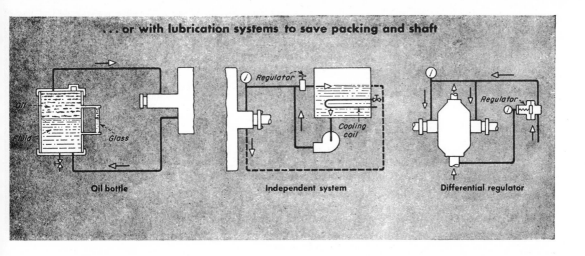

Oil bottle   **Independent system**   **Differential regulator**

packing needs to be lubricated from outside

pressure to inject sealing oil, needs no regulator or pump. Just watch gage glass, keep bottle filled with oil. This system uses little oil. When pump is shut down, drain water from bottle, fill with oil, and it's ready.

Circulating system has relief regulator set a few psi above casing pressure. To carry away heat, line from stuffing-box bottom is piped back to tank (dashed line). Fluid in tank can be cooled, if necessary. Differential regulator system keeps a set pressure between casing, gland.

Corrosion. Before injecting sealing liquid, you should know that fluids trapped in stuffing box often cause corrosion. This is known as crevice corrosion, often takes place with fluids that don't corrode metals in an open vessel,

but cause damage when trapped in clearance between packing and metal. Packing materials suspected of causing this action are copper and brass reinforcements, also graphite. Oily packings are less likely to give trouble.

One solution for packing that's used with water or aqueous solutions is to install it just before machine is used. Then remove it if unit is to be idle for long periods.

Applications that normally need a sealing liquid include:

1. All packing against gases—they leak excessively because of low viscosity. Also vacuum units, steam engines and turbines, h-p autoclaves where agitator enters gas space at top.

2. All packings for liquids of low viscosity or high volatility (such as gas-

oline) used at high speeds and high pressures. Examples: Lube oil injected into stuffing boxes of pumps handling propane; oil between double axial seals on water pumps, etc.

3. Packings for liquids carrying large amounts of dissolved or suspended matter that tends to drop out in passage through packing. Examples: Water-sealed boxes on centrifugal pumps handling caustic soda and on hydraulic turbines operating on muddy or contaminated water.

4. All packings for fluids of any kind that are poisonous, corrosive or too valuable to lose. Examples: Pure gasoline injected into boxes of pumps handling anhydrous hydrogen fluoride in the alkalization process; oil injected into ammonia-compressor rod packing, etc.

Should you swab rods, and oil new packings?

Metallic packing on reciprocating rod may need oil feed, needs no swabbing. Excess oil from swabbing many kinds of steam packings may do harm—isn't needed. You can swab V-type lip packings, but usually even this isn't necessary. Lubricant put into most asbestos packings is all that is needed on steam jobs. Here condensate has lubricating qualities. Laminated-type duck-rubber packings around rams of large hydraulic bloomer-mill shears in steel mills are lubricated by spraying ram with viscous, high-temperature oil every hour or so. But here radiant heat from hot steel is problem.

As for oiling new packing, there is up to 30% saturant in most jam-type packings—so why oil them? Main thing is to see that shafts are not rusty, not covered with abrasive grit and generally in good condition before applying new packing. Like any bearing, shaft should be oily and not dry. If in doubt, wiping the shaft with oily cloth may be good insurance when repacking a stuffing box.

Q—Does pressure affect gaskets?

A—We have very little trouble that way; today's gasketing materials can contain just about all pressures involved. But keep in mind that flange or joint design is the key Poor flange conditions due to warpage, incorrect design or improper bolt tightening are *deficiencies*. Don't assume that they can be overcome by changing gasket types or thicknesses—go to the root of the trouble.

Q—How important is gasketing material to the final temperature of a system?

A—Temperature is an item which we can easily measure. Materials are graded for certain temperature ranges and rated for certain jobs—it's obvious that asbestos and rubber will have different specifications. But the packing specialist often recommends a material for temperatures that seem far beyond its strength. He does this because he knows that some decomposition at high temperatures is not necessarily fatal—the products of decomposition may even play an important role in the gasket.

For example, many rubber gaskets are combined with products such as asbestos and the combination handles far above what rubber alone will stand. Take compressed asbestos sheet where the prime gasketing material, the asbestos fiber, is bonded with an elastomer. Your expert knows that the rubber is only a binder, a means of manufacturing the actual gasket. There's so little elastomer that when it decomposes and is confined between metal surfaces under pressure we still have a very adequate gasket.

We must also be aware that temperatures in a pipe or a pump may be somewhat lower in the gasketing or packing areas—sometimes as much as 25% below that of liquid or gas handled. This is good since the actual temperature of the fluid could completely decompose a packing or gasket.

Q—Do engineers give enough attention to how the sealing material affects the medium being sealed?

A—No! Since all liquids and gases handled must be packed or contained or stored in some manner, every aspect of the packing materials must be considered. We must also think in terms of the fluid or combination of fluids we're working with. I find that it's common for packing-recommendation requests to make a great point, for example, of the fact that an acid problem exists. Yet they may not even mention that the acid content of the fluid is only 2 or 3%. So the expert devotes his time and thought to the acid side of the case. He's left completely unaware that a solvent is present in a much higher percentage and is really the tough part of the problem.

Not only should we consider the content of a fluid or combination, but we must know what happens during leakage. Of course we know that the fluid is supposed to leak a little to lubricate its packing. But during the process, a leaking liquid often evaporates as it flows to atmosphere. And if evaporation leaves a residue of crystals in the packing, we have a really difficult problem. Yet this abrasion factor is often overlooked when we

consider packing. Here again, a lantern ring with a 5- to 10-psi pressure increase will help cancel the abrasion.

Q—How does fluid leakage affect packing?

A—A fluid film between packing and moving shaft is a must if any packing is to give good service. If we depend on lubrication from the fluid being contained, we'll have many points to consider. Some fluids will be good lubes and some won't. But in any case, pumpage of vacuum pumps and gas pumps or compressors will have no inherent lubrication. In these cases lube will have to be applied from an outside source or from the packing itself.

In this area, Teflon has provided our greatest advance among packing materials. Teflon and graphite are both considered solid lubricants, but Teflon is more versatile and more efficient.

Any packing problem brings up two important considerations. What is available to act as a lubricant? How can it be maintained? Lubrication in a packing is just as important as it is in a bearing.

Q—How do you pack against high shaft speeds?

A—Shafts and plungers can move slowly or fast. In a reciprocating plunger, rubbing speeds of over 250 fpm are critical and they're often tough to pack properly. Length of stroke also is important, because high pressure, high speed and short stroke give you friction, and heat builds up rapidly. This heat can't be carried away if strokes are short. So localized heating occurs quickly and scorches the wearing surface of a packing.

With rotating shafts, it isn't rpm alone that counts. Product of rpm and shaft circumference establishes rubbing speed. While the ranges aren't as critical as they are in reciprocating equipment, certain types of packing are very poor performers at rubbing speeds over 2500 fpm. Below this range, many types are OK.

Leakage rates are also important, to provide adequate cooling fluid for high speeds. A rotating piece of equipment tracks the packing in the same spot. The reciprocating pump continually makes new surface contact with the packing. For this reason molded V-type packing doesn't properly seal rotary shafts, though it's ideal for reciprocating motion.

Q—How does clearance around a packing affect its life?

A—Clearances designed into a piece of equipment are enlarged by wear and misalignment—packing life drops as clearances increase. Clearance problems are compounded by a shifting shaft or a gland follower. Annular clearance might be only 0.010 in., but it takes just a slight shift to compound this clearance to 0.020 in. because of misalignment or wear.

Misalignment, called "postholing," kills your packings. Misaligned or off-center shafting will posthole packing by beating it back. Soon leakage becomes excessive and packing life shortens.

An off-center shaft will also become a wobble pump as it postholes through the packing. With each revolution, the fluids being sealed will flood the clearances. Then as the shaft rotates it pushes fluids toward the low-pressure side.

Q—Is there any relationship between packing life and plant personnel?
A—Yes! Packing life depends a lot on maintenance habits and procedures followed and these in turn depend on plant personnel. That's why a change of personnel may reduce the life of a packing, shaft, or plunger.

Q—Does better surface finish of machined elements have much effect on packing life?
A—You'll have to think in terms of the type of finish you want, and the type you can tolerate costwise. Although a fine finish is always desirable, its profile is often a greater consideration than its finesse. A sharp saw-tooth profile, even in the better finishes, can file down the packing.

A so-called "rough" finish which has been touched with abrasive cloth to knock off the sharp edges can be ideal. I have seen reciprocating plungers with grooves as deep as 1/32 in. worn into the plunger. Yet if alignment and clearances are OK the plunger doesn't leak badly.

Q—How does stuffing-box depth affect sealing efficiency?
A—Depth of stuffing boxes varies greatly with different types of equipment. Stuffing boxes for centrifugal pumps can vary from 4 rings of packing to as many as 12. Both operate well, but they'll need different adjustment and breaking in.

A stuffing box with many rings has a long throttling effect on the leakage. But if temperature caused by friction expands the packing, that may shut the leakage off quickly. Result is scorched packing and scored shaft. A stuffing box with fewer rings may react the same way because greater adjusting pressures must be placed on the packing. This will limit adjustment, and again you're risking a quick shutoff of leakage. Use five rings of packing with or without a lantern gland.

Q—How many basic types of packings are there?
A—Some seven groups or types of materials are used to manufacture packing and gaskets, but they're made into many combinations for varied services. While the types of service can be listed under a very few categories, packing combinations on the market for these services are legion.

Q—Do organic fibers fill a useful need in packings today?
A—Organic fibers, flax and cotton, were popular many years ago. They're gradually fading out of the packing picture—especially cotton. The greatest limitation on these two materials is their narrow range of temperature resistance. They were popular mainly because both were easily available and both had great absorbing qualities. So many types of lubricants could be combined with them.

But even when they're used on shafts or plungers where temperature is normally within their operating range, you always run the danger that these temperatures

will be exceeded because of friction and high rubbing speeds. We find no logical reason to combine organic fibers with Teflon, because of the temperature factor.

Q—What are the limitations of inorganic fibers like asbestos and glass?
A—Asbestos, both the white and blue types, is the most-used material in packing and gaskets. Asbestos is basically a rock. Fortunately it can be processed into usable forms. Because of its great temperature and chemical resistance it's an excellent material for packings and gaskets. But asbestos isn't indestructible and it is affected by various acids to varying degrees.

Heat drives off the water of crystallization, reducing the asbestos to a fine powder with no structural strength. Around 1000 F, asbestos is on its way to decomposition but still functions as a packing material. Asbestos yarns and cloth are reinforced with many types of wire to increase their tensile strength and let them withstand high temperatures and pressures.

Asbestos is very common as a friction material in brake linings and clutch facings. These require opposite frictional characteristics to those needed in packing. Modification for low friction combines asbestos with special materials and lubricants.

Spun into yarns and woven into cloth, asbestos contains varying amounts of cotton depending on service required. Commercial-grade asbestos used in packing contains about 25% cotton.

Q—Are the elastomers holding their own as a packing or gasket material?
A—Today the synthetic rubbers are considered much more efficient than natural rubber for packings and gaskets. Natural-rubber compounds have a weakness: at relatively low temperatures the rubber reverts to a soft gooey mass even though it is vulcanized. This causes rapid deterioration of packing and gaskets because the rubber loses its ability to bind—hold together—a packing structure. However, rubber is still a popular gasket-sheet material for low-pressure steam and hot water.

Our synthetic-rubber materials usually don't go through this reversion. Silicone rubbers have the extended physical characteristics needed for higher temperatures. These have brought about a breakthrough, jumping limitations from 275 up to 500 F—a great step for the packing industry.

Rubber and synthetic-rubber materials are either used in sheet form or converted into cement.

Q—What jobs do the metals—lead, aluminum, brass and others—handle?
A—Although some packings are made entirely of metals such as lead and aluminum foil, the metals are not universally accepted packing materials. But they are often used together with asbestos. Metals contribute to extrusion resistance and low friction in many packing situations; packings which must perform an auxiliary bearing function are combined with or made wholly of metals.

What you should know about design and application of mechanical seals

HARRY TANKUS, extreme right, demonstrates mechanical seal components to plant operators during one of many seal seminars

Here are the seventeen questions most frequently asked by engineers attending the author's technical seal seminars given across the country. These answers from an expert will help you get top performance from your mechanical seals

1 Why use mechanical seals rather than packing?

A There are two good reasons: economy and practically no leakage. Besides, mechanical seals lend themselves to long, uninterrupted equipment usage with minimum leakage. Teardowns are less frequent, and you don't have to replace shaft sleeves or metalize shafts as you do when using packing. While packing requires the human *touch* for proper installation, the mechanical seal comes as a unit and is installed mechanically.

2 Do packing and mechanical seals prevent leakage in the same way?

A Packing smothers the fluid. The tighter you squeeze it, the less leakage. But you must always permit a visible amount of leakage so that the packing does not burn up and excessively wear the shaft or sleeves. On the other hand, mechanical seals utilize hydraulic principles in containing fluids. While there is leakage, it is generally invisible to the naked eye. (See *Mechanical Seals*, POWER, March 1956, 24 pp.) The accepted leakage ratio between packing and mechanical seals is about 100-to-1.

3 How do seal and packing costs compare?

A While this depends on the unit and operating conditions, the initial cost of seals is generally higher than packing. But economy is still on the side of end-face seals because of their lack of downtime; less maintenance.

4 Is there a universal seal for all jobs?

A No. Such a seal would have to be manufactured for the most hazardous and exotic application and would therefore be too expensive for less sophisticated use. Engineering a specific seal for a given unit and set of operating conditions permits application of the most economical seal design and materials.

5 What are inside and outside mounted seals?

A These terms refer to the seal's position relative to the stuffing box. Inside seals are mounted in the box, while outside seals are mounted outside the box. But the difference goes beyond this physical fact.

Inside seals are advantageous because hydraulics place all seal materials in compression rather than tension. Thus fluid flow to the faces is generally directed through a bypass line from the discharge jetted in close proximity of the faces. This permits better lubrication and also has thermal advantages. Dirt and oxides have a tendency to be thrown centrifugally away from the faces.

Outside seals are easy to assemble and maintain. Since the seal is completely exposed to the atmosphere, no cracking of the box or auxiliary gland plate is necessary for seal removal or adjustment. But hydraulics place the seal in tension and eliminate safety factors on many otherwise desirable materials. Liquid flow is to the outside diameter of the faces and goes through closely restricted areas that do not allow heat dissipation. This makes face lubrication more critical and bypass use not as effective.

Dirt and foreign matter further inhibit lubrication of the faces and present a clogging hazard through close areas on the inside diameter. In some cases, clogged areas can change balance characteristics of the seal design. Since fluid tends to come through the faces centrifugally, leakage can become more pronounced.

6 How do balanced and unbalanced seals differ?

A Balanced seals generally require steps in the shaft or sleeve—and more complicated parts. Therefore, they should be used only when necessary. Balanced seals permit a hydraulic pressure gradient across the bearing faces which cancels a portion of the face load. Their use is called for when the combination of fluid, pressure, temperature and speed adds up to a poor lubricant.

On the other hand, if fluid, pressure, temperature and speed permit a good lubrication film between faces, it's practical to use an unbalanced seal.

7 Can seals handle abrasive fluids?

A Certainly. But handling abrasives usually calls for a seal system, rather than the seal itself. The best solution is to inject a clean fluid through the stuffing box and then use a metering throat-bushing. Double seals are also used against abrasives.

8 Do mechanical seals have limitations? If so, what can be done about them?

A Basic limits are on temperature, pressure and rotative speed. These, coupled with the lubricity features of the fluid, finally dictate the PV (pressure-velocity) limitations of a given seal. A PV rating of 500,000 is becoming common, but 1,000,000 will soon be possible.

9 What is maximum shaft speed, temperature?

A Speeds up to 100,000 rpm have been reached, but this must be qualified by shaft size for an ultimate fpm (ft per min) factor. Temperature environment of up to 1300 F has also been claimed. But the lubricity of the sealed fluid must be known at this temperature.

10 What physical changes take place at higher operating temperatures?

A For the components, the seal designer must consider the expansion coefficients of the materials used, especially as related to critical clearances between relatively moving parts. Also, corrosion effects are important on materials at these temperatures. But most important is the lubricity of the fluid at extreme temperatures, especially as it comes across the seal faces using the pressure gradient technique. A physical change of the fluid can take place. It's not uncommon to have friction of the faces, coupled with the hydraulic and mechanical pressures involved, modify fluid to a solid at the faces. This is normally called *coking*. Since abrasives at the faces is the enemy of any end-face seal, it follows that coking leads to early seal failure.

11 How about higher pressures?

A Although we have covered temperatures, speeds and fluid characteristics, we must also consider the pressure to which the seal will be exposed. Pressure plays a vital part in finally dictating the type of lubricity (pressure) gradient that will come across the face. Pressures up to 3000 psi have been successfully held at standard pump speeds by end-face seals. This has been possible using fairly well-known fluids at nominal temperatures. But pressures in this range are still in the exceptional class, whereas 2000-psi seals are becoming commonplace.

If, one day, a breakthrough on temperature or pressure takes place, it will be the result of (1) the concentrated research by seal manufacturers now under way, and (2) industry-sponsored programs for outside research activity.

12 How do seal materials effect reliability?

A The predictable life of any seal relates back to the question of materials reliability. Designs are proven through extreme tests in the field and the laboratory. But without consistent quality of materials, all sophisticated designs will suffer. That's why quality control programs must be maintained.

13 Are better face materials being developed?

A It is axiomatic that basic materials used as the end-face bearing combination will play an important role in the success or failure of the overall seal. Considering design and workmanship equal, face combinations can be selected with thermal absorption in mind. At this stage, work must be done by basic material research laboratories to develop materials that can dissipate generated heat faster—and to a higher degree while still remaining stable. To date, special grades of carbon graphite have shown the most desirable properties.

Selection of face materials must be made in the light of an understanding of surface finish requirements and distortion effects. Surface roughness will have a significant effect on the function of a given face combination, while distortion of the face materials will modify balance characteristics with detrimental results.

14 Can you predict mechanical seal life?

A This generally depends on the application. Mechanical features of the equipment—such as shaft run-out, end play, vibration, and fluid characteristics in terms of quantity of foreign material or abrasives, lubricity, temperature and corrosion characteristics—all dictate seal life. Under commercially accepted conditions, a two-year seal life with 24 hour-a-day operation is common. Under ideal conditions, seals have worked beyond 15 and 20 years.

15 Are mechanical seals temperamental? We hear about millionths of an inch, light bands, hospital cleanliness, etc. So, are seals practical?

A These references are generally used in seal manufacturing techniques rather than by seal users. It is true that more care has to be taken with these seals than with packing. But reliable service depends merely on following the seal manufacturer's instruction.

16 Is it true that when mechanical seals fail they fail suddenly, allowing gush leakage?

A Certainly not! A failure of this type is extremely unusual. Even if one sealing member broke completely, which is highly improbable, there would be labyrinth-type leakage. Besides, for critical service, safety devices can be placed on the atmosphere side of the seal. There are quench glands with vent lines that direct any undesirable leakage to a safe area. Failure is most often gradual, thus giving an operator enough time to place replacement.

17 Can seals be repaired on job sites?

A Yes. Only a few critical parts are normally replaced. Carrying spare elements allows almost immediate repair or replacement. There are also special seal-cartridge designs which permit removing a complete cartridge and replacing it with a new one. Then the old cartridge can be taken back to any convenient machine shop, repaired and placed back on the shelf, ready for re-use.

Here's one way to convert

Follow these steps___

1 Dismantle pump; remove burrs, sharp corners from shaft, also from the shaft threads. Shaft must be free of defects at O-ring location, see sketch

2 Machine the shaft sleeve according to drawing furnished by seal firm. If your pump's shaft has no sleeve, seal will be supplied to fit your present shaft

3 Here are mechanical seal parts, also gland backing-flange machined to fit the gland. Sleeve is machined and on the shaft, ready to receive the seal

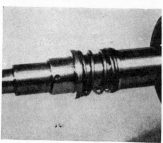

7 Slip seal-ring with O ring over the shaft; don't harm O ring. Turn ring so it spreads the drive-spring open, is held firmly in place as shown, sketch

8 Shaft sleeve here has drive-spring with seal-ring, all rotating parts. Seal-ring face is lapped to within a few millionths of an inch; don't harm face

9 Place stationary unit (gland, gland backing-flange, seal-ring, gasket) on shaft. Take care not to harm lapped sealing faces; keep these parts clean

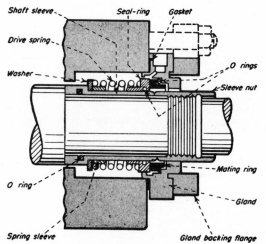

Shaft sleeve · Seal-ring · Gasket · Drive spring · Washer · O rings · Sleeve nut · Mating ring · Gland · O ring · Spring sleeve · Gland backing flange

How seal works___

▶ MECHANICAL SEAL has come to mean two flat surfaces with sealing faces at right angles to the rotating shaft. This makes it an axial seal, also. Because these two rings take all the wear, they save the shaft. One ring is fastened to the shaft and rotates, while the other is stationary. Usually one ring is metal with a hardened face, while the other is a composition carbon ceramic, or metal. Instead of a metal ring, some small pump shafts have a lapped shoulder that the carbon ring seals against.

One ring is mounted so it moves axially with the shaft. That allows its face to remain in contact with the stationary ring, regardless of shaft endplay, sealing-face wear or sealing face run-out. So this ring, usually called the seal ring, must be mounted flexibly.

The opposing ring, known as the mating ring, is mounted rigidly to the gland or pump housing. This rigid ring is backed up with a gasket or O ring to prevent leakage.

your pump to mechanical seals

4 Mating ring (carbon) is inside gland here and gland is on gland backing-flange. Drive spring is also assembled with seal ring to show how unit looks

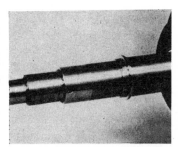

5 Slide washer against sleeve shoulder, then slip spring-sleeve against the washer. Sleeve is split so drive spring will hold it tightly against the shaft

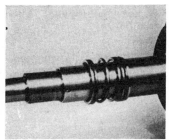

6 Place spring over split spring-sleeve with turning motion. Turn right-hand spring clockwise, left-hand spring on shaft's other end counterclockwise

10 Install pump rotor assembly in the pump casing. Center impellers in the pump casing recesses. Lock impellers, sleeves and nuts in place with screws

11 Cut pump-casing gasket so gasket extends beyond the stuffing-box face. Bolt down casing head, then cut off the gasket flush with face at stuffing box

12 With pump closed, gasket flush with casing to prevent leakage between upper and lower halves, assemble back-up flange and bolt the gland evenly

There are many types of mechanical seals—stationary, rotary, balanced, unbalanced, bellows, diaphragm. These and others on the market are covered in detail in POWER's 24-page report on Mechanical Seals.

While mechanical seals are relatively new in industry, most of us use them without knowing it. The water pump in your car and the one in our hot-water heating system at home are the best examples. You don't know you have them because they don't leak and so do not have to be repacked or adjusted.

One disadvantage of a mechanical seal for a large industrial unit is that it must be installed over the shaft's end. That means breaking the coupling and removing the motor. If a seal is installed for the first time, as in these photos, then the pump impeller must be removed as shown. This is best done during the usual annual overhaul, when pump is opened up anyway. Then this job can be done correctly.

An advantage is that these seals don't usually leak visibly. What leakage there is evaporates. But when they do start leaking, it's ordinarily the first sign of trouble. For pumping abrasive fluids, two seals are often used on a shaft, with a clean fluid circulated between them at a slightly higher pressure than the abrasive fluid. This keeps the harmful solids out of the sealing faces.

Seals are now built for shafts revolving at over 40,000 rpm. Fluids sealed include crude oil, gasoline, acids and all kinds of corrosive chemicals. Pump shown handles oil, kerosene and gasoline. Leakage here can mean a fire or asphyxiation in closed spaces. For pumps lifting liquids, leakage through packing on suction side often draws in air, reduces vacuum and increases pumping time unless a sealing fluid is used in lantern ring. With mechanical seals, stripping time is reduced because there is no air leakage.

Courtesy, Sealol Corp, Providence, R. I.

Sealing hot water with mechanical seals?

Answers to ten basic questions cover critical importance of operating temperature, effect of total dissolved solids present and the special role of water treatment chemicals

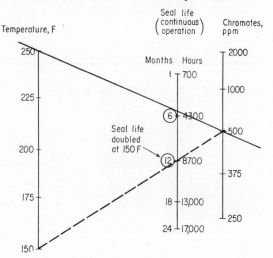

Data above based on following factors:
a–Up to 2-in. diameter shaft
b–Up to 100 psi
c–Up to 3600 rpm
Factors of assembly, mechanical malfunction, or foreign material in system are not included

Q What are the biggest problems when sealing hot boiler feedwater with mechanical seals?

A Hot boiler feedwater presents many unique sealing problems not encountered with other hot liquids. These problems have a definite relation to (1) water temperature, (2) chemical water treatment used, and (3) chemical properties of the water itself, which vary considerably depending upon the source of supply.

Q Which problem causes the greatest headache?

A Temperature is the biggest headache. Reason is that water at a temperature above its boiling point (212 F) is prevented from changing to steam at that temperature when under a higher pressure, as in a pump. In the area between the faces of any seal. a pressure drop (pressure gradient) takes place. Then as pressure drops, the water is allowed to form steam. And the steam opens the seal faces and thus allows leakage.

Q If the boiling point of water is 212 F, why must the seal faces be kept at 160 F, or less, to prevent leakage and ultimate failure of the seal?

A Experience dictates that a maximum temperature of 160 F at the seal faces be set because of the frictional heat generated at these contact faces. For example, if maximum water temperature is 190 F, and if 25 F of friction heat is generated at the faces, the resulting 215 F will generate steam. Of course frictional heat may be much higher than 25 F, depending on the installation.

1 Reducing seal face temperature from 250 to 150 F doubles seal life. Chromate content of water is also an important factor

Q Where does lubricity (viscosity) of water come into the picture?

A From the preceding question, we can see that the higher the water temperature at the seal faces, the poorer is the water's lubrication value. Mechanical seals must always have a pressure gradient film of liquid between their sealing faces. Water is not generally considered a good lubricant. The viscosity of water, at 70 F for example, is about one centipoise. And water gradually reduces in viscosity as the temperature rises. At 212 F, its viscosity is reduced to only 0.28 centipoise. At this low rating the water has lost almost all lubricating qualities and the seal faces will run dry in direct contact (loss of lubricating film gradient). Under these conditions the seal actually operates in a vapor, or steam phase. From this it can be seen that the lower the temperature of the water in the stuffing box, the longer the seal life.

Q If the temperature of the sealing faces must be maintained below the boiling point of water, how can hot water above the boiling point be sealed?

A Above 160 F the temperature deterioration is accelerated until the boiling point of water is reached at the contact mating faces. This takes place regardless of the pressure above vapor pressure in the system. The answer to longer seal life is to cool the sealing faces so they remain below 160 F. Four methods are shown right.

try these stuffing box cooling hookups, then carefully control chemical treatment

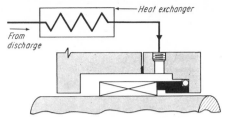

2 Hot water from pump discharge is cooled to the required temperature in heat exchanger before entering the stuffing box

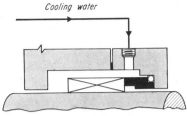

3 Cooling water at higher than the pump's gland pressure is forced into the stuffing box from outside source, but must not fail

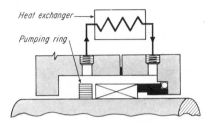

4 Pumping ring on pump shaft in stuffing box sends hot water from pump through heat exchanger before it cools the seal faces

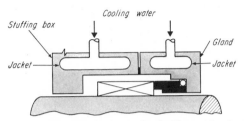

5 Jackets in the stuffing box and in the gland keep sealing faces cool without cooling water mixing with water inside the pump

Q What is the second biggest sealing problem?

A The physical properties of the water itself cause many sealing problems. For one thing, formation of abrasive mineral is increased as temperature increases. Here is another good reason why water temperature must be cooled to a safe maximum. Aim of cooling is to prevent formation of minerals such as lime, calcium and black iron oxide particles at the seal faces. Black iron oxide is not only abrasive, but forms a hard black deposit over the seal parts and shaft, thus hindering seal performance.

Q What role does chemical treatment play?

A Chemical treatment is fed into the water to prevent formation of scale, pitting, etc, inside the system. While these chemicals are necessary for this service, they are often detrimental to seal life. Specifically, use of chromate treating compounds can be tough on seal faces. The nomograph, left, indicates expected seal life under conditions indicated. Obvious answer is to keep the chromate, or any chemical, at a level consistent with good treatment practice, yet not high enough to cause excessive wear of the seal faces.

Q Since most boiler water contains particles of iron oxide, how are these particles prevented from reaching the faces of the pump's mechanical seal?

A The stuffing box of a boiler feed pump is a natural trap for the collection of iron oxide, as well as for other abrasives. These particles are extremely harmful because some of them are so small that they migrate across the seal faces in the lubricating film. Others find their way into clearances of sliding parts and under the gaskets.

Installing a magnetic separator into the cooling water will remove most of the iron oxide particles. The separator contains Alnico (permanent) magnets. Water enters the separator through one of the taps (direction of flow is unimportant), impinges against a baffle, flows over a second baffle and leaves the separator through the outlet.

Q Since stellite is so extremely hard, won't a stellite sealing ring, running against carbon graphite, prevent excessive wear when sealing hot water?

A For lower pressure systems, stellite has been used. But if the pump should lose suction, even momentarily, stellite, when used for sealing water, will heat check and the seal will probably fail completely.

Q What are four common methods of keeping the wearing faces of mechanical seals below 160 F?

A There are many schemes; above are four popular ones. First method is to bleed discharge water from the pump through a small heat exchanger and run this cooled water into the flushing tap on the gland (Fig. 2).

Another way is to run cooling water from an outside source into the stuffing box (Fig. 3).

Third method (Fig. 4) circulates water with a pumping ring attached to the shaft in the stuffing box. Hot water circulates through a heat exchanger and returns into the packing gland to cool the sealing rings.

Some hot water pumps and boiler-feed pumps have water jackets cast into the stuffing box (Fig. 5). Here packing gland also has a cooling water jacket. Because stuffing box is a dead end for water forced past the neck bushing into the stuffing box, jackets do a good job.

Treat Your Compressor Seals Right

▶ Shaft seals on refrigeration compressors prevent leakage of (1) gas and oil out of the crankcase and (2) air into it. Seals on modern high-speed compressors are often the spring-loaded type, Fig. 1. Another popular design, Fig. 2, uses tempered-steel diaphragms. In both, friction surface is much less than in a stuffing-box-type seal. In Fig. 3, unit has a bellows element.

Coil Spring. In this type, Fig. 1, the entire seal assembly rotates with the compressor shaft. Contact between the seal-face surface and its mating stationary seal seat forms the running seal joint. Both surfaces are highly finished so they are smooth and flat.

The seal face loosely encircles the shaft and is held tight against the stationary seat by the coil spring. Any gas or liquid pressure in the crankcase also acts on seal face, pressing it to the seat. But the spring furnishes main force.

Seal-face back contacts front of the resilient friction ring. This ring is closely fitted to shaft and prevents leakage at this point. Expansion of the ring is restricted by the friction-ring band, keeping seal tight at all times.

Diaphragms, Fig. 2, give a positive balance between internal crankcase pressure and external atmospheric pressure. Case-hardened steel collars retain the diaphragms, which are immersed in oil. Total friction surface is said to be less than 5% of the conventional stuffing-box-type seal.

Bellows. Seal nose, Fig. 3, carries a narrow ring with an accurately finished surface that is made to bear against a hardened-steel collar on the rotating shaft. Collar sealing surface is also highly polished. Flange mounts the seal assembly and the bellows provides a flexible and gas-tight connection between flange and seal nose. Spring maintains a firm contact between seal nose and shaft collar.

Leaks. Shaft seals seldom develop leaks if the compressor is run right. In good operation one of the most important items is the right amount of lube oil in the compressor crankcase. Keep the oil clean at all times and be sure it is the grade recommended by compressor manufacturer.

High-speed Freon compressors squeak or squeal when shaft seal is dry or scored. Stop the compressor as soon as you hear noises like this that do not come from drive belts or other parts of the compressor. Check crankcase oil level immediately. A damaged seal can lead to expensive repairs.

Once a week, during normal running, stop compressor and check seal for oil leaks. If much oil is found on seal exterior, test for refrigerant leaks, using a halide torch.

Seals like that in Fig. 1 may develop a slow leak when compressor is first started. Long storage or lack of oil between sealing surfaces is usually the cause. On most jobs, crankcase oil reaches the seal immediately after unit starts, establishes a tight seal again.

If leak doesn't stop within a short time after starting the compressor, shut the unit down and open seal for inspection. Handle its parts carefully because seal faces must be perfectly smooth to be effective. Deep scratches or grooves usually mean that seal must be replaced or reconditioned.

Choosing Seal. Life of a compressor seal depends on the basic design, proper finish of sealing surfaces and right installation. Seal must be suitable for shaft speed because there may be many variations in a unit suitable for 400 rpm and one built for 1200 rpm. So be most careful when replacing a seal to see that it is right for the speed, pressure, temperature, fluid conditions on your job.

Maintenance. Your first rule in caring for seals is cleanliness, just as it is for ball and roller bearings. Don't unwrap a new seal until you have your work area spotless. Store seals only in a clean, dry place. Once the seal is installed, check it every week for liquid and gas leaks. Seals require little attention other than this; they'll run for years without trouble if properly lubricated.

---

1 **SHAFT SEAL** fitted with coil spring to maintain pressure between rubbing, stationary faces. Friction ring, closely fitted to shaft, revolves with it

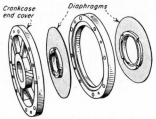

2 **BALANCED** packless shaft seal's two steel diaphragms run in oil

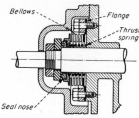

3 **BELLOWS** in shaft seal provides a flexible and gas-tight connection

13
Power
transmission

You <u>can</u> stop

Watch belt tension for longer belt life _____

RELAXED BELT has straight side, as shown by ruler. Rubber tries to bulge in pulley groove but sheave sides keep it straight

BULGED BELT is bent as when going around sheave. Bulge gives belt the added friction it needs to grip sheave, keep from slipping

Here's how you can diagnose belt failures _____

Oil deterioration

Oil saturation ruined the rubber in this belt. For complete protection against oil install splash guards. Where oil cannot be avoided replace your belts with ones that are resistant to oil

Cover fabric ruptured

Cover fabric, left, was ruptured by prying belt over sheave during installation. Wood or metal falling into groove will do same thing. Avoid by moving motor up until belt fits into groove

Slip burn

Because this belt was too loose it didn't move; friction against sheave burned rubber. When belt finally did grab sheave, it snapped. So check tension. If OK, check drive for overloading

Snub break

Here's what happens when belt gets crack-the-whip action. Cover wear shows that belt was slipping badly, while clean break indicates a sudden snap. Keep belts taut — check tension

Abrasion wear

Sidewalls of this belt were worn away from foreign material and rust in the sheave until belt dropped to bottom of groove. Install dust guards. Belt must be kept taut in dusty places

Base cracking

This belt has been running against severe back-bend idlers, has been stored improperly or has been operating under high temperatures. Avoid ambients over 150 F and check storage

these V-belt headaches

Follow these points for trouble-free operation

First rule for long belt life is to maintain right belt tension. Loose belts slip, causing belt and sheave wear. Snapping action of loose belt adds sudden stress, often breaks belt

To test for tension press down firmly on each belt as shown in the photo at left. If there is proper tension you can depress the belt an amount equal to the thickness of the belt for each four feet of center-on-center distance. This is an important fact to remember

Keep sheave aligned or you'll have excessive belt and sheave wear. Unparallel shafts are common, cause belts to work harder on one side. In such a case belts wear faster and you have to replace entire set. Misalignment in sheave only is indicated by cover, sheave wear

Sheave grooves must be smooth. Dust, oil and other foreign matter cause pitting and rust. "Dished out" sidewalls ruin belts fast. A shiny pulley bottom shows that either sheave, belt or both are badly worn. Big headache with badly worn groove is that it causes belt to ride lower than others on the same sheave. Result: "differential driving," meaning that the belts riding high in the grooves are traveling faster than the low-riding belts. Here one belt on the sheave works against the other

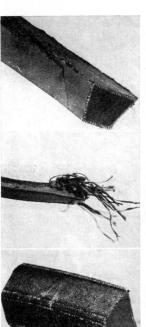

Ply separation

Too small sheave caused the split along pitch line on this belt. Remember: every belt size has a minimum sheave diameter. Always check the drive design and install right one

Rupture

Many things can cause this condition. Rocks or tools falling into sheave grooves will rupture cords in the grommet or plies. Belts loose enough to twist in the groove can rupture cords

Worn belt sides

When the sides of your belt show wear, it may be from normal causes. But look for misalignment. Grit or dirt will also cause abnormal wear. Replace belt before the wear is excessive

Install belts right

MOVE DRIVER UNIT forward so V-belt can be slipped into the groove easily, as shown

DON'T FORCE BELT into sheave with screwdriver or wedge. Fabric, cords rupture

Answers to Your V-Belt

Here are 15 frequently asked questions.

What is a V-belt drive?

Drive that uses sheaves (really grooved pulleys) and one or more V-belts. Where the speed ratios are large (3 to 1 or above), it is often possible to have the larger sheave flat-faced instead of grooved. The smaller sheave, however, must always be grooved.

What is a multiple V-belt drive?

A drive that has two or more parallel belts working together on same drive.

Should a V-belt touch bottom of sheave groove?

No. Wedging-contact between *sides* of belt and sheave transmits power. If belt touches bottom of groove, even slightly, less wedging action results.

Why not standardize on one cross-sectional size, ganging in parallel for multiple drives?

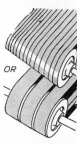

This could be done. But it would mean adopting as a standard one small belt that would be practical for smallest drives in industry. Then when you got to the big fellows, the drive would become impractical because of large number of belts needed. One belt of large cross section transmits 10 to 20 times more hp, taking only several times the space of one small belt.

What are standard cross-sectional sizes for the industry?

There are five sizes as shown. They are labeled *A, B, C, D, E.* Each is progressively larger and stronger. Belt widths along the tops' sides are as follows: A, $\frac{1}{2}$ in.; B, 21/32 in.; C, $\frac{7}{8}$ in.; D, $1\frac{1}{4}$ in.; E, $1\frac{1}{2}$ in. Sheaves are available that will take one or more belt sizes.

Why are pitch measurements of belts, sheaves used rather than regular outside measurements?

First let's define terms. *Pitch line* of a belt is a line about halfway between belt's top and bottom. Theoretically, it

is in the neutral axis between tension and compression sides of belt. *Pitch length* of belt is lengthwise measurement taken along pitch line. *Pitch diameter* of sheave is diameter of circle whose circumference coincides with belt's pitch line.

Pitch measurements are used for greater accuracy. Not all V-belts and sheaves have the same difference between pitch and outside dimensions.

What are practical speed limits?

Anywhere from a low of 1000 fpm to a high of 5000 fpm depending on diameter of driving sheave. Here's why: Hp transmitting capacity drops off fast at less than 1000 fpm because of the high torque needed for any given hp. As belt speeds go over 5000 fpm, centrifugal force stretches belt. This throws it away from grooved sheave, thus reducing wedging action.

What are minimum recommended pitch diameters of sheaves?

The standard endless V-belt is designed to bend in a normal easy arc as it turns a sheave. Naturally, the belt arc corresponds to the sheave pitch diameter. If the belt is used with a sheave of too small diameter it will distort, cutting its expected life. Here are recommended maximum pitch diameters of sheaves for standard V-belt cross sections:

Belt section	Min pitch dia, in.
A	3.6
B	5.4
C	9.0
D	13.0
E	21.6

As with the recommended minimum and maximum belt speeds, future improvements in construction may make it possible to change these limits.

What is the *optimum* pitch diameter of a sheave?

Most favorable pitch diameter possible. That is, not less than minimum recommended diameter yet large enough to produce highest belt speed within the 5000 fpm maximum. Belt has a greater hp capacity at higher speeds.

Questions

Texrope Drive Section, Allis-Chalmers Mfg Co

What are *stepdown* and *stepup* drives?

Stepdown drive is used where driven shaft runs slower than driver shaft. Therefore, smaller sheave is used on prime mover, larger sheave on driven shaft. Most multiple V-belt drives are stepdown.

Stepup drive has its driven shaft running faster than its driver. In this case the larger sheave is used on the motor or prime mover, and the smaller sheave on the driven shaft. Watch out for excessive belt speeds!

What's service factor, where does it fit in belt picture?

Motor hp x SF = drive rating

An adequate drive must be able to handle normal hp loads, plus reasonable temporary overloads. Type of service and characteristics of the prime mover and driven machine must be considered in determining how much overload capacity to allow. V-belt manufacturers' catalogs list the service factors for various driven machines and prime movers or electric motors. To determine the hp of V-belt drive, multiply the rated hp of driving unit (nameplate rating) by service factor.

Why must arc-of-contact be considered when you are engineering a drive?

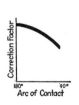

Hp ratings for different size V-belts are based on 180-deg arc-of-contact between belt and sheave. If one sheave in any drive is smaller than the other, it has less than 180-deg contact. Sheave will have less gripping area, hence can transmit less power. Use the arc-of-contact correction formulas in your manufacturers' catalogs to calculate the effective hp transmitting capacity in belt. Each belt, of a multiple V-belt drive, will have this same percentage of relative horsepower effectiveness.

What is the correct center distance for a V-belt drive?

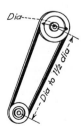

Ideal lies between 1 and 1½ times the diameter of large sheave. Actually, centers may be as short or as long as conditions dictate. But remember, the V-belt drive is inherently a short-center drive. You'll get best results when centers are as short as reasonably possible. The short belts cost less and require less maintenance.

What are *stock sheaves*?

Originally sheaves used in V-belt drives were made to order. As use of V-belts became common it was noted that certain size sheaves were used more often than others. With passing of time these more popular-size sheaves were handled by jobbers as stock items.

Today stock sizes of sheaves are fairly well standardized among all manufacturers. Most drives up to 150-hp capacity can be handled with stock sheaves. That 150-hp figure holds for getting proper sheave diameter and right number of grooves. In exceptional cases and for heavier hp requirements, sheaves are made to order. Wherever possible, use stock sheaves rather than "specials."

Why do stock sheaves have interchangeable bushings?

Practice of using interchangeable bushings in sheaves originated as a convenience to sheave manufacturers and dealers. To have each size sheave available in every popular bore size meant carrying a large number of sheaves in stock. But through use of interchangeable bushings, any particular sheave can be furnished with bore size required. This means carrying fewer sheaves with a choice of different bore-size bushings for each standard sheave.

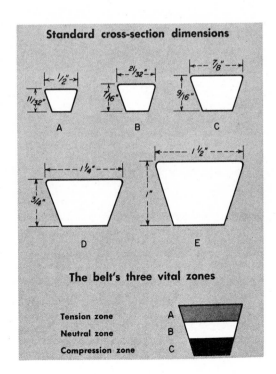

Standard cross-section dimensions

The belt's three vital zones

Tension zone   A
Neutral zone   B
Compression zone   C

Handy Aspirin for

SLIPS AND SQUEALS?———————— STRETCHES?————

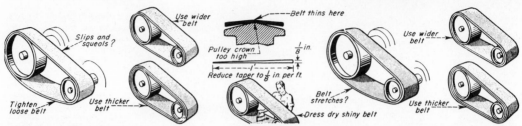

TIGHTEN loose belt, use wider or thicker belt. Keep crown taper to ⅛ in. per ft or belt thins in center and slips

USE WIDER or thicker belt. Excessive stretch usually means that belt used is not strong enough for pulling the load

RUNS OFF PULLEY?————————————

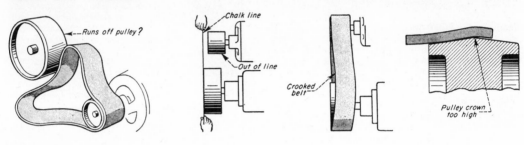

ALIGN PULLEYS with chalk line if they are out of line. Do this by holding line against rim of both pulleys as shown.

On short centers, use a straightedge. Check belt's **side** for straightness. If straight, the pulley crown is probably high

WHIPS AND FLAPS?————————————

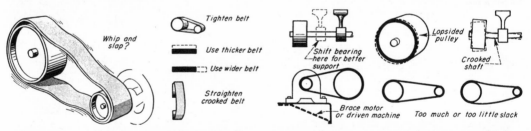

TIGHTEN BELT, use thicker or wider belt, or straighten belt. If none of these, support shaft bearing right or brace motor

or driven machine. Look for lopsided pulley, crooked shaft or too much or too little slack. This is the usual trouble

CRACKED OUTSIDE PLY?———— CRACKED INSIDE PLY?————

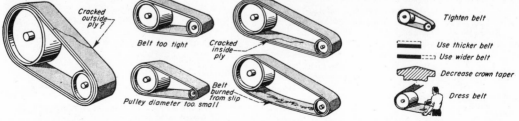

LOOSEN tight belt by moving pulleys closer together or by splicing belt. Or use larger pulley to increase radius

TIGHTEN belt, use thicker or wider belt, decrease crown taper or dress belt. One of these will stop inside ply cracks

Leather Belt Headaches

RUNS CROOKED?

CROOKED RUNNING is hard on belts. Never force belts over pulley or over high pulley crown. That stretches it badly on one side. Take up evenly on both sides of clamp when joining the ends. Align the pulleys and then keep them so

RUNS TO ONE SIDE OF DRIVEN PULLEY?

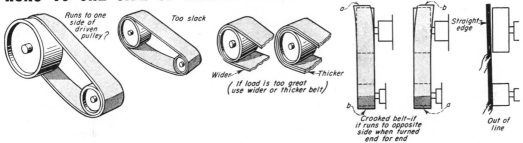

TAKE SLACK out of belt. Or you may need a wider or a thicker belt. Quick way to check if belt is crooked is to turn it around. If it runs on opposite side of pulley, that's your trouble. And here again, check the pulleys for alignment

WEAVES BACK AND FORTH ACROSS PULLEY?

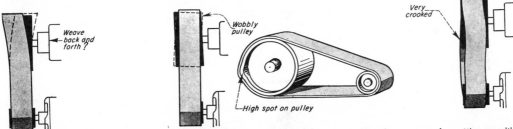

STOP WOBBLE by making pulley bearing on shaft snug fit. Or true up pulley in lathe, and remove high spot that causes belt to climb over it. There's no excuse for putting up with crooked belts, yet they cause weave that makes belt wear

PEELING GRAIN?

PREVENT belt slipping and squealing, clean belt with a commercial solvent, stop chemical fumes, scrape off loose grain, use right belt dressing and keep oil off belt.

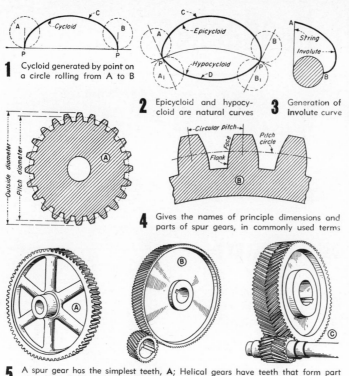

1 Cycloid generated by point on a circle rolling from A to B

2 Epicycloid and hypocycloid are natural curves

3 Generation of involute curve

4 Gives the names of principle dimensions and parts of spur gears, in commonly used terms

5 A spur gear has the simplest teeth, **A**; Helical gears have teeth that form part of a helix, **B**; A herringbone gear is a combination of two helical gears, **C**

Know Your Gears!

If you don't, here is your opportunity to get wised up on the simple principles on which they are designed, the more common types and how they differ in construction and arrangement

▶ GEARS are among the oldest power-transmission devices used by man, dating back thousands of years. The early designs were made of wood and very crude affairs compared to modern gears built to precisions of one ten-thousandth of an inch. Modern gears are built in many forms and from materials ranging from silk to heat-treated alloy steels.

Gear Teeth. To mesh properly, gear teeth must be formed according to certain natural curves, such as cycloid, epicycloid, hypocycloid, involute and modifications of these. A cycloid is a curve generated by a point on a circle rolling along a plane. For example, in Fig. 1, if a circle in position *A* makes a complete revolution to position *B*, point *P* on the circle will follow path *C*.

If a circle is rolled along the outside of a circle, Fig. 2, the curve *C* followed by point *P* is called an epicycloid. A hypocycloid *D* is described when the small circle is rolled along the inside of larger one, Fig. 2.

An involute is a curve followed by the end of a string as it is wound onto a cylinder. For example, in Fig. 3, if the string is kept taut and wound onto the cylinder by swinging its end to the right, the end will follow a curve *AB*, which is an involute.

Gear Dimensions. Gear dimensions are expressed in a number of terms. First, outside diameter is the over-all diameter measured to the ends of the teeth. Pitch diameter is the diameter of the pitch circle, Fig. 4A, and is the diameter used in making the teeth calculations.

Circular pitch of a gear is the distance from the center of one tooth to the center of an adjacent tooth measured on the pitch circle, Fig. 4B. Diametrical pitch is the number of

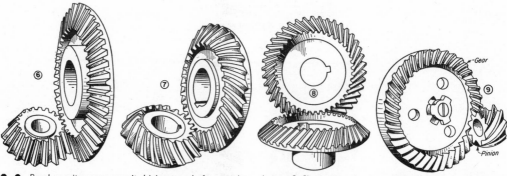

6-9 Bevel or miter gears permit driving two shafts at right angles. **7** Spiral bevel gears are an improved bevel design. **8** Skew gears have straight teeth skewed at about the same angle as a spiral-bevel design. **9** Hypoid gears

teeth per inch of pitch-circle diameter, usually a whole number.

Spur Gears. A spur gear is the simplest type, having straight teeth cut parallel to its shaft, Fig. 5A. In this type not more than two teeth are in mesh at any one time and load is transferred from one tooth to another very quickly. Because of this action, spur gears are limited to comparatively slow speeds and are inclined to be noisy in operation. They are used, however, very extensively on many applications where a reliable slow-speed drive is required.

Helical Gears. Helical gears, Fig. 5B, have teeth that form part of a helix. In this type several teeth are in mesh at the same time. The teeth mesh at one end and gradually roll together and break contact at the other end. Because of this action helical gears are generally quieter than spur gears. On the other hand, an end thrust is produced by one gear tending to push the other axially.

Herringbone Gears. End thrust on the bearings of helical gears can be avoided by combining two gears, one with a right-hand lead and the other with a left-hand lead, Fig. 5C. This type is known as a herringbone gear.

Properly designed and cut herringbone gears are excellent, but must be kept in perfect alignment. Any slight endwise movement of one gear by the other throws them out of proper mesh in a way that unloads one side and overloads the other.

This difficulty may be overcome by mounting the gears in bearings that will hold them in rigid alignment, and by connecting them to the power source and load with flexible couplings that prevent external thrust being transmitted to the gears.

Another method of keeping the gears in proper mesh is to permit either the pinion or the gear to float so that one may follow the other. The V-shaped teeth then mate up correctly and equalize the load. Herringbone gears should not be operated in a vertical position, because any wear in the bearings tends to throw the teeth out of proper mesh.

Bevel Gears. Bevel or miter gears, Fig. 6, permit driving two shafts at right angles to each other. In this drive the axes of the two shafts must be in the same plane. The gear teeth are straight and placed so that their center lines form a cone with its apex on a projection of the shaft axis.

Because of the difference in the inner and outer periphery of the gears the teeth are thicker at their outer than at their inner ends. In their action the gears operate like spur gears in that they never have more than two teeth in contact at any one time.

To get a bevel gear approaching the performance of a helical design, the spiral bevel, Fig. 7, was developed. It is evident from the figure that the teeth are curved and do not point toward the center of the shaft. Bevel gears have teeth that are sections of a cone; in Fig. 7, the teeth are part of helices around the surface of a cone.

Skew-bevel gears, Fig. 8, have straight teeth, but skewed at about the same angle as the spiral-bevel design. A hypoid gear, Fig. 9, is a form of bevel design that has the axis of the pinion below or above the gear axis.

Worm Gears. Where large speed reductions are required in a single step, or a right-angle drive is desired, worm gears are frequently used, Fig. 10. The worm always connects to the power source and the gear-wheel shaft to the load. With a single thread on the

worm the gear wheel moves one tooth for each revolution of the worm. A single-thread worm meshing a 50-tooth gear therefore gives a 50-to-1 speed reduction. A worm and gear inherently gives a high-speed reduction.

To obtain lower speed reduction from a worm-and-gear, multiple-threaded worms are used. Fig. 10 shows a worm with 10 threads meshing into a 32-tooth gear wheel to give a speed reduction of 3.2 to 1. If we continue the process of increasing the number of threads on the worm until they become equal to the teeth in the gear we get what is known as a spiral gear, Fig. 11, also used as a right-angle drive.

Internal Gears. What are called internal gears, Fig. 12, have been used quite extensively. In this design the main gear has its teeth cut on the inside of the rim instead of the outside. These gears have the advantage of being more compact, run smoother, and their teeth are stronger than in comparable external spur gears. Both gears also operate in the same direction, where external gears run in opposite directions.

Planetary Arrangement. Helical or spur gears may be assembled in a planetary arrangement, as in Fig. 13. The center or sun gear is keyed to the motor shaft and driven by it. Supported on bearings in a cage that connects to the output shaft, the three planet gears mesh into the fixed ring gear.

If the sun gear turns in a clockwise direction, the planet gear will turn counter clockwise and rotate around the ring gear clockwise. That is, both the input and output-shafts turn in the same direction. Another feature of this type of gear unit is that the input and output shafts have their axes in the same straight line.

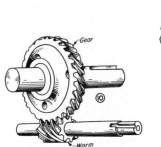

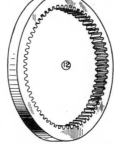
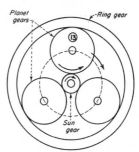

10-13 Worm and gear. **11** Spiral gears are used where right-angle drives are required. **12** On internal gears, the main gear has teeth cut inside the rim. **13** Outline of spur or helical gears assembled in planetary arrangement

1 Single-strand finished-steel-roller chain runs on cut-tooth sprockets

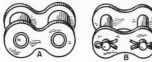

2 Steel-roller links. A: roller or inside link. B: pin or outside link of a chain

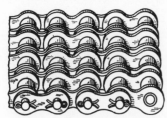

3 Multiple - strand finished - steel roller chain is used for increased power

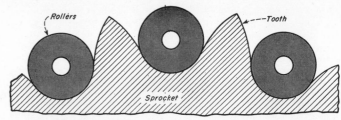

4 Steel-roller-chain sprocket profile approved by the American Gear Manufacturers' Association, takes care of increased chain pitch due to wear of chain links

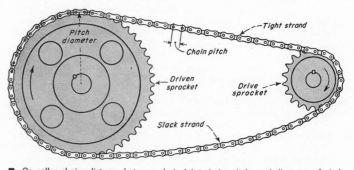

5 On roller chain, distance between chain joints is its pitch, and diameter of circle through the chain joints in mesh with sprocket is pitch diameter of sprocket

Put Chain Drives to Work

Here's the low-down on finished steel roller-chain drives—types available, number of teeth used in sprockets, center-distance between sprockets, how to select chain drive, how drive position affects chain operation and a lot of other practical know-how

▶ POWER-TRANSMISSION CHAIN is applied to practically all kinds of power jobs. Its great strength permits transmitting large power with small cross section, and it can run in temperatures that would quickly destroy belts or ropes. It is developed in many forms, but here we will consider only finished-steel-roller designs.

Roller Chain. Such a chain drive is simple in construction, consisting of two or more toothed wheels, "sprockets," mounted on parallel shafts and connected by an endless chain. Center distance between two adjacent joints, Fig. 1 and 5, is called the chain's pitch. Diameter of a circle drawn through center of chain joints in mesh with the sprocket, Fig. 5, is the pitch diameter of the sprocket.

Finished-steel-roller chain on cut-tooth steel sprockets is a highly developed type that can run at high speeds and be built to transmit large power. It consists of roller links connected by pins and side links, Fig. 1. A roller or inside link, Fig. 2, is an assembly of two rollers and two bushings, whose ends are pressed into holes in the two side links. A pin or outside link has two pins,

whose ends fit tightly into two pin link plates.

Number of Strands. Roller chains are built in single- and multiple-strand designs, Fig. 1 and 3. Multiple-strand allows using a shorter pitch chain without sacrificing high power capacity. Because the shorter pitch allows more teeth for a given sprocket diameter, along with a lighter driving, it gives quieter and smoother operation. Multiple-strand chains may be run as fast as a single strand of equal pitch. Theoretically the power that can be transmitted by a multiple-strand chain is equal to the capacity of a single-strand times the number of strands. Actually, this figure is reduced about 10%.

Extreme care used in manufacturing some roller chains makes them suitable for speeds up to 4000 ft per min. But

unless the chain is enclosed and lubricated, don't exceed 1400 ft per min.

Sprocket Standards. To get full advantage of modern roller chain run it on sprockets having accurately machined teeth, the shape of which has been approved by the American Gear Manufacturers Assn. and other organizations. This shape, Fig. 4, is made of circular arcs designed to allow for the increase in pitch due to natural wear. It provides maximum efficiency during sprocket life.

Sprocket Teeth. Even though as few as five or six teeth have been used on roller-chain sprockets, and as many as 250, experience and research show that (on the basis of speed and efficiency), the number of teeth should be kept well

*This article is based on information supplied by the Power Transmission Council

14
Bearings,
lubricating oil

ABC's of Oil

WHY GROOVE IS NEEDED

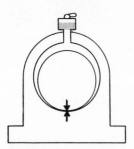

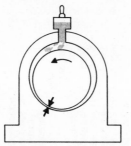

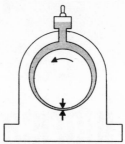

AT REST, with oil cut off, oil leaks out of bearing. Weight of shaft squeezes oil film until shaft and bearing come into direct contact as shown

DRY SHAFT starts climbing up on bearing's side when shaft starts turning. Always make sure oil is flowing to bearing before turning shaft

PUMP-ACTION of shaft starts as speed increases. Then oil is forced between shaft and bearing. Oil separates shaft so it slides to bottom

HYDRAULIC PRESSURE creates oil wedge when shaft turns rapidly. This pressure is so great that oil wedge forces shaft up against opposite side

WHERE TO GROOVE

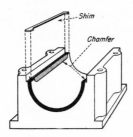

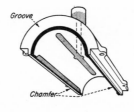

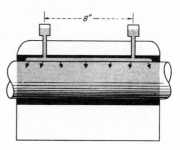

SPLIT BEARINGS must be chamfered at edges. That holds true for both halves. Shims must be cut back to keep from scraping oil off shaft and should almost touch shaft near ends, to keep oil in

HEAVY-DUTY split bearing (especially if bearing is long or uses heavy-bodied oil) needs extra attention. Besides chamfering split edges, longitudinal groove at top assures even oil supply all around

LONG BEARINGS must have more than one source of oil supply to feed long distributing groove at top. For every 8-in. length of bearing, add another oil supply or bearing will starve for lack of oil

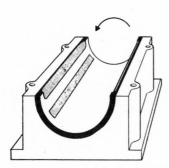

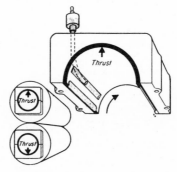

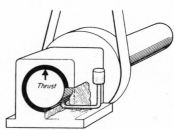

HEAVY-DUTY bearings turning at slow speeds need an extra oil groove in lower half just before pressure area. This gives heavy oil supply just before pressure area, where support is most needed

PRESS-ROLL bearings have thrust **up** on top bearing and **down** on bottom bearing. Oil distributing groove must be placed on side before pressure area. Then there's good oil-wedge from hydraulic action

BELT PULL fixes oil-groove location. Oil-distribution groove is on bottom so pressure area on top is supplied with even oil wedge, not broken up with grooves. Watch this bearing for leakage at top

Grooving Bearings

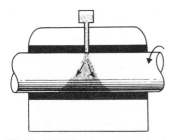

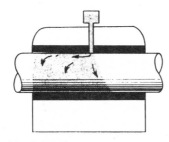

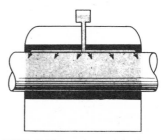

NO OIL GROOVE in bearing is bad. It causes oil to be carried around shaft directly under oil-supply opening. Then because shaft gets no oil near ends, bearing runs hot and finally burns

OIL FOLLOWS least resistance. If bearing has no oil groove, but is out of line as shown, oil flows along clearance between left side of shaft and bearing. Right side runs dry because oil doesn't spread

RIGHT WAY is to have oil-distributing groove shown here along top of bearing. Oil from supply flows into groove to within a half inch or so from bearing's end. As shaft turns, oil is carried down

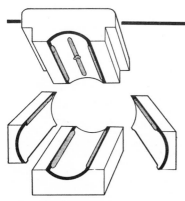

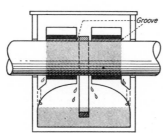

FOUR-PART BEARINGS must have distributing groove in top. All quarter-box edges must be chamfered to keep from scraping oil off shaft with sharp edges. Cut shims to fit along chamfered edges

RING-OILED one-piece bearing needs groove at top only. Ring carries oil from sump to this distribution groove. Hydraulic action of oil carried by shaft forces oil-wedge between shaft and bearing

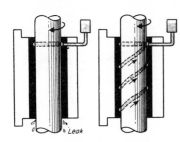

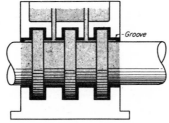

VERTICAL BEARINGS need groove around top to feed oil down by gravity. If leakage is bad, spiraled groove against shaft rotation helps shaft force oil up. Thrust bearing at bottom takes load

THRUST-BEARING needs oil fed to shaft and not to collars at top. Centrifugal force throws oil away from shaft so that it oils the collars' thrust sides. Oil fed to collars' top would not feed shaft

DON'T . . . .

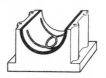

. . . CIRCLE-GROOVE pressure area of any bearing. Oil escapes through groove into low-pressure area—then won't support shaft

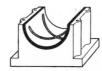

. . . CROSS-GROOVE pressure area of any bearing. Grooves cut down bearing area besides helping oil film to leave the area of wear

. . . CHAMFER split-bearing to end of bearing. Oil leaks before it can be forced under shaft by hydraulic action of oil wedge

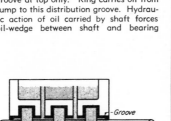

HOT BEARINGS—

WHY BEARINGS GET HOT

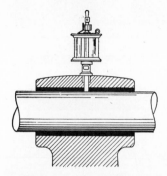
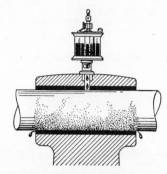
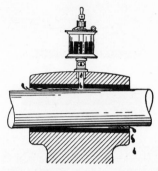

1 **NO OIL** or not enough oil is common cause of hot bearings. Check lubricators and bearing at set periods to be safe

2 **DIRTY OIL** causes undue friction and bearing soon heats. Keep oil supply clean and stored away from dusty air

3 **OUT-OF-LINE** shafting and bearings concentrate load on small area. Then bearing runs hot. Line up the bearings

FIRST STEP TO COOL

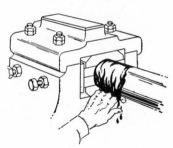

1 **FLOOD WITH OIL** by screwing up on cup adjusting nut. Or unscrew oil cup and pour oil into the bearing

2 **REMOVE WICKS** if wick fed. Then pour oil into box to flood bearing. Idea is to have oil wash through bearing

3 **FEEL SHAFT** near bearing with back of hand for heat. If temperature goes up after flooding, try to find the **cause**

EMERGENCY STEPS TO COOL

1 **DON'T GET EXCITED,** that won't help. If engine bearing is sizzling hot, throw the load off engine quickly

2 **GET ASSISTANT** to throttle engine to slow speed, **but don't stop.** If engine is stopped bearing may seize the pin

3 **REMOVE CENTER-OILER** (if this type). Flood bearing with oil can while engine turns slowly, until bearing **cools**

Causes & Cures

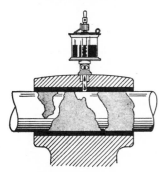

4 WRONG-GRADE OIL for bearing won't support shaft. Oil film collapses —shaft and bearing metals make contact

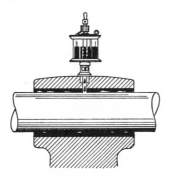

5 UNEVEN BEARING or journal surface has same effect as out-of-line shaft. Reduced bearing area causes heating

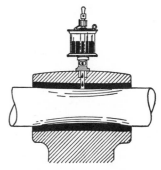

6 TIGHTLY KEYED bearings allow no space for oil between bearing and shaft. Loosen bearing bolts until shutdown

4 EXAMINE OIL on back of fingers like this. If oil is gritty and dirty keep flooding bearing to wash out grit

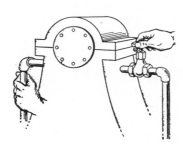

5 INCREASE WATER-FLOW if water-cooled bearing. Feel discharge line to make sure water is flowing and not plugged

6 POUR OIL on shaft if temperature keeps climbing. This will help carry away some heat so the bearing can cool

4 NEVER POUR COLD WATER on bearing or shaft. Bearing will contract and seize shaft. Shaft may even crack.

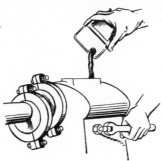

5 SLACKEN SETSCREWS or bearing cap nuts and flood bearing with cylinder oil. Heavy oil stands heat much better

6 SPEED ENGINE SLOWLY when bearing cools and keep checking temperature. If okay, throw on load.

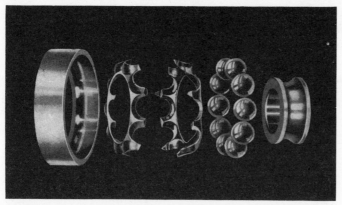

1 Here are the basic parts of a ball bearing. Going from left to right we have the outer ring or race, retainers to position the balls and then the inner race

4 This is most popular single-row ball bearing; carries a fair thrust load

Pointers

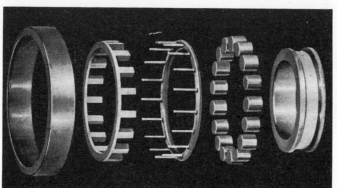

2 Working surfaces of a tapered-roller bearing's cup, cone and rollers are tapered for true rolling action. This type has high thrust capacity in one direction

Every operating man needs a little dope from time to time on anti-friction bearings. Here's a concentrated dose* telling of popular types with a dash of maintenance info

▶ STARTING FROM SCRATCH, let's define an *anti-friction* bearing. It's simply a group name given to ball and roller bearings. The fundamental idea behind them is to change sliding to rolling friction. This is handled by simply putting a ball or roller between the stationary and rotating parts of the bearing.

Advantages. Low friction, high reliability, little maintenance and close clearances that remain throughout the bearing's life are the chief advantages of anti-friction bearings. Keep in mind that there is but little difference between the "running friction" of anti-friction and plain bearings when maintained properly. Remember that the starting friction of anti-friction bearings is much lower than that of plain bearings. Why? Well, picture the shaft in a plain bearing when it is at rest. The lubricant is squeezed out between the point of contact at the load zone. Now as soon as the shaft starts to rev up, you have metal-to-metal contact between shaft and bearing. It's at this time that you get greatest wear.

Likewise a ball or roller bearing will

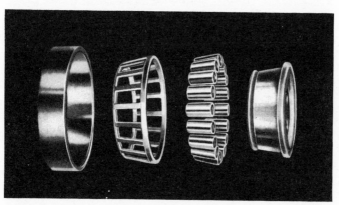

3 Straight-roller bearings are suitable for loads at right angles to the shaft (called radial loads). Needle and hollow-spiral roller bearings are in same boat

382

5 Filling-notch bearing is suitable for high speeds

6 Double-row type takes big radial and thrust loads

7 If shaft is not in line, use self-aligning bearing

8 Felt-seal type holds the lube in and keeps dirt out

on Anti-Friction Bearings

show but little wear during many years of service if properly selected, installed and maintained. Remember that bearing housings can be sealed snug to hold in oil or grease. This explains why anti-friction bearings don't need relubricating very often.

Bearing Parts. Most all anti-friction bearings you will come across have four main parts: an outer ring, inner ring, balls or rollers, and the retainer or cage, Fig. 1, 2 and 3.

The retainer or cage has the simple job of spacing the rolling elements. If a retainer were not used, the balls or rollers would not stay equally spaced as they enter and leave the point of maximum bearing load. Then again, the retainer reduces any friction that would be present if the balls or rollers were allowed to touch one another. Some types, however, mainly needle bearings, are designed to be used without a retainer for certain applications.

Type Bearings. While there are many types of bearings on the market, we'll just talk about those most generally found in the small plant. The most popular single-row bearing is shown in Fig. 4. It's sometimes called the deep-groove, non-filling slot, no-notch, or continuous-raceway bearing. This bearing will carry high thrust loads in either direction since the grooves are deep.

The four general classifications of roller bearings are tapered, square, journal and needle rollers. Each is available in designs suited to specific applications. Probably the most common tapered-roller design is the single-row bearing shown in Fig. 9. This particular design is suitable for heavy

radial loads, heavy thrust loads in one direction, and for use at moderate speed. When you come across heavy thrust loads in either direction you will probably find the double-cup tapered-roller bearing used, Fig. 10.

A worthwhile point to remember in connection with straight-roller bearings, Fig. 11, is careful ordering of replacements. To make sure identification is correct, get the following information: (1) manufacturer's name (2) bearing number if possible (3) bore (4) outside diameter, width, number and diameter of rolls.

Maintenance. No, we're not going to go ahead and describe the features of all bearings shown on this page. They're to let you know the popular types. Your job as an operating engineer is to keep them in top-notch shape. In a nutshell, this means keeping them clean and well lubricated. Follow the advice of the manufacturer regarding

lubrication. He'll specify the kind of grease or oil to use, how often to renew it and how much to use.

While talking about lubrication, remember that more than one operating engineer has caused more harm than good by over-lubricating. This can easily happen if you are applying grease with a hand gun. Avoid this trouble by opening the plug located on the bottom of most bearing housings. Keep this out for a short period after starting up machine; this gives any excess grease a chance to drain off. When you're sure that all excess grease has been removed, replace plug.

As a parting shot, remember: Dirt is a bearing's worst enemy. So work with clean tools in a clean place.

When wiping bearings, use only clean lint-free rags. Then, before putting new bearing in place, wipe housing to prevent any dirt present from slipping into the clean bearing.

9 Tapered-roller style usually mounted in sets to take heavy thrust

10 The double-cup tapered job handles thrust in either direction

11 Here's straight-roller bearing without inner race; runs direct on shaft

Your Centrifuges: How

The Disk-Type Bowl --

▶ MIXTURE of oil, water and dirt is separated by centrifugal force into three layers in oil purifier bowl. Dirt, the heaviest, is thrown against bowl's rim. Water, the next layer, discharges continuously. Oil, being the lightest, collects and is discharged nearer center.

Disk-type centrifuge has larger diameter. Liquid is thrown about same distance as depth of liquid layer. In gravity settling, shallow vessels work faster and better than deep ones. Reason is that more time is needed if particles to be separated have greater distance to travel to bottom. Same is true in centrifugal bowl.

For best results under these conditions, and with lower centrifugal force because of larger diameter, this bowl has series of disks to separate liquid into thin layers. That creates shallow settling distances between disks.

Centrifugal purifier makes use of force thousands of times stronger than gravity. If separator is designed right, centrifuge does work that gravity separation alone can't—and much faster.

Disk-type bowl travels up to 7200 rpm, is designed to take advantage of well-known principle of thin-strata distribution. Oil is fed into the bowl through a strainer cup in top, and passes down through hollow shaft to bow's bottom. From there it rises into the bowl.

Oil is thrown out toward rim. Heavy impurities hug rim. Water is thrown out against dirt, travels up and out as shown. Clean oil floats up to the top. Liquid is divided into thin layers of 0.050-in. space between disks. Disks increase speed and quality of purification. Oil passes between disks, up through holes in disk, to outlet near center. Balance between columns of oil and water is attained by using discharge ring of right size. Light oil takes larger opening in the ring than heavy oil—usually.

1 Apply brake until bowl stops. Remove cover. Lock bowl, loosen bowl ring

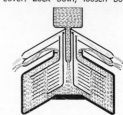

MORE DISKS cause solids to be forced against them sooner, doing better job

The Tubular-Type Bowl --

This is the second type of liquid separator. A tubular bowl is hung from a flexible spindle, which is driven by belt and pulleys from an electric motor.

Dirty oil enters bowl at bottom, is separated, and oil and water discharge from the top into separate covers—the lighter oil into the upper cover and the heavier water into the lower. Sediment accumulates within the bowl and is periodically removed.

The bowl is designed to pass liquid from inlet to outlet with a minimum of turbulence that might interfere with separation.

Bowl turns at 15,000 rpm, developing high centrifugal force of 13,200 gravity. Bowl length, and so liquid travel, is many times settling distance. A simple one-piece, 3-wing device within the bowl brings the liquid up to speed smoothly, and keeps it from slipping within the bowl. Sketch shows straight-through path of liquid in bowl, and how particles of sediment are thrown against bowl wall.

Stress on a centrifuge bowl is proportional to peripheral speed. The smaller the bowl diameter, the greater the centrifugal force it can safely develop. The centrifugal force increases because there is greater change in the liquid's direction in the smaller bowl in a given time. There is a "sharper corner to turn."

Impurities commonly found in lubricating and fuel oil are water, sludge, dirt, carbon, bits of metal, and acid. All but acid and sludge, which are oil soluble, can be removed with either type of centrifuge.

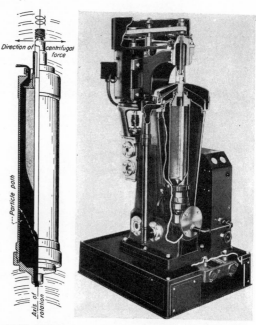

LIQUID TRAVEL is up through long bowl. Cutaway section shows the working part of centrifuge and fins in tubular bowl

To Know and Clean Them

How to Clean...

2 Unscrew bowl cover. Remove shaft and disks as unit, put in pail of solvent

3 Scrape bowl shell with scoop to loosen and remove dirt collected against rim

4 Replace disks (wet with solvent) in bowl. Much dirt is already washed off

5 Reassemble bowl. Start machine without admitting oil. Run "dry" 5 min

6 Centrifugal force throws dirt off disks. Seal with water—turn on oil

How to Clean...

1 Remove the cover to get at the collected solids inside

2 Screw extension sleeve onto the bowl's bottom

3 Lift bowl with extension sleeve above block—drop

4 Remove sleeve, pull out caked sediment as here

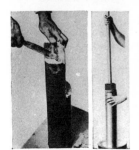

5 Remove solids from fins, brush dirt from the bowl

6 Roll parchment-paper liner loosely into long bowl

7 Push in fins (3-wing assembly) as is shown here

8 Seat fins against bowl. Screw on—tighten bottom

How important is lube life?

Many plant engineers fail to recognize the growing importance of lubricant life—and not simply as a means to trim their lube bills. Here's the reaction we got to questions asked of field experts...

Q. Why is lubricant life more important today?
A. Mainly because of labor saving says the user, and the oil supplier agrees. Even though automatic lubrication has come a long way, there are plenty of bearings and gear housings that need manual lubrication. Stretching the time between lube applications here can save the user real money.

Q. What about lubricant cost—is it important?
A. Yes, but not as important as the labor saving. The oilers' payroll in at least one plant is two-thirds higher than the lubricant bill, and that doesn't include storage and handling costs.

Q. How do you go about boosting lube life?
A. First, get some basic data from the oil suppliers and equipment manufacturers on lubricant life under good operating conditions. Dig back into your plant records and check lube schedules for your actual working conditions. Cross check lubrication frequency in the different areas of your plant to see if bearings and gears working under similar circumstances are getting the same treatment. Talk the problem over with engineers from other plants with the same problem.

Tie all this information together and judge for yourself what your frequency rates should be. Stick in a safety factor and don't double the rates right off the bat; take it in steps.

The oil supplier goes along with this approach, stressing the caution needed. A program designed to extend application periods on a large scale in a plant needs close supervision since the trouble caused by oversights will cost more than the saving in lubes and manpower.

Q. What are some of pitfalls in extending lube life?
A. Parts failure. Oil and grease breakdown caused by overwork leads to serious repair bills.

For example, in a recirculation system, oxidation brings on deposits which can clog lines or filters, resulting in reduced or cut off oil supply. Recirculation oil systems have a twofold purpose—to lubricate and cool. Cutting down lube flow might not hurt lube action, but reduced cooling effect can have fast, serious results.

Leaving oil in too long increases viscosity and boosts heating of the parts. Also, chemical reaction rates double every 18 F rise, and experimenting with a high-temperature application is risky.

Q. How can you check on lube performance?
A. If it's an oil, you can have it analyzed. This is costly, but about the only sure way to tell if oil is breaking down. Visual inspection of many oils won't tell you much until it's too late.

With greases, just looking at them will usually let you know whether they need to be replaced. They turn dark, get stiff and may show a separation of the oil and soap.

Q. What do you check before changing frequency?
A. If high temperatures, moisture or leaks are causing frequent lube applications, there isn't much point in doing anything until these troubles are cleared up. Often the choice of a different grease may give you longer life automatically, since it will resist the effects of moisture or heat.

Q. Where has extending lube life paid off?
A. The user we talked with got the best lube frequency change with motor bearings, pillow blocks and gear housings. User greases his motors and certain pillow blocks once a year and that's it.

He uses the expendable theory on some pillow blocks figuring that it's cheaper to replace them now and then rather than to lose product because of over greasing.

Oil supplier believes rolling-contact bearings and gears are easier places to extend lube life than sleeve bearings. He says having an all-loss or recirculating varieties to do the job.

Q. Are there any new greases or oils on the market that promise longer life?
A. Both user and oil supplier say "yes." User has switched over to EP grease entirely for all but the very special applications, allowing them to use one grease now where it previously took six different varieties to do the job before.

These greases cost more than ordinary grease but they cut oiler mistakes since there's only one grease to use. They have a tougher film, giving an added operating safety factor. As far as giving longer life, the oil supplier says the EP's can do two jobs: (1) Tackle the tough assignment and extend frequency rates from, say, once a day to once every three days. (2) Replace many types and do a good job over a wide range of applications. Either way the user can cut both his lubrication bills and the time needed for the oilers to do their job.

STEEL DRUMS carry by far the major share of industrial lubricants entering the average plant. Careless handling or storage can completely ruin them and their contents

Here Are Storage and Handling...

POINTERS ON LUBES*

▶ MODERN OILS AND GREASES for lubricating services are refined and manufactured to meet rigid standards. Many oils carry special additives to equip them with precise properties for particular duties. Careless or even haphazard handling and storage at your own plant can destroy the very values you may have paid a premium to get.

Problems. For example, exposure to abnormal temperatures or contamination from moisture or dirt—two strong possibilities—can bring on leakage or produce deterioration that destroys a lubricant's suitability. We'll see why later. Of course, careless handling of the containers themselves may result in dents and split seams that take their toll in direct loss. Here are some simple rules for storage and handling.

Abnormal Temperature. Both oils and greases experience considerable change under temperature variations. Their effectiveness as lubricants is reduced by hot locations. These make heavy-bodied oils thinner, tend to evaporate the required moisture content of (1) soluble cutting oils, leaving behind a jelly-like mass (2) certain greases where oil separates out from the soap.

Cold areas, on the other hand, make lubricating oils and greases more difficult to handle. The oils flow more slowly, and the greases become stiffer. Oils containing fatty materials congeal them out of solution under low-temperature conditions. In addition, soluble oils, cold enough to freeze their moisture content, prove useless.

Somewhat along the lines of temperature effect is spoilage in long storage. Soluble cutting oils suffer most from this cause through a loss of moisture. But preventing this loss is more a problem of stock control than of storage or handling.

Contamination. An improperly protected lubricant supply is almost certain to experience dust and dirt contamination. No plant man worthy of the name needs to have pointed out to him the dangers of dirt in lubricating oils and greases.

Moisture contamination is a major problem, too. Oils are more likely to be contaminated than greases. Some oils are designed to function as a lubricating emulsion in the presence of moisture, or even small amounts of water, but water contamination during storage can render them useless for later service. This holds true for most steam-cylinder oils and similar ones meant for duties like lubricating the bearings of the wet ends of paper machines. Nonsoluble cutting oils and heat-treating oils suffer complete destruction if contaminated by appreciable quantites of water.

Still another form of contamination develops. This is the introduction or unwanted mixing of different oils when dispensing them within your plant. It stands to reason oils of different qualities when blended produce a third, entirely new oil. You must take certain precautions, and then exercise definite care to keep original oils effective.

(More on next page)

*Abstracted from *Technical Bulletin on Handling, Storing* and *Dispensing Industrial Lubricants*, by Socony-Vacuum Oil Co, Inc.

Draw on these easy-to-use tips based on in-plant...

STORAGE

Outdoor storage is never too good. Brand markings and labels wash off or become obliterated under weathering. Future dispensing is needlessly endangered and other lubricant contamination becomes more possible. Temperature changes from season to season can produce enough expansion and contraction stresses over a period of time to develop leaks in the container seams.

Moisture contamination is a potent threat with outdoor storage. Rain water collects readily inside the chime top, see first sketch. It is gradually sucked past the bung by the drum's breathing during alternate hot-and-cold periods.

If you must store outdoors, spread tarpaulins over the drums or erect a temporary shelter. Then if mechanical means are available, turn each drum over so the bung end is down. If you can't do this, lay the drums on their sides, and line them up so the bungs make up a horizontal line across the drums' middle.

Actually, there are four major recommendations that apply to outdoor storage: (1) Provide at least temporary shelter against the weather. (2) Place drums so they cannot breathe through the bungs. (3) Make sure bungs are made tight if you are going to move drums. (4) Before opening, dry and wipe bungs thoroughly as well as the surrounding surfaces.

An oil house is the ideal and logical place to store and dispense oils and greases of all sorts, as well as to carry out the cleaning of dispensing equipment. Size varies with the individual plant. Some plants require only a single oil house or even a part of a full-package warehouse. But no matter what its size, every plant needs an oil house of sorts.

Experience indicates the oil house should be closed to all but authorized personnel. Locate it as centrally as possible, yet in an area free from possible process contamination, as from cement dusts, chemical fumes, coke dust and similar byproduct wastes.

Storage takes many forms. Mechanical aids, in sketch below, are one. You can buy the storage racks commercially. Take steps to prevent any harm to personnel from toppling or rolling drums, barrels and containers. Further, it's smart to see that the drums are grounded electrically, second sketch, especially with flammable volatile solvents.

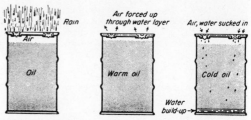

OUTDOOR STORAGE, never desirable, makes drum contents an easy prey to contamination by an outside moisture like rain

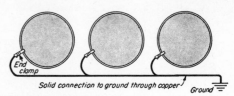

ELECTRICALLY GROUNDED connections on drums, especially if they hold volatile solvents, is a smart measure for safety

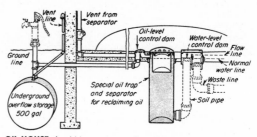

OIL HOUSE should be set up so all its facilities are permanently installed, and ample waste drains supplied as best storage bet

HANDLING

Handling the containers, when lubricants are both coming into storage and being dispensed, goes best with certain mechanical aids. They can be anything from the simple hand truck, sketch top of facing page, to chain falls and trolleys, sketch at right, or electrically operated elevators or lift trucks. Some companies, however, just roll tht drums from delivery point to storage.

By far the greatest volume of industrial lubricants arrives in drums shipped by railroad or motor truck. The load consists mostly of a single tier of drums, each standing on end. Unloading platforms, level with railway car or truck, make it easier to handle the drums and reduce possible damage risks to containers. Wooden or metal ramps serve well where there are no unloading platforms. Slide the drums endwise

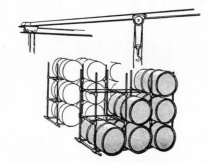

MECHANICAL AIDS like chain falls and trolleys make handling of drums into and out of storage comparatively easy

to the ground. Never drop an oil drum from a truck or railway car onto a cushion of discarded automobile tires.

Major transporting activity of oil drums should be confined to trips from unloading area to the oil house. This method involves the handling of bulk quantities only. It entails stacking and storing, plus dispensing and disposal of waste oils and empty containers.

MANUAL LABOR, particularly if helped by hand trucks, can do effective job in the average small-to-medium-sized plant

DISPENSING

Hand oilcans, large dispensing safety cans, oil wagons, grease guns, lubricating cars and sump pumps comprise the more important dispensing and transporting equipment for grease and oil.

Clean this class of equipment regularly. But don't limit your cleaning to mere wiping with solvent-soaked rags. Instead, use generous quantities of solvent for both cleaning and rinsing. Baths of safety solvent (flash above 100 F) meet the usual safety regulations where ventilation is adequate and correct covers are provided. First sketch at right shows a good arrangement with two baths, one for cleaning and one for rinsing.

Actual transfer of oil from container to dispenser may entail nothing more than a hand-operated drum pump, shown in sketch, driven into the bung of the oil drum. Such a pump is a positive-delivery type capable of supplying measured oil quantities.

You can get pumps, however, with spring-action closable returns, and in air-operated or electrically driven designs. Drum faucets also prove efficient and economical transfer devices. They come in different sizes for fast, slow oils.

Because of its consistency, grease presents some special problems in transfer to dispensing equipment. A good metal paddle does a fine job particularly for heavier greases.

But the major problem comes from dirt and other contaminants that land on the surface of open drums and exposed grease. Take every care to keep drums closed when not in use. Extra precautions against stray dirt should apply while the grease is being removed.

Air-operated pumps do a fairly good job in pulling grease out of drums without exposure to the air. But you cannot always count on them for greases harder than No. 2 or 3 consistency.

Sump Pumps. Used oil has to be periodically removed from reservoirs and sumps. The sumps can be anything from the small reservoirs for textile-spindle and machine-tool-spindle housings to large systems like hydraulic machines.

Sump drainers can be homemade. One such example is shown in second sketch, right figure. It employs a heavy steel drum, a siphon, and compressed air to draw from sump.

Often you'll find it desirable to purify certain types of lubricating oils for reuse. Extent of the purifying equipment depends entirely on how much of a saving it gives you. A very simple arrangement is pictured in the second sketch, left figure. If your wants are greater, it may pay to set up a purifying section within your oil house.

All in all, though, lube oil-and-grease handling is a relatively simple operation that gives little trouble if it is planned and controlled properly.

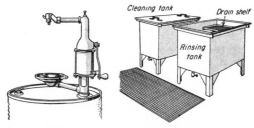

HAND-OPERATED drum pump, *left*, is a good dispenser of lubes. keep it and other dispensers clean by solvent baths, *right*

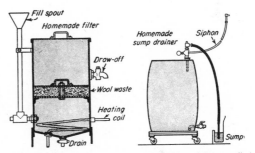

SPENT OILS can be reclaimed in simple filter, *left*, or pulled from sump by compressed air siphon, *right*, for disposal

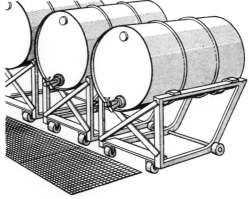

AS DISPENSING METHOD that is both quick and easy, use rocker-type racks and a set of drum faucets, as is shown here

Guess I Forgot To Oil It

You sure did, Jocko, and there goes a week's top-priority space production shot to blazes. But we chiefs can't hold a guy like Jocko responsible—not if we have all our marbles.

That's why the trend today is to put the brightest mechanic in charge of lubrication and not trust it to a grease monkey. Why? One little slip, and *zingo*, the whole plant can be knocked out—at least temporarily.

So smart engineers in larger plants see that this key spot is filled by a man who keeps on his toes about lubrication practices. Instead of monkeying around, he keeps log sheets of all oil and grease changes, watches supplies so they're not contaminated, and so on. In smaller plants, engineers assign the oil can to a well paid mechanic who at least knows how to overhaul a bearing. A guy like that makes each drop of oil strike home—and he never misses squirting that drop to the right place, at the right time.

If one bearing failure can be so expensive, what cheaper insurance can a chief buy against breakdowns and costly wear than smart, well paid help?

Monkey business and lubrication don't mix. It's either one or the other, but never both.

15
Shopwork

Yours for Better Filing

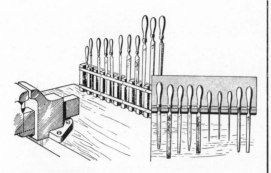

1 File racks like these get more mileage out of files. NEVER THROW FILES IN A DRAWER. You wouldn't throw your razor blades around and expect a smooth shave. Files have numerous sharp edges. Save them by handling as a delicate tool

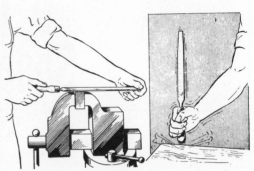

2 Filing is an art. Have work at elbow level. Take long even strokes, keeping file level. Ease up on return stroke. NEVER RUB FINGERS OVER FILE. That makes file slide. Tap file handle on bench every few strokes to knock out chips

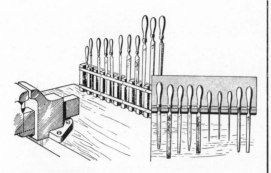

3 Lathe filing is dangerous. A good mechanic learns to file with either hand to get away from revolving chuck. Move file forward and sideways over turning piece—never hold in one place. Moving helps file clear chips and avoids ridges

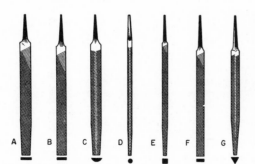

4 Draw-filing makes flat surfaces like valve faces, valve chests, smoother than "straight filing." Hold file in both hands and push, then pull across work. Use mill file for most draw-filing jobs. Clean file with file card often

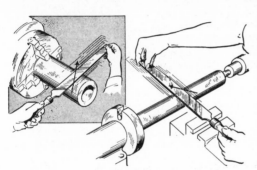

5 Machinists' files: A, FLAT, cuts all around; B, HAND, for finishing flat surfaces; C, HALF ROUND, for curved surfaces; D, ROUND for openings; E, SQUARE for keyways, slots; F, PILLAR for slots; G, THREE SQUARE for angles

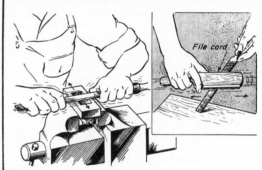

6 Single-cut and double-cut files have angles shown. Larger file cuts are designated as: rough, course, bastard, second-cut, smooth, and dead smooth. Small files are numbered, as 00, 0, 1, 2, etc. Courtesy, Nicholson File Co

HOW TO

Handle Your Hacksaw

USED PROPERLY, the hacksaw is a useful, long-lasting tool. Secure blade firmly in rigid, nonbending frame. After use, protect sharpness by oiling teeth lightly

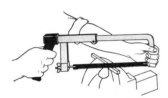

MOUNT BLADE with teeth pointing away from frame handle. Use long, smooth strokes. Lift blade slightly on return stroke. That prevents dragging teeth through material you cut, wearing teeth

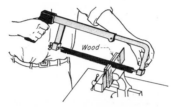

SANDWICH very thin stock between wood blocks. Then stock won't tear, and you can cut faster without harming saw blade. General rule is: Use blade of 32 teeth per in. for 1/16 in., or thinner stock

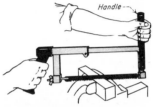

HEAVY STOCK takes greater pressure. Auxiliary handle, mounted on forward end of hacksaw frame, is useful. Slot a broom handle, drill and bolt to frame as shown. Keep this frame for heavy work

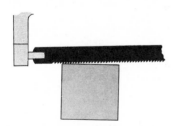

WHEN STARTING, reduce stroke angle to minimum so at least two or three teeth contact the work. Good cutting speed is 40 to 60 strokes per minute. But go slower and lighter on material that's softer

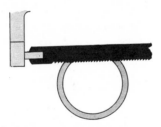

USE BLADE of 24 teeth per in. for metal of 1/16 to 1/4 in. thick, on round, as here, or on straight stock. Use blade that has enough teeth so at least two teeth always contact thin material

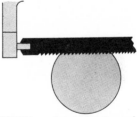

EIGHTEEN teeth per in. is good for general shop use. Stock over one in. thick, also softer materials, can take a 14-teeth-per-in. blade. Fine blades on thin stock prevent loading one tooth

TO MAKE cuts of any depth in thin material, use this special rig. Plate for the mounting blade is rigid enough to withstand pressure. Either rivet or weld plate to the long screw shown here

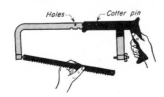

CHANGING BLADES is nuisance because frame tries to collapse. Drill holes in frame as shown. Lock frame with cotter pin for size blade usually used. If you change blade length, just move pin

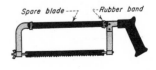

SPARE BLADES are handy to carry, fastened to frame as shown. Use rubber bands and slip enough blades under so you won't waste time running back and forth when supplies are at a distance

Grind Drills Right

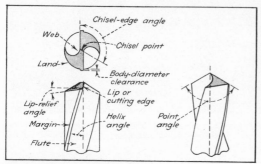

1 All operating engineers use drills. For best results, certain basic facts should be known. Lip-relief angle, chisel point, cutting-edge angle, body-diameter clearance and web of drills are important. Drills are only as good as the user

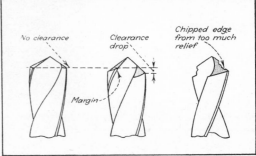

2 Lip relief is important. With no clearance, cutting lip can't cut into metal. It only rubs. By having right clearance drop, cutting edge digs into material. Use less relief at margin than at center. Edges chip from too much relief

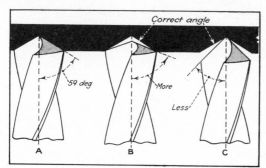

3 For normal all-around jobs, a 59 degree angle is best. Skilled machinists grind drills to about this angle without measuring. More angle makes drill "walk" and hard to center. Less angle gives long cutting edges, strains drill

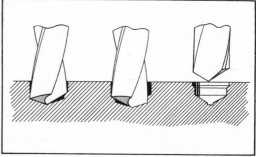

4 Grinding cutting edges right is important. If angles are different, one lip cuts more than next. If lips are off-center, they will cut larger hole than drill diameter. If off-center and off-angle, hole is larger and drill is strained

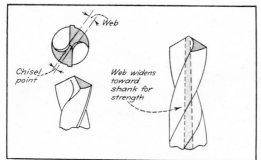

5 Chisel point of drill widens as drill is ground shorter because web widens toward shank for strength. This wider web should be thinned so drill's cutting edges can dig into metal without being held back by wide chisel-point rubbing

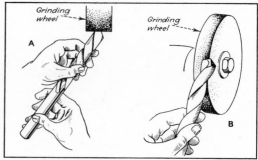

6 To grind drills right (A) learn to hold at right angle when starting to grind. Then swing drill downward with a slight twist to lessen relief angle near margin. To thin web (B) hold near wheel edge. Never grind on side of wheel, it may break

Practical Drilling Hints

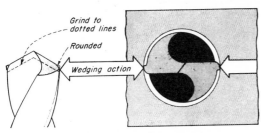

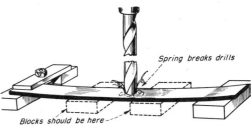

1 Always check drill's cutting edges before using. If corners are rounded, grind down to dotted lines shown. Rounded corners wedge into metal, causing friction. Then heat anneals edge. Grind away roundness, no matter how small

2 Drills break like this when work springs against it. The blocks should have been nearer drill. When the drills point cuts through metal, spring action forces metal against drill, causing greater "bite." Drill can't take it and snaps

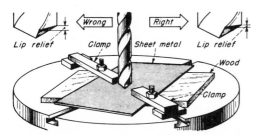

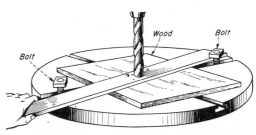

3 Drilling thin metal is extremely dangerous if not clamped down right. When drill's point cuts through, cutting edges dig in. Drill either spins metal, hurting operator, or drill breaks. Grind drills with slightest lip relief for thin metal

4 It's OK to hold work being drilled with stop bolts placed so they keep work from turning. Nuts on bolts keep work from climbing up when drill cuts through, hurting operator or breaking drill. For safety, clamping work down is better

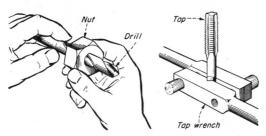

5 Quick way to find nearest drill size for tap is using new nut like this as drill gage. Largest drill just slipping through nut is right size. Of course best method is looking up size on chart. In hurry, nut takes place of chart or gage

6 Oiling drill is important. Oil above work near shank so oil feeds onto work (for slower speeds). Use oil when drilling steel, wrought iron, monel metal or malleable iron. No oil is needed on cast iron, copper or most bronze and brass

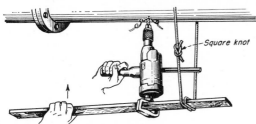

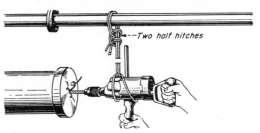

7 Rope is handy when using portable drill like this. That's why OE should know knots. Any knot that's easy to untie when tight is OK here. Pushing heavy portable drill overhead is hard work unless easily adjusted fulcrum gives leverage

8 Another way to have rope do hardest part of drilling job is by using like this. Fasten to drill and then throw ends over any convenient pipe. Two half hitches are simple knots anyone can make. They don't get tight, untie easily

LAYOUT AND DRILL FLANGE

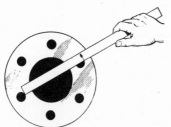

Rub chalk on surface to show scribe marks

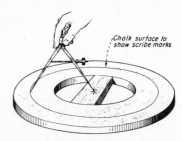

Chalk surface to show scribe marks

1 Laying out and drilling flange is important because holes must line up. This is right way to layout and drill holes. Measure old bolt circle from the edge of opposite holes like above. This is easier than trying to guess at center of holes

2 Saw off piece of wood to plug flange center and rub chalk over surface. Hermaphrodite is right tool to find center. Scribe four arcs on chalked wood 90 deg. apart like this. Then prick punch center for dividers to find the true center

3 With center punched in wood, it's easy to scribe bolt circle with dividers. Set dividers to one half of circle diameter found in Fig. 1. Chalk flange face and use sharp-pointed dividers to scribe a clear circle. Good tools save time

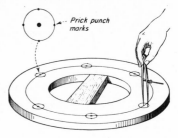

Prick punch marks

4 Circle must be divided in two exact halves. Place machinists' scale on the flange so it splits center mark. Then scribe line on flange like this. Machinists' scriber is best tool. Learn to use the right tool for each operation

5 Center line is shown dotted here. Since this is a six-hole flange, distance between holes is equal to radius of bolt circle (distance from center to bolt circle). Prick punch circle at line and scribe four arcs shown on bolt circle

6 After prick punching each arc on circle, set dividers to half of drill diameter. Then scribe six circles around flange. Next, punch four marks around drill circles. These marks are important. All machinists lay out holes this way

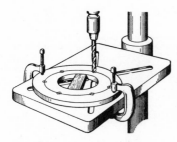

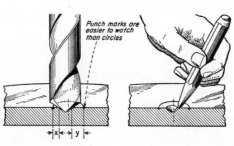

Punch marks are easier to watch than circles

7 In clamping flange to drill press, make sure drill lines up with a slot in table or place wood under flange to protect table. C-clamps, used like this, are OK. Making table and flange tight helps drill do better job and prevents accident

8 Drills seldom follow prick punch. They have tendency to "walk." Then distances **X** and **Y** aren't equal. As soon as drill point "spots" flange as shown, check for being centered. If not centered use prick punch as chisel and cut groove from spots top to bottom shown by dotted line, center sketch. Drill this groove out and check again. Size of groove depends on distance drill is off center. Be sure to bring drill back so the four center punch marks are split and distances **X** and **Y** are equal

HOW TO INSTALL DOWEL PINS

REASON FOR PINS. Keep machinery from moving by pinning it to metal base with dowel pins. Holding-down bolts are slack in holes and often loosen because machinery vibrates. Then there's lateral movement under bolt head. Dowel pins help holding-down bolts by preventing any side movement. Or if machine is removed for any reason, it's only necessary to tap pins back into place without realigning the machine.

Dowel pins can be either straight or tapered. Tapered pins hold firmly and seldom loosen if right reamer is used.

You can buy the pins in standard sizes, with reamers for all sizes. Or they can be made from round stock found around most plants. Manufactured pins have a taper of ¼ in. to the foot. They come in ¾-in. to 6-in.

lengths. Reamers for these pins are proportioned so each size 'overlaps' the next smaller size by about ½ in.

Handy use for small tapered pins is on motors and pumps connected by flexible couplings. This equipment is usually screwed to an iron base. Dowel pins are handy for pinning flat machined pieces, also for pinning cross-head knuckles to pump rods, shafts, etc.

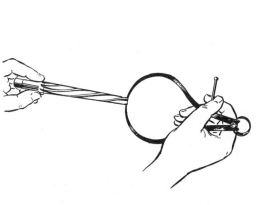

MEASURING FOR TAPER

Homemade tapered pins are easy to make but taper must be same as on reamer used in drilling hole. This taper is ¼ in. to the foot on standard pins. Measure widest diameter across reamer flutes at both ends with caliper. Turn pin same size

DRILLING HOLE

Use a drill that's same size as small end of reamer. If the hole is blind run the drill deep enough to clear pin's end

REAMING

Go easy on tapered reamer. Turn only in cutting direction —never backward. Always use oil on drill and reamer

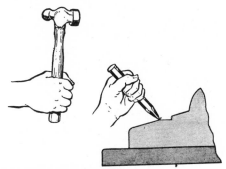

INCLINED SURFACES

On some machines best place to install dowel pin might be at tapered surface. It's easy enough to level a small area around location with a chisel or a portable grinding wheel before using a prick punch for the drill. Use at least two pins

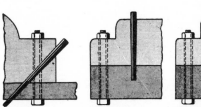

ANGLE . . . VERTICAL . . . SPECIAL

Angle holes take care of pinning needs where there's no other way . . . Whether angle or vertical, leave one or both

pin ends sticking out so it can be removed . . . Special pins have threaded end with nut. Clean hole; graphite the pin

Centering Stock for Lathe

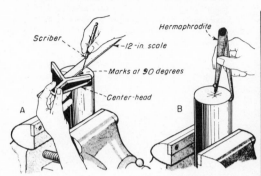

Scriber

Hermaphrodite

-12-in. scale

Marks at 90 degrees

Center-head

A   B

1 There are many ways to center round stock for turning between lathe centers. First, chalk stock ends. Use center-head and scribe marks at right angles, **A.** Hermaphrodite is handy, **B.** Scratch four marks at right angles. Punch center

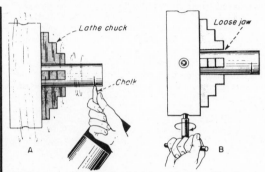

Lathe chuck

Loose jaw

Chalk

A   B

4 To center stock in lathe, chuck it and start machine. Steady hand on compound rest, **A,** and hold chalk so revolving stock barely touches. Stop lathe, **B,** loosen jaw opposite chalk mark and tighten jaw nearest mark. Repeat until stock is true

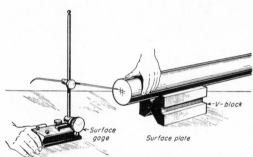

V-block

Surface gage

Surface plate

2 Surface plate, surface gage and V blocks are a third way to center round stock. Place stock to be centered on V block or blocks, depending on stock's length. Scribe lines as with hermaphrodite. Hammer deep centerpunch mark with lines

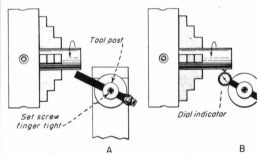

Tool post

Set screw finger tight

Dial indicator

A   B

5 Another way, **A,** to center stock is by placing tail end of tool holder against stock, with set screw tightened finger tight. Don't run lathe but turn chuck by hand, tightening or loosening jaws. The quickest way is with a dial indicator, **B**

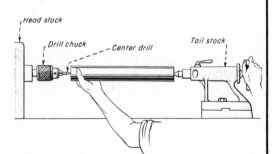

Head stock

Drill chuck — Center drill

Tail stock

3 With centers punched at both ends, place round stock against lathe center in tail stock and support other end with hand to take weight off center drill. Feed slowly against drill. Center drills are delicate. Oil drill, feed carefully

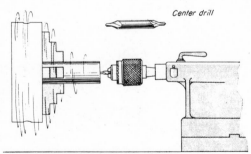

Center drill

6 In Fig. 5 **B,** set dial indicator in tool holder and move compound rest against work until dial reads zero. Turn chuck by hand and move jaws with key until reading is the same all around. When stock's true, oil drill, run it slowly into stock

Measuring Instruments

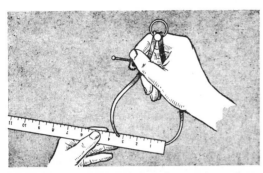

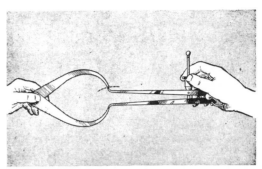

1 To fit replacement parts accurately, operating engineers must know how to use machinists' measuring instruments. An outside caliper is a precision instrument in hands of trained machinist. Set it right by hooking to end of scale like this

2 Never set inside caliper from scale, but from outside caliper like this. First set outside caliper as in Fig. 1. Then rest inside caliper on outside caliper and swing back and forth. Adjust until the contact can just barely be felt

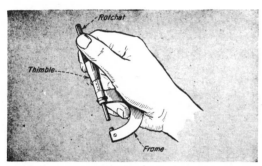

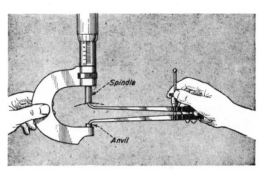

3 There is only one way to hold an outside micrometer. Rest frame in palm of hand and press with tip of ring finger. Adjust thimble or ratchet with thumb and index finger of same hand. Now other hand is free to hold piece being measured

4 Easiest and most accurate way to set inside caliper is with outside micrometer. Rest one leg on micrometer anvil and swing other leg against spindle like this. For right setting, you should feel for slightest contact with mike

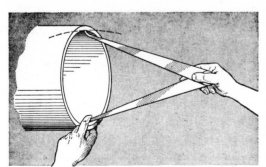

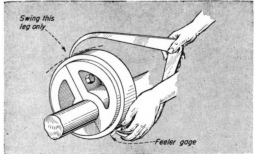

5 In measuring inside diameter of large cylinder, hold caliper's bottom leg against surface to keep it from moving. Then find exact opposite point by swinging caliper slightly sideways as well as back and forth. Set caliper by tapping leg

6 Eccentrics wear out-of-round. Wear in thousandths of inch can be found by caliper. First caliper largest diameter. Then hold feeler gage under caliper leg to find smallest diameter. Try different feeler gages until exact wear is known

Making Condenser Plugs

▶ Because operating engineers are responsible for machinery repairs, knowing how to run a lathe is important. A small repair shop in any plant saves time and money. But engineers who haven't learned the machinist trade pick up many *wrong habits* without expert advice. Then they never learn the correct way and do faulty work.

This series on machine-shop practice is intended to point out important details by starting readers on the right track. Instead of showing how to do various things on a lathe we're using an old trick we used in teaching trade school. That's showing how to do jobs that come up in most plants. A good machinist uses either hand so he can file and saw facing chuck as in Fig. 6, to keep from leaning over revolving chuck.

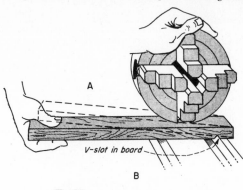

V-slot in board

A

B

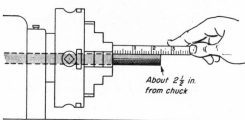

About 2½ in. from chuck

1 The facing page shows in detail how to plug condensers. Here's way to make plugs, cutting taper with compound rest. Use board to rest chuck on, **A.** Place ¾-in. brass stock through headstock with end near chuck to prevent chatter, **B**

Lathe chuck

Chalk

Chalk marks

A

B

2 To center stock in chuck, rest hand on lathe and touch chalk to revolving stock, **A.** Move stock in chuck by loosening and tightening jaws until chalk makes even ring around stock. Mark two jaws, **B**, to use for next pieces

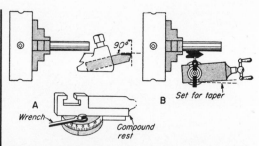

A

Wrench

B Set for taper

Compound rest

3 Loosen compound-rest nuts, set rest to 4 degrees and tighten nuts. Grind tool for brass at 90 deg so top surface has no rake, **A.** Set tool with center gage against stock after compound rest is set for taper. Oil all the lathe bearings.

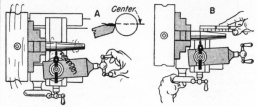

A Center

B

4 Set tool tip to center of stock, **A.** Set lathe speed for about 200 ft per minute. That's surface speed of stock, or about 1000 rpm for ¾-in. stock. Now move lathe carriage so tool tip is 1¾ in. from stock end, **B**, length of the taper

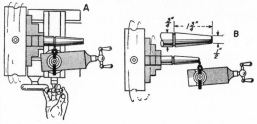

A

B

5 No oil is used on work when turning brass. With tool set for plug's length, run in crossfeed to make deep groove, **A.** Now run compound rest out by hand and knock sharp edge off plug with tool, **B.** This does neat job and prevents cut hands

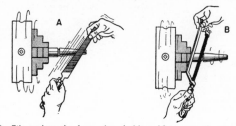

A

B

6 File tool marks from plug, holding 12-in. smooth-cut mill file as shown. Push file towards work and towards chuck lightly, **A.** Use hacksaw in groove with lathe turning same speed as before. Saw shown askew here but keep in line

Finding Condenser Leaks

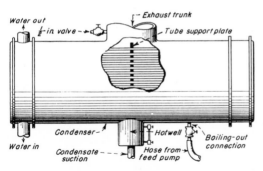

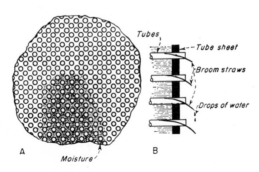

1 Finding leaks in tubes of surface condensers, oil coolers and other heat-transfer apparatus isn't easy. First fill condenser's steam side with hot water from feed pump through boiling-out connection, until it reaches exhaust-trunk vent

2 Tiny pinholes in tubes will keep tube sheets moist over large area, even with 160-F water inside condenser, evaporating moisture, **A.** Stick broom straws, **B,** into tubes suspected leaking. Water dripping from straws will show leaky tubes

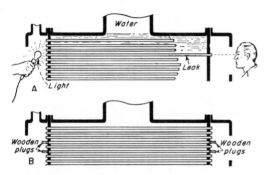

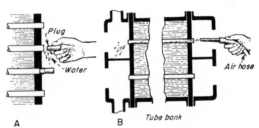

3 One way to find leaks inside tubes is to look through tubes against light at other end, **A.** Larger leaks show up easily. To be sure about smaller leaks, stick soft pine tapered plug into both ends of all tubes appearing to leak, **B**

4 After 15 or 20 minutes, remove one plug, **A,** from one end of a tube. If it leaks, water that filled tube will flow out, proving tube has hole. If compressed air is handy, another way to find leaks is with air, **B,** then look as in **3A**

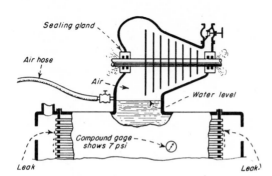

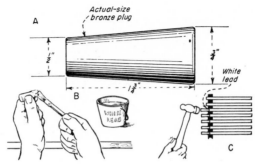

5 On large condensers connected directly to steam turbines, connect air hose to vent valve or to turbine-casing drain. Build air pressure in exhaust trunk and turbine until compound gage shows 7 psi. This pressure helps show up leaks

6 After leaks are found, plug both ends with bronze plugs of size shown. Never use wood because wood often leaks before tubes can be renewed. White lead the plugs and tap into tubes. After 10% of tubes are plugged, install new ones

Don't Shut Down for A Lost Spring
Wind One Like This

1 Because springs get lost easily when taking machinery apart, operating engineers should know how to wind them, especially when there's no time to order a new one. Use a mandril of slightly smaller diameter than inside of spring. Drill small hole through mandril for wire. Wind first few coils by turning lathe slowly. Wind expansion spring turns close together. Start compression spring same way but spread coils proper distance apart

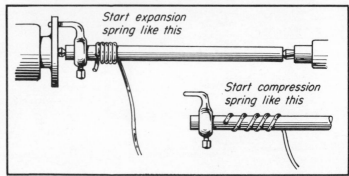

Start expansion spring like this

Start compression spring like this

2 Make clamp of hard wood or flat metal stock with jaws of soft wood or fiber. Tighten clamp enough to wind compression coils tight against mandril. Hold clamp with one hand and control the mandril speed with the other by having it on the starting button or belt shift. Make sure coil of wire is fed into clamp without binding. The safest way is to cut off approximate length of wire for spring and have assistant hold other end

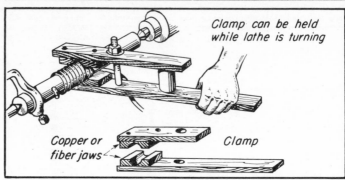

Clamp can be held while lathe is turning

Copper or fiber jaws

Clamp

3 Winding heavy compression springs like this assures a good spring. Rest clamp on offset tool holder. Replace hardened toolholder set screw with brass one and tighten lightly against wire. Then set lathe for same number of threads as turns per inch in compression spring. As lathe feeds wire along revolving mandril, oil wire to keep it from scoring. Loosen clamp and tool holder slowly when spring is wound to prevent violent recoil

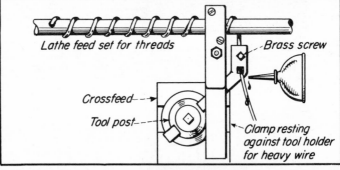

Lathe feed set for threads

Brass screw

Crossfeed

Tool post

Clamp resting against tool holder for heavy wire

4 Small springs of 1/4-in. outside diameter or less can be wound by placing fiber jaws in a parallel clamp and holding clamp by hand. The faster the mandril turns the better "life" your spring will have.

If you don't have a lathe, a drill press can be used for turning the mandril. Any good grade of spring wire stock of brass, monel or steel will do, depending on the service the spring is made for. Wear goggles when handling spring wire

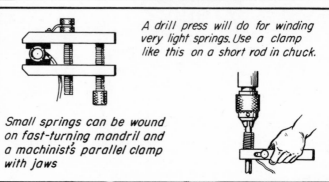

A drill press will do for winding very light springs. Use a clamp like this on a short rod in chuck.

Small springs can be wound on fast-turning mandril and a machinist's parallel clamp with jaws

Dry Ice Makes Expansion Fit

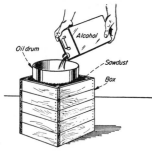

1 Use drum having at least 2-in. of space around liner. Place drum in box and insulate with sawdust. Add industrial alcohol to cover liner

2 Dry ice is sold in 10-in. cubes. Place a cube in a burlap bag and break it into small pieces by pounding with a wood mallet or a hammer

3 A metal disk and hook are handy but not essential. A disk like this one might be needed if the liner has to be tapped in with a hammer

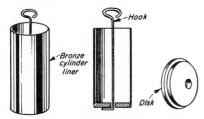

4 Place liner in the alcohol bath and pour in the dry ice. The drum should be at least 6-in. above the liner because the alcohol will "boil"

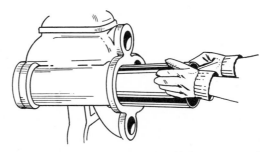

5 Dry ice and alcohol cools liner to about 80 F below zero in about 15 minutes. This 6-in. diameter liner shrinks about 0.006 of an inch. Alcohol lubricates liner and prevents frost forming on it

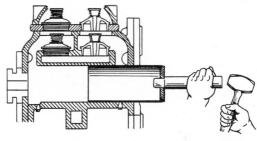

6 Work fast after removing liner from bath. Delay warms liner quickly. It might seize when part way in cylinder. If the liner won't slide in by hand, cool again and force it into cylinder like this

▶ DRY ICE is a quick practical chilling medium for making expansion fits, because its temperature is 109 F below zero. It works better with alcohol because alcohol prevents frost and acts as lubricant. Dry ice is solid carbon dioxide, (CO_2) and it changes to gas when it absorbs heat. It's sold in 10-in. cubes, weighing fifty pounds each. While 97 F below zero is possible, the liner is cooled between -50 F and -90 F. That's because it warms rapidly when taken from bath. Count on about 0.001 in. per inch shrinkage for each inch of liner diameter. This holds for most metals. A water-tight container, same shape as liner, is best. Alcohol should cover liner or parts to be shrunk and container should be higher than liner to allow for alcohol boiling. Putting bag of dry ice in liner while fitting keeps it from expanding too fast. Caution: CO_2 displaces air. *Ventilate working area.*

Shrink and Expansion Fits

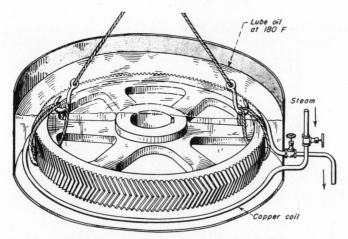

1 Gear is lowered into lube oil, heated to about 180F (do not boil) till hole expands. Steam coil or electric heater does trick. Oil should just cover gear

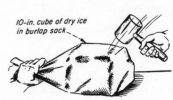

2 Break dry ice by placing cube in burlap bag and hitting with mallet

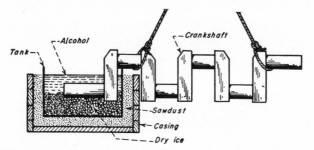

3 Fill sawdust-insulated tank with industrial alcohol to cover crankpin. Tank should be 6 in. higher than alcohol level. Keep adding dry ice to keep "boiling"

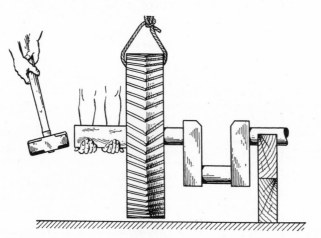

4 To fit expanded gear hub over shrunk crankshaft, have tools ready and work fast. Tap gear lightly using wooden block and move crane to help slide gear on shaft

▶ COMBINING SHRINK WITH expansion fit does tightest possible job on two pieces to be held together. This method has been used where machinery works under severe conditions.

Lubricating oil heated to about 180 F expands female part about 0.001 in. per inch of diameter. Dry ice shrinks male part about the same amount. When parts, machined to allow for diameter change, are expanded and shrunk and quickly fitted together, fit is twice as tight as "one-method" job.

Dry ice is a quick practical chilling medium for making expansion fits, because its temperature is 109 F below zero. It works better with alcohol because alcohol prevents frost and acts as a lubricant. Dry ice is solid carbon dioxide and it changes to gas when it absorbs heat. It's sold in 10-in. cubes, weighing fifty pounds each. While 97 F below zero is possible, piece is cooled between −50 F and −90 F. That's because it warms rapidly when taken from bath. Count on about 0.001-in. shrinkage for each inch of diameter. This holds for most metals. A watertight container, slightly larger than piece being cooled, is best. Alcohol should cover part to be shrunk and container should have high sides to allow for alcohol 'boiling'. Ventilate working area.

Removing Cylinder Shoulders

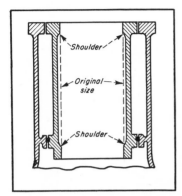

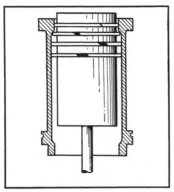

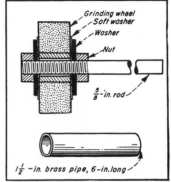

1 Internal-combustion-engine cylinders wear from pistons and rings. Dotted line shows exaggerated wear, with shoulder at top and bottom of liner. Shoulders form even when counterbored

2 Effect of liner wear on piston rings is shown by ring and clearances. New ring at top is closed but new rings at worn section are opened against walls. Excessive openings make new rings leak

3 To remove shoulders before installing new rings use grinding rig like this made by threading round stock and securing small grinding wheel. Brass pipe is for holding wheel against cylinder wall

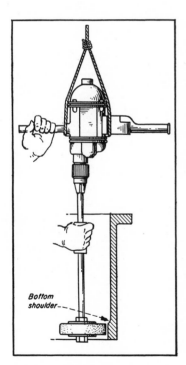

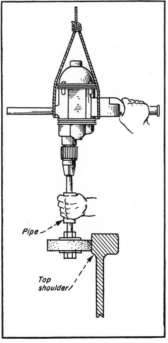

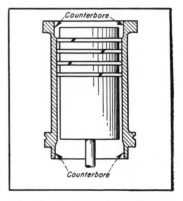

6 This is the cylinder liner after top and bottom shoulders have been ground and counterbores (shown exaggerated) ground. Larger rings can now be used with minimum gap openings. This saves fuel

▶ NEVER fit new rings into diesel-engine cylinder liner without removing shoulders at top and bottom. New rings fitted over shoulders aren't much better than old rings because they open in worn part, Fig. 2. Shoulders don't have to be removed by reboring liner; they can be ground out with home-made rig if necessary. Caution: Keep grit out of crankcase by spreading canvas.

4 Bottom shoulders can be removed by using electric drill with long rod hung above cylinder. Place oil on rod so it slides easily within brass pipe while you are holding wheel against shoulder

5 For grinding top shoulder a shorter rod can be used if there is little head room. Pull wheel against shoulder as shown. Caution: Don't grind worn cylinder wearing surface. **Always wear goggles**

Hand Boring Small Cylinders

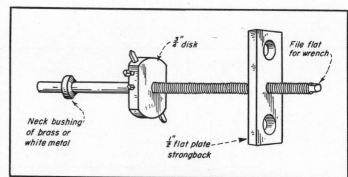

1 Small cast-iron cylinders are easy to bore with hand boring rig. Make boring bar 2½ times cylinder length of same diameter stock as piston rod. Make brass-bushing guide to fit stuffing box and rod snugly. Cut 40 threads to inch on bar and in flat disk to fit. Drill bolt holes in strongback and tool and screw holes in disk. Tap holes and file flats on bar.

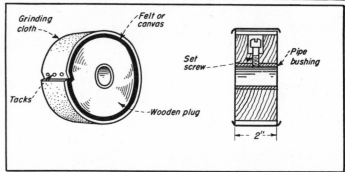

2 Replace old packing and neck bushing with new neck bushing. Hold it in place with gland. Oil rod and bolt boring rig in place after centering bar with inside caliper against counterbore. Don't start boring until bar is perfectly centered. Adjust pointed tool. If tool chatters, insert second tool on opposite side. Take finish cut with round-nosed tool. Complete each cut to end of cylinder.

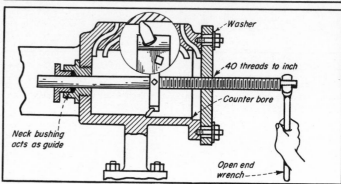

3 Make wooden plug for polishing tool marks from cylinder. Cut plug from 3-in. block in lathe with parting tool. Tack on No. "0" grinding cloth as shown. Bore plug for piece of pipe to be pressed into center. Pipe bushing can be loose fit on bar. Turn outside diameter of plug so it is snug fit when felt or canvass pad and grinding cloth are tacked to rim.

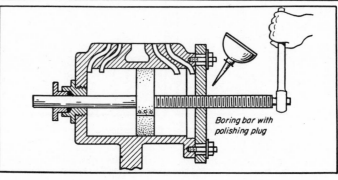

4 Remove tool disk from bar and attach polishing plug. Bolts can be left loose against strongback, as plug will center itself. If grinding cloth doesn't fit snugly enough to remove tool marks, wrap another strip of felt or canvass cloth underneath. After cylinder is polished bright, remove rig, clean and coat cylinder with beeswax. This gives cylinder a hard glazed surface for low friction.

Removing Broken Nipple

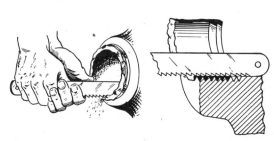

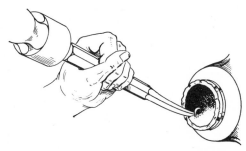

1 Neatest job is to saw broken nipple with a hack saw blade like this. Be sure to stop sawing when you reach threads

2 If saw can't be used, sharpen a diamond point chisel and split nipple all way through like this. Don't cut threads

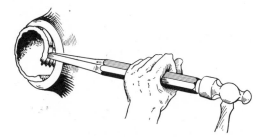

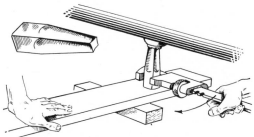

3 Split nipple is easy to collapse with chisel if it's split right. Caution: Don't knock nipple through hole to inside

4 If broken nipple is in a close place and you have no "easy-out", hollow grind hard stock and work with lever and wrench

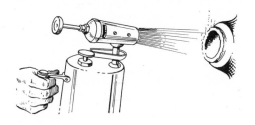

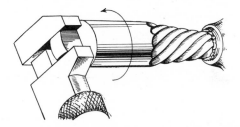

5 Don't let nipple that's rusted solid stop you. Heat with blowtorch after it's split. Then it will collapse easily

6 This "easyout" won't always work. If the nipple is badly frozen and won't budge, don't spoil threads by wedging

▶ A BROKEN NIPPLE can lead to a major shutdown and a big maintenance job. Whether it's made of cast iron, steel or brass, there is a right way and a wrong way to go about removing it. If it's a cast-iron nipple, the easiest way is sawing to the threads with a hacksaw blade. Then chip pieces out of it by collapsing near the cut. With a little patience, the nipple can be removed so there are no marks and it won't even be necessary to clean the threads with a pipe tap. The nipple can be split with a chisel, but it's easier to botch up the job. Then again hammering may loosen fittings unless they're properly braced near the nipple. If an "easyout" won't remove the nipple on the first try, don't spoil the threads by putting your beef on it and wedging this tool deeper into the nipple. Heat it with a blowtorch. Heat expansion usually breaks rust or dried thread sealing compound so that "easyout" will remove nipple. A hollow ground square file end—turned with a wrench and held in place with a lever often does the trick in tight spots. Using the *right* method for *each job* makes it quick and easy. Caution: Don't knock chips into line.

Repair Your Valves

MACHINING, RESEATING and GRINDING

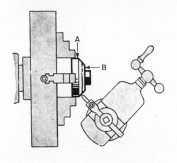

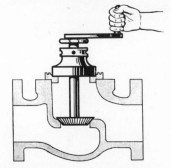

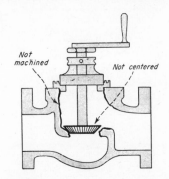

TAKE light cut on valve disk after truing up with a dial indicator at *A* and *B*. Set the compound rest to valve's angle

RESEAT valve body with reseating tool, or by truing in a lathe chuck. Remove only enough metal to clean up the pits

CAUTION! Do not use reseating tool if valve body has no machined finish. Only way then is to machine seat in lathe

NO LATHE or RESEATER? USE HAND TOOLS

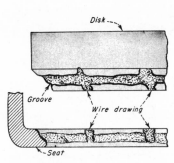

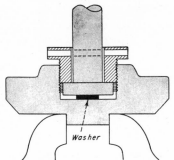

BADLY GROOVED and wire-drawn seat and disk look like this. The sealing surfaces can be restored with hand tools

REMOVE DISK from stem and place washer or small piece of gasket under stem to lock parts solidly so disk won't turn

REMOVE metal on both sides of groove with file (after giving light grind to find bearing surface), keep turning disk

TESTING

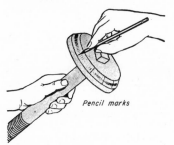

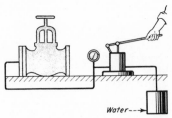

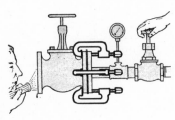

LIGHT PENCIL-MARK the disk. Place stem in bonnet, and turn disk few degrees against seat to see that all marks rub

LEAK-TEST for normal work pressure with small hydraulic pump like this, after assembling valve. If not tight, grind again

NO HYDRAULIC PUMP? Use outlet from feedwater discharge for test. C-clamps handle all size valves on the test flange

Before They Give Trouble

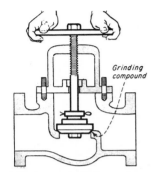 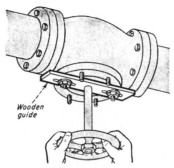 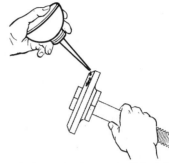

GRIND valve and disk by locking disk to shaft loosely with split pin. Use bonnet as guide. No nuts; no packing

IN-PLACE grinding of large valves. Use wooden guide if upside down. Then you won't have to support the heavy bonnet

FINISH by cleaning compound from disk and valve. "Grind" with oil, clean, and again grind with oil until seat shines

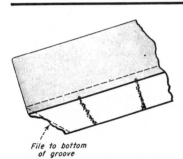 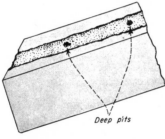 

GRIND with coarse compound each time you file around once. Don't touch ground surface with a file—don't file deep pits

FILE EVENLY all around by turning disk so you don't file flat in disk. Idea is to remove metal that would be machined

FINISH FILING carefully with fine mill file. Give grind with fine compound and grind out all pits. Then give oil rub

Don't Forget These . . .

 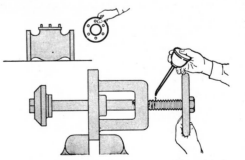

OVERHAUL bonnet assembly before finishing job. Shine up valve stem with 00 emery cloth. Replace old packing in bonnet. Replace old gasket with new. Clean threads in Stoddard's Solvent, oil. Work stem full thread length

Tips On Making

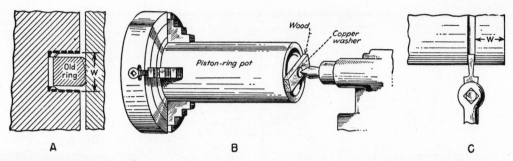

A   B   C

MANUFACTURED RINGS ARE BEST—but sometimes you have to "roll your own". Start by chucking cast-iron ring pot in lathe as at **B**. Use dead center on wood block to prevent "chattering" of long pots. Rough-turn outside until diameter equals cylinder diameter plus 12 thousandths in. per in. cylinder dia. (to allow for cut in making joint) plus ⅛ in. (to allow for final finishing). Then rough-bore inside so thickness is 1/30 cylinder diameter plus ¼ in. for finishing. Now lock carriage to lathe bed, face edge of pot with facing tool and cut off ring **C**. Width of ring **W** is determined by width of most-worn part of piston

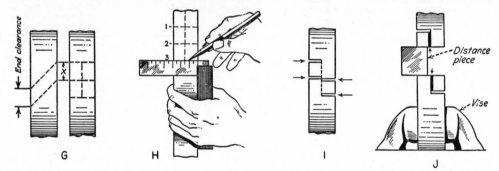

G   H   I   J

FOR ALL JOINTS the critical dimension is the end clearance **X**, sketch **G**. This allows for expansion and should be 3 thousandths in. of cylinder diameter when ring is closed. (Make cut a little larger). We allowed for this by adding to the diameter when rough-turning ring pot. To make step cut, take scriber and square and lay off as in **G** and **H**. Then make four cuts with hacksaw as shown at **I**. Note that middle cuts are made off-center. Now spread gap with a distance piece **J** to allow room for saw to finish cuts. Cut along heavy black lines indicated by arrows. Note: these two cuts are also off center.

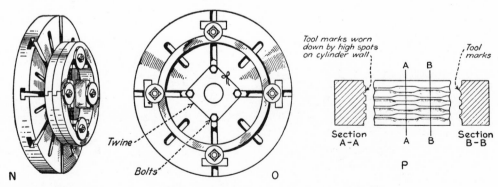

N   O   P

MAKING THE SECOND CUT. With ring bound, clamp it to face plate as in **N**. Any number of rings may be trued up at the same time. Sketch shows two. Use distance piece under washers on the inside for uniform purchase. Note step joints are directly under washer. Overhang gives washer enough pur-chase to hold ring closed. Remove wire and turn ring down to exact cylinder diameter less desired clearance. Then clamp ring from outside as at **O**. Don't let ring open while clamping. Inner bolts can't be removed, so tie them to center with string to keep out of way. Turn inside of ring until thickness is 1/30

And Fitting Piston Rings

Angle Joints

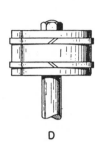

D

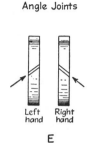

Left hand Right hand

E

Step Joints

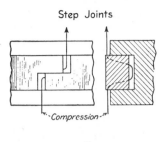

Compression

F

groove **A** as groove will have to be turned to this dimension to prevent rounding of new ring. After cutoff, finish other side of ring with grinding cloth on face plate. Now we are ready for the joint. What kind of joint? Angle or step. An angle joint is easier to make, but a step joint makes a better ring. Some

people say combine left-hand and right-hand angle joints **as** in **D** because it makes for turning action to keep cylinder round. Actually it encourages blowby. Other people claim a step **joint** can not leak, yet gas can leak around the back as at **F** if **ring** is loose. Still **step** joint is best. Let's make one to find **why**

K

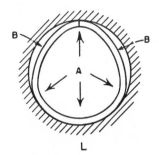

L

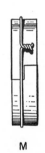

M

If cuts are made right, the joint will look like **K**. The indicated over-hang is very important. Now comes a step a lot of people overlook. The ring is not yet a true circle. Put it in the cylinder and it will leak at **B** and bear at **A**, sketch **L**. Why? Well, cut a piece out of a circle and put the ends together. The result

will have an egg shape—not a true circle. To make the **piston** ring a true circle it is necessary to make a second cut. **The** second cut can be made only with a step joint. Without **the** overhang in **K** the ring can not be held closed while **turning** it to a true circle on the lathe. Next, tie ring closed as at **M**

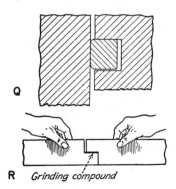

Q

R Grinding compound

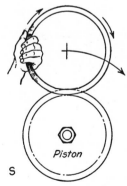

Piston

S

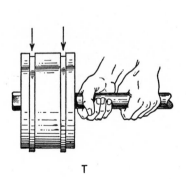

T

of cylinder diameter, more if piston groove is deep. Note: Always leave tool marks on ring as at **P**. They help ring wear to cylinder quickly and prevent blowby. Next, slightly round ring as at **Q** to let it slide over lubricant without scraping it off. Then remove overhang at step joint by grinding inside

surfaces as at **R**. Never file the outer edges. **S** and **T** show quick checks for fitting to piston. If ring is too tight, **turn** piston groove, not ring. Groove has to be turned down to **heavy** dotted line in sketch **A** anyway before new ring will fit. These tips aren't the whole story. Practice will tell you the **rest**

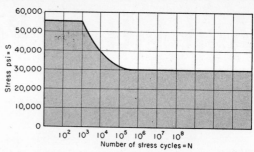

1 Typical S-N curve for a low-carbon steel shows that one application of 55,000-psi stress will cause sure failure

'Stop Holes' Don't Stop Crack

Laboratory tests show stop holes at end of crack weaken boiler plate. Drilling holes crowds stress lines around them, weakens metal. Then because nature of load on plate is such it has made a crack, holes in tiered metal may be disastrous

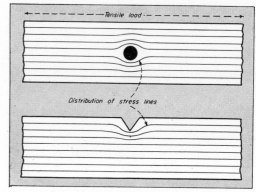

2 Stress lines crowd around hole in plate as shown above. With notch in loaded plate at bottom, lines crowd same way

3 With stop holes drilled, failure may occur at 100 cycles and 22,500-psi stress as metal around holes is weakened

▶ OLD-TIMERS REPAIRED small boiler cracks by "sewing" or "stitching." That is, they drilled a series of overlapping holes, with round soft iron or brass plugs driven in and peened over. Method was common before fusion welding.

When welding, certain types of cracks might be veed out for full thickness and electric welded. But some welders still drill holes at each end of crack to prevent it spreading. This is bad practice as we shall see.

To understand problem, let's see how a fatigue crack develops and what happens when "stop holes" are drilled, as shown in Fig. 3.

Fig. 1 is a typical "S-N" curve for low-carbon steel, plotted from fatigue test data. It shows relationship between number of stress cycles of different magnitudes that produce a fatigue failure. Curve shows that one application of 55,000-psi stress magnitude causes fail-

ure. Then, if stress magnitude is reduced to 40,000 psi, it takes about 10,000 applications before fatigue failure. In Fig. 1 this is 10^4, meaning 1 with 4 zeros. At stress magnitudes of 30,000 psi or lower, number of cycles is theoretically infinity, with no failure taking place.

Fig. 2 shows stress lines distributed through cross section of a plate having uniform thickness. Note how they crowd up at a hole or a notch. It's this crowding of stress lines that produces a unit stress in that specific location far higher than average stress throughout cross section. And, it's this crowding of stress that reduces limit of fatigue endurance. So premature failure at hole is a probability, unless surrounding area is reinforced or thickness is increased to reduce unit stress.

To structures subject to static loading, this localized stress concentration may

not be too serious. But, if loading's nature demonstrates ability to cause a fatigue crack, a hole in "tired metal" may be detrimental.

For example, instead of standing up for 10,000 cycles at 31,500 psi, failure may occur on only 100 cycles at 22,500 psi.

When repairing a crack, first eliminate fatigue cause, such as vibration. Then install an oval-shaped patch after cutting out the defective plate for at least an inch beyond each end of crack. Design the cutout and patch so longitudinal axis of cutout is not more than one-half of circumferential axis. Make patch same thickness and material as shell plate. Overlap on *inside* by at least one-half inch. Use full fillet welding for the attachment.

First have repair approved by a commissioned boiler inspector, so it doesn't conflict with local regulations.

HOW TO EXPAND BOILER TUBES

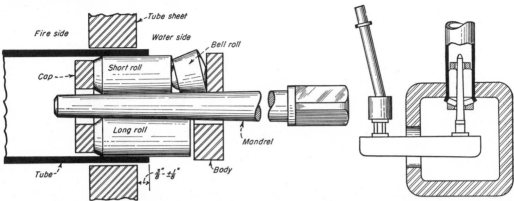

1 Boiler tube holes are slightly larger than the tubes' outside diameter. To make tube ends tight in holes, they are cold-rolled into tubesheet. This is done with an expander made with a tapered mandrel and rolls. The tube is expanded about ¼ to ½-in. past the tubesheet. This gives tube a slight bulge near tubesheet in form of a collar, which helps hold tube in place, making tighter joint

5 With handholes 90 degrees to tubes, use expander with mandrels and ratchet or gear drive. Always clean the tube ends with a fine grade abrasive cloth. Remove all burrs from the hole

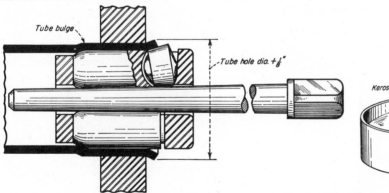

2 There are many types of expanders. This one rolls the tube and bells tube's end in one operation. It's non-adjustable and rolls are right length for tubesheet thickness. Greatest danger in rolling tubes is from over-rolling. It's better to under-roll. Then, after making hydrostatic test, leaky tubes can be re-rolled. Over-rolling enlarges tubesheet hole, thins tube and causes failure

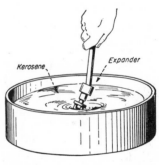

6 Use vegetable oil on expander. It's easier to clean from boiler than mineral oil. Wash expander often in kerosene to keep clean. Expand bottom tubes first so oil does not run into holes below

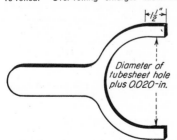

3 When rolling a number of tubes, gage is handy for measuring tube bulge. Bulge is usually about 0.020 in. larger than the hole diameter in the tubesheet

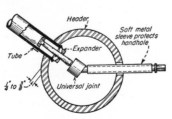

4 Extend tubes from ¼ to ⅜ in. past hole. Drive brass wedge between tube and hole at one end while rolling other end. This is for drums and headers

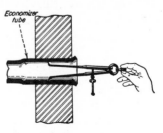

7 Measure economizer tubes from the inside. They need more expanding, or about 0.035 to 0.040 in. plus the 0.025 in original clearance around tube

413

READY FOR ACTION: to roll wide range of tube sizes. Whether ferrous or nonferrous metal, this electropneumatic tube-rolling control does the job, without over- or under-rolling. Each tube is rolled to a set dimension, and all of the tubes are uniform

1 Measure diameter of hole in tube sheet for right setting of electrical control. Only one setting is needed for job

2 Set the control unit, based on hole diameter of tube sheet and OD of tubes to be rolled. This is a simple operation

HOW TO take
'guess' out of tube rolling

▶ Now, you can roll in your tubes with air and control their tightness so they are neither over- nor under-rolled. Tool is a handy portable electropneumatc unit. The advantages claimed are:

Air-operated motors are better for driving. They allow one motor for wide range of tube sizes: up to 6-in. diameter have been rolled.

Electronic brain controls amount of expansion to tube sheet, tightness and the holding strength to exact values, thus doing away with "weepers" or "leakers." Cold-work stress is minimized and distortion or fracture of ligaments in tube sheets prevented.

Operator does not need to have right "feel," as all tubes are rolled to set dimension.

Courtesy, Crane Packing Co, Chicago, Ill.

3 Insert expander into tube and activate air gun with rotating-handle switch. Valve's circuit controls air to motor

4 At right point of tube expansion, torque load on air motor increases, air shuts off and rolling stops automatically

Sheet-Metal Layout

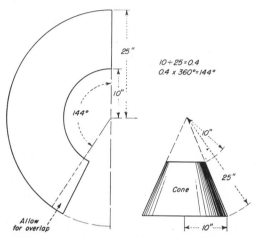

FIRST, draw desired cone and measure indicated radii. Use these to draw arcs on paper or metal sheet. Allow for lap

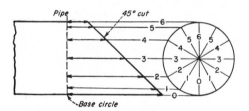

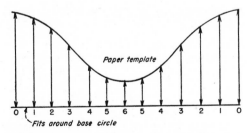

DIVISIONS of template base line correspond to 12 divisions of pipe circumference. At each division measure distance to cut

▶ ALL PROBLEMS in sheet-metal layout are based on principles of descriptive geometry, but you can go a long way with a combination of mechanical horse sense, some knowledge of drawing and a bit of practice. For a starter try the following two easy examples of layouts often found in actual maintenance and construction.

Better get out your drawing board and follow through, step by step.

CUTOFF CONE (FUSTRUM)

First lay out a flat shape for this cutoff cone, which is a common sheet-metal shape. As a practice example you draw it small scale. If the problem is actual, you should lay out the flat full scale on a wrapping-paper template or directly on the metal sheet.

To draw the full-scale circle you can easily improvise a beam compass as long as needed. Just drive a nail (for a center) near one end of a long stick. Then drill holes (for pencil) 10 in. and 25 in. from nail.

The only real trick is to know how much of the arc to use. Merely take the same fraction of the whole circle (360 deg) as radius 10 is of radius 25. That's 144 deg. Clearly mark end of 144 deg arcs for matching purposes. Then add what metal you need for overlap.

PIPE CUT 45 DEG

Next take common case of pipe or round duct sliced off at 45 deg—either to enter a flat surface or to pair with another 45 deg to make a 90-deg turn.

End view of pipe is the circle shown. Divide this into even parts. Twelve 30-deg divisions, as pictured, are convenient and usually close enough. Project these left and then on to the base circle. Base circle is drawn at any convenient place.

Next measure pipe circumference with string or tape and lay it out full scale as the horizontal base line of the wrapping-paper template. Divide this into 12 equal divisions, and number them as shown to correspond with the divisions of the pipe circumference. On side view of pipe measure distance from base circle to cut line at each division and transfer to verticals on template. Draw smooth curve.

Wrap template around until base circles match. Then transfer template curve to metal, and cut.

SEAL ANCHOR HOLES with asbestos paper held with friction tape, after melting old babbitt. Wood blocks the oil-ring slot

BABBITT IS RIGHT TEMP when soft pine stick chars. Use shell and mandrel set, fill the shell completely with babbitt

STAND SHELL ENDWISE on asbestos paper and seal bottom with asbestos cement. Put warm mandrel in center of bearing

"HOW TO" TIPS ON BABBITTING

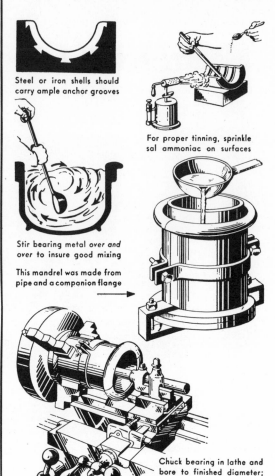

Steel or iron shells should carry ample anchor grooves

For proper tinning, sprinkle sal ammoniac on surfaces

Stir bearing metal over *and* over to insure good mixing

This mandrel was made from pipe and a companion flange

Chuck bearing in lathe and bore to finished diameter; scrape to true fit on shaft

Line That Bearing!

Bearings shot? Can't be replaced with new ones from suppliers stock? Stop worrying about it.

▶ Is ON-THE-JOB-rebabbitting becoming a lost art? The answer is yes! A recent industrial survey showed that many operating engineers not only never poured a bearing, *but had never seen one poured.*

That's because conveniently located mill-supply houses, with large stocks of bearings and pillow-blocks make it easier to replace than to rebabbitt.

What's wrong with that picture? Nothing at all. You can't improve on a good factory-made bearing. But, in case you can't get a bearing in a hurry to avoid a delayed shutdown, you might have to pour your own. That's where this article will help you.

Types of Bearings. There are many types of bearings and boxes. Solid journal bearings; rigid pillow blocks, ample pillow boxes; gibbed rigid pillow blocks; common flat boxes and others.

Babbitting procedure may vary, depending on type of bearing to be poured, but methods are fairly general and standard. Solid journal bearings and boxes are babbitted in one pour, while the individual halves of split bearings and boxes call for separate pouring. Speeds and loads have much to do with the grade and type of babbitt used, but the sequence of operations is usually the same when babbitting bearing.

Removing Old Babbitt. First step is to remove the old babbitt. This might be done with a hammer and chisel or by applying heat. Whatever method used, see that all old babbitt is completely removed from the anchors or dovetails in the bearing. These bearing grooves serve to hold babbitt tight and make it a solid part of the bearing.

In very old split bearings, staggered holes may be drilled part way into the casting. These holes serve as babbitt locks or anchors. In the upper half of some bearings, two holes—either cast or drilled, and same distance from ends and center, serve dual role of anchor and babbitt pouring holes. Whatever method used for holding babbitt, see that all old babbitt is removed before pouring new babbitt. Make sure the bearing is thoroughly clean and dry.

Aligning Shaft And Bearing. Next, align shaft—whether horizontal, angular or perpendicular—perfectly. Make sure shaft runs absolutely true and that it's perfectly smooth through bearing section. Then center bearing radially around shaft (usually with inside calipers) before tightening down. This insures equal babbitting from shaft to bearing, all around shaft. See that center line of bearing coincides with center line of shaft, if that is easier.

Preparing For Babbitting. Method to use depends on the bearing. A one-piece vertical bearing is easily babbitted because it's easy to seal the lower end and pour from top. Vertical-type split bearings are also top poured, but take-up shims should be in place between bearing halves. These shims should just touch the shaft to form a "parting line" when babbitt is poured. This type bearing is also bottom sealed and poured from top separately through each side of shim-divided halves.

Horizontal solid bearings can be poured through pouring holes on top, or from lip formed in the sealing compound at either of both ends. The lower-half ends of horizontal split bearing are sealed. With the upper half removed, the bearing is poured to the parting line. If poured too high it may be easily dressed down with a babbitting file. Then insert takeup shims. After top half has been installed, tighten down and seal end. It also may be top or end poured same as solid horizontal bearing.

Some operating engineers pour babbitt through the lubrication hole in the bearing. This is common practice, but means redrilling babbitt from oil hole. Or if it's a tapped hole, it must be re-tapped after babbitt is drilled.

Sealing Compounds. While clay or putty is commonly used to seal the ends of bearings for babbitting, there are some good flexible compounds on the market. These compounds stick close and tight. They can be reused indefinitely without preparation or re-mixing. Because they are moisture proof, they eliminate accidents from blowouts.

Approved compounds are much superior to clay, dough, putty, or similar improvised plastics. They effectively seal every seam and crevice in and around the bearing, preventing molten babbitt from seeping through.

Asbestos lead runners are also used to seal and pour bearing ends. But most operators usually set up collars or cut disks of tin or cardboard—either in one piece or halved—to tightly fit shaft and extend beyond area to be poured. These retainers are also held in place with sealing compound.

Pouring The Bearing. Before pouring the bearing, select right grade and type of babbitt. There are many kinds of bearing metals and supply houses usually have the right metal for most jobs. Some metals melt at low temperatures, yet have ductility, tensility, toughness and elasticity. They also can take severe shocks without crushing or deforming.

Magnolia anti-friction metal contains graphite and has peculiar property of embedding particles of dirt and grit below bearing surface where they cannot score or cut the shaft. This metal stands up under tough conditions and requires minimum lubrication. Select the right metal for each job.

A fire pot, blow torch or other heating medium to melt the babbitt or bearing metal is needed. Take care that the metal is not overheated. Pyrometers are especially suited to measure babbitt temperature accurately. If no instrument is handy, use white pine shavings. When molten babbitt starts to turn shaving brown, it's right temperature for pouring.

Melting ladles come in various sizes. They are either single or double lipped. Double-lipped ladles permit either right- or left-hand pouring. Bottom-pour ladles pour only clean metal from bowl's bottom, without skimming. They are best.

Heat the bearing before pouring—not only to make it perfectly dry, but also to prevent "babbitt chill," and retard "flowability." Check sealing compound after heating bearing. Then after skimming the refuse from molten babbitt, pour the bearing.

Pour babbitt slowly at first to see compound holds tight, then pour quickly and in a steady stream. As bearing fills, slacken flow gradually and continue to pour until bearing is completely full. Be sure metal solidifies before removing compound and retainers.

Fitting Bearing. Before putting a re-babbitted bearing "on the line" check it. If necessary, relieve high spots by scraping. Adjust shims on split bearings and lubricate. If bearing was poured through oil hole, redrill and re-tap. If not, remove sealing compound that plugged hole.

Some operating engineers wrap paper around the shaft when pouring one-piece bearings. Paper is tight and wrapped evenly. It may be secured with Scotch tape or asbestos string. The paper insures clearance for removing shaft and prevents sticking.

Safety. Wear babbitting masks when pouring babbitt. Mask protects you from accidental blow-out when there's moisture in bearing. With a little practice, any operating engineer can pour a first-class bearing in a tight pinch.

Did You Know . . .

. . . bearings can be keyed up in an emergency without taking leads or without removing the bearing-bolt nuts? If the bearing knocks badly and must be keyed up in a few minutes time, stop the machine and loosen the bearing nuts a few turns. Then remove a thin shim from each side of the bearing. Tighten the nuts and see if shaft turns freely. Keep removing shims until the shaft is hard to turn. Then add shims equalling at least 0.001-in. for each inch of the shafts diameter. A few thousandths extra will assure enough clearance without the danger of burning up the bearing.

This method doesn't allow inspection of oil grooves or bearing condition. But will keep the machine running quietly again until the next shutdown.

READY FOR BORING. These large diesel crank bearings were relined by welding babbitt with acetylene torch. Ready now for machining, then many hours of service

MACHINED BEARINGS. Metal is free of blow holes, securely bonded and can't crack loose. Time was saved and a better job done because torch heated the shell

Mandrel Babbitting—Goodby

▶ SPLIT BEARINGS can be relined without a mandrel—and a better job done. But let's talk about bearings first. Bearing maintenance has been a major problem to operating engineers for years.

Corliss main bearings are heavy cast shells, dovetailed to secure the babbitt. The babbitt is often ½ in. to ¾ in. thick. On large units, these four-piece bearings weighed from 300 to 400 lbs. But modern turbine, uniflow and diesel engines have almost entirely replaced these old units.

Now high speeds require less weight in reciprocating and rotating elements. Cast-iron shells have to be replaced with thin steel in many cases. These shells are too thin to hold dovetailing for securing the babbit lining. So a method of securely bonding the babbitt to the shell is needed. Many engine builders buy their bearings from manufacturers who specialize in bearings.

Centrifugal Casting. Machines have been designed for centrifugally casting the babbitt. By this method, the shell is rotated, the centrifugal force holding the molten metal against the shell. The speed isn't reduced until the metal has cooled sufficiently to be held in place. Claim for this method is that the more dense elements of the babbitt metal are deposited near the shell. Then the lighter components, which make the better bearings, are nearer the surface.

From my 50 years experience this claim is questionable. Especially with babbitt linings ranging from 1/32 to ⅜ in. in thickness. Relining these thin shells has baffled many operators, because they cannot make a successful job by the old method of casting the metal on a mandrel.

Tough Problem. During the last war we were faced with the problem of repairing 9-in. bearings. Their shells had only about ⅛ in. of metal. As the bearings were not manufactured by the engine builders, they would not reline them. They suggested sending the shells back to the bearing manufacturers. This would make long periods of delay. So we developed a scheme of doing this work ourselves. And I think it's interesting to operating men because it has proved so successful.

Any operating engineer can get the hang of it. With a little practice, relining bearings of any size or with any thickness of metal is easy. The right equipment is a bottle of acetylene, compressed air and a small welding torch.

Welding Torch Method. If the bearing must be relined completely, place it on a piece of sheet iron on your work bench. Melt out all the old metal

418

REMOVE METAL, REBABBITT AND PATCH CRACKS WITH BLOW TORCH

MELT BABBITT from old bearing shell like this. Start at one end and run torch from top down, removing metal in rows

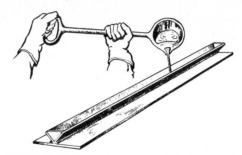

CAST BABBITT for use as welding rods in trough made of angle iron. Cast stick is easy to remove from the trough

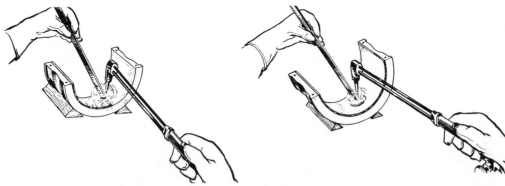

WELD BABBITT to shell in layers after thoroughly tinning it. Keep moving shell so lowest part takes the melted babbitt

REPAIR CRACKS by welding in patches of babbitt. For quick repairs this method puts the machinery right back in service

and burn off all traces of oil and grease. The shell is now ready for tinning.

If you know the composition of the old metal and if it's in good condition, add sufficient new metal of the same kind. But never mix metals of unknown composition.

Melt the babbitt in a pot and get it hot enough to pour easily. Then cast the babbitt in strips about 18 in. long, by pouring into a 1 x 1-in. angle trough. These strips can be used like welding rods. Next tin the shell thoroughly with the same metal. There are many tinning compounds on the market. Some are not suitable for this kind of work. One that we use is the Farco tinning compound manufactured by the Farrelly Co in Philadelphia.

Hold the shell in position with two wooden blocks and float on a thin layer of metal all around. The shell is now ready for building up.

With a torch and babbitt welding rod, weld on light ribs of metal same as if reinforcing a section of steel plate with an iron rod. Put on one complete layer over the entire shell and repeat until you have the right thickness. When

a very thin lining is needed, one layer of beads will give ample thickness for machining.

Keep the heat localized. If the finished areas have tendency to soften and slide, back up this section with a little wet asbestos cement. After you get a little practice you will be surprised how easy this job is.

Advantages. The metal is securely bonded to the tinned surface in a permanent weld. It doesn't shrink and the metal is free from small blowholes. I found repairs with this method to old-type cast-iron shells far superior to casting on a mandrel. Peening is not necessary. With the mandrel method, the thicker the babbitt the more it shrinks. This headache is eliminated by this method. Besides quick repairs can be made to cracked sections by welding a little metal in the affected areas.

Oil Grooves. Method of cutting oil grooves depends entirely on engine design and type of lubricating systems.

Many builders of diesel and high-speed vertical compressors provide pressure lubricating systems. Oil is pumped directly to main bearings or

through drilled crankshafts to crank bearings. From there, through connecting rods to wrist-pin bearings. In this type bearing it's customary to cut one circumferential oil groove to hold the oil in the bearings and keep side clearance to a minimum. Otherwise so much oil is lost at this point that the wrist pins don't receive enough lubrication.

Babbitting Tips. Remember — don't mix metals of different or unknown compositions.

I've had excellent results with metal of the following composition and properties. Copper 4.5%, tin 91%, antimony 4.5%. This metal has a Brinnell hardness of about 16 at 70 F and is good for crankpin service on diesel engines because it does not crack easily. That's important.

Another good metal has 3½% copper, 88% tin and 8½% antimony. This metal is harder, has higher bearing values and will not pound out as easily. Both metals are excellent for modern high-speed work. And they lend themselves well to this method of relining bearings because they sweat on easily, due to high tin content of the alloy.

Aligning Couplings and

Why You Must Check Couplings_____

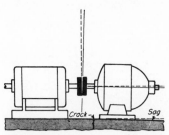

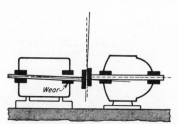

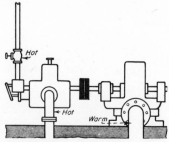

FOUNDATION SAG may not be evident to the naked eye, but vibration, building settling, etc, lower or change level of machinery, wearing the flexible couplings

BEARING WEAR may be uneven. Then most accurately aligned equipment gradually gets out of line. That wears contact surfaces of even best-designed coupling

EXPANSION or contraction from hot, then cold piping often pushes equipment out of line. Lining up cold machinery does not mean it will be aligned when hot

Handy Uses for Indicators_____

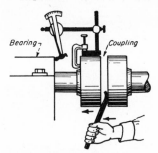

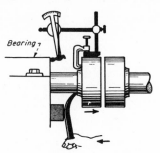

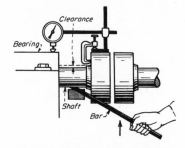

THRUST CLEARANCE or end play in most rotary equipment is highly important. If this clearance is excessive, vital parts, like rotors of turbines or pumps, rub on casing, causing costly damage. To check end-play without taking equipment apart, break coupling, then move shaft both ways while the indicator is set up as shown here

BEARING CLEARANCE (also wear) is easy to check. Break coupling, clamp indicator to coupling and rest the indicator on bearing. Not necessary to take leads

Simple Checks_____ | ## How to Align Couplings_

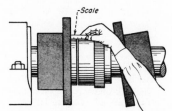

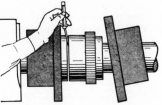

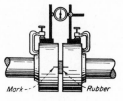

Angular check

SCALE or straightedge placed across two coupling halves at four places around coupling is visual check. But indicator is much more accurate, saves costly wear

FEELERS between coupling halves at four places around circle show if shafts are aligned. But this naturally cannot be as accurate as using a dial indicator

1 Place rubber between shaft ends (not near rim) to hold halves apart. Mark halves, attach indicator as shown, turn the halves together for angular check

Other Handy Indicator Checks

New Machine Has to Be Level

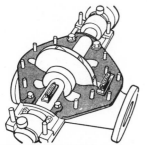

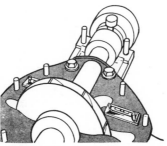

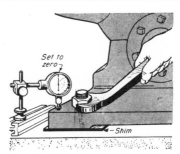

HORIZONTALLY ALIGN equipment with machinist's precision level by placing level on shaft or on machined surface. This helps parts to wear as intended

HORIZONTAL alignment also means that machine must be level in all directions. This helps in right setup of the piping and other equipment hooked to machine

INDICATE machine base by setting gage to zero. Tighten down most rigid unit first (pump). If base springs, loosen and place shims until there's no spring

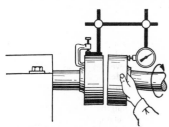

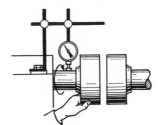

BENT SHAFT or coupling half not square is found by turning one half, while indicator is on the other. Don't assume couplings are true on shaft. Check them

CROOKED SHAFT shows up with indicator anchored to bearing while pin touches shaft. If shaft is true, check coupling half for being true on shaft (right). If not true, hole may be eccentric to rim, or coupling may not be snug fit on shaft. Soon as you know that shaft and coupling's half are true, you can line up coupling (below)

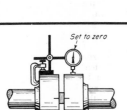

Rim check

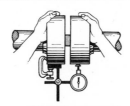

Figuring rim misalignment

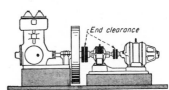

2 For rim check, change indicator as here. Turn two halves together, watch dial. Raise machine legs same amount with shims if angular check is perfect

3 With 0 reading at top and 0.020 in. at bottom; 0.020 in. divided by 2 equals 0.010 in. that one machine is out. Raise low machine with 0.010 in. shims

4 End clearance between the coupling halves must be right. If not, the worn thrust bearing or heavy rotor may load lighter motors, causing serious damage

421

PROPER ALIGNMENT of flexible couplings can be speeded up with right indicator assembly using two dial indicators. This rig is easy to clamp to one half of coupling in most designs

Use two indicators for alignment

Misalignment of rotating machinery crops up in new and old installations. Building settling is a common cause. Aggravating the problem, many maintenance men have an idea that a flexible coupling is also a universal joint. Reading the equipment catalogs, they see that the couplings will withstand so much runout and angular misalignment. True, the couplings *will* handle the misalignment —but rotating-equipment bearings wind up taking a terrible beating.

Machinists' stethoscope is a handy tool for quick misalignment checks on ball-bearing-equipped machinery. After a little practice listening to these bearings you can usually recognize flexible-coupling misalignment. Ball bearings make a roaring sound when alignment isn't right, a clear ringing sound when the flexible coupling is properly aligned.

Bearing temperatures are another quick check for detecting misalignment. Indicating pyrometer, right, gives an instantaneous temperature reading. If temperature is

above normal, suspect misalignment.

Don't overlook the engineer's precision level. I once checked an air compressor each month for a year. During that time the concrete base settled more than an inch, throwing the coupling way out of line.

Flexible couplings are often aligned during an emergency shutdown with nothing more than a straight edge. Then months later maintenance engineers complain about bearing failures, blaming the manufacturer.

My experience shows it's better to take a few extra minutes to make the alignments and do them right. Second and third failures cost big money. But proper alignment starts with the proper tools.

Basic tools you'll need are two good dial indicators and the necessary jigs to install them on the machine. There are several good makes; many have standard attachment connectors. For my work, I take 5/16-in. dia drill rod, then make up a series of attachments in lengths I think I'll need. I turn down one end

and thread for No. 10-32 machine screw threads. In this way I can select whatever mounting length the gages need to fit any job. For the gage extensions I use 1/4-in. drill rod, make up a series of these attachments in various lengths, photo, lower right.

A good clamping device can be made of 7/8-in. key stock. Use two pieces, each about 12-in. long. Drill them to take 3/8-in. hold-together rods. Two of these rods are needed. The key stock is drilled on two sides and tapped at 1-in. intervals for the No. 10-32 threads of the 5/16-in. drill-rod extensions. This makes it easy to find a suitable combination of holes for positioning the indicator.

To align, open the flexible couplings and set the distance between halves according to manufacturer's instructions. Then push a piece of fairly soft rubber between the two halves, sketch at right. The two halves must be marked so they can be rotated and kept together during the test. A through bolt between the two halves on some couplings will

TWO INDICATORS show angular runout between two rims of coupling at one quick glance

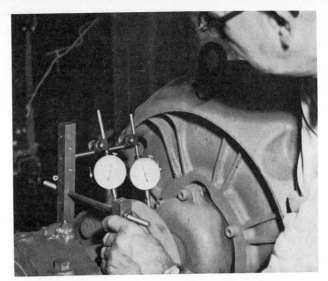

TURN INDICATORS 180 DEG to opposite side, making sure both halves of coupling turn together. Check these readings with those taken before

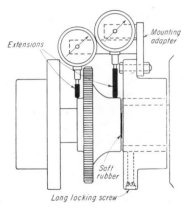

EXTENSIONS of various lengths help you align every size and shape of coupling

INDICATING PYROMETER of this type gives instant temperature reading. Use on bearings to check for abnormal temperature rise: sign of misalignment

keep both halves turning together.

To check runout, set two dial indicators to straddle the coupling half. Set each indicator as near the end of each rim as possible. In this way the indicator farthest from the shaft end measures the angular runout. Distance between the two extensions is the radius of this angle.

After machine is aligned so these gages read within 0.002 in., rotate the gages 180 deg and take readings again. Angular alignment should be held within 0.001 in. per in. distance between the two indicators.

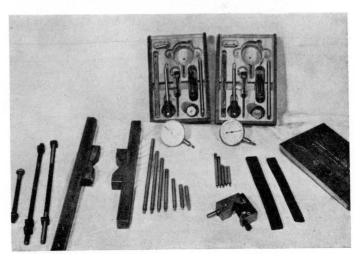

PARTS for dial indicators are made of key stock, drill rod and other standard materials found in most plants. Use indicators as shown above to save time

HOW TO BEND SMALL TUBING RIGHT

Wire Wound Around Tube Makes a Neat Bending Job

WIND IRON or steel wire tightly around the section of tubing that is to be bent

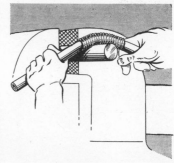

PLACE A LARGE PIPE or bar in vise and then bend the tube over it like this

If Wire Solder Fits Snugly, That Keeps Pipe Round

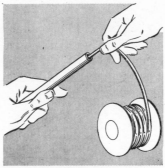

BEND VERY SMALL tubing by inserting a length of solid wire solder inside it

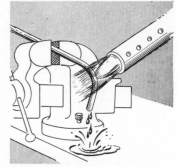

IF THE WIRE SOLDER cannot be pulled out, heat the tubes outside with a torch

Sand for Larger Tubing Holds Job Round at Bend

FILL LARGER TUBING with pitch (easily removed by melting) or with dry sand

HAMMER TAPERED wooden plug into the end of tube to hold tightly packed sand

Anneal Tubing So It Bends, Doesn't Crack

ANNEAL THE TUBE before bending. This is sure way of getting a much neater job

TO ANNEAL COPPER, heat to cherry red, dip in water. But cool iron pipe slowly

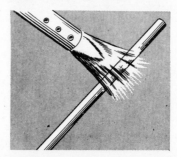

1 Assemble clamp around leaking pipe. Have loose enough to slide over leak

2 Lift clamp and slide over leak. Be very careful if the fluid is toxic

3 Tighten nuts a turn at a time until cylinder's base is secure on the pipe

PRESSURE is tamed as clamp leakage brings it under control

HOW TO plug bad leaks

▶ Now YOU CAN BRING costly and dangerous leaks under control with one of many clamps designed for this purpose. As photos show, head of clamp is iron cylinder. Its base has vents or ports on either side. Cylinder has rubber-faced metal piston that is screwed down against leak, same as the closing valve.

To prevent burning the gasket while shutting off leak, a brass band protects its lower rim. This also keeps rubber from being blown out through vents. Bottom of gasket is formed to fit pipe. Piston is keyed in cylinder so it doesn't turn while closing.
—*Courtesy, M B Skinner Co, South Bend, Ind.*

4 Tighten thrust screw until piston covers the leak and vents do not blow

Stress bolts right to make up tight flanges

MANUAL IMPACT WRENCH builds up spring power; number of blows indicates bolt tension

Initial stresses for American Standard steel flanges

Size of alloy-steel bolt-stud (note 1)	Average stress applied manually, lb per sq in. (note 2)	Approximate torque to obtain stress, ft-lb (note 3)	Elongation, in. per in. of effective length (note 4)
¾	52,000	175	.00173
⅞	48,000	255	.00160
1	45,000	370	.00150
1⅛	42,500	500	.00142
1¼	40,000	665	.00133
1⅜	38,000	860	.00127
1½	36,500	975	.00122
1⅝	35,000	1285	.00117
1¾	34,000	1700	.00113
1⅞	33,000	2200	.00110
2	32,000	2350	.00107

Notes: (1) Coarse-thread series, 1 in. and smaller; 8-pitch-thread series, 1⅛-in. and larger. (2) Average stress applied by maintenance man in assembly, using a lever and wrench or sledging. (3) Based on well-lubricated threads. (4) Based on modulus of elasticity of 30,000,000
Courtesy, Crane Co, Chicago, Ill.

Today more than ever before, greater speeds, heavier loads, higher temperatures and pressures demand accurate flange tightening. In selecting the best fastener for each unit and specifying torque-wrench settings, designers consider and balance factors such as (1) tension needed to keep bolt tight under maximum stress (2) smallest adequate bolt to reduce weight and cost (3) type of bolt material (4) class and type of thread (5) safe maximum tension and clamping to prevent distortion and breaking.

All this adds up to a truism: after such careful engineering analysis went into selecting the component, it's only good sense to follow proper procedures when tightening the assembly during manufacture or maintenance.

When a flanged joint is assembled, each component comes under varying tensile or compressive stresses. During service, especially at high temperatures, the flanges transmit heat to the stud or bolt, causing it in turn to expand at the rate of its coefficient of linear expansion. But normally this is such an infinitesimal difference that it can only be calculated on paper or measured under laboratory conditions. You can ignore it in the field.

Initial bolt stresses for flanged joints must be high enough to insure a tight joint. But at the same time you'll want a reasonable safety factor, to be sure stress doesn't come too close to the yield point of the bolt material. Usual practice in field-erected flanged joints is shown in table at left. According to Crane Co's survey, these initial stresses are customary when alloy-steel bolt-studs are pulled up by hand. Experience has proved these stresses OK for American Standard steel flanges. Values in this table are graduated by size—if you can't apply higher torques, it's best to keep initial bolt stresses at about one-third of the material's yield point.

Stress can be applied by one of three methods: (1) figuring proper torque and pulling up the well-lubricated bolt-stud by that amount (2) figuring elongation needed for the desired stress and pulling the bolt to that length (3) prestressing bolt with a pneumatic tensioner. If you're

BEAM TYPE TORQUE WRENCH flexes as torque increases, changing relationship between the pointer and the dial near the handle

PNEUMATIC IMPACT WRENCH has integral torque control. Its setting can be adjusted to any value between 500 and 1000 ft-lb

using the second method, you will need to calculate total elongation of the bolt. Do this by multiplying the value given in the table for elongation (in. per in.) by the effective bolt length (distance from center of nut at one end of the stud to center of nut at the other end).

Most operating engineers work with seals, know the effects of gasket flow, bolt pressure and flange-surface finish on sealing action, and the consequences of improperly tightened bolts. In this connection, remember that properly specified new gaskets and flanges *should not leak*. If they do, it's a signal that they were tightened improperly.

Torque is a twisting, the effort we apply in turning a nut, bolt or cap screw. It is measured in units such as ft-lb or in.-lb—in other words, length multiplied by weight. Let's say you have a ft-long socket wrench extending horizontally from a nut. If you hang your own 200-lb weight on the wrench's end, you're applying 200 ft-lb or 2400 in.-lb of torque to the nut.

Last month we said a bolt stretches about 0.001-in. per in. of length for each 30,000-psi tensile stress. Torquing is the common way of getting this stretch. How much torque do we need? A torque chart supplied by the wrench or equipment manufacturer—table at left—will give the approximate answer.

Three basic torquing tools are the (1) spanner (2) impact wrench (3) torque wrench. Bolts can also be expanded by heat or with tensioners, but in the strict sense neither of these methods is torquing.

(1) Hand-tightening with a spanner or a box, socket or monkey wrench is an old hit-or-miss, unscientific way. But it's still used. To make matters worse, someone is likely to add a long piece of pipe to the wrench for leverage, or try a hammer as a "convincer". Results are overstretched bolts or stripped threads, making the flange assembly useless and sometimes dangerous.

(2) Impact wrenches are pneumatic- or electric-power driven, or spring-actuated. There are many models on the market, each with its own distinct features. Photo, above right, shows a heavy-duty pneumatic-operated impact wrench with integral torque control. This wrench shuts itself off once it has reached the proper torque. Torque

setting can be adjusted to any value between 500 and 1000 ft-lb, simply by turning the adjustment screw with an Allen wrench. A lockout button lets tool do the work of conventional impact wrench with full 1½-in. bolt capacity.

Calibrators are used to determine proper settings for many types of air-pressure-operated and torque-control impact wrenches. They are especially necessary when you're installing high-strength structural bolts. Calibrator gage reads directly in lb of tension. Actual bolts, nuts and washers from the job are tightened in the calibrator—a new set of these should be used each time the wrench is applied. Adjust wrench setting until a series of three bolts can be tightened to more than the needed minimum tension, and no bolts break.

Spring actuated wrench, photo at left, is self-contained, hand-operated—a rachet type impact wrench. It uses principles of the hammer, lever and spring; automatically delivered rotary hammer blows or impacts produce the torsional force. This wrench will apply brute force to loosen a frozen nut or it can be switched to the precise task of tightening a bolt to desired tension with little manual effort. Torquing is based on the relationship between number of impacts and bolt tension—a click clearly indicates each blow, and mechanic estimates the tension by counting the clicks.

Sensory torque wrench sounds a distinct click, then imparts a strong impulse to the operator's hand the instant a preset torque has been applied to the nut. Operator can also read torque on the dial.

Center photo shows a torque wrench which releases automatically when proper torque is reached. To set it, operator turns the handle until the micrometer adjustment indicates desired torque. Then he locks the handle, attaches the socket, applies his wrench to the nut or bolt and pulls smoothly until the wrench releases.

Torque readings, do not always mean that the load has been placed in the stud. With torquing devices, the true load can only be determined by actually measuring elongation of the bolt in question. You can do this with an extensometer, measuring over the full length of each bolt. Or you can use a measuring device.

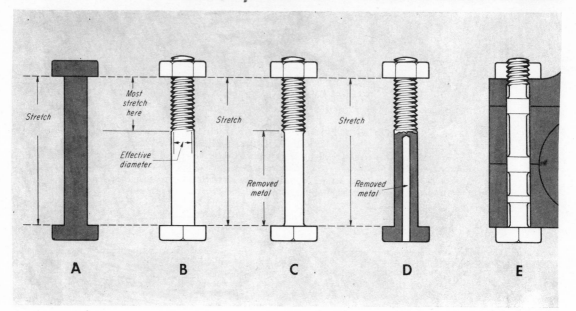

Let's talk about bolt stretch and measurement

Tightening flange bolts uniformly and to the right tension is a vital chore. Experience shows that when you're working with high-temperature flanges, uneven bolt loading causes real trouble. Not only do gaskets deform, but flanges bend, bolts relax and finally the flanges start leaking. More loading may twist bolts or strip threads.

Many articles have been written that discussed prestressing bolts with hydraulic bolt stretchers instead of tightening them with wrenches. Now let's see how bolts act when they're pretensioned or torqued. We'll look at some typical and special-purpose bolts, see how to get the most life out of a bolt and how much it can be loaded to

do most efficiently the job for which it was designed.

Elongation of the bolt is an effective measure of the stress it's carrying. Lab tests show that any steel is elastic, will stretch about 0.001 in. per in. of its length under a load of 30,000 psi *if* load is kept below the yield point. Soon as load is removed, steel returns to its original length. This happens regardless of the steel's hardness, whether it is stainless, carbon or alloy.

External forces acting upon a material produce tension, compression, bending, shearing or torsional stresses. And these stresses cause deformation. But as we mentioned before, if the stress isn't too great the material's elasticity returns it to its original shape soon as the stress is removed. If the metal doesn't take its original shape and dimensions, the *elastic limit* is exceeded. Then the bolt is *dead;* dangerous to use, it should be scrapped.

You can easily find the elastic limit or *modulus of elasticity* by dividing load in psi by elongation, in. per in. of unit length. Bolts made of steel average 30,000,000-psi modulus of elasticity, a figure to remember.

What bolt load should each stud carry? According to Edward Valves, Inc, bolting used with steel valves for pressures over 300 psi is usually alloy steel, having a yield strength of 105,000 psi. For best initial tightness in flanged joints, they recommend bolting loads under half this value, around 45,000 psi.

Ideal bolt would be tightened by upsetting the end like a rivet, sketch **A.** If it could be ideally loaded, entire bolt between the two heads would stretch evenly because the shank is minimum diameter over its entire length. But

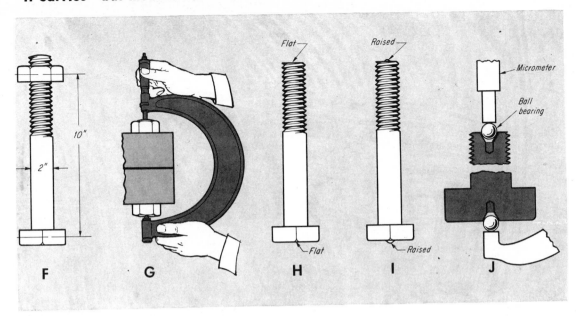

most bolts are constructed like **B.** Their weakest part is the threaded section from the bottom of the nut to the end of the threaded shank. Effective diameter is measured from the thread roots as shown, so most stretch of the loaded bolt takes place in only a part of the length. Of course, parts above and below nut ends are not stressed uniformly, and threaded portion carries a varying stress. But rule of thumb for any bolt says total stress is divided equally along portion of bolt between nut centers.

Reducing shank is one way to increase the strength of a bolt, especially to withstand shock loads. Besides the remarkable increase in impact strength, modified bolts with reduced shank allow greater stretching or elongation if overtightened. That gives the nut less chance to loosen because of vibration. Resistance to fatigue failure also goes up and the bolt lasts longer.

Most of bolt **C**'s shank is turned down to the same diameter as the smallest diameter through its threads. Shank is reduced to a given optimum size, which depends on type, method of manufacture, etc. Interesting thing about this bolt: although it is modified to weigh less, it can absorb several times more impact energy than conventional bolt **B**, which may surprise you.

Another way to reduce shank area or theoretically make it equal to effective diameter through threaded section would be to drill metal from shank's center, bolt **D.** This would have same strengthening effect as removing metal from outside the shank. Bolt **E** is a special-purpose model, used for the connecting rod of large diesel engines, or for standard railroad-car carriages, large rock crushers, etc.

How to figure loadings? **F** is a 2-in.-dia bolt 10-in. long between nut centers. How much should this bolt be loaded? Answer is no mystery.

Let's assume we want stress of 45,000 psi. Divide 45,000 by the modulus of elasticity, 30,000,000. That gives us 0.0015 in. Since the bolt is 10-in. long between nut centers, multiply 0.0015 x 10 in. This means we'll stretch the bolt 0.015 in., which is to be added to the bolt length over ends. If all bolts in the flange are elongated or stretched exactly this amount, and taken up in sequence by tightening bolts opposite each other, flange and gasket should give maximum service for life of installation. This method is important and should always be followed.

Method of measuring bolt stretch with a micrometer is shown in **G.** But if both ends of bolt are not parallel, readings will be off. You could grind or face off ends at right angles to the bolt shank by chucking in a lathe, **H.** Another way is to machine a small raised tit or bump on both ends as shown in **I.** Still another way: use a center drill and countersink both ends, **J.** Then you can set in ball bearings, measure with the micrometer.

Any measuring device that reads in thousandths of an in. will do. This might be an outside micrometer, vernier caliper or a dial indicator as used with tensioners.

If you're using a torque wrench, remember that friction between the threads and bottom nut will vary. So there will be a difference of elongation between bolts even though torque indicator reads the same for each. More-accurate method: predetermine the elongation, then, through sequence tightening, set all bolts to this value.

Test your fastener know-how

Russel, Burdsall & Ward Bolt and Nut Co

Q 1—In a gasketed joint, protection against leakage increases proportionately as the strength of the bolt is increased.   T F

Q 2—Rated clamping force indicates the strength of a bolt only if it is tightened properly.   T F

Q 3—Undertightening a bolt is as serious a mistake as overtightening. In other words, a minimum torque rating is as important as a maximum one.   T F

Q 4—A jam nut should always be installed above the full size nut.   T F

Q 5—Severe overtightening of either a fine- or a coarse-thread bolt will lead to the same results; the bolt will break.   T F

Q 6—Mathematically, a fine-thread bolt has a greater stress area than a coarse-thread one. It is, therefore, safer.   T F

Q 7—A bolt with no head marking has an ultimate tensile strength of about 70,000 psi.   T F

Q 8—Head markings should always include the manufacturer's identifying symbol.   T F

Q 9—A bolt should always be replaced with one of an identical head marking, or with one known to be stronger.   T F

HEAD MARKINGS on hex head bolts, indicate their strength level. For safety, you must know the identification symbols

Energy-systems engineers know that you can't replace one bolt and nut with another simply because it fits. The safety factor built into threaded fasteners works only if the right class of bolt is selected and the assembly is tightened properly. Industry standards have therefore been established to help users with fastener selection and proper installation procedures. Unfortunately these standards are often misunderstood or misapplied.

How about you? Do you know all that you should know about fasteners? Let's review your answers to the True or False questions above. If you work with fasteners and miss more than two answers, you had better read carefully the discussions following each of the answers.

A 1—FALSE. Bolt-users who experience joint failures often go to larger bolts. But these do not solve their problems, because the bolts are not torqued to their proper loads. Besides, using larger bolts for greater tightening often results in bowing of the flange. This was the problem of a manufacturer of sealed compressors designed to contain a toxic gas. The larger bolts resulted in leakage. Why? Because larger, stronger bolts caused more bowing of the flange and still greater leakage. We solved the problem by recommending harder gaskets to minimize compression, and slightly smaller bolts tightened with less torque. The bolts were also spaced closer together to help achieve a more uniform seal area.

A 2—TRUE. One of the most misunderstood points about threaded fasteners is their rated clamping force. Rated clamping force indicates the strength of a fastener *only* when the unit is properly torqued. The more a fastener is torqued, the stronger it gets—within, of course, the limit of the fastener material's inherent strength. Thus, clamping force of a threaded fastener depends on two factors: (1) inherent strength of the material, and (2) tightening procedure. If a fastener has a rated clamping force of 20,000 lb but is tightened to develop only 5000 lb, the extra inherent strength does nothing but resist shear.

A 3—TRUE. Recently I noticed stenciled NAS (National Aerospace Standards) numbers alongside some screws fastening the skin of a jet plane. They read "maximum torque 20 in.-lb." This worry about overtightening is typical. But what some people don't realize is that undertightening can be as damaging as overtightening. In what way? Undertightening can cause external premature failure caused by a joint loosening under applied load. On the other hand, severe overtightening—which can cause the bolt to stretch beyond its elastic limit—will prevent its safe reuse after the joint is disassembled. The point to remember is that listing a minimum torque is as important as listing a maximum one. So don't overlook the importance of torque.

A 4—FALSE. Jam nuts help prevent loosening of nut and bolt combinations. It seems to be traditional to place the thinner jam nut on top of a standard hex nut. Yet from a technical point of view, this is incorrect. The jam nut, when placed on top, merely adds frictional resistance at the interface with the standard nut. But if the position is reversed, not only are the same things accomplished, but frictional resistance is also introduced in the thread flank. The standard hex nut, being stronger than the jam nut, stretches the bolt thread off lead. (Lead is the distance a threaded part moves axially in one complete rotation with respect to a fixed mating part.)

A 5—FALSE. Ten years ago, the automotive industry used 80% fine-thread fasteners. Today, the industry uses 80% coarse threads. Why the switch? Because, mathematically, a fine-thread fastener has a greater stress area than a coarse-thread one. But when a nut is tightened on a fine-thread bolt, full strength cannot be developed in the stress area. If a nut is overtightened on a coarse-thread fastener, the bolt fractures. With a fine-thread unit, overtightening will cause the thread to fail in progressive shear. This results in a difficult-to-detect stripped connection. For this reason, one automotive company now uses only coarse-thread fasteners in all critical connections. If a coarse-thread bolt is overtightened (torque wrench out of adjustment), it breaks. In effect, breaking becomes an automatic inspection procedure, and the bolt is removed and replaced. If fine-thread bolts are used, the torque wrench could kick out as usual and the machine might have a de-

ASTM and SAE grade markings for steel bolts		
Grade marking	Specification	Material
	SAE – Grade 1 ASTM – A 307	Low carbon steel
	SAE – Grade 2	Low carbon steel
	SAE – Grade 5 ASTM – A 449	Medium carbon steel, quenched and tempered
A325	ASTM – A 325	Medium carbon steel, quenched and tempered
A490	ASTM – A 490	Alloy steel, quenched and tempered
BB	ASTM – A 354 Grade BB	Low alloy steel, quenched and tempered (medium carbon steel, quenched and tempered may be substituted)
BC	ASTM – A 354 Grade BC	Low alloy steel, quenched and tempered (medium carbon steel, quenched and tempered may be substituted)
	SAE – Grade 8 ASTM – A 354 Grade BD	Medium carbon alloy steel, quenched and tempered

*Manufacturers identification symbol required

fective stripped-thread connection. This of course could be critical and lead to a major breakdown and costly repairs.

A 6—FALSE. The reason for increased use of coarse-thread fasteners in the automotive industry is safety. Don't be misled by the fact that fine-thread fasteners have a greater stress area. This is true theoretically, but in practice coarse-thread fasteners are safer, especially in installations in which fasteners are tightened automatically.

A 7—TRUE. Identification markings on fastener heads must be understood for safe and proper use. No grade identification marking on a bolt head means that the bolt is an SAE Grade-2, or its ASTM equivalent, and has an ultimate tensile rating of 69,000 psi. This classification is now being raised to 74,000 psi.

A 8—TRUE. One of the head markings (SAE Grade-5) of the type shown in the table above indicates an ultimate tensile strength level of 120,000 psi in sizes of ¼ through 1 in. A bolt with such a head marking cannot be replaced by an unmarked bolt because the latter would not meet the required strength level.

A 9—TRUE. Many markings are used to indicate strength levels of bolts. The main ones are shown in the table above. The manufacturer's identifying symbols are also part of head marking requirements. That is why it is important that bolt users check these markings carefully and that nothing should be done to alter the physical properties of the bolts. Posting these markings pays dividends. ●

You can spot breakdowns

Some plant uses for this handy tool—the reflectoscope

Crankshaft may have flaws

Ultrasonic testing of this new diesel crankshaft shows up any hidden flaws. This is good insurance before placing new parts in machine. Here, old crank cracked through web. Operator is placing crystal (of reflectoscope) on crank web. Crystal projects a mechanical vibration or wave of extremely high frequency into the metal. Wave is projected through length of web, in path directly under the crystal, tests only segment of steel directly in its path. Keep moving crystal.

Crankpin tested in place

Here instrument is set up near machine, test made to find flaws in engine crankpin. If a defect is present, wave is reflected from it. If this defect happens to extend in a direction parallel to the wave, it presents practically no obstruction to the wave and won't be detected unless crystal is changed to new location. To detect such a defect, move crystal to another surface so wave travels at angle to defect, thus being partially or completely blocked in wave's path.

Flywheel shaft

Fatigue cracks in axles or shafts, which are hidden from view by flywheels, bearings or crank disks, show up distinctly under reflectoscope inspection. Besides, size and location of such cracks may be found. That helps operators to use the shaft as long as possible, yet alerts them to remove it from machine before failure from any defect growth.

Metal thickness

Lost metal is detected on location. Extent of both internal and external storage-tank corrosion is found without emptying—or entering tank. Pitting plates has same effect on sound waves as defects or cracks. Thickness of plates can also be measured with reflectoscope. You learn if plates have become dangerously thin with time.

Pipe for flaws

Critical-high-temperature, high-pressure pipe is inspected for dangerous defects. Here the angle search unit is used. Sound ricochets off inner and outer walls, circling the pipe until it returns to the search unit. Defects in the path of sound reflect back and appear as an intermediate signal on screen, which operator watches.

432

quickly <u>before</u> they happen

Here's how the reflectoscope works

▶ IF YOU CAN'T AFFORD costly breakdowns, answer is preventive maintenance. One way to prevent work stoppage is to "look" inside metal for flaws, replace such parts before they fail. Instrument coming into common use for this job is the reflectoscope. This is a portable electronic instrument producing pulses of ultrahigh sound. These pulses are induced into an object under test by a crystal search unit.

Nondestructive tests can be quickly made for fatigue cracks, internal flaws, corrosion, wear and imperfect welds. You not only don't have to drill or cut open part being tested (for loss of metal) but you often don't even have to remove part from machine, as these photos show. That saves time and costly labor.

Two beams. You can use two search beams: a straight beam and an angle beam. With a straight beam, sound waves travel straight through the object. They bounce off the opposite side, and are picked up again by the search unit. This gives you thickness measurements. This beam also detects any defects in the material that are in the wave's path, doing a twofold job.

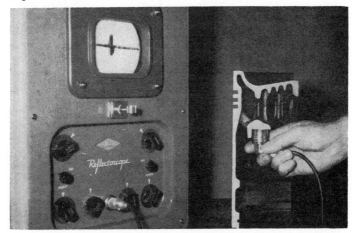

With an angle-beam search unit, sound waves don't bounce straight back to the search unit, but ricochet off all sides and through all areas of the object. This search unit is especially useful for checking welds, pipe and tubing—also, other objects where a straight unit might prove impractical or difficult. Instrument penetrates and indicates reflections through 30 ft of aluminum or steel. In porous metallic and nonmetallic materials, maximum penetration varies—depending on wave length to within fraction of an inch. *Courtesy, Sperry Products, Inc, Danbury, Conn.*

Let's test your pipe lines for flaws and thickness

1. Sound bounces off inside walls

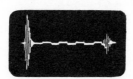

2. Flaw shows on viewing screen

3. Crack sends the sound back

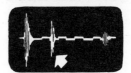

4. Screen shows crack like this

5. Thickness is measured here

Typical maintenance test is of high-pressure steam piping. Here sound is beamed into pipe wall at an angle with search unit. Sound ricochets off inner and outer pipe-wall boundaries in its travel around pipe, and returns to search unit, Fig. 1. Uninterrupted flow and beam's return are indicated on reviewing screen, Fig. 2. Any discontinuity in path of sound beam reflects a portion of energy back to search unit, Fig. 3. This shows on screen, Fig. 4.

Thickness measurement. Here the sound is beamed into pipe wall with a straight beam and reflects from inner face, Fig. 5. Length of time needed for sound to reach inner face and return to search unit varies with thickness of material. Thickness is indicated on screen. You can thus measure degree to which bent pipe (for example) thins on outside and thickens on inside of bend, assuring full specified thickness in high-pressure work. It has many other plant uses.

20 ways smart maintenance management can get their money's worth out of epoxy resins

Almost across the line in the energy systems, epoxy resins are stepping in as versatile maintenance tools. Epoxies are thermosetting materials. Converted by a hardener and cured by heat, the resins become hard, infusible plastics; some harden at room temperatures.

Epoxies are naturals for coatings, castings, laminates—some are tough enough to cast drop-hammer dies. They are superior adhesives for most metals, unaffected by a wide range of acids, alkalies and solvents. A filled-epoxy-resin system has virtually zero shrinkage.

But on the other side of the coin, epoxies aren't elastic. They won't bond to everything. To help you skirt the pitfalls, here are answers to 20 questions on best use of epoxies for industrial repairs.

1 I know epoxy compounds do a good job, but I've heard they are toxic. How about that?

A True, epoxy compounds can cause skin irritation, and they should be well ventilated. But if your men are properly instructed, resins can be used without fear. A pamphlet about safety is available from most manufacturers.

2 Why do epoxy-resin repair compounds come in two parts? Mixing is a problem. Why can't they be packaged in one part like the cold solders?

A Epoxies harden chemically, rather than drying through solvent evaporation. The best of them contain no solvents, so they are 100% reactive. Result: very little shrinkage; however thick you apply them, they harden throughout the mass, not just on the surface. Mixing is certainly a nuisance. But when you want maximum performance, it's worth the trouble.

3 I notice that the makers of epoxy-resin cements want us to wait, sometimes up to 24 hours, before using a component that's been repaired. But often we have to get a machine back to work in a hurry. What can be done about this?

A Some of the so-called *room-temperature-setting* epoxy cements harden in a few hours, some need a day. In general terms, the cements with the best chemical or heat resistance require a longer time. But any of them can be hardened in minutes with a little warmth. Heat from any source will do the trick: the sun, lamps or a heat gun.

4 According to some claims, epoxy resin bonds everything to anything. Is that 100% true?

A No. Nothing we know of does everything equally well except maybe money. Epoxy-resin cements adhere best to cold-rolled steel, aluminum, magnesium; to glass (but not Pyrex); to concrete and plaster (if they are dry); to wood (if it's roughened); to other epoxy materials, phenolics and polyesters. They do fairly well with copper, brass and low-nickel stainless steels. But they don't adhere well to nickel, tin or zinc, nor to thermoplastics like polyethylene, vinyl, styrene.

5 Is it necessary to clamp parts while bonding them?

A No. Contact pressure is all that's needed.

6 The manufacturers of epoxy cements stress proper surface preparation. Sometimes this is easy, sometimes not. Coming down to brass tacks, what preparation is really basic?

A The area *must* be dry, oil-free, clean, and that's basic. Beyond this point, adhesion is greatly improved by sandblasting or hand sanding; this not only assures that the area is clean and bright, but increases the area of bond. Epoxy repairs should be made in a dry atmosphere.

7 Should I weld, or use epoxy instead?

A Depends on your need. Welding is stronger and takes heat, which epoxy won't. But epoxy is often good enough.

8 How permanent is an epoxy repair?

A This question is asked frequently, but it's impossible to answer; there are too many variables. Epoxy resin is a plastic. Whatever you put in it—steel, iron, aluminum, sand—the epoxy is doing the work. You can repair steel with epoxy, but epoxy doesn't *replace* steel. Why? Because steel has many times the tensile strength of any epoxy-resin formulation.

9 How do I know what epoxy compound to use where? Aren't they all more or less the same?

A Epoxy-resin compounds are formulated for a given job. A "general-purpose" resin is a compromise of the possible properties: speed of set, heat resistance, chemical stability. Any epoxy cement on the market, if it's any good at all, will do run-of-the-mill repair jobs. But if you have one particular property in mind you have to buy a formulation which emphasizes that property. Heat resistance and chemical stability are gained at the expense of speed, for instance. Rapid hardening is obtained at the expense of mixing life. Putty-like qualities are achieved by sacrificing wetting ability.

10 I used an epoxy compound to fill seams in an old wooden tank on our roof. It stood up well for a while; then the epoxy fell out. Why?

A Even so-called *flexible* epoxies are not elastic. A polysulfide-rubber caulking compound should have been used here, and for similar repairs.

11 I tried to bond some galvanized metal with epoxy and got a very poor bond. Why?

A That's because epoxy bonds poorly to zinc, tin, nickel. Trick here is to remove the zinc coating by sanding. Then you'll have a good bond.

12 We repaired a steam line in our plant with epoxy, and it didn't take long for steam to start leaking again. Are epoxies overrated?

A Ordinary epoxy cements, the kind you buy at the hardware store, can be ruined by live steam. You'll have to try again with a special epoxy compound formulated to stand up to steam. Spread it on glass cloth and wind the cloth about three times around the (dry) hole. Let the repair harden 24 hours, then turn on your steam. You'll be happy. Of course superheated steam is another story.

13 I understand you want me to turn off the water in the system before I make an epoxy repair. But I can't turn it off. What can I do?

A There are various ultraquick-setting compounds. Some people insert wooden slivers—the wood swells, holds up long enough. Have glass tape cut and ready to apply. Spread your epoxy cement on the tape and wrap it around on itself three or four times. Set the epoxy up rapidly by applying 175-F heat.

14 Should we use epoxy or grouting compounds for expansion joints in our cold-climate plant?

A Epoxies, even the flexibilized epoxy compounds, are too rigid—not suitable for applications where there is a lot of movement. Use polysulfide sealants instead.

15 Will epoxy resin grout heavy machinery?

A Yes, but choose a very heavily filled compound. Or you can fill it yourself. Start with epoxy resin and a flexibilized hardener. Mix these and add dry aggregate, about five parts of aggregate to one of liquid material by volume.

16 I have an old eroded condensate tank made of concrete. The water is somewhat acid and around 200 F. Can I repair this tank with epoxy materials?

A No. That's a mighty rough environment for an organic compound. Instead, cut out all the weak concrete, dry the tank thoroughly, paint on an epoxy-Thiokol tie coat, and, soon as this is tacky-hard, trowel or pour a thick coat of portland-cement concrete.

17 We are changing the process in our chemical plant, exposing our many electric motors to highly corrosive acid fumes. Will coating the windings with epoxy resins protect them?

A Epoxy resins are noted not only for adhesion and insulating value, but also for chemical resistance. This includes most solvents, alkalies, and acids except for the strong oxidizers (for example, nitric acid). But to be on the safe side, the windings should be thoroughly impregnated as well as coated on their exterior.

18 We may change the hydrocarbon lubricating oil in our large steam-turbine generating units to a newer synthetic phosphate-ester fire-resistant lubricant. Would an epoxy coating prevent insulation on the electric-generator end from softening —if some fluid gets by the shaft seals?

A Epoxy-resin insulation will not be adversely affected by the phosphate-ester type lubricant. But whether an external application of epoxy will succeed here depends on the type of existing insulation. Ideally, generator windings should be completely impregnated, then coated.

19 Piping, valves and other equipment around the cooling towers in our petrochemical plant are corroding badly because of the humid atmosphere. Could we spray them with an epoxy-base paint?

A Yes, epoxy coatings provide excellent resistance to corrosion. Despite a tendency to fade and chalk, they are dependable and longer-lasting. But epoxy compounds are just that—compounds. Even the 1-component high-temperature epoxies contain a latent hardener. Choice of the proper hardener is at least as important as choice of the proper epoxy resin.

20 Can we use epoxy to seal a multiple-wire electric connector?

A Epoxy will work, but an elastic polysulfide or silicone rubber would probably do a better job for you.

ROPE for general industrial use is handy because it's flexible, light. But use a rope with safety factor of 5 to 1 for the load

"Pick It Up?"

Rigging jobs come up often around plants, but usually few landlubbers know too much about them unless they have sailed aboard ship. It takes know-how to work safely with manila rope, wire cable, chain hoists, while saving costly manhours

▶ "PICK IT UP, it ain't spit!" Sounds like something old Marmaduke Surfaceblow might bellow at a well-intentioned but unhurried wiper boy on the SS *Suretosink*. But how many of us with plenty of plant experience know all we should about how to "pick it up"?

Safety committees and insurance companies put up enough posters on safe manual lifting so we ought to know the important things: Bend your knees, keep your back straight, push with your legs instead of lifting with your back—and if it's too heavy for one man, get some help. If it's still too heavy, then you've got to resort to some kind of rigging. *(More on next page)*

Handy Table of Rope Sizes and Strengths

3-strand rope, standard lay; use at least 5 to 1 safety factor

Threads	Circumference In.	Diameter In.	Net wgt of 100 feet Lb	Minimum length in one lb (net wgt) Ft	Approximate gross wgt full coils Lb	Minimum breaking strength of good rope Manila Lb	Sisal Lb
6-Fine	9/16	3/16	1.47	67.9	50	450	360
6	3/4	1/4	1.96	51.0	50	600	480
9	1	5/16	2.84	35.2	50	1,000	800
12	1 1/8	3/8	4.02	24.9	50	1,350	1,080
15	1 1/4	7/16	5.15	19.4	63	1,750	1,400
18	1 3/8	15/32	6.13	16.3	75	2,250	1,800
21	1 1/2	1/2	7.35	13.6	90	2,650	2,120
	1 3/4	9/16	10.2	9.80	125	3,450	2,760
	2	5/8	13.1	7.65	160	4,400	3,520
	2 1/4	3/4	16.3	6.12	200	5,400	4,320
	2 1/2	13/16	19.1	5.23	234	6,500	5,200
	2 3/4	7/8	22.0	4.54	270	7,700	6,160
	3	1	26.5	3.78	324	9,000	7,200
	3 1/4	1 1/16	30.7	3.26	375	10,500	8,400
	3 1/2	1 1/8	35.2	2.84	432	12,000	9,600
	3 3/4	1 1/4	40.8	2.45	502	13,500	10,800
	4	1 5/16	46.9	2.13	576	15,000	12,000
	4 1/2	1 1/2	58.8	1.70	720	18,500	14,800
	5	1 5/8	73.0	1.37	893	22,500	18,000
	5 1/2	1 3/4	87.7	1.14	1,073	26,500	21,200
	6	2	105.0	0.949	1,290	31,000	24,800
	6 1/2	2 1/8	123.0	0.816	1,503	36,000	28,800
	7	2 1/4	143.0	0.699	1,752	41,000	32,800
	7 1/2	2 1/2	163.0	0.612	2,004	46,500	37.200
	8	2 5/8	187.0	0.534	2,290	52,000	41,600
	8 1/2	2 7/8	211.0	0.474	2,580	58,000	46,400
	9	3	237.0	0.422	2,900	64,000	51,200
	9 1/2	3 1/8	264.0	0.379	3,225	71,000	56,800
	10	3 1/4	292.0	0.342	3,590	77,000	61,600
	11	3 1/2	360.0	0.278	4,400	91,000	72,800
	12	4	426.0	0.235	5,225	105,000	84,000

Wire Cable. How can you check a wire cable? You *don't* do it by running your hand over the wire—that's one of the easiest ways to pick up some nasty metal slivers. You *do* look the wire over completely. If more than 10% of the wires in one running foot are broken, the cable is unsafe.

Chain Hoist. A weak chain is not easily detected. It may look just a little old and rusty, yet may have reached its fatigue limit and be full of invisible cracks. Then it will suddenly break without warning. So, chain hoists and chains used for lifting should be inspected and tested for 150% capacity. Inspect them at least every 12 months. If you notice a kink in a chain when making a lift, lower the load at once and take out the kink. This is important.

Best bet for precision rigging work is a chain hoist on the hook of a crane lift. Then when you get close to landing the load, further lift or lower only

with the chain hoist, never with the crane itself. We learned this the hard way.

We were landing the water-end head of a large condenser on a chain hoist on the big hook. It was let down nice and easy, and two bolts got caught. But to line up rest of holes, we had to lift it up just a hair and tap it over a bit. OK so far, but an eager beaver in the gang signaled the crane man to hoist. He did—with a vengeance. One bolt sheared off. The chain-hoist hook slipped out of the eyebolt. The chain whipped through the air while the big plate slid down and hung on the other good bolt. We were lucky. No one was hurt. But it brought home two more lessons:

1. Always mouse a hook if there is the slightest chance that the load might slip off.

2. Have one responsible man, and only *one*, give the necessary signals to the man handling the crane.

How to Take Care of Your Rope

Uncoil Rope Properly

Unwind right-laid rope counter-clockwise. Lay coil flat — pull end up through the center of coil

Dry Rope If It Gets Wet

Don't ever store wet rope. First dry thoroughly to avoid short life, loss of strength and even mildew

Overloading Is Risky

Safety factor is ratio of minimum breaking strength to load. Make 5 to 1 on new rope

Slack Off The Guys

Guy lines and other supports exposed to the weather tighten up from wetting; keep them slack

Sharp Bends Are Very Bad

Extra strain on outer fibers from sharp bends reduce rope's tensile strength. Pad corners as shown

RIGHT

WRONG

Reverse Rope End-for-End

Localized wear in short section is bad. Reversing rope ends helps all sections get needed equal wear

Keep Rope Kink-Free

Twisting and kinks harm rope structure. Avoid these if you want most service from your good rope

HOW TO TIE THE RIGHT KNOT IN A JIFFY

CLOVE HITCH

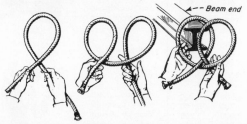

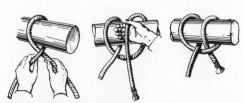

FOR MAINTENANCE WORK around the plant, the clove hitch is one of the most useful knots. Make single loop with two hands. Then make a second loop with right hand. Now move second loop over first loop and slide over beam end or load

TIE CLOVE HITCH IN PLACE around pipe or beam like this. Throw loop over pipe, cross rope end underneath, then over pipe. Stick end of rope through cross-under loop. This knot won't slip and yet can be untied with no trouble at all

BOWLINE

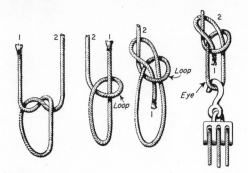

HERE'S A MUST. To tie bowline quickly, tie an ordinary overhand knot. Flip end 1 over to form loop. Tuck end 1 behind end 2 and down through loop. One of many uses is to form eye on end of line as shown. Knot unties very easily

STOPPER HITCH

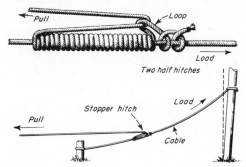

FOR TIGHTENING LINE already under strain, the stopper hitch is best. Coil rope around line to be tightened and allow for loop, with two half hitches as shown. This knot stays in place on cable or rope and won't slip on hard surface

SHEEPSHANK

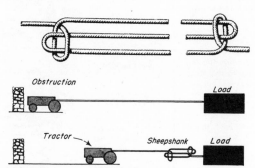

TO SHORTEN a long line without untying or cutting rope, use the sheepshank. For example, if a long line is needed to tie on to a load and then, because of obstruction, line must be shortened, back up tractor and tie a sheepshank

SLING SHORTENER

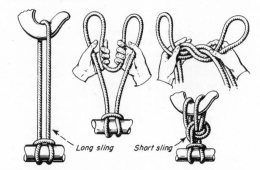

SHORT SLING can be made from long sling easily when there isn't enough head room. Form into two loops and tie overhand knot. Then place two loops over hook. This knot saves having too many slings of different sizes around the plant

HOW TO USE RIGHT KNOT FOR JOB

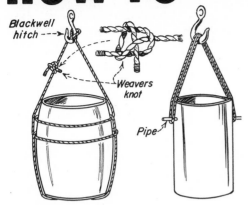

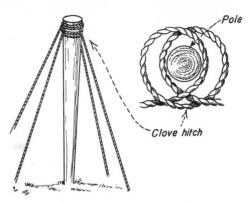

1 Knots are useful for operating engineers. Hoisting oil drums or barrels is safe with rope under barrel, two half hitches around barrel and a weavers knot. For hoisting small parts, use steel drum and sling on pipe shoved through drum

2 For hoisting heavy equipment in or out of plant, gin pole is easy to rig. Fasten guy ropes to any suitable anchor near base. At top, clove hitch around pole will hold. Some states require 4 guy lines on this job, others 5 guy lines

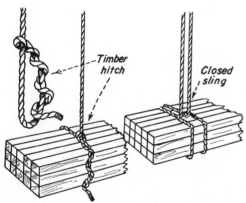

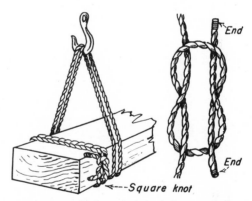

3 Hoisting piping or lumber saves time with this timber hitch. Load material over single line and wind end back around as shown. It won't slip after strain is taken and is easy to undo. Closed sling around load is another easy way

4 Another way to lift heavy loads is winding as many turns around load as necessary to be safe. Tie ends with square knot. This is only right way to tie square knot. It won't slip. No matter how tight knot gets, it's easy to untie

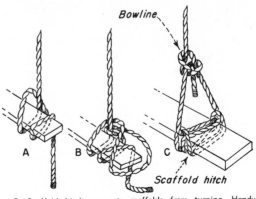

5 Blackwell hitch on hook keeps load from shifting and prevents rope falling off hook if strain is suddenly taken off rope. Bowline on hook doesn't slip and never gets tight. This is one of most useful all-around knots to use anywhere

6 Scaffold hitch prevents scaffolds from turning. Handy when painting plant walls and ceilings, or working on high suspended pipes. **A, B** and **C** show easy steps. Be sure to tie loose end above scaffold with bowline knot as shown here

SOLID ANCHORING, with base embedded or surfaced,

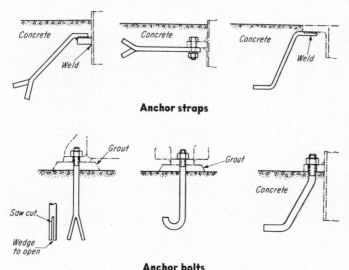

Anchor straps

Anchor bolts

Let's look at some specific engineering practices. One of them may be just the method you need to anchor your machine.

First three sketches show how *anchor straps* are fastened to machine parts embedded in concrete. These straps are all easy to make. Applied as shown, they'll do a reliable job.

Anchor bolts of various types hold down embedded or removable parts. Top end fastens the machine, while the lower end takes one of a number of special shapes to secure the bolt firmly in concrete. Big point to watch here: the bolts must be spaced to fit the machine's base perfectly. There's no margin for error, because once the concrete is poured, your bolts are frozen in place for keeps.

The *spring-rod anchor bolt* is one way of coping with this headache.

MACHINE LEVELING can be done before the final concrete

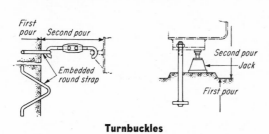

Turnbuckles

Turnbuckle design, far left, helps position the machine, and it preloads the installation before the final pour of concrete. Design at near left is a variation of the same principle; it uses a jack for preloading.

The *leveling screw* at right is an old standby, stepping in where you need to adjust your machines precisely on uneven floors. In recent years, *wedge blocks* have picked up the load where heavy and long machine beds must be kept straight. You can shift these blocks around from machine to machine as you need them.

One thing to keep in mind: These devices are all

VIBRATION ISOLATORS are resilient materials, springs, or

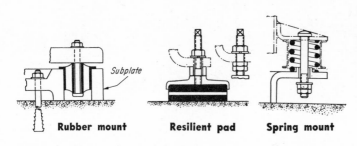

Rubber mount   **Resilient pad**   **Spring mount**

Rubber mounts or cork mounts are available in an almost endless spectrum of varying designs. They are especially effective for small- and medium-sized machinery, where they help reduce noise, shock, stress and strain. *Spring mounts* can be made very soft, and then you can alter the spring tension to suit the load. Bare springs provide the best vibration isolation because they absorb almost no energy; rubber will absorb about

uses grouted studs or easily removable hold-down bolts

Although you cannot remove this bolt from the foundation, either, its spring permits slight adjustment to suit the machine's position. Another variation is the *removable anchor bolt*—here, the protruding platforms of the foundation should be ribbed.

Embedded base plate is designed to prevent horizontal shift and to anchor the machine firmly at the same time. *Expansion bolts* come into the picture when specs call for mounting machinery on existing foundations or floors. Bolt holes of proper diameter and depth can be drilled and expansion bolts inserted.

Here's a tip to remember: cleanliness around the foundation is essential. Leaking oil lines, for example, should never be tolerated. Animal fats and vegetable oil will actually disintegrate your concrete.

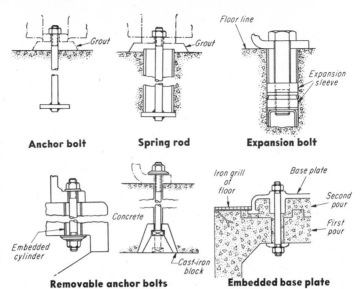

Anchor bolt **Spring rod** **Expansion bolt**

Removable anchor bolts **Embedded base plate**

pour or anytime after you've set your machine in position

machine positioners. They will not isolate or damp vibration. In order to stop vibration transmission between machines and building structures, and also to make it easier to live with the machine you are anchoring, you may want to consider using vibration isolators under the machine base. Power's 16-p special report, *Vibration Isolation*, August 1960, goes into more details on the vibration problem and what you can do about it, gives charts to help with your calculations. Now, let's pick up a few pointers on some types of vibration isolators, and see how they should be installed for best service.

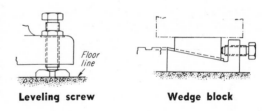

Leveling screw **Wedge block**

for high-mass machines, isolated bases

27% and natural cork 6 to 11%, in a range of 800 to 2000 fpm. To retain the elasticity of cork and rubber, the static load should not exceed 70 lb per in.—cork will harden under excessive loads. Mounting on *resilient pads* combines several valuable features: the pads absorb shock and vibration; machines can be leveled. To avoid shock transmission, design an *air gap* into your foundation or build in some soft material.

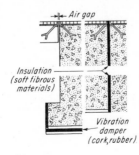

Air gap

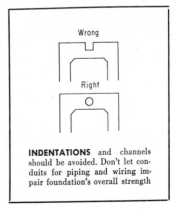

INDENTATIONS and channels should be avoided. Don't let conduits for piping and wiring impair foundation's overall strength

NET, woven to protect steam turbines, is inspected by Art Cannon. Ropes are strong enough to support workmen, materials. This is also good insurance against outage

SAFETY NET with dropcloth is spread above machinery before painters work on 70-ft-high ceiling in power plant. Paint splashes and dropped tools kept from machinery

Soak Up This...

► "SAVE THE SURFACE and you save all" is good advice, but just *how* do you save the surface? For an answer to this and other plant-painting questions, I visited Art Cannon of Oliver B Cannon & Sons, in Philadelphia. With 32 painting crews about the country, Art Cannon is our leading industrial painting contractor.

Regardless of who does your painting, you should know how this topnotch industrial specialist goes about a job. Then you will see that painting takes plenty of know-how.

Staff Meetings. Art Cannon has staff meetings each month. His key men fly in from the field. Sales, estimating, production and testing personnel attend. An expert from a paint manufacturer, one from a chemical firm, or a specialist on corrosion, etc, is usually present.

Current jobs are discussed and solutions arrived at. Art's own paint chemist gives lab findings on new products. He also advises on tough surface-protection problems and licking some unusual local condition.

Everyone has a chance to put in his oar. So supervisors go back into the field ready to tackle their problems from a new slant. Do these meetings pay off? They helped put this firm at the head of the list, competing with local contractors everywhere.

Another reason Cannon does so much work in power plants is that his experts do critical areas, without a shutdown. The plant's regular painting crew does routine safe areas that almost any painter can do.

Every Cannon supervisor has ten years

SAFETY SIGNS tell plant employes that men are working above them here

PROTECT running machinery with dropcloth over frames so no shutdown's needed

IDENTIFY pipes with removable plastic strips, either lettered or plain colors

of previous training with the firm. Safety is a big item, especially when working high above and around machinery, as in power plants. A green painter always works with a trained man until he knows the ropes.

Estimating is important. Condition of each building or section is listed as good, fair, poor or very poor. When to paint, next scheduled repainting, quotation and comments are listed. Here are some typical remarks I copied from an $85,000 estimate:

"(1) Buildings can wait until rehabilitation program is scheduled, then interior and exterior painting can be done. (2) Buildings viewed by public need painting for public relations and appearance. (3) Can wait until more essential buildings are finished. (4) Building important to production should be done before more replacements are needed."

Scheduling. All preparations are completed *before* plant is shut down. That means putting up protection, having enough men for shutdown equipment so work isn't held up, etc.

His average power-plant team of painters is about 40 men. Idea is to put as many men as possible on a job at one time as overhead costs the same for 10 men as for 100.

Safety. Each painting super checks scaffolds, ladders, rope and other equipment before sending them to a job. Safety nets are homemade because Art isn't satisfied with ones on market. Every rope, scaffold, ladder, etc, is numbered and recorded on an office card. Age, repairs and all other data are

jotted down. All rope is scrapped after one year's service, regardless of condition. "CAUTION, men working overhead" signs are placed below painters. Because of an excellent safety record, this firm has very low insurance rates.

Doing the Job. Here are a few rules on painting: Wire-brush, scrape or chip all rust and corrosion spots. Vacuum-clean to prevent loose dust, flyash, etc, getting into machinery.

Spraying can be more expensive than a brush job as it needs more protection. But, usually, spraying gives a thinner coat, does not protect as well.

Never red-lead new metal exposed to weather. When building outdoor metal tanks and steel structure, let them rust for a year or two until all mill-scale rusts off. Coat of primer and paint lasts on such a job. If red lead is painted over new metal, mill scale peels off with time, taking new paint with it.

In our electric-generating plants, 35% burn coal, 35% burn oil or gas, 30% are hydroelectric. Newest plants can switch from coal to oil, whichever is cheaper. These dual-design boilers present temperature problems.

Ducts get hotter when oil-fired, need a heat-resistant paint to cover changing temperatures. Aluminum is often used —one coat. Aluminum's varnish-vehicle stacks up with other enamels. If not aluminum, use a light pearl gray, a semi-alkyd oil-base paint, with a semi-gloss finish. There is no good heat-resistant paint in green or blue. Here are a few color schemes for power plants.

Turbine Room, ceiling—oyster white for right light reflection. Ceilings are

usually 60 to 70 feet high. Light green or cream adds good character to this room. Walls—light gray. Machinery —soft green enamel.

Boiler room, ceiling—oyster white. If walls are brick, well struck and pointed, or of good monolithic concrete, leave unpainted. Don't paint unnecessarily if surface is attractive and needs no protection. Once painted, a surface must always be painted. Use heat-resistant aluminum or a gray paint for boilers and adjacent piping, boiler doors, buck stays and ducts.

Piping. Paint the same color as boiler. Paint all insulated piping connected with boiler feed system a contrasting enamel of dark gray or green. On other machinery, such as pumps, drivers, coal pulverizers, coal feeders, use a gray hard-drying gloss surface. Colored plastic bands are best for piping. If over 12 colors are used, have labeled bands. Adopt U.S. Bureau of Ships color scheme.

Crane. Paint high-visibility yellow for safety.

Railing. Paint top rail black, so soiling from hands won't show. Paint middle rail safety yellow, and the rest gray. Yellow middle-rail makes stairway "stand out" for quick access to trouble spot.

Fire Lines. Paint red or use red bands to identify them.

Stacks. Paint with special coating to withstand temperature. Paint aluminum stacks at top with broad black band to camouflage smoke discoloration.

"Remember," Art Cannon reminded me as I left, "It costs more not to paint."

CORROSION is tough on steel sash if putty fails between glass and the sash

CALK steel sash with gun after digging out old putty—before painting sash

SPRAYING: not usually as good as brush work, calls for mask in close quarters

Here's what you should know before preparing steel surfaces for painting

- Make sure surface preparation is properly done
- Select the right contractors
- Since weather and temperature are key factors, use to advantage
- Reblast surfaces again if cleaned surfaces get wet
- Supervise and check jobs thoroughly to assure quality

Since every type of energy systems equipment from turbines to boilers and pumps is being installed outdoors today, surface preparation of metal is critical. Life of a protective coating depends on its long-term attachment to the surface. Thus, in preparing surfaces for painting you must remove mill scale, rust, grease, etc.

Specifications. Start with the proper set of painting specifications, then follow them out carefully.

Surface preparation. Sandblasting is usually the most reliable and is often used on new work. Three types are (1) Class A (white surface blast)—removal of all mill scale and rust, (2) Class B (commercial blast)—removal of mill scale, and (3) Class C (industrial)—removal of loose scale only (rust stains are firmly attached to mill scale and are still present).

Remember that type of sand, equipment used and techniques in the blasting sequence are most important for adherence of coatings. That's why we use only contractors or personnel with reputations for high-quality work.

Make sure safety regulations and company policies are understood by contractors. Enforcing them saves time and eliminates injuries. You'll find that cheap one-shot deals such as power brushing or light scraping, followed by cheap paint, are costly. Prices for sandblasting and a three-coat system vary, depending on locality and wage structure. Always call in contractors and indicate work involved. Have them inspect job, then note recommendations.

On completion of painting, list date completed, types of coating, thickness of each coat and weather encountered. Keep records of cost per sq ft. Repainting of metal is generally done after eight years when the intermediate coat usually starts to show.

New work. A definite program is followed: (1) Sandblast (Delmonte 20-mesh sand or equivalent) to a Class B finish for painted surfaces, (2) Apply prime coat of quality primer of 400-sq ft coverage per gallon, (3) Apply intermediate prime coat and tint slightly with aluminum powder to insure complete coverage of first coat, (4) Apply finish coat of green enamel, aluminum or other.

Applying the paint. Use spray, airless spray, brush or roller. Rollers give excellent coverage and maximum thickness without sagging. Don't sandblast in areas where paint is being applied. Be sure to paint blasted surfaces the same day and before contact with moisture. If there is rainfall before blasted surfaces are coated, reblast again to prevent rusting or "bloom" on steel.

Maintenance painting program should include the following: (1) Wash with soap and rinse all surfaces to be painted; (2) On completion of washing, check all surfaces—rust spots must be scraped, primed and given one complete coat; (3) Apply one coat of good quality paint with roller. Don't let undercoats become exposed to weathering, since this permits fumes, vapors and moisture to penetrate pores of undercoats.

Weather and temperature are key factors in any painting program. Be sure to finalize your program, material and personnel prior to favorable painting weather. Where rainfall is a problem, utilize wet days for washing vessels. Temperatures of 70 F are excellent for application and proper drying conditions.

Jobs to be done should be viewed in the field and recommendations noted in relation to types of paint, expected life, thickness of coats and equipment to be used. Some large jobs can use 5-gallon pails, others 1-gallon cans.

Supervise the job to insure quality work. Don't be afraid to ask for quality or to check the odd jobs carefully. While most contractors try to do good work, some workmen are marginal and require constant checking. If the job is unsatisfactory, talk it over with the contractor. Gaining his confidence will assure quality.

HOW TO HANDLE BALL BEARINGS

Courtesy of Marlin-Rockwell Corp.

DON'T LEAVE BEARINGS EXPOSED OVERNIGHT

PLACE UNWRAPPED BEARINGS on a clean newspaper, never on dirty bench. Oily bearing surface collects dust particles from air if left open overnight. Dust shortens bearing life. Always cover open bearings with cloth during maintenance

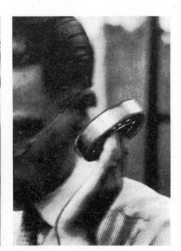

CLEAN BALL BEARINGS LIKE THIS

WASH IN CLEAR KEROSENE while moving inner ring back and forth to loosen grease. Blow with air, but don't let bearing turn or it may scratch. Flush again, then blow. Oil to keep from rusting. A clean-oiled bearing runs very quietly

REMEMBER THESE TEN POINTS

1 Don't take bearings from box until ready to install
2 Keep hands and tools clean
3 Don't wash packed grease out of new bearing
4 Keep grease can covered
5 Use clean rags; they are cheaper than bearings
6 Keep assembly bench clean
7 Lay bearings on clean newspapers
8 Get shaft and housing seats perfectly clean
9 Paint housings inside to seal core-sand, chips
10 Cover exposed bearings if left overnight

1 Proper size is important. Hoist capacity must at least equal load weight

2 Rig the chain block and load carefully. Never use the load chain as a sling

3 A metal sling carries the load safely. Never use rope to tie load to the hook

4 Tail-chain anchor pin must be free of defects. Inspect it before using hoist

5 Load brake must be in good operating condition, never slipping or dragging

6 An open load hook is dangerous. It may drop load. Replace it immediately

7 Stand aside. Never get under the load. Why ask for a crushed or broken head?

8 Handle all chain blocks carefully for good service. Never drop or throw them

9 Report defective chain blocks promptly. Never use one; it may drop the load

9 steps to chain-block safety

Sketches, above, illustrate some of the essentials of safe chain-block operation. If your men follow them, they will obtain maximum safety and keep chain block maintenance to a minimum.

Additional points to consider include the following: (1) Always suspend chain block from the saddle, not the tip, of the hook. (2) After hoist is suspended, check load chain for twisting or kinking. Never put a strain on a kinked chain. Take up the slack slowly and see that every link seats properly. (3) Never use the load chain for a sling. This can damage the load chain. (4) Use hoists of larger capacity than the regular work requires. (5) Tag each sling to indicate the maximum load it can

lift safely. Remember, increasing the angle between legs of a chain sling and the vertical increases the strain on the sling.

Follow manufacturer's instructions for maintenance and repair. Inspect all parts periodically. Replace worn, corroded or damaged parts. Inspect upper and lower load hooks. Replace if they are distorted, opened, elongated in shank, or damaged in any way.

Inspect load chain with a magnifying glass. Replace if worn, corroded or damaged. Look for bent links, cracks, scores or markings, corrosion pits, lifted fins. Lubricate unit as recommended by manufacturer.

16
Welding,
hard surfacing

A Preparation of joints is most important step. Once the surface is clean it can be wire-brushed and **made** ready for first welding **pass**

B Remove slag from each welding pass by chipping surface and then wire-brushing. When it is clean you can make additional passes

C After preparing the surface, by chipping and wire brushing, weld the reinforcing pass. In the sample case it is the fourth welding run

D Last step in the welding operation is to chip out the weld root and prepare its surface, as above, for the final overlaying weld pass

1—Gas or Electric Fusion Welding

Where Each Welding Method

All-welded boiler drums and welded tubes indicate growing importance of understanding welding fundamentals and where the operating engineer can apply them.

▶ WITH SIX DIFFERENT TYPES of welding, each important in boiler and power-plant equipment repair, an operating man needs to know which one to use. The six methods are: (1) gas (2) electric arc (3) electric resistance (4) thermit (5) forge (6) brazing.

Gas Welding. Gas Welding is a form of fusion welding. An oxygen-acetylene torch heats the parts to be joined. Once these parts reach melting temperature and while still molten, metal is added from a welding rod. This addition of metal is performed in such a manner that upon cooling the parts become as one. Heating up of surrounding areas is one limitation.

Electric Arc Welding. This is still another form of fusion welding. A welding rod or electrode applied to the joint transfers its metal by electricity. That is, the parts to be joined are connected to one terminal of a special high amperage-low voltage generator. The electrode connects to the other terminal. Scratching the electrode in the "grounded" metal next to the joint produces a short circuit.

Withdraw the end of the electrode a fraction of an inch and an arc forms. Directed into the joint between the parts, this arc develops melting temperatures. The electrode melts and its metal combines with molten parent metal to fill the joint. Moving the electrode along as the joint fills completes the welding operation.

Because bare rods and electrodes proved hard to control coated rods came into use. Over and above the control of the arc are two other advantages. As the coating melts off it (1) gives off a blanket of inert gas that keeps out oxygen (2) forms a slag over the welded joint that lengthens cooling period. A too rapid cooling of the joint can cause trouble.

The slag coating must be chipped off and the surface wire-brushed before applying another pass of welding. Otherwise slag melts into the joint and sets up weak points that could crack.

The high melting temperatures in this form of welding develop so quickly that surrounding areas don't heat up too greatly. This is a big advantage.

Preparation of joints for fusion welding is important. Fig. 1 shows a number of steps for a complete weld.

Resistance Welding. Electric resistance welding for boiler plant service applies almost entirely to re-ending or safe-ending fire tubes. Each tube piece connects to a low voltage-high amperage source. Mechanical pressure brings the joints together until they

4—Fundamentals of Brazing

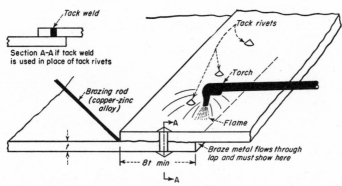

Brazed joints have temperature limitations that affect application. In brazing apply heat to one side, rod to other until braze metal shows all through lap

Edges of the above joint are scarfed back and heat is applied until plastic temperature range is reached; press together to make the bond

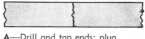

A—Drill and tap ends; plug

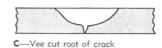

C—Vee cut root of crack

B—Grind sides 2 in.; down t/2

D—Fill completely by brazing

2—Forge Welding   3—Braze Welding Cracks

Fits the Operating Engineer's Job

make a short circuit. They reach melting temperature almost at once.

No metal is added, but the fluid ends under pressure squeeze together to form the bond. While the metal remains in the plastic-temperature range the tube is rotated, with a mandrel inside and pressure rolls on the outside, to work the upset metal and form a joint.

Thermit Welding. This is rare in boiler joints. It is produced by damming at the joint a mixture of finely powered aluminum and iron oxide. On "priming" or igniting with magnesium, the mixture burns rapidly at very high temperatures, bringing the metal to a fluid state. This forms the joint.

Forge Welding. Forge welding once served for re-ending tubes and even for making up some joints in boiler drums. The edges of the joints are scarfed back, Fig. 2, heated to plastic temperature, then pressed or hammered to make the bond while still plastic.

Brazing. Brazed joints have temperature limitations that rule them out of boiler-construction service. They do apply in unfired pressure vessels where the plate is lapped over eight times its thickness and secured by tack welds or tack rivets. Heat is applied to one side and the braze rod to the other until braze metal shows all the way through the lap, Fig. 4.

Brazing is not used on steel boiler repairs because an acetylene torch applied to the plate heats up large areas and may cause distortion. Temperature limitation applies to repairs as well.

Cast-iron, non-pressure parts or cast sectional boilers sometimes can use brazing in repairing a crack. Grinding the joint back to prepare for braze welding Fig. 3, adds strength.

WHERE TO USE DIFFERENT TYPES OF WELDING

Where Used	Fusion Gas	Fusion Electric	Electric Resistance	Thermit	Forge	Brazing
Joints and drums	No	Yes	No	Rare but sometimes in small drums	Rare but sometimes in small drums	No
Piecing tubes, new installations	Yes	Yes	No	No	No	No
Piecing tubes for repair	Yes	Yes	Yes	No	Yes	No
Forming braces or stays	No	No	Rare	Rare	Yes	No
Building up corroded areas, pits	No	Yes	No	No	No	No
Seal welds, riveted girthwise joints	No	Yes	No	No	No	No
Seal welds, ends of fire tubes	No	Yes	No	No	No	No
Joining steam pipe	Yes	Yes	Rare	No	Rare	No
Joints of unfired-pressure vessels	Yes	Yes	No	Rare	Rare	Yes
Boiler patches where permitted	No	Yes	No	No	No	No
Non-pressure parts, doors, grates, etc.	Yes	Yes	Rare	Rare	Rare	Yes
Filling cracks where permitted	No	Yes	No	No	No	*No

Table above summarizes various welding types and where they do and do not apply. Electric welding applies in most instances and is one form of welding that has grown in importance for general boiler-room maintenance.

Larger plants often carry electric-welding mechanics as a normal part of the maintenance crew. Some appreciation of the application and value of this maintenance tool has become almost a requisite for power-plant operation.

NOW Low-Temperature

EUTECTIC* PRINCIPLE

*EUTECTIC means easily melted. Pertaining to an alloy or solution having its components in such proportion that the melting point is lowest possible with these components

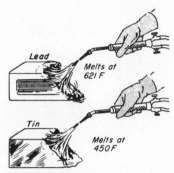

1 Every pure metal has its own melting point, 621 F for lead, 450 for tin

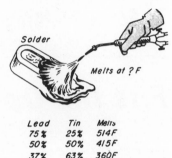

Lead	Tin	Melts
75%	25%	514F
50%	50%	415F
37%	63%	360F

2 Melting point of solder varies, but may be much less than for tin or lead

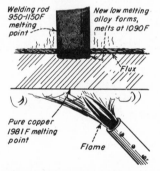

6 New low-melting alloy forms and melts under rod at 1090 F; sinks in copper

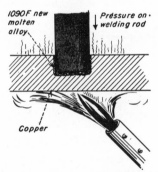

7 Keep pressing down on rod. As new alloy forms, rod keeps sinking deeper

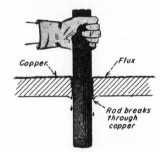

8 Rod breaks through. You did not have to heat copper to its melting point

WELDING ALUMINUM

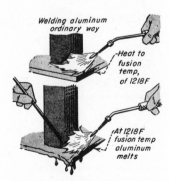

1 Aluminum, welded usual way, needs fusion heat of 1218, can damage

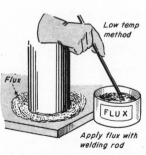

2 To low-temperature weld, first apply flux to area that is to be welded

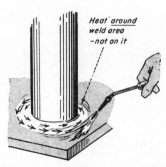

3 Play torch around area to be welded, not directly on it. That's important

450

Welding Saves You $$$

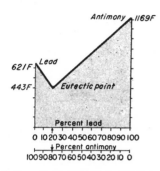

3 Eutectic point of both is 443 F, but lead melts 621 F—antimony 1169 F

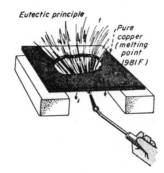

4 Flame breaks through a sheet of pure copper at its melting temp of 1981 F

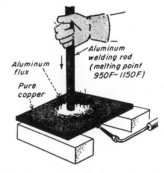

5 New flux fuses aluminum rod to hot copper. Press down on aluminum rod

WELDING CAST IRON

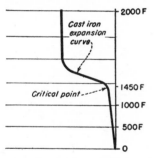

1 Critical point for cast iron cracking is 1450 F. It expands fast there

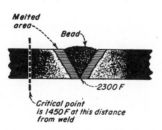

2 Fusion welding heats metal above its critical point of 2300 F, beyond weld

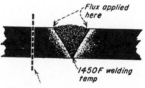

3 Heat to only 1450 F with low-temp method; keep below critical point

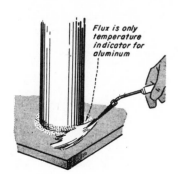

4 Melted flux (950 F) is temperature indicator. Apply the welding rod now

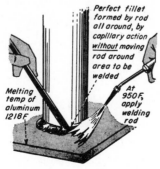

5 Perfect fillet is formed all around by holding welding rod in one spot

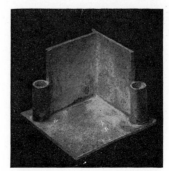

6 Low temp did not thin metal. Fillet is even and formed at all the joints

HOW TO TEST ACETYLENE SYSTEM

1 With working pressure inside, hold hose under water. Use soap suds to test hose ends near blowpipe and regulator

2 Hold blowpipe tip in water, with blowpipe valves closed and pressure on. If there are bubbles, valve seat leaks

3 Leaks are fire hazard wasting oxygen and acetylene. Test eight spots shown with soapy water under oxygen pressure

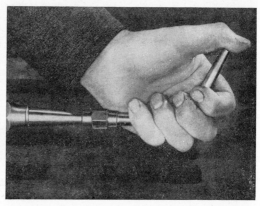

4 To place oxygen passages under pressure, open blowpipe oxygen valve and hold your thumb over end like this

Courtesy, The Linde Air Products Co

5 Never test leaks with a match or flame. To stop gland leak, tighten the acetylene cylinder-valve-packing nut

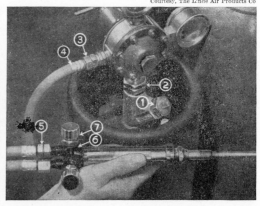

6 With acetylene system under pressure, test these seven spots with soapy water. Make tests at regular intervals

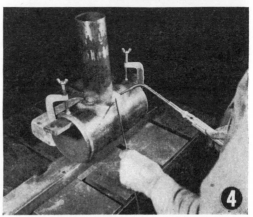

Joining Little Pipe to Big

▶ WHERE A SMALL SIDE OUTLET is welded into a much bigger pipe the difference in wall thickness may cause some trouble. Here's how to insure a good job.

MARK THE HOLE

1. Place the nipple in position and mark the hole on the header with a piece of sharpened soapstone. Center punch mark the soapstone circle with about ¼-in. spacing. Heat from the torch quickly destroys soapstone, but the center punch marks will remain visible.

CUT THE HOLE

2. Cut out the circle by holding cutting torch nozzle square with the pipe surface. Try to cut a good round hole just big enough to admit nipple after all oxide is knocked off the cut edge. Bevel the edges to a 45 deg angle, leaving a shoulder at the bottom end of cut about ⅓ the pipe wall thickness.

TRIM THE NIPPLE

3. Trim end of nipple to follow curve of large pipe. To mark off the nipple end, insert it in the hole resting against a bar clamped below inner surface of header. Edge of nipple should come just flush with lower end of bevel at side of hole, as pictured. Mark this curve permanently with center punch, and trim the end to fit with a cutting torch. Cut edges square; beveling nipple edges is not necessary.

WELD THE NIPPLE

4. Draw up bar until it rests snugly against inside surface of header and put nipple in position for welding. Tack weld it on one side. Straighten it into exact alignment by tapping with a hammer before making the second tack.

What's a little water leak, say a faucet dripping at the rate of one tumblerful every minute? Suppose this goes uncorrected for a whole year—525,000 minutes. Allowing 16 tumblers per gallon, that figures more than thirty thousand gallons down the drain.

1 Start out by giving your plates double-vee butt, ⅛- to 3/16-in. clearance

2 Heat of first welding pass may buckle up plates. It shows stress is present

Weld bead

Straight edge shows direction of buckle

3 Turn plates over after the third pass; then clean and make ready for welding

4 First pass on the second side of plates pulls the plates back towards balance

5 Finished plate looks like this. Peen weld hard for any slight unbalance

Welding Plates Made Easy

You don't have to worry about heavy steel plates buckling under welding. Here's how to do it right

▶ TOO OFTEN a plant maintenance man thinks that welding two flat steel plates together is a simple job. All you need do, he thinks, is slap the two together and weld away. Actually to weld a plate of any size, say 10 ft by 4 ft by ½ in. thick, to another of the same size and keep them straight is a matter of knowing stress and strain.

But there is a definite way of going about the job to both do it right and avoid undue stress. Plates must have a double-vee butt, Fig. 1, on the facing sides. Lay the plates flat, with a ⅛- to ⅟₁₆-in. space between them. This spacing makes sure you burn through to make a pure metal job.

Tack the plates together and proceed with the job. A straight edge will help you see which way the strain is pulling. Here's how you proceed:

1. If you have a dc welding machine, use straight polarity all through the job. An ac machine is simpler to handle, because you don't have to worry about any magnetic blow. Magnetic blow comes as a result of magnetizing the metal so much at one spot that the welding material is repelled there and won't stick. Welding from the other direction cancels this magnetic field.

2. Lay your first pass through the entire bevel with a ⅟₁₆-in. mild-steel electrode, Fig. 2. It is safer to weld two feet and stop. Then move up about two feet and repeat. Now go back and weld the open spaces. This method, known as step welding, spreads heat evenly through entire plate.

3. Once you complete the first pass, clean the weld thoroughly with a wire brush and hammer. Now lay a continuous bead with a ¼-in. mild-steel electrode starting from left to right. Weld the whole length unless you come to a magnetic blow. Then start on the opposite end and weld until you complete the entire pass.

By this time, any stress in the two plates caused by the heat of the depositing metal will have buckled up the plates, Fig. 2. Clean the second pass and apply the third just as before.

4. After the third pass, turn the plates over, Fig. 3. With an air gun and a sharp gouging tool, such as a flat chisel, clean the weld until you see clear metal, and get rid of all the slag, excess metal and dirt deposited with the first pass.

5. Now you can make the first pass on this side. Use more heat than usual for the ¼-in. mild-steel electrode to make sure you penetrate well into the metal of the plates to form a solid piece. When this pass is finished, you'll find the strain has pulled the plates back towards balance, Fig. 4. A straight edge will help you decide. If the plates haven't pulled too much, lay a second pass. (Clean all welds thoroughly before making additional passes). Second pass should bring plates in line.

If plates have stayed pretty well in line, you may apply a third pass. But if they are so close that additional heat will pull the plates past balance and buckle them the opposite way, peen the weld. To do it right, use the peening tool in an air gun. It evens the strain and helps hold plates in balance. You can now make a third pass.

6. After the third pass, check plate balance with straight edge. If they balance, you are ready for the next move. If they have buckled past balance, peen the weld hard to bring plates back in line.

You can then turn the plate back to its original side and make passes four and five to finish that side.

7. Turn the plate over again and finish the reverse side. Metal already deposited by this time should be enough to hold plates even and no unbalance will develop. When finished, Fig. 5, you have a straight plate 20 ft by 4 ft by ½ in. thick with a welded seam as pure as the plates' metal.

NEEDLE VALVE for tough use finished with alloy of cobalt-chromium-tungsten

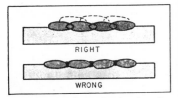

OVERLAP hard-facing beads, then lay second-layer beads between first beads

SUBMERGED ARC is semi-automatic. Bare electrode wire is power-fed through blanket of alloy flux. Alloy pickup from flux gives up to 62 Rockwell C hardness

Hard Surfacing

1—Electric-Arc Welding Method

Two major hard-facing tools are the electric arc and the oxyacetylene torch. The electric arc deposits alloys fast so heat doesn't penetrate deep. But steel with more than 0.3% carbon may crack unless preheated. Because arc heat is intense, base metal melts near surface. This dilutes the first layer of hard-facing alloy with base metal—more than when a torch is used. It may take two or more layers to get undiluted hard facing.

Electric arc forms when two conductors of an electric circuit touch, then separate, and there's enough voltage to maintain current flow across the gap. Arc is struck between a carbon or metal electrode connected to one terminal of a generator, and the metal to be welded, which is connected to the other terminal. Arc size is varied by changing distance across gap, altering current output, or putting resistance in the circuit. Distribution of heat in the arc depends on current density in the electrode.

For *straight polarity*, connect electrode to generator's negative terminal; for *reverse polarity*, tie to positive.

Hard-facing electrodes are both bare and coated. The flux coating reduces spatter loss, assures good penetration, prevents oxidation of deposited metal and helps stabilize the arc. Flux makes metal "wet" easily so weld metal flows and diffuses through melted flux. It also helps carry flame-produced oxide to the surface so it doesn't remain to contaminate the hard facing.

Current Settings. If you don't have electrode maker's data, use the table below as a handy guide:

Rod-size, in.	Current, amp
1/8	80 to 110
3/16	150 to 200
1/4	200 to 350

Low or medium current with a short arc of about 3/16 in. prevents loss of alloying metals. It also reduces dilution of alloy by giving minimum penetration. Increasing current increases penetration, softens deposit, and increases dilution with base metal.

Electrode size depends on nature of work and current that it is to carry. Because the electrode acts as an electrical conductor, be sure to use right current for each size electrode.

Good rule is to use largest convenient rod for the job. Large electrodes permit faster work because more metal is deposited. They are cheaper and faster and their arc stability is often greater. But if material is crack-sensitive, use a small-dia electrode. That reduces heat input and also danger of distortion.

Use only dry electrodes. Best way is to make sure they are kept clean and are stored properly in a dry place. When using, be sure to avoid trouble by following the manufacturer's directions. Better hard-facing jobs will more than pay you for your troubles.

Submerged arc (photo) increases depositing speed 3 to 4 times because high current density with small wire gives high burnoff rate. For example, 600 amp on 5/64- or 3/32-in. wire is equal to 100 amp on 5/16-in. electrode. Alloy pickup is from granular flux, containing alloy needed, as flux pours from funnel and hides arc. Inexpensive wire can thus make hard-facing deposits of 36 to 62 Rockwell C hardness.

Twin carbon arc has two carbon electrodes in holder, gives softer arc and will even hard-face thin materials without burning through. Reason is that current flows from tip of one copper-coated and cored carbon to other, not through metal being surfaced.

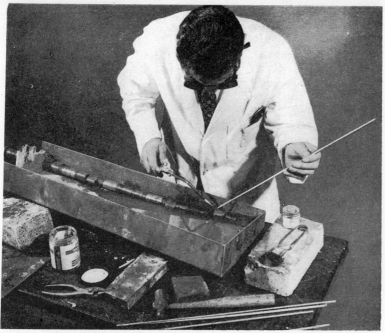

WORN CAMSHAFT LOBES are built up with torch. Water keeps shaft from warping by carrying away heat. This repaired job will last much longer than a new shaft

DIESEL VALVE turns slowly while hard-facing rod is deposited around its seat

2—Oxyacetylene-Torch Method: It's Handy for Parts

Big advantage of surfacing by oxy-acetylene torch is that surfacing alloy can be applied by flowing alloy onto the base metal with a technique known as sweating or tinning.

These terms describe formation of a lower-melting-point element that enables alloy to flow onto base metal while hardly melting it.

Term "sweating" refers to action that takes place when applying high-melting-point ferrous, nickel and cobalt-base alloys because of using an excess acetylene flame. This flame carburizes the surface of base metal, which in turn lowers melting point of the film. Alloy flows easily over this thin liquid film.

Term "tinning" describes action when surfacing with bronze or copper alloys. Only here it's more like a cleaning action caused by flux or flame that prepares base metal ahead of the welding action. That allows capillary flow of low-melting-point alloys onto the higher-melting-point base metal.

Flame Types. To burn acetylene gas completely takes $2\frac{1}{2}$ times its own volume of oxygen. But about equal volumes of the two gases are fed through the torch. Other volume and half of oxygen is gotten from the air. The combustion of acetylene in oxygen produces carbon monoxide and hydrogen. Oxygen for burning the CO and H_2 to carbon dioxide and water vapor comes from the surrounding air.

Hard-facing flames usually have an excess of acetylene. Sketch shows a neutral flame—equal volumes of acetylene and oxygen going into the torch. Increasing the acetylene builds up a white flare caused by unburned white-hot carbon particles. These free-carbon particles tend to carburize the surface of the base metal and also to reduce any oxides present to the metal form. So excess-acetylene flames are also called carburizing flames and reducing flames.

The usual hard-surfacing flame has a white flare about three times as long as the inner cone. This flame gives good sweating action because the free-carbon particles are absorbed into the base-metal surface, lowering its melting point. Melting of an extremely thin surface layer provides a liquid medium on which molten facing alloy flows. Carburized base metal, to a depth of about 0.001 in., is absorbed by alloying with the molten hard-facing layer. So gas-welded deposits can be made with penetration less than 0.010 in. into the base metal.

How Much Acetylene? Excess-acetylene flame also prevents loss of carbon content from welding rod and may even increase carbon content of weld deposit from 0.2 to 0.4%. Because some alloys absorb carbon from flame, hardness of deposit depends on excess acetylene used, and on cooling rate of finished work. But long excess-acetylene "feathers" are unsatisfactory because they are not as hot, put too much carbon into deposit and make poor welds.

Always keep inner cone from touching molten metal already deposited. Inner cone contacting puddle puts in too much carbon. That hardens deposit, which might cause cracks. Keep inner-cone tip 1/16 to $\frac{1}{8}$ in. away.

A slight reducing flame is used with rods having fairly high iron content, where dilution of facing alloy with base metal is not serious. When working on cast iron, watch corners and sharp edges, because cast iron has a lower melting point than many hard-facing rods.

When hard-surfacing with the gas torch, use only a slight excess of acetylene. Use flame to keep air from contact with molten metal. This prevents oxidation. Never remove flame completely until metal starts to solidify.

Which Way? Deposit hard-facing alloys with either forehand or backhand technique, sketch. Forehand is usually

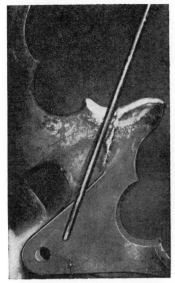

EXCESS hard-facing metal is wiped away from this sprocket wheel while plastic

NO GRINDING needed; contour is formed by die while the alloy is still plastic

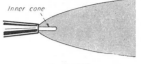

To Be Finished to Close Tolerances

recommended because penetration is controlled easier, rod is deposited faster with less scale in molten pool.

Backhand technique shown is also common with the sweating method. Direct excess-acetylene flame at a 30- to 60-deg angle to the surface. As you deposit alloy, move flame back to remelt rough spots, get a smoother finish.

Typical Job. Torch method is handy for doing parts that will be finished to close tolerances by machining or grinding. Cam shaft of internal-combustion engine, photo on facing page, forms a typical job. To repair worn lobes, partly submerge shaft in water so heat is carried away and shaft doesn't warp.

In photo, shaft is held in place with a plastic compound at end of pan. Get this at a supply house or substitute loose asbestos mixed with lube oil. Heat can also be removed by packing wet asbestos around shaft, building it up to area being surfaced.

Set torch to carburizing flame and hold against cam lobe until metal sweats. Extent of sweating area depends on size of welding tip, but it should extend a distance of $\frac{1}{4}$ in. around excess-acetylene flare.

Withdraw flame enough so end of rod is between flame's inner cone and the hot base metal. Apply the rod to sweated area, with tip of inner cone just touching rod. As rod melts a pool forms, flowing over sweating area. Remember: Base metal is too cold if first few drops don't flow evenly.

Remove rod from flame and flow molten pool where wanted by playing flame on it. Return rod to flame and melt off more metal into pool. Keep flame on front edge of molten pool so flame keeps it molten and also sweats base metal in front of pool. Move rod in and out of flame as it moves steadily forward. Float off scale specks, don't cover them up.

Keep supplying the pool from the rod as the pool flows forward. Use rod so it deposits molten metal uniformly at a steady rate over area to be covered. On round work, this is easier if work is turning. Play the flame over the deposited alloy from time to time to remelt and smooth out any high or low spots, left behind.

Plastic Working. The sprocket wheel in the photo above is built up with an alloy that can be worked in the plastic state. Rod is used to wipe away excess alloy. Then, while alloy is still plastic, it's die formed. This establishes the contour without need for grinding. The alloy used is 75% nickel, 25% chromium and boron and wears very well.

Right Flame and Welding Methods for Job

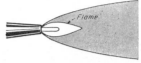

Neutral

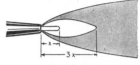

Reducing

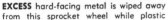

Hard-facing

CARBURIZING. The flame welds but hardly dilutes facing alloy

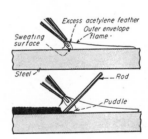

SWEATING (top): Relationship of the tip, rod and base metal

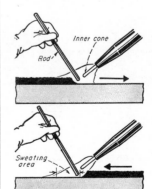

BACKHAND (top) or forehand depends on welder's preference

3—Spraying Methods: They Use Alloys in Powder or Wire Form____

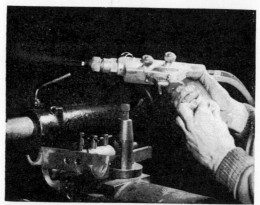

FIRST, adjust powder and flame with knobs on top of the gun. Then, open gun's trigger as shown. Powder is now spraying.

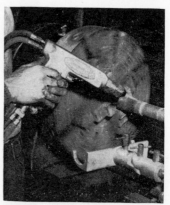

SECOND, build up corners of undercut by hand, while shaft revolves in lathe

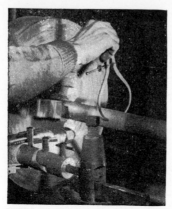

THIRD, run gun back and forth by feed of lathe. Shaft builds up to size here

FOURTH, second stage is fusing alloy to shaft with large gas torch at right temp

MANY ALLOYS may be sprayed on by one of several methods. Spraying gives a thin, uniform coating—from a few thousandths thick up to 0.60 in. Contours and shapes are easily coated. Spraying also handles alloys not easily applied in rod form.

In one method, alloys are sprayed from a suitable gun, then fused with a torch. In another, powdered alloy and plastic binder form a wire that's fed through a typical metalizing gun. Resulting deposit is then fused. Still a third way uses a special torch to apply a coating and fuse it in one step. Let's take a look at each scheme:

Spraying Alloys Available in Powder Form

Powder spraying is done with a gun roughly like that for metalizing, but with a powder hopper and a carburetor for mixing powder with air. This applies alloy but does not fuse it. A separate torch is used to fuse sprayed alloy to base metal. It can also be done in a furnace or by induction heating.

Typical Job. Conveyor shaft, in photo, of cold-rolled steel, wears badly in center. First, undercut it so there is at least 0.030 in. for hard facing all around. Never make sharp corners in groove, or oxide film will develop porosity when fusing and the coating will peel back.

Blast with angular-steel or aluminum-oxide grit, not round shot. Caution: Don't touch base metal. If it must stay in lathe overnight, cover it.

In lathe, blow shaft with compressed air while it revolves, then wire-brush. Adjust to neutral flame of 13-psi oxygen and 11-psi acetylene. Have 75-psi air pressure to propel powder.

Spraying. Turn on powder, then open gun's trigger and build up corners of undercut, holding gun in hand. Next, set gun in holder attached to toolpost and adjust so spray is against shaft's center. Start thread-cutting feed and reverse at end of each stroke. This sprays even layers of alloy. Spray on enough alloy to allow for shrinkage. This depends on coating thickness. Check with set calipers.

Fusing. With shaft built up to right diameter, use large torch with neutral flame. Play back and forth over revolving shaft, heating evenly. Then hold flame at one end of shaft, photo, until shaft has a transparent appearance. Move flame slowly along shaft as alloy shines under flame. Now alloy is fused to shaft. With asbestos gloves, remove part from lathe and drop into a box of powdered insulation for slow cooling overnight. This step is important or deposit will crack.

Unless cooled gradually, hot alloy shrinks faster than inside of shaft, and cracks. If alloy is inside a sleeve, fast cooling pulls alloy away from wall.

Spraying Alloys in Wire Form

Besides the many wires made for various metalizing jobs, hard-surfacing wire is made of powdered alloys extruded with plastic binder. Plastic evaporates in metalizing-gun flame, so only the alloy deposits on base metal. Alloys may be same as for powder spraying—only difference is that wire is deposited with usual metalizing guns.

After spraying, deposit is similar to metalizing job, which means it's not fused. There are various ways of fusing: For small to medium work, use an acetylene torch. For larger work, use two or more torches, or put in a furnace.

Liner Job. Photo shows hard surfacing inside of a gas-engine cylinder liner. This isn't a common application and is probably one of the toughest to do right. But cylinder will have from three to four times more life than new.

First bake casting at from 450 to 500 F to burn oil out of cast-iron "pores." This is important on old oil-soaked castings. Baking takes about an hour, depending on heating method. On liner, Tempil-stick mark shows when right baking temperature is reached. This is a temperature-indicating stick that melts at given temperature.

Bore out to get rid of ring steps, fluted wear and scoring. Finish bore with a No. 18 or 20 thread final cut for anchoring alloys. Then grit blast. Preheat from 200 to 300 F. Keep gas torches on outside as shown to maintain this temperature while spraying.

Spraying. Metalizing tool shown is fed back and forth by lathe for smooth, even deposit. Spray cylinder up to original

WIRE BEING SPRAYED inside gas-engine liner is a powdered hard-facing alloy bound by a plastic that vaporizes in flame

diameter, plus about 0.030 in. for finishing. Because alloy used here is over 50 Rockwell C, and hard to machine, cylinder will have to be finished by grinding.

Fusing. After spraying, bring up to shiny red heat for fusing. For this job, a furnace is best, as it will give even heat at all parts of the liner. Then allow to cool gradually over a few days until room temperature is reached. If liner cools too fast, or unevenly, alloy coating pulls away from the liner. For most jobs, but particularly for one like this cylinder liner, slow, uniform cooling can't be overemphasized.

Spraying Powdered Alloys and Fusing Them in One Operation

This method feeds powdered hard-facing alloy onto the base metal through an oxyacetylene flame, as when spraying powder and wire. But here the process differs.

Powder, with flux if needed, is projected through the flame by an auxiliary gas, such as carbon dioxide, nitrogen or hydrogen, by means of a spray gun that is specially designed for the purpose.

Continuous Fusing. Temperature of oxyacetylene flame is adjusted so deposit is continuously fused or braze-welded to the base metal. This makes the process something like oxyacetylene hard surfacing, as the base metal is prepared and must be preheated and treated in the same way.

One advantage is that coatings as thin as 0.010 in. may be deposited. Up to 0.150 in. of coating has been applied. There is a full bond with base metal, as the deposit is completely melted without a second fusing operation.

Control. Accurate control of base-metal temperature is obtained by increasing or decreasing amount of flame, then proportioning the projecting gas (and alloy), and the processing gas. Both base-metal and alloy temperatures are controllable, all the way from melting temperatures down. This gives a high-quality metal-bonded hard surface.

Equipment. Photo shows torch, control box with air- and gas-pressure regulating valves, and hopper for powdered metals. Torch resembles usual welding torch but has three control knobs on the handle. These are part of a total of 16 controls on the apparatus for accurate regulation of all oper-

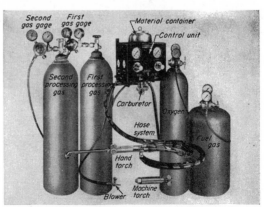

POWDERED ALLOY FUSES to base metal as deposited under regulation of controls that help operator make needed adjustments

ating conditions. Torch also applies plastics or synthetics.

Alloys containing beryllium, bismuth, carbon, cobalt, molybdenum, tungsten, etc, may be deposited with this equipment (not yet commercially available to the industry).

COKE CHUTE, hard faced with chromium-boride crystals applied in paste form, is expected to last about eight times longer

4—Spread Alloy in Paste Form as if Icing a Cake

Some hard-facing materials come in paste form. One of these is chromium-boride, which reads 68 to 72 on Rockwell C scale. It has good resistance to extreme abrasive wear found in materials-handling chutes, fan blades, pump impellers, scraper blades, crusher rings, screw conveyors, etc.

In addition to resisting abrasion, chromium boride is good against corrosion and impact, has hot-hardness qualities. It's easy to apply, by carbon arc or oxyacetylene torch.

Applying Paste. Clean base metal thoroughly. If there's scale or rust, remove by grinding, shot-blasting or sanding with abrasive cloth. Amount of preparation needed depends on previous service of material to be coated. Watch out for problems like oil-soaked castings. If not baked out, oil can cause hard-surfacing job to fail.

Ordinarily, preheat the work to about 100 F. Play oxyacetylene flame or blow torch back and forth across piece so it heats up slowly and evenly.

Paste comes in cans, mixed with water. If it's too thick, add water and stir to the right consistency. Then spread it on metal with a putty knife, a stiff brush or a piece of wood. Coating thickness runs about 1/16 in. Putting on the paste is much like icing a cake. Next, let paste dry. The preheat evaporates the water quickly.

Sweating. If metal to be coated is 14-gage or less, use oxyacetylene or carbon arc torch. Spread paste about 1/64 in. Heat base metal up to sweating temperature. Fusing of base metal and paste will show as a dark line ahead of the torch flame. Move torch across work, keeping the dark line ahead of the flame.

For metal of over 14-gage, use standard carbon electrode of 3/16, 1/4 or 3/8 in. dia, depending on thickness of base metal. Have straight polarity with just enough current to maintain arc. Hold carbon about 80 deg to the vertical.

Using Carbon Arc. Work with a quarter-moon weaving motion, not over 2½ times diameter of carbon. Point carbon in pencil sharpener or by holding against grinding wheel. If handled right, carbon melts so point stays as it should. Have about 3 in. of carbon sticking out of holder.

Keep black spot of carbon floating on the surface just in front of the point. After sweating-on first weave, pass across work and return with about half the width of the second weave overlapping the first. If you get porosity on the edge of the weave, pay no attention. It will disappear when you overlap the next weave.

Use paste made of chromium-boride crystals only on rough work that is not to be finished.

Sweating-On Powder. Chromium carbide comes as a powder, mixed with flux. Sprinkle it on the base metal and fuse with a carbon arc or carbon arc torch for a smooth, dense, abrasion-resisting surface, which can be as thin as 0.025 in.

Hardness depends on mixing with base metal, runs about 54-61 Rockwell C for one layer and 57-63 for second. Deposit develops full hardness in resisting scaling at high temperatures. Corrosion resistance is like that of stainless steel. Don't use this type of alloy where impact is great.

FOLLOW THESE EASY STEPS

1 Preheat base metal with torch to only about 100 F

2 If paste is too thick, mix with water to thin out

3 Spread on about 1/16 in. thick, as if icing cake

4 Let paste dry—the preheat does the job quickly

5 Use about 1/4-in.-dia carbon, pointed like pencil

6 Use straight dc polarity and hold carbon as shown

7 Weave quarter moons, and use lowest amperage

8 Reverse, overlap the previous stringer, as shown

1 Grit-blast the tungsten-carbide inserts to remove the surface oxides and other films. Machine surface of the base metal, then do a thorough degreasing job

2 Flux surface of the base metal with brush. Before fluxing, clean in carbon tetrachloride. Degreasing, cleaning and grit-blasting must be done properly

3 Position silver-solder strips after fluxing. Place one strip on base metal, a copper screen on the solder strip and second solder strip on copper: flux

4 Put the tungsten-carbide inserts in place, on top of "sandwich" of silver-solder strips and copper screen. Six inserts are butted together on this job

5 Heating to 1350-1400 F by the propane-air burners against underside of base metal bonds the inserts. Test bond by tone of a hammer blow as shown

6 Attaching hard-faced peg to moving rotor of coal pulverizer. Coal gets pulverized by impact and attrition between moving and stationary pegs shown

5—Brazing Tungsten-Carbide Plates

Diamond-substitute alloys of Type III, A, used here, come as inserts of various shapes. Some contain 90 to 95% tungsten carbide. Remaining 5 to 10% is nickel, cobalt, iron or other elements to give the desired properties.

Braze these carbide plates to wearing surfaces because brazing heat (torch flame) will not melt the carbide materials. They do "wet" easily, so can be bonded to metal surfaces.

Because stationary and moving grinding pegs of the coal pulverizer shown take heavy punishment from impact and abrasive wear, they are hard faced with inserts of tungsten carbide. Here are steps in typical application:

Operations. (1) Machine surface of base metal to be hard-faced. (2) Degrease with trichlorethylene, clean with carbon tetrachloride. (3) Grit-blast tungsten-carbide inserts with silicon-carbide grit to remove surface oxides and other films.

(4) Flux machined surface of base metal with brush. (5) Cut strip silver (silver solder) to size and dip in flux. Then put in position on base metal. (6) Dip a copper screen or copper sheet in flux and place on strip silver. Because insert's coefficient of expansion is not the same as the base metal's, always put copper between. This also takes impact shock, as inserts are hard, with no "give." (7) Flux second strip of silver alloy and place on copper. (8) Put inserts in position on top layer of strip silver. (9) Braze by directing flame under base metal until both strips of silver melt. (10) Test bond by tapping inserts with a hammer. Tone indicates the quality of the bond.

Tungsten carbide can also be used to tip your maintenance tools. Technique is same as explained on this page.

More About the Electric-Arc Method: How to

Getting Ready

WEAR on this shaft comes from highly abrasive coal dust working into bearing

WHICH ELECTRODE?

Build up shaft with iron-base rods having less than 20% alloying elements. For a surface resistant to abrasion, finish with Type I, A, 4 iron-base rods having more than 20% alloying elements

Electrode dia, in.	Current, amp straight	Voltage, arc volts
1/16	20 to 40	17 to 20
5/64	25 to 50	17 to 20
3/32	30 to 80	17 to 21
1/8	70 to 120	18 to 22
5/32	120 to 170	18 to 22
3/16	140 to 240	20 to 24
7/32	170 to 300	21 to 25
1/4	200 to 350	22 to 26
5/16	250 to 450	23 to 27

Electrode dia, in.	Current, amp reverse	Voltage, arc volts
1/16	20 to 40	17 to 20
5/64	20 to 60	17 to 21
3/32	30 to 80	17 to 21
1/8	80 to 130	18 to 22
5/32	120 to 180	18 to 22
3/16	140 to 250	20 to 24
7/32	170 to 300	20 to 24
1/4	200 to 400	20 to 24
5/16	250 to 500	22 to 26

ELECTRODE size, amperage and polarity

Applying the Facing

4 Ventilate well if the welding must be done inside some confined space

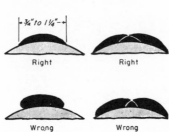

5 Bead must spread as shown. Be sure it overlaps to avoid porosity, slag

6 Run two or three beads around circumference few inches at shaft's end

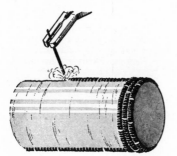

10 Run next bead on opposite side of shaft to distribute the heat evenly

11 Use bead sequence shown to prevent shaft from distorting permanently

12 Stagger the bead ends all around the shaft, as here, to strengthen pads

Put It to Work Hard-Facing a Worn Stoker Shaft

1 Clean shaft with safe solvent, such as trisodium phosphate; rinse and dry

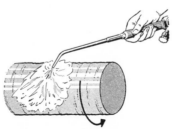

2 With a gas torch, preheat part to be coated—just enough to remove chill

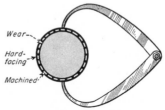

3 Set calipers for finished size plus 20% for finish machining, shrinkage

7 Turn shaft around and run some beads on other side to spread heat evenly

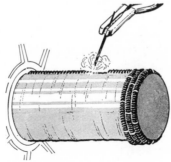

8 Run bead along shaft. Start from outer edge and work in from the end

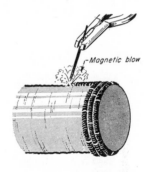

9 Stop if you have magnetic blow (this happens with ac); start at other end

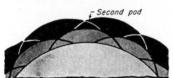

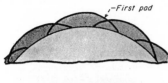

13 Laying second pad, with Type I, A, 4 alloy: put beads between first ones

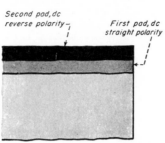

14 First pad takes dc straight polarity, second pad, dc reverse polarity

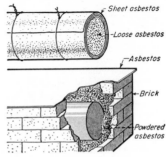

15 Cool slowly by covering shaft with sheet or powdered asbestos, as shown

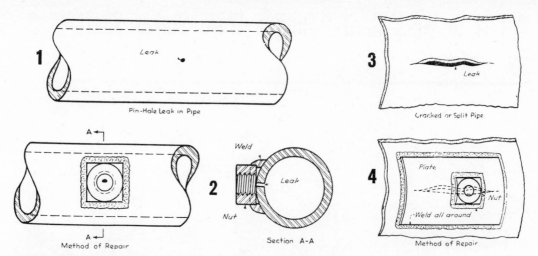

Around the pin hole shown in **1** arc weld an ordinary machine nut **2**. Leakage will continue through nut center, but will not disturb welding operation. Then shut off nut leakage with a machine bolt and packing compound. To seal a pipe crack under pressure (exaggerated in **3**), first weld a machine nut over a small hole drilled in a large patch. Next weld the patch over the crack as shown in **4** and finally seal the remaining leak through the nut center with bolt and sealing compound, as before.

You Can Weld Leaks Under Pressure

Arc welding can easily seal a machine nut around the leak. Then screwing in a short bolt plugs the flow. Avoids plant shutdown

▶ ANY KIND of a piping-system leak is bad, but when this leak occurs in a high-pressure line, it is really serious. This is particularly true if the line cannot be bypassed. If bypassing is impossible, usually the only thing left is a plant shutdown, and every operating engineer knows only too well the embarrassing situations that accompany unexpected shutdowns.

When welding high-pressure lines became accepted practice, many expected that some of the worries connected with leaks would vanish. But, operators ran into trouble when they tried to make repairs without shutting down.

Welding Difficulties. Oxy-acetylene welding was generally unsuited for making such repairs for several reasons, the two most common being: (1) Cooling effect of the steam, water or gas in the line usually prevented proper fusion of weld metal to the pipe, or (2) if it was possible to heat the pipe to fusion temperature, the pipe, weakened by heat penetration, was likely to blow.

These two difficulties were much less with metallic arc welding, but, nevertheless, it was nearly impossible to repair one of these leaks by arc welding without taking the pipe section out of service. It was usually found that a patch could be welded to the pipe except for the final closing of the weld. At that point, nearly every attempt to complete the weld failed, because pressure would build up under the patch and blow away the final sealing drop of weld metal. The leak remained, but shifted to the edge of the patch.

Here's the Trick. This difficulty can be overcome by combining mechanics with welding. To close a pin-hole leak (Fig 1) weld an ordinary machine-bolt nut over the leak. This allows any water, air, etc, to escape through the hole in the nut during welding, making it possible to complete the weld.

The result is the simple yet effective patch shown in Fig 2. To stop the leak simply screw a bolt into the nut with sealing compound.

What about a crack? A cracked or split pipe, see Fig 3, can be closed in the same way. In this case, size of the leak is usually too large to be covered by a nut, so a piece of plate stock is used for the patch. Always cut a small hole in the plate after it has been shaped to the pipe. Then weld a nut over the hole and weld the plate to the pipe, (Fig. 4).

Although this type of repair is usually only a temporary measure, it can be considered permanent if the bolt is finally seal welded in the nut.

The repairs described here have been used a great deal to stop leaks in high-pressure air and water lines. I know of no reason why the same methods cannot be used on lines carrying other gases or liquids, providing there is no safety hazard. Such a hazard might exist in attempting repairs on high-pressure steam lines.

17
Fire fighting, accident prevention

Know How to Use the

TYPE OF FIRE

CLASS A: Fires of wood, paper, cloth, rubbish, etc. Use pump tank, soda acid or just water

CLASS B: Fires of oil, gasoline, grease, paint, etc. Use CO$_2$, dry-chemical or foam type

CLASS C: Electrical-equipment fires. Put out with CO$_2$, dry chemical or vaporizing liquid

Five Handy Fire Fighters

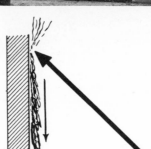

SODA ACID OR WATER. Use on class A fires only. Start at top of wall and work down, wetting surface as you go to knock down flame. Then go back and drench thoroughly to prevent reflash. Range of stream is from 30 to 50 feet

CARBON DIOXIDE (CO$_2$) is for class B and C fires. Direct the gas across the flaming area and sweep the fire before you. CO$_2$ is a nonconductor, leaves no residue, and does not affect equipment or even foodstuffs. Range is 3 to 6 feet

How to Operate

SODA ACID OR WATER. Turn upside down, bump on ground to create pressure. Main extinguishing agent is water, so keep away from the electrical equipment

CARBON DIOXIDE (CO$_2$). Unhook hose, pull out lock pin, squeeze trigger or turn handwheel. Discharge at fire base. Keep the stream range at about 3 to 6 feet

Right Fire Extinguisher

FOAM. Use n class B, alsoo on class A fires. Back the stream of foam off the opposite side of the wall (if possible) to build up a solid blanket of foam without splashing the burning liquid and without spreading it. Range is 30 to 40 feet

DRY CHEMICAL. For class B and C fires. Direct stream into nearest corner of fire (if possible) and sweep across flame. Chemical releases a smothering gas on fire; fog of dry chemical shields operator from the heat. Range is 8 to 12 feet

VAPORIZING LIQUID. For class B and C fires. Play stream against side of container or wall to break it into droplets so liquid vaporizes quickly. Remember, it's the vaporized *gas* and not the liquid that puts out fire so gas must contact fire

FOAM. Turn over, bump on the ground. Aim at base of wood fire, float foam on top. The foam is a flowing blanket of durable bubbles filled with an inert gas

DRY CHEMICAL. Remove ring pin, free hose, push lever down, squeeze nozzle-handle—or, if another make, unhook hose, pull lock pin and squeeze trigger

VAPORIZING LIQUID. Unlock by turning handle. Pump stream on burning material. Range of stream is 20 to 30 ft. V-L is nonconductor, won't damage equipment

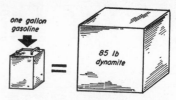

2 You wouldn't monkey with this much dynamite in your plant, would you?

1 Heavy vapors spread quickly along floor. Then open flame can destroy plant. Light fumes are bad, they intoxicate.

Clean That Crankcase- But . .

National Safety Council

> *Don't use gasoline as a solvent*
>
> *Don't use carbon tetrachloride*
>
> *Don't use kerosene*
>
> *Don't store solvent in open container*
>
> *Don't dump used solvent in sewer*
>
> *Don't store volatile solvents at temp over 100 F*
>
> *Don't do job without face shield and rubber gloves.*

▶ WHY WAS THE pipefitter dead? Because he was fitting pipes in the small basement under the machinery room.

What's wrong with that? Nothing, only that some jerk was cleaning a refrigeration compressor crankcase and motor windings with carbon tetrachloride, directly above the pipefitter.

And why was he a jerk? Because he didn't know that carbon-tet fumes are heavier than air and poisonous.

Cleaning Crankcase. Winter or early spring is a good time to check, overhaul and repair refrigeration compressors, as load is lightest then.

One of the first things to do is to drain lubricating oil from crankcase. Change crankcase oil during this annual check-up, also several times during the year. After draining, flush the crankcase to wash out all sludge and carbon residue. Then fresh lubricating oil won't become contaminated.

Gasoline Is Bad. A common but dangerous flushing liquid is gasoline. Gasoline is dangerous because: (1) vapors are explosive (2) vapors have toxic effect on respiratory system (3) it burns hands and skin, causing infection or dermatitis. Chemical action dries out skin's natural protective oils.

Most dangerous hazard is explosion of gasoline vapors in crankcase. Hazard also exists when other low-flashpoint solvents, such as naphtha or benzol are used. Gasoline, naphtha and benzol have low flashpoint of 25 F.

Gasoline, when vaporized in air, has about three times explosive energy of TNT. To make this powerful explosive, *vapor must be mixed with air.* It explodes at different mixtures, but most dangerous is six per cent gasoline mixed with 94 per cent air, by volume.

Beware! A small residual coating of gasoline left in crankcase can generate enough explosive vapor to blow up the machine. And low-flashpoint solvents form heavier-than-air vapors. When gasoline or any low-flashpoint solvent is used, it's very hard to remove gas vapors that settle in lowest part of crankcase. While using those solvents, and especially gasoline, either inside or outside crankcase, vapors escape along floor to settle in low spots. They flow into drains where they may be ignited.

As an example, pour gasoline in bottom of an ordinary wash-tub. Almost instantly tub is filled to brim with gasoline vapors. Then gas vapors pour over tubs brim same as if water overflowed tub. But invisible heavy vapor spreads over the floor *ten times faster than water.* If a spark is near, there's a terrible explosion.

To see how gasoline vapors spread, we took an open trough, 30 feet long, and put a lighted candle at the lower end, Fig. 3. Other end was raised three feet. We poured gasoline in at raised end. Before gasoline ran 3 feet, fumes reached candle, flashed back and lighted what gasoline had not evaporated. Air seems to offer no resistance

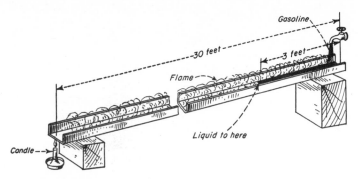

3 Before gasoline runs three feet, its vapors reach candle flame twenty seven feet away, ignite and flash back to liquid. That's why gasoline is dangerous

to gasoline fumes—they travel fast.

So What? A little gasoline used to flush a crankcase may spread an explosive mixture over refrigerating plant. Remember! One gallon of vaporized gasoline in air has explosive energy of 85 pounds of dynamite.

Low-flashpoint vapor remaining in crankcase tends to break down quality of lubricating oil. That might even damage machine bearings.

Vapors given off from these solvents are highly toxic. Breathing them in enclosed space could overcome workmen in short time. Toxic effect also tends to make workmen dizzy or drunk.

Gasoline splashed on hands can badly burn or blister them from chemical effects. Gasoline may cause skin breaks to become infected and hard to heal.

How About Kerosene? In some refrigeration plants kerosene is substituted to flush sludge and carbon residue. Kerosene doesn't have explosive potential of gasoline, *if kept below* 100 *F*. But it does create a fire hazard, if room

temperature is high and a spark or open flame is near. In engine-room where temperature is above 100 F, kerosene gives off vapors similar to gasoline.

Carbon Tetrachloride. This is often used as solvent to remove sludge from metal parts, but it's a killer. Carbon tetrachloride evaporates rapidly, and has a very toxic vapor. In a confined space, with no direct ventilation, vapor quickly overcomes any worker.

To avoid flammable or explosive hazard of low-flashpoint solvents and kerosene, carbon tetrachloride is often mixed with them. While this reduces fire and explosion hazard, it does not remove toxic-vapor hazard. It's difficult to ventilate confined space around a compressor to eliminate toxic vapor unless you use a forced-circulation blower.

Safe Solvents. There are several petroleum solvent products, also chemical solvents, with a very high flashpoint for safely flushing out compressor crankcases. But these solvents are flammable. Keep away from open

flames or electrical sparks. These safer solvents are cheaper and can be obtained easily. Nearly all oil companies advertise and sell them.

A good safe solvent is 25 percent methylene chloride, 70 percent Stoddard solvent, and 5 percent perchloroethylene.

Storing. When draining solvents from crankcase, drain into a closed positive sealing container. Then remove container from plant and dispose in a safe place. Never drain solvents from crankcase into drain or sewer.

When storing these solvents in plant, store in compliance with all local fire regulations, as to amount which can be stored, etc. Container must be approved by the Underwriters' Laboratory. Keep container at temperature below 100 F and away from open flames or electrical contacts.

Other Solvents. There are solvents that are good cleaning agents, easy to use, and cheaper than ones mentioned. These are alkaline chemicals. They dissolve in water, make excellent solutions to clean and flush crankcases. Solutions are made by dissolving soda ash or trisodiumphosphate in water. For exact proportions, chemical manufacturer gives correct formula.

After flushing crankcase, rinse thoroughly with warm water to prevent corrosive effects. Drain water from the crankcase and dry it out.

Protection. Workmen using or handling strong alkali solutions should wear protective clothing, such as face shields and rubber gloves.

Where a strong alkaline solution is objectionable to flush out a crankcase, a weak solution of dissolved alkali will work. First fill crankcase with weak solution of alkali and water and heat with a steam hose in water until it reaches 180 F. Hot alkali removes heavy crankcase sludge and it's easily disposed of by draining to a sewer.

DRY-ICE SAFETY

WEAR GLOVES when handling dry ice

▶ Dry ice (solid carbon dioxide, temp about −109 F) can cause frostbite similar to a severe burn. In a tight container, the pressure can cause an explosion. Here are some good rules to follow when handling dry ice:

1 Wear gloves when handling either dry ice or metal that has been in contact with it.

2 Never put dry ice in any drink; a small piece can cause frostbite.

3 Never place dry ice in any tight

container except one designed for it.

4 When sawing dry ice, see that guards are in place. Use a wooden strip to push ice against the blade.

5 In a small enclosed space, gas from dry ice can cause suffocation. If warned by quickened breathing, dizziness or ringing in the ears, get to fresh air immediately.

6 Learn to use artificial respiration.

*From National Safety Council's SAFETY INSTRUCTION CARD NO. 417

How Electricity Affects The Human Body

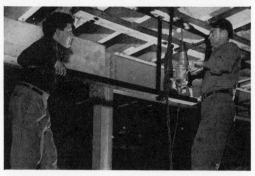

SLIGHT GROUND may tickle, but it can throw lads off scaffold

Man-made electricity killed its first victim in 1879. Today in the United States and Canada, electricity kills seven per million population each year. About half the reported electric-shock accidents are fatal.

Here are main points when considering effect of body shock: (1) circuit voltage and frequency, (2) body resistance between points of contact (3) current path.

Deaths have occurred on commercial frequency circuits at voltages as low as 46 v. It is generally considered that voltages in the order of 24 and below are safe. This may be true from a purely electrical standpoint, but they may cause accidents because of involuntary muscular contraction.

Body Resistance. The outer layer of skin (the gray-beards call this epidermis) is the body's best layer of protection against electric flow. When this outer skin is dry, its resistance is relatively high. For instance, the calloused palm of a lad in your plant will measure about 15,000 ohms per sq in. But when this outer skin is soaked, either with water or sweat, its resistance may drop to below 150 ohms. Once the current sneaks by the outer skin (epidermis) there is but little bodily resistance to the flow of electric current.

In addition to changing its resistivity when wet, skin resistance will increase during sleeping and it is high in sections of the body that are paralyzed.

On top of that, bodily resistance varies with applied voltage. For example, measuring body resistance from hand to hand, using 60-cycle ac source, gives the following values: 10,000 ohms at 50 v, 1200 ohms at 500 v, and 1000 ohms at 1000 v.

Low voltage brings up the blister hazard. When blisters form, the outer skin loses its protective high resistance. And heat resulting from contact with a 120-v circuit will form blisters within a few seconds.

Because of the marked lowering of skin resistance by moisture take extra precaution when sending men out to work with electric tools in damp locations (read what Dr Schweisheimer says about electrical hazards because of sweat . . . the doc's story is right before this piece).

Ohm's Law Holds! Value of current flow through body is determined by our old friend, Ohm's Law. This means that current in amperes equals applied voltage divided by resistance of circuit in ohms. Remember, it is the quantity of electric current flowing through the body that determines to a large extent the resultant injury.

And path the current takes in the body is important. In general, if there are no vital organs in the way, resulting injury is usually confined to burns.

One milliampere of 60-cycle ac generally produces a tingling sensation at point of contact. Of course, this precise value varies with individuals and with frequency. This minimum "tingling" current holds to 300 cps. The tingling decreases as frequency climbs. At 100,000 cps there's more heat than tingling sensation. These minor current sensations are not in themselves dangerous. But they can readily be a direct cause of a serious accident. For instance, the fellow who receives a minor electric shock then falls off a ladder because he jerks away from the circuit.

Let-Go Current. When current flows through the body, muscles in its path contract. This contraction may be so severe the victim cannot release his grasp on the live line.

We define the *let-go current* as that value a man can stand without harmful effects. At least, for the time required to release his hold on the circuit. At commercial frequency (60 cps) the let-go current is 9 milliamperes for men.

Freezing Current. Using a similar yardstick, we define *freezing current* as the value that will contract the muscles to the point where the victim cannot let go the circuit (about 20 ma). Of course, the freezing current is naturally greater than the let-go. This freezing current is mighty important since it requires only a few seconds for blisters to form. These blisters are generally in that portion of the body in path of current. When they form, the epidermis loses its protective resistance and current increases in accordance with Ohm's law. And if someone isn't around to release the poor fellow from the circuit, the result could easily be fatal.

Maximum current that any man can tolerate for a short period without losing consciousness lies somewhere between the freezing current and the current that would stun the heart (ventricular fibrillation). This maximum current lies anywhere between the freezing current of 15 to 25 ma, and the fibrillating current of 75 to 100 ma.

Ac or Dc? To answer this time-honored question, as to which causes most serious shock—ac or dc—let us first review two fundamental facts. Muscles contract when contact is made with an electric circuit. This contraction of the muscles is more severe on commercial frequency circuits than on dc. Where exceptionally good contact is made, or where high voltage is involved, the victim may be thrown clear of the circuit. Therefore, dc has less of a shocking effect on the human body because of (1) bodily reaction and (2) the usual low voltages involved.

Electric Burns. Often burns are the direct effect of an electric shock. These are generally of two types: (1) Heat burns, which result from the electric arc. These are generally formed at the point of contact. (2) A straight electric burning of the tissue through which the current passes.

The true electric burn forms a pinkish spot on the skin's surface. Many such burns, however, may penetrate deeply and take a long time to heal. Burns or blisters may not even be present on the skin after an electrical accident. Unless the skin is soaked with water and the contact area is large, a fatal shock may not leave the slightest blemish.

The electric-shock victim may lose consciousness. In some instances he may recover immediately and in others only after the application of artificial respiration. There are cases on record where men have been knocked unconscious from contact with a circuit, at two points on the same leg or arm, when there was no burning of the tissues. In such cases no current passed through any of the vital organs. It is generally agreed that unconsciousness was caused by severe shock to the system.

National Safety Council

"Green to green, red to red, perfect safety—go ahead," is a basic traffic law for ships; and it's also a good safety rule to remember when . . .

Handling your oxy-acetylene outfit

▶ THE OXY-ACETYLENE PROCESS is one of industry's most versatile tools for welding, cutting and heating. Efficiency and safety in its use go hand-in-hand with careful observation of precautions and safe practices. The "green to green, red to red" theme fits for two reasons: (1) The oxygen-pressure regulator and oxygen hose are always colored green and the acetylene-pressure regulator and acetylene hose are always colored red. The two should never be interchanged. (2) Oxygen and acetylene are different animals and require different handling. Consequently, if we think of "green" precautions when we're dealing with oxygen, and "red" precautions when we're handling acetylene we'll be on the safe side.

"Use no oil" is probably the most important thing to keep in mind when using oxygen. Never let oil or grease in any form come in contact with oxygen under high pressure because they may ignite violently. Every piece of equipment through which oxygen may pass (from cylinder to blowpipe) must be kept entirely free from oil and grease. Too, oily or greasy hands, gloves, clothing or other material must never be exposed to oxygen or air that's rich in oxygen. Even if your clothing is not oily, if it becomes saturated with oxygen, a tiny spark could turn you into a human torch.

Never use oxygen from a cylinder unless through a (green) *oxygen* pres-sure-reducing regulator. Regulator and oxygen cylinder valve outlet both have righthand thread connections. Don't connect your "green" hose to anything else. Don't use any pipe compounds, oil or grease for making connections, and *don't* force connections that do not fit. Use only hose and connections that are made especially for oxygen service.

Always call oxygen by its proper name—"oxygen." It should never be called "air" nor confused with compressed air. Don't ever use oxygen in pneumatic tools, oil preheating burners, to start internal-combustion engines, blow out pipelines, "dust" clothing or workpiece, for pressure tests of any kind, or for ventilation. Don't store oxygen cylinders where oil or grease from overhead belts or cranes might fall on them. Oxygen, itself, will not burn, but it supports and accelerates combustion, hence causes oily materials to burn with great intensity.

Acetylene is a highly flammable and explosive gas and should be respected as such. But make a habit of calling it "acetylene" and not "gas." It's far different from city gas or furnace gas. Air and acetylene mixtures between 2.6 and 80% acetylene have been shown to be explosive if ignited. Under certain conditions acetylene may ignite at a temperature as low as 650 F. Never "crack" an acetylene cylinder valve nor release acetylene to the atmosphere near other welding or cutting work, or near sparks, flame or any

other possible sources of ignition.

Never use acetylene from a cylinder except through a (red) *acetylene* pressure-reducing regulator. Acetylene regulator and cylinder valve outlets have lefthand thread connections. As with oxygen, only use hose and connections made especially for acetylene service. Again, don't use pipe dope, oil or grease for making up joints and don't force connections.

The oxy-acetylene flame itself is rarely the cause of a fire—probably because it's obvious that flame and combustible material should be kept away from each other. Big danger comes from the heat of the metal being welded or cut and from flying sparks. Another fire hazard lies in using acetylene hose that's leaking or in poor condition. Acetylene-hose rupture could produce a messy situation.

A deep respect for the oxy-acetylene outfit, backed up by firm knowledge on how to handle, use and store oxygen and acetylene, is your best bet for avoiding trouble. Rules and recommendations, based on the combined experience of oxy-acetylene equipment users and manufacturers, are available from the National Fire Protection Association, National Safety Council and International Acetylene Association. Other instructive literature can be had from your oxygen and acetylene supplier. It's all for one purpose—*your* safety.—*Courtesy Linde Air Products Co, a division of Union Carbide Corp*

How's your accident IQ?

There are ten basic causes of accidents. Use this checklist as a guide in evaluating the safety program in your own plant

I have found that when a plant encounters operating troubles, management always takes one course of action. "Find the cause of the problem." Will the same approach work in Safety? Yes! But let's look at what happens when an accident occurs and someone is injured. The injured employee's supervisor is required to investigate and fill out an accident report form. The usual report form has questions on location, equipment concerned, action taking place, and protective devices used.

Primary cause. Usually the supervisor must answer the question, *Primary cause of accident?;* and usually it is answered incorrectly.

Why is this the case? Since many supervisors do not investigate in depth, they come up with what they think is a "suitable answer." Another reason is that a supervisor may tend to "cover up" for the employee or for himself.

As a result of a program of "in depth" accident investigations, a list of ten basic causes of accidents was drawn up, and is reprinted below. These were used to train supervisors in *accident investigations*. Please note the ten items concern "people." This is because our study indicated that most accidents are caused by people. And an equipment failure in most cases is the result of some person not doing his job. Thus the greatest benefit in a *safety program* is obtained by concentrating on the "people" aspect. Only then (in most cases) will equipment failures take care of themselves. Here is our list of the ten basic causes of plant accidents:

1.—**Disobedience of rules.** Most plants, large and small, today have *safety rules.* These may be plant-developed or supplied by the insurance carrier and vary from a few broad guide lines to a complete list of detailed rules. In either case, failure to comply will cause many accidents.

2.—**Ignorance of rules.** A new employee usually receives indoctrination which covers basic safety rules. But this teaching must be continuous, as employees will forget. And new equipment, new processes with new hazards create a pressing need for continuous updating of safety rules.

3.—**Inattention to work,** or failure to keep alert. Employees become preoccupied and their minds tend to wander. This may be due to personal problems at home, or perhaps thinking about a forthcoming hunting or fishing trip. Did you know that accidents actually increase when a company and union approach a strike deadline?

4.—**Lack of complete instructions.** Here the supervisor fails to realize the need for detailed instructions. He may assume the employee "knows how," and not realize that a few added words will help the man to perform his assignment safely.

5.—**Not following instructions.** Many times the supervisor will give detailed instructions; yet they are not followed. The man may feel he "knows how" and not realize the importance of following the instructions as they were outlined.

6.—**Interference by others**. This may vary from simple horseplay to the case where another employee closes a switch or opens the wrong valve. Signs and locks can help prevent this kind of accident.

7.—**Not using protective devices** or safety equipment. Goggles, masks, gloves, aprons, suits and other protective equipment are not comfortable to wear, but wearing them is certainly the lesser of two evils. The use of such protective equipment requires education, training and supervision.

8.—**Lack of skill or experience.** In any manpower pool there exist experience and skill in varying degrees. Thus the supervisor should take care to select personnel with the required skill and/or experience.

9.—**Poor judgment**. The supervisor and employee must visualize any potential hazards and take necessary precautions. This often is based on judgment. And we know that many accidents are caused by poor judgment.

10.—**Taking chances.** "I thought I could get by" is too often the answer of an injured employee. He truthfully tells that, being in a hurry, he "took a chance"—and lost.

When supervisors begin to dig a little deeper looking for the real cause, they will improve their accident IQ. Note that the word careless covers all 10 items. This is one good reason why it should never be used in a report. And remember that "be careful" is a meaningless statement in job instructions; be careful of *what?* ●

18
Maintenance management

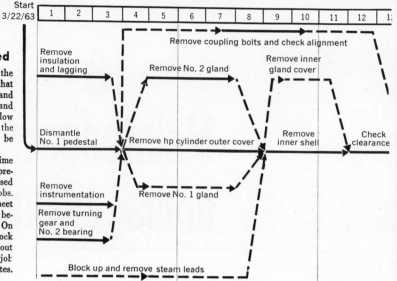

Here's how a simple critical-path chart is used

Solid straight line running through the center is the critical path; shows that the turbine must be buttoned up and ready to run in 50 days. Dotted and solid straight lines above and below are noncritical jobs and indicate the many other operations that must be going on at the same time.

Solid portions indicate *estimated* time to do the work. Dotted portion represents extra *float* time that may be used for various reasons on noncritical jobs. But lines above and below *must* meet the target dates indicated, or they become critical and extend outage. On this job, work goes on around clock because each day of outage costs about $30,000. This chart is posted on job site so everyone is informed of dates.

Major turbine overhauls in record time

A new approach involving sophisticated instrumentation, critical-path charts and improved turbine operation. All save time—and money—for Con Ed

The larger and more efficient steam turbine units of the past few years have called for an updating of our repair procedures. Now we monitor and chart each unit as to output, efficiency and reliability. Our repair and operating experiences have taught us that we need more complete information about the internal conditions of the large units before we open them.

The economics now include (1) the cost of the outage itself, of course, and (2) the difference in cost of operating much less efficient units during the outage period. We have found that complete instrumentation—essential to good operation—can, with proper analysis, help us better evaluate the unit's internal condition for the purpose of efficiently scheduling overhaul outages.

Instrumentation. At first, the connected instruments were regarded as an obstacle to overcome before getting at the *meat* of the job. They were often damaged by careless removal, then put aside and forgotten during the overhaul period. And, because they were not needed to run the unit, in the rush to complete the job, some in-

struments were not reinstalled. In addition, some of those that were installed were not always operative. The fault? *Not scheduling time* for taking proper care of these instruments at removal and during outage, or for properly installing and calibrating them at reinstallation. This problem of looking after the instruments must be recognized and specifically provided for, since proper care insures safe and efficient operation of our units. We now have an instrument group within the operating department of each station.

A technician from this group is assigned to work *with the repair department* at the start of each job. He is responsible for proper removal and tagging of instruments, whether pressure gages, thermocouples, supervisory instruments or recorders. After overhauling as necessary, these instruments are stored for the repair period. During reassembly, the technician sees that each instrument is properly reinstalled by the repair department and then calibrated as necessary.

We, as a repair group, requested this. Why? Because

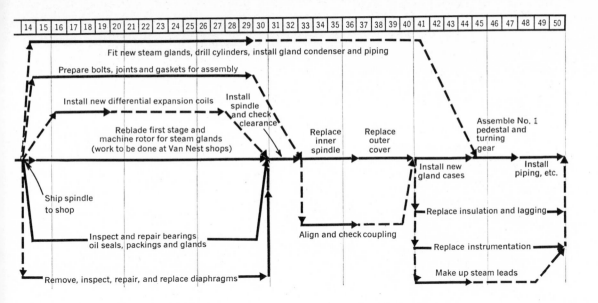

we also use data from this instrumentation to help us determine the turbine's internal conditions, and plan a realistic overhaul time schedule.

We developed tables, based on our records of actual experience, to show how instrumentation helps decide when to schedule unit outages. Data from instrumentation also averted serious damage that could have occurred if operations had continued. In *all* instances, there were no other signs of distress, such as squealing, smoking, pedestal or floor vibrations, or excessive bearing oil temperatures. But since efficiency and reliability instrumentation represents added cost, its many advantages must be clearly understood by system managers, repair departments and plant operators.

Scheduling outages. At Con Ed we must schedule turbine outages within somewhat narrow time limits, dictated by system load requirements. On our newer large units, overhaul time is longer than for older ones.

To just *open* an hp-turbine section — which is very complex—takes a minimum of eight hours turning-gear roll, preferably 16 hours. Then about four days are needed to remove instrumentation, inlet piping, insulation, outer cover bolts, outer cover, inner covers, blade rings and seal casings; to uncouple; check clearances and alignment; and remove spindle. At this time the cylinder joints and spindle are still uncomfortably hot.

To close the unit, ready to run, takes another five to six days, working on a 24-hour basis. Of course, any additional work, such as replacing seals or blading, will extend this outage time. That's why we use the critical-path chart shown above. It's most useful in planning and in following the progress on major repair jobs. This approach is an excellent control tool for both the manager

and individual supervisor and has the advantage of releasing them from a multitude of details.

Today we only open the turbine section that indicates trouble; and open other sections only if further trouble or damage is indicated. We follow this policy because, in the past, where no serious defects were found by an inspection opening, we did more harm than good.

One of our topping unit turbines has not been opened in 13 years. On some of our larger compound units, the low-pressure sections have not been opened for over 10 years. They still operate efficiently.

Mobile repair group. Our repair department has a turbine overhaul group that travels to all our power houses. They do major overhauls on turbines, generators and rotating auxiliaries associated with the turbine. These highly skilled men do all the mechanical work associated with the job. Their major task is adhering to the scheduled time for the outage.

Our central shop does all the machine work, reblading and rebuilding necessary for the proper overhaul and repair of turbines, generators and auxiliary equipment.

This central shop also has a welding group, electric shop and electric field repair group. Along with the machine shop, they also provide highly skilled mechanics on job site for special welding, generator, exciter or large motor repair, rewinding or special machine work. They also reblade turbines, when it is more economical to do so, on the job site rather than at the shop.

Having our own central shop gives us full control of job progress and priority—essential in getting a major unit overhaul done in scheduled time. Also, we can make immediate decisions on repairing and rebuilding the equipment where an unusual repair must be made.

BOILER INSPECTION REPORT calls for specific conditions of all critical components being examined by the inspector

Inspection reports improve PM programs

Maintenance engineers have set up an inspection and repair program with scheduling handled by a computer and costs accumulated and studied. But the actual inspection can be a hidden weak link unless reports are well designed and executed.

Let's assume that maintenance programming, scheduling and cost studies and repairs are carefully thought out and supervised. But what about carrying out the actual inspection?

Does your plant have a systemized inspection procedure? If it doesn't, inspection can be the weakest link in the maintenance chain. Inspection of a major piece of equipment is usually done according to a set procedure, by trained and experienced personnel. Certain readings or measurements are taken and a written report is prepared.

But what about the multitude of auxiliary equipment which can also fail and interrupt production, and increase maintenance costs? One common example is the electric motor used in most plants to drive blowers, compressors, pumps and other process equipment. The failure of an electric motor can cause a process upset, loss of production and unnecessary costs. But this kind of trouble can be prevented by thorough, periodic inspections which point up needed repairs *before* failure takes place.

The periodic part of the problem is solved by scheduling handled by a computer. But a successful program depends on a great many variables. First step is to prepare questionnaire-type checklists and train your personnel to use them properly. The kind of questions included is extremely important. It is only after key questions about each machine are answered that the inspection is effective.

The answers to about twenty questions can help determine the condition of an average electric motor. This should

INSPECTION REPORT — Engines, Pumps, Compressors

ITEM	ITEM (Continued)
KIND OF INSPECTION (Regular, Special or Survey)	AIR COMPRESSOR
SCOPE OF INSPECTION (Dismantled, Operating or Shut-down)	Unloader Trips—Lbs. per sq. in.
KIND OF EQUIPMENT	INDEPENDENT OVERSP. STOP
	Quick Closing Valve—Operation (Defective or Free)
RATING	Tested Tripping Speed—R.P.M.
Manufacturer	GOVERNOR
Cylinders—No. and Diameter	Governor Stop—Operation When Gov. Lifts or Belt is Removed (Sluggish or Drops Freely)
Cylinders—Kind (Air, Ammonia, Steam, etc.)	Response to Load Change (Sluggish, Hunting or Suitable)
Speed—R.P.M. ENGINES	Belt, Chain or Gear Drive (Poor, Fair or Good)
Throttle Pressure—Lbs. per sq. in.	CYLINDERS AND PISTONS
Steam Temp.—Deg. Fahr.	Liners or Cylinder Bore—Wear (Elliptical, Scored or Smooth)
Horsepower COMPRESSORS, PUMPS	Piston Rings—Wear (Excessive, Tolerable or Small)
Suction Press.—Lbs. per sq. in.	Pistons—Assembly (Relaxed or Tight)
Discharge Press.—Lbs. per sq. in. PUMPS	Piston Glands—Leakage (Excessive, Tolerable or Small)
Capacity—Gallons per minute	ENGINES
Head—Feet of Water or Fluid DEEP WELL PUMPS	Moisture Entering—Amount (Excessive, Tolerable or Small)
Casing Depth—Feet	Items not satisfactory are identified as H.P., L.P., etc., under "Recommendations"
NUMBER	MAIN BEARINGS, CRANKS AND CROSSHEADS
Owner's	Main Bearings—Wear (Excessive, Tolerable or Small)
Manufacturer's Serial	Main Bearings—Temperature (Excessive, Hot or Normal)
LOCATION OF EQUIPMENT	Cranks—Wear (Excessive, Tolerable or Small)
Building	Cranks—Temperature (Excessive, Hot or Normal)
Room	Crossheads—Wear (Excessive, Tolerable or Small)
Department	Crossheads—Temperature (Excessive, Hot or Normal)
DRIVING OR DRIVEN MACHINE	Items not satisfactory are identified as H.P., L.P., etc. under "Recommendations"
Manufacturer	FLYWHEEL, PULLEY
Kind	Rim—Alignment (Dangerous, Tolerable or True)
Size—H.P. or Kw.	Wheel Sections—Soundness (Cracked or Sound)
Owner's No.	Teeth—Wear (Gear Wheel) (Excessive, Tolerable or Small)
Serial No.	Keys, Links and Bolts—Assembly (Relaxed or Tight)
LOAD—TIME OF INSPECTION	Belt, Rope or Chain Drive (Poor, Fair or Good)
Speed—R.P.M. ENGINE	LUBRICATING SYSTEM
Kilowatts (If Connected to Gen.)	Oil Lines and Fittings—Support (Poor, Fair or Good)
Throttle Pressure—Lbs. per sq. in.	Lubricant—Supply (Scant or Ample)
Steam Temp.—Deg. Fahr.	Lubricant—Quality (Hydrous, Dirty or Clean)
Back Pr.; Vac.-psi; In. Hg.	FOUNDATION
Jacket Water Temp.—Inlet	(Dangerous, Yielding or Rigid)
Outlet	
Exhaust Temperatures COMPRESSORS, PUMPS	
Suction Press.—Lbs. per sq. in.	
Discharge Press.—Lbs. per sq. in.	
Relief Valve Set.—Lbs. per sq. in.	

INSPECTION REPORT — Rotating Electrical Machines

ITEM	ITEM (Continued)
KIND OF INSPECTION (Regular, Special or Survey)	VENTILATION
SCOPE OF INSPECTION (Dismantled, Operating or Shut-down)	Cooling Air—Quantity (Scant or Ample)
KIND OF MACHINE	Cooling Air—Quality (Injurious, Stagnant, Tolerable, Good)
RATING	BEARINGS
Manufacturer	Wear—Driving or Driven End (Excessive, Tolerable or Small)
Kilovolt Amperes	Wear—Opp. Driving or Driven End (Excessive, Tolerable or Small)
Kilowatts	Temp.—Driving or Driven End (Excessive, Hot or Normal)
Horse Power	Temp.—Opp. Driving or Driven End (Excessive, Hot or Normal)
Power Factor—Per cent	Lubricant—Supply (Scant or Ample)
Volts—Line	Lubricant—Quality (Hydrous, Dirty or Clean)
Phase and Frequency (Number) (Cycles per Second)	Wear In Mils If Machine Is Dismantled Temp. In Deg. If Indicator Is Provided
Amperes—Line	COMMUTATOR, COLLECTOR RINGS AND BRUSHES
Amperes—Field (Synchronous Machines Only)	Commutator—Surface (Rough, Grooved, Flats or Smooth)
Speed—R.P.M.	Commutator—Sparking (Excessive, Tolerable or Slight)
NUMBER	Collector Rings—Surface (Rough, Grooved or Smooth)
Owner's	Collector Rings—Sparking (Excessive, Tolerable or Slight)
Manufacturer's Serial	Brushes—Wear (Excessive, Tolerable or Small)
LOCATION OF MACHINE	Insulation Bet. Commutator Bars (Burnt, High, Flush or Undercut)
Building	LINE SWITCH
Room	Contact Surfaces (Burnt, Blistered or Smooth)
Department	Temp.—Current Carrying Parts (Excessive, Hot or Normal)
DRIVING OR DRIVEN MACHINE	Switch Oil—Quantity (Scant or Ample)
Manufacturer	Switch Oil—Quality (Sludged, Hydrous, Dirty or Clean)
Kind	Bushings and Terminals (Cracked, Chipped, Dirty or Norm.)
Owner's No.	PROTECTIVE EQUIPMENT
Serial No.	Relay—Setting (Amperes or Per Cent of Full Load)
LOAD-TIME OF INSPECTION	Relay—Setting (High, Low or Suitable)
Kilovolt Amperes	Fuses or Cut-outs—Rating (High, Low or Suitable)
Kilowatts	GROUNDING
Power Factor—Per cent	Machine Neutral (Ungrounded, Defective or Adequate)
Volts—Line	Machine Frame (Ungrounded, Defective or Adequate)
Amperes—Line	FOUNDATION OR SUPPORT (Dangerous, Yielding or Rigid)
Amperes—Field (Synchronous Machines Only)	INSULATION RESISTANCE (Megger Test Voltage 500 D.C.)
Speed—R.P.M.	Stator Winding to Frame - Megohms (Field If D.C. Machine)
WINDINGS	Rotor Winding to Frame - Megohms (Armature If D.C. Machine)
Stator—Insulation (Soft, Brittle, Broken or Normal)	Time Between Shut-down and Test
Rotor—Insulation (Soft, Brittle, Broken or Normal)	
Temp.—Stator Winding (Excessive, Hot or Normal)	
Foreign Matter Present (Moisture, Oil, Lint, Dust, Etc.)	
Temp. In Deg. If Indicator Is Provided	

MACHINERY INSPECTION DATA when form is completely filled out leave no doubt as to the exact condition of the equipment

cover: nameplate data and location; base foundation; power leads or conduit; frame ground; bearings; lubrication; air gap mechanical parts; vibration or misalignment; ventilating air passages; windings insulation resistance; commutator or collector rings; temperatures; fire protection; control equipment circuit; protective devices; overload protection; voltage readings; ampere readings; and observation of any unusual conditions.

In order to get useable information, specific questions should be asked. For example; under the category of *Windings* ask if there are any cut-out coils. Ask the inspector to describe any adverse conditions such as dirt, oil, or foreign material; any frayed, broken or brittle insulation; any loose connections or signs of heating; any loose or missing wedges; any poor contacts between rotor bars and end rings. In the case of *Control Equipment,* make sure the report asks the inspector to describe fully burned, pitted or loose contacts, also loose connections, signs of heating, improper or frayed insula-

tions, and other specific conditions.

The inspection reports reproduced above show the kind of information that is required. Check-off lists like these will improve your PM program because it's very difficult to fill out this kind of form without performing a thorough inspection. And anyone reviewing the report will know the condition of the equipment. Comparison with previous inspection reports will often disclose a trend, even when the inspections have been made by different engineers in large organizations. ●

Instruction books and *PM* check sheets aim to cover every contingency

On these two pages, we'll sample the various types of paperwork. Every component of every unit under the care of Tom and Bob is cataloged, described, itemized; for each unit, records include (1) comprehensive description of the equipment (2) mechanical drawings with numbered components (3) instruction sheets detailing each job in step-by-step sequence. For example, instructions show how to start up a unit that has been shut down for a long time, and also the procedure for daily startups.

These instructions leave nothing to chance. The only way a maintenance man can possibly pull a boner is if he fails to follow the instructions he takes to the job. But even that is pretty well covered, since he must indicate on his check list that each operation is complete.

Instruction books condense manufacturers' recommendations

Service Schedule:	D – Daily, W – Weekly, M – Monthly, Q – Quarterly, S – Semi-annual, A – Annual, R – As Required by Operating Conditions.		SCHEDULE
ITEM	DESCRIPTION	SERVICE REQUIREMENTS	
1.	Fresh Air Intake. Standard sheet metal weatherproof louvers with bird screen - Size 5'0" x 7'0"	Check to make sure the opening is not partly blocked by foreign materials sticking to screen.	D
2.	Minimum Section Fresh Air Damper. Standard louver type. Johnson D-203, Johnson Service Company, Seattle - SE. 5213	Check operation. Clean, adjust and reset or repair as necessary.	A
3.	Minimum Section Fresh Air Damper Operator. Johnson D-251. Johnson Service Company, Seattle - SE. 5213	Check operation. Clean, adjust and reset or repair as necessary.	A
4.	Maximum Section Fresh A...		

1. Routine Starting

Follow the steps in sequence as listed below:

a. Check crankcase oil levels.

b. Check oil level in force feed lubricator reservoir. Fill as required. Use Gargoyle DTE-103 oil.

c. Start one circulating pump. Make sure water is flowing through both cylinders, intercooler and aftercooler.

d. Start water cooler fan and circulation pump.

e. Make sure there is sufficient evaporative water in cooler for proper operation.

valves to regulator are open. These
...to regulator.

Heating system

Maintenance and service procedure for heating and ventilating system is a 17-p set of instructions, ending with a list of 18 manufacturers' service manuals that pertain to various pieces of equipment in the system. And for each energy system at Boeing, a similar master set of instructions is drawn up and kept handy.

Air compressors

Routine starting of large air compressor is covered in a brochure including the responsibilities of those who operate and service compressors, a description of equipment, detailed operating information, inspection and servicing requirements. This info mastered, the instructions and inspection charts follow: daily log sheets; monthly, semi-annual and annual inspection sheets.

Inspection sheets check off step-by-step procedure

Item No.	Item Description	Service Requirements	Condition						
			Mon	Tues	Wed	Thurs	Fri	Sat	Sun
22a	Air Filter	Inspect to see that indicating system shows normal operation. Both red and clear lamps should be on for normal operation and off when a short circuit exists. Advise electrician if malfunction exists.							
26a	Filter and Pressure Reducer for Supply Air to Johnson Controllers	Check filter assembly. Blow down filter. Air pressure should be 15 psig leaving reducer							
28	Expansion Tank 24" x 72" Tank	Check water level in tank. If water is low, check feed water system for faulty operation and/ or hot water line for leaks. Feed water is admitted automatically through a pressure reducing valve set at 20 psig. Advise plumber of malfunction.							

Daily . . .

inspection of heating, ventilating and air-conditioning system gives marching orders for cleaning, lubricating and adjusting. The maintenance man is told, "state what was adjusted; new unit or part required; explain what failed, why it failed and what was done." There's a space which he uses to warn the next maintenance man about anything unusual that might need attention.

Item No.	Item Description	Service Requirements		Condition	
23a	Air Conditioning Unit for Telecomputing Room	1. Check water temperature leaving condenser. Should not exceed 95° F.			
		2. Wipe off dust and dirt in unit.			
		3. Check for loose tubing connections.			
		4. Check evaporator coils and lines for icing.			
		5. Check ...			

Monthly . . .

inspection sheet lists eight items to be checked in the a-c system proper and eight more items in its cooling tower. Like every other sheet, this must be signed by the mechanic making the inspection and by his foreman, who must approve each item.

Plumber's check list covers such items as the oil-filter pump, steam traps, strainers for circulating and condensate pumps, circulating pump and motor and condensate pump in system.

Item No.	Item Description	Service Requirements	Condition
12a	Steam Traps	Check operation of trap and blowdown strainer.	
		South Mechanical Room	
		Central Mechanical Room	
		North Mechanical Room	
		Hangar Unit Heaters	
		Control Tower Air Conditioning Unit	
34a	Strainers for	Blow down until water runs clear.	

Item No.	Item Description	Service Requirements	Condition
11b	Steam Control Valves	Inspect valves. Tighten packing only enough to prevent leaks. Replace any leaky diaphragms. If valve does not close tightly, replace the valve disks.	
		South Mechanical Room (4 valves)	
		Central Mechanical Room (2 valves)	
		North Mechanical Room (4 valves)	
		Hangar Unit Heaters ...	

Semiannual . . .

inspection sheets are again drawn up for the air-conditioning-system plumber. These sheets list various items to be checked twice a year on the steam-control valves and mixing valves in the system.

Annual . . .

inspection sheet for a-c system lists 27 separate points that need once-a-year checking to give optimum performance. Here again, every possible problem is spelled out. For example: "set sensitivity as high as possible (5 psig per F) without hunting. Reset thermostat at 40 F" leaves no question about what is to be done. The 6-p form lists about 100 specific check points.

Vault-inspection list (prior to date of de-energizing) details 13 points to be checked. Accent is on safety—and when a problem comes up, other department heads must be kept informed.

Item No.	Item Description	Service Requirements	Condition
9	Supply Intake Air Mixture Temperature Thermostat (Direct Acting)	Check operation. Move setting manually. Control pressure should equal supply pressure when thermostat setting is moved $1\frac{1}{2}°$ F. below actual mixture temperature. Control pressure should be 0 psig when moved $1\frac{1}{2}°$ F. above. Adjust or repair as necessary. Set sensitivity as high as possible (5 psig/° F.) without hunting. Reset thermostat at 40° F.	
		... Room	

11				Condition		
	1. Test transformer fluids. (See page 8).			✓ - Satisfactory X - Unsatisfactory		
	2. Procure and test new fluid as required.			Insp.by	As found	As left
	3. Write letter to departments which will be affected by vault shutdown.					
	4. Arrange for required accessories:					
	a. Compressed air b. Vacuum cleaners c. Emergency power d. Cleaning fluids and cloths					
	5. Calibrate and adjust relays ... needed.					

Fred Wendel is shown checking piping installation he constructed on Bermuda

Can you answer these important questions before . . .

Buying buried piping systems?

First you must know the exact condition of your soil; then
get familiar with water table and drainage conditions in the area. Piping,
thermal insulation, flexible joints and manholes must also be right.
During construction phase, you can assure a first-class system.

Q Where do you start when designing an underground piping system for steam, hot or chilled water energy systems?

A First is a thorough inspection and analysis of the ground itself. This means knowing the water table and also the history of the ground in the immediate area. Ask neighbors if there are old drainage trenches, or buried piping systems already installed on the property. If so, be sure to obtain old plans of the piping and sewage lines previously laid. Some people will remember earlier usages. If the land was used by a chemical plant, be sure to have a chemical analysis made. If soil is found to be acid, or to contain any corrosive elements, you must take precautions as this condition is very tough on buried piping systems.

Was the ground used for buildings? Is the soil filled in? If so, with what? Is it filled-in swampland? Are there buried boulders or rubble? Trace out the length of the proposed piping, then take test soundings or borings every hundred feet or so. Remember that a plot of ground can vary greatly, even over short distances.

Q Is there an inexpensive way to take soundings?

A Yes. Drive a length of 2-in. pipe into the ground at various places where the piping system will be buried. Then with a rod punch the soil out of the pipe. Besides showing you the nature of the soil, this method will give some idea of the soil's resistance. Check the resistance at various levels; at 2 ft, 4 ft, and at 6 ft depths below the planned bottom of the conduit. Also dig up samples from various depths. Is there muck? Sand? Gravel? What? There are firms which spe-

cialize in testing ground conditions. It is good business to retain such expert help if your budget permits.

Make sure you check for ground water—especially during the wet season. Also know annual rainfall in the area. Then investigate drainage of the land. If drainage is poor, you must provide for extra protection against water leaking into the system, which may ruin piping insulation, rust the piping and flood the manholes. You may need a deeper trench. For better drainage, the trench may have to be lined with stone or broken masonry.

Since drainage can be such a headache, provision must be made against infiltration of surface water from sewer backups, overflowing creeks, streams, etc. The local water commission can often be helpful. Neighboring plants may have engineering drainage records.

Check rock formations common to area. If filled area, bottom may not be consistent because of the fill materials used. The bottom may not even provide a uniform base to properly support the piping, especially if previously filled with rubble, boulders or blocks of broken concrete.

Q Are there any standard recommendations for underground heat distribution systems?

A Yes. The Corps of Engineers published Procedures for Establishing Acceptability of Heat Distribution Systems, which set down test requirements that system suppliers must meet to be acceptable for military construction. Since water is the major problem, two basic sites are defined: Class A and Class B (see next question). Conduit systems specified for all sites must have these three characteristics: (1) They must be capable of being drained in place. (2) Piping insulation must be able to withstand lengthy exposure to boiling water and return to its original physical characteristics. (3) The system must have the ability to regain original operating efficiency following flooding of the casing.

Q How may the water table in the area affect the design of the buried piping system?

A If the water table is expected to be above the bottom of the conduit at any time, the system is called Class A, and the conduit must be watertight. It should also have an unobstructed passage so the insulation can be dried out should it become flooded. Provision must also be made to allow air pressure testing of piping for leaks at any time. And the system must be capable of being drained in place, if it becomes wet.

But if the water table does not rise above the bottom of the conduit (based on soil survey), the system is called Class B and must still have an unobstructed air passage suitable for drying out the insulation. Here again, provision must be made for draining in place, if insulation and piping become wet.

Q If the system is installed in a waterproof conduit, why must provision be made for draining?

A Because very costly experience shows that eventually all underground piping systems become wet. This condition takes place from internal pipe breaks, water infiltration from ground water. Normal water table may vary with time, manholes become flooded sooner or later, or flash floods and melting snow may turn the pipe trench into a catch basin for water.

Q What items are important to keep in mind when making drawings of the system?

A The plant's engineering office should have comprehensive drawings of the plant piping system to be served by the buried piping. These drawings should have legends identifying all piping. The drawings of the new buried system should show when piping was buried, elevation, location of all joints, all fittings, valves, outlets, etc. Also show size and type of pipe, and be sure to plot to proper dimensions. If the system is an extension of present buried system and sections have been discontinued from use by blanking off, this information must be indicated on drawings, along with dates. When system is complete, insist on *as built* drawings, to include changes invariably made during construction.

Q What data is needed to design the piping system?

A Pitch of pipe versus capacity of the pipe is of major importance, especially for drainage. Also know your hydraulic gradient (maximum difference in elevation at inlet and outlet ends of piping system). Make sure you select the correct insulation, especially if there is any danger of ground water leaking into the pipe tunnel. Seriously consider using pre-insulated, hermetically-sealed piping. Selecting proper manholes, pipe guides and expansion joints usually calls for previous experience.

Consider possibility of extending or adding to system in future. This may affect type of burial and size of pipe. How about access to piping in case repairs are needed? This must be provided during installation.

Q What considerations should be given manholes?

A Never "cheat" on size of manholes. Be sure to make them big enough for safe and easy access. Otherwise, there won't be enough room for maintenance. If manholes are to be poured concrete, be sure the foundation for the expansion joints housed inside is cast integrally with the manhole. For best results, the manholes are to be poured concrete, be sure the foundation is cut off at the underside of the manhole roof and the other extends almost to the manhole floor. Both extend above grade and terminate in a U bend. Pipe should be 4-in. diameter or larger. Provide for positive and reliable means of draining the manhole. If steam ejector is used for draining, remember that the float controlling the ejector must be inspected at set periods. If ejector fails to work, piping insulation will be saturated. If lay of land is favorable, best way to keep manholes dry is by gravity. Where gravity drainage is used, be sure to install a backwater valve (check valve) in drainage lines from manholes to any stream or body of water whose level may rise due to flash storms. This will prevent water backing through the drain line. Make sure gravity drain line from sump is minimum of 3-in. diameter. Stick to gravity drainage if at all possible, for where a sump pump is used in each manhole, price of electric wiring and maintenance can be costly. Today, completely piped manholes (prefab) are built in one unit, ready to install. These should also be considered.

Q During construction, what can be done to assure a reliable system?

A Cap off all open piping at end of each day to prevent debris being placed inside, either carelessly or maliciously. If practical, eliminate flanges and slip joints in favor of loops. Loops need no maintenance, although they do take up more space. Keep in mind the inherent flexibility of the pipe itself, especially during cycling periods. Piping must be allowed to move (expansion and contraction) while in operation. Remember that guiding the piping into the expansion joint is critical. If done improperly, expansion joints (especially bellows or slip type) can become damaged and totally useless. Soon as piping is installed (before covering system with dirt) blank off and test with air pressure from portable system. If factory insulated and encased piping is used, be sure to put a separate air test on the metal casing. This phase of testing is the one usually (and deliberately) omitted. The entire system must be watertight.

Q What considerations should be given vibration?

A Visit neighboring plants and study their vibration problems. Some soils are springy; vibration of passing locomotives or trucks will be carried to the buried piping and cause trouble. There may be drop hammers or other shock-producing equipment in the immediate area that can cause problems if soil is springy. In such areas the trench may have to be lined with sand to change soil characteristics to help absorb vibration.

'Here's My Idea, Boss'

THIS PHOTO MAKES ME THINK of an old engineer friend who worked so successfully with people that he eventually ended up as the firm's president. When asked why people always tried to break their necks doing things for him, he summed up his philosophy as follows:

1. Always ask a subordinate how *he* thinks a job should be done before assigning the work.
2. Pay the full price for work, whether in money, materials or manpower.
3. Face issues squarely. Putting them off only stores up trouble, and is a bad example, besides.
4. Use new methods. Don't be afraid to pioneer just because no one else has done it before.
5. Hire consultants who have solved problems similar to yours. Cash in on their experience without wasting your time to find out the answers they already know.
6. Put all instructions in writing to make them definite and specific.
7. Start and finish one big job at a time.
8. Snap decisions are usually bad. If a "yes" or "no" is demanded, usually answer "no."
9. Share honors and bonuses for department with all in department, right down to last fireman. You can't have a "team" by treating only the bosses as prima donnas, or your team will let you carry the ball by yourself. This last is the master secret.

I've seen a lot of success formulas, but this one strikes me as having the slightest margin of error. If you are a youngster just starting, take my advice and paste this in your hat.

19
License examination
calculations

Figuring Efficiency

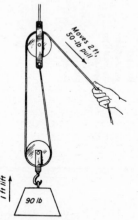

1 Man pulls two feet to lift load one foot. Here the efficiency is 90%

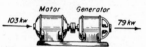

2 Efficiency of this motor-generator set is 76.7%

3 Mechanical efficiency of steam engine is 92.9%

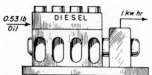

4 Overall efficiency of this diesel generating set is 33.4%

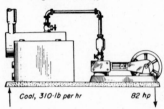

5 This small steam plant, exhausting to waste, figures only 4.98% efficiency

▶ ONCE YOU GET the basic idea, efficiency is easy to figure. I mean almost any kind of machine efficiency, with a few exceptions that I won't try to cover. If a machine could give out as much as went into it the efficiency would be 100 per cent. That never happens although big generators and hydro turbines get close to this perfect mark.

A machine that delivers half its input has an efficiency of 50%. One that delivers three quarters has an efficiency of 75%, and so on.

The input and output being considered may be work, heat, mechanical power or electrical power. The important thing is to figure both the input and output in the same kind of units, whether they be horsepower, kilowatt hours, Btu or foot pounds.

EXAMPLES

Input and output are in like units in Fig. 1 and 2. In Fig. 1 both input and output are measured in foot pounds. The man has to pull 50 lb to lift 90 lb and he pulls 2 ft for every 1 ft the load rises. Then:

Input=50×2=100 ft lb
Output=90×1=90 ft lb
Efficiency=90÷100=0.90=90%

In the mg set, Fig. 2. both input and output are kilowatts. By feeding 103 kw to the motor you get 79 kw out of the generator. Then:

Efficiency=79÷103=0.767=76.7%

In Fig. 3 the two units might appear to be different since input is *indicated* horsepower and output is *brake* horsepower. Yet both are mechanical horsepower. The indicated horsepower is the mechanical power delivered by the steam to the piston and the brake horsepower is the mechanical power delivered by the engine to the load. So the efficiency of the engine in converting cylinder power into useful power (called the mechanical efficiency) is:

39÷42=.929=92.9%

The case of the diesel generating set, Fig. 4, is somewhat different. Here input is heat and output is electrical. For every kilowatt hour delivered this engine burns 0.53 lb of 19,300-Btu oil. Then the input per kilowatt hour is 0.53×19,300=10230 Btu.

Table shows that one kilowatt hour equals 3413 Btu, so:

Input=10,230 Btu
Output=3413 Btu
Efficiency=3413÷10230=0.334=
  33.4%

For a final problem, assume that the simple steam plant shown in Fig. 5 blows all exhaust to waste and that, when delivering 82 hp, the fuel burned is 310 lb per hr of 13,500 Btu coal.

This figures 310÷82=3.78 lb coal per bhp hr, so:

Input=3.78×13,500=51,030 Btu
Output=2544 Btu (see Table)
Effi.=2544÷51,030=0.498=4.98%

Conversion Factors

1 kilowatt=1.3415 horsepower
  =738 ft lb per sec
  =44.268 ft lb per min
  =2,656.100 ft lb per hr
  =56.9 Btu per min
  =3413 Btu per hr

1 horsepower=0.7455 kw
  =550 ft lb per sec
  =33.000 ft lb per min
  =1,980,000 ft lb per hr

1 horsepower=424 Btu per min
  =2544 Btu per hr

1 Btu=778.3 ft lb

1 kilowatt hr=3413 Btu
  =1.342 hp hr

1 horsepower hr=2544 Btu
  =0.7455 kw hr

Torque, Speed

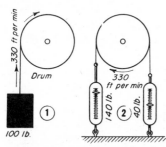

1 One horsepower does useful work
2 One horsepower delivered as heat

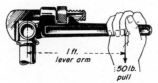

$F = force$

Radius or lever arm is R feet or r inches

$N = rev. per min.$

3 Rim speed is 6.283 × R × N

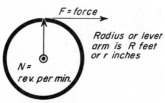

1 ft. lever arm

50 lb. pull

4 Here torque is 50 ft lb or 600 in. lb

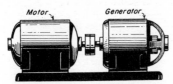

Motor   Generator

5 At 1750 rpm, 36 in.-lb is 1 hp

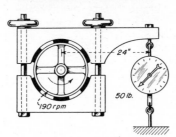

24"

50 lb.

190 rpm

6 24 × 50 × 190 ÷ 63,000 = 3.6 hp

▶ MORE OFTEN THAN NOT, power is delivered through rotating parts rather than in a straight-line push. In either case horsepower is the force in pounds times the distance through which it acts in one minute, divided by 33,000, because one horsepower is 33,000 ft lb. per min.

Fig. 1 shows a hoisting drum turning with a rim speed of 330 ft per min, winding up a weight of 100 lb at the same speed. Work done is 330 × 100 = 33,000 ft lb per min. or 1 hp.

Now look at Fig. 2. The two spring balances pull on the leather-strap brake. The net drag on the drum surface is 140 − 40 = 100 lb. The surface moves 330 ft in one minute, so the work done is 330 × 100 = 33,000 ft lb per min. Again this is one horsepower, but the power is delivered as heat rather than as "useful work."

TORQUE

Now let's consider torque and the relation of torque and rotative speed to power delivered. Torque is nothing but twisting effort measured as the product of the pull by the length of the lever arm of the pull, Fig. 3.

The lever arm may be measured in feet, in which case the torque is in foot pounds. Often it is more convenient to measure the lever arm in inches, in which case the product of lever arm in inches by the force in pounds gives the torque in inch pounds.

For example, in Fig. 4, if man pulls 50 lb at a point 12 in. from the center of the fitting the torque is 1 × 50 = 50 ft lb, also 12 × 50 = 600 in. lb.

TORQUE AND POWER

Suppose (Fig. 3) that we have the radius of a wheel and the rotative speed in RPM. In one revolution every point on the rim moves 2 × 3.1416 × R feet, where the R is the radius in feet. This is 6.283 × R feet per revolution.

If N is the rotative speed in *RPM*, the distance traveled by the rim in one minute will be 6.28 × R × N feet. If the force at the rim is called F the horsepower will be 6.28 × R × N × F ÷ 33,000.

Note that this boils down to R × N × F ÷ 5250. Now R × N is nothing but the torque in foot pounds. Call this T. Then horsepower is T × N ÷ 5250.

If the radius is measured in inches the product of radius and force is the torque in inch pounds (call it t) and horsepower will be t × N ÷ 63,000. For convenient reference the torque-speed-power formulas are repeated in the box. Let's apply these to the one-horsepower motor generator set pictured in Fig. 5. The set runs at 1750 rpm. Find the torque. Then t = 63,000 ÷ 1750 = 36 inch pounds.

It may be worth remembering that when any machine's shaft, delivering one horsepower, is running at the common electrical speed of 1750 rpm, the torque or twist in the shaft equals a one pound pull at the end of a wrench a yard long.

Fig. 6 shows a "prony-brake" measuring the power of a small engine. Here torque t = 24 × 50 = 1200 in. lb. At 190 RPM the brake power is 190 × 1200 ÷ 63,000 = 3.6 hp.

Formulas for Rotary Power

R = radius of torque arm, feet
r = radius of torque arm, inches
F = force at end of torque arm, pounds
T = R × F = torque, foot pounds
t = r × F = torque, inch pounds
N is rotative speed, rpm
P is horsepower

Then:

P = T × N ÷ 5250
P = t × N ÷ 63,000

Figuring Force, Work and Power

FORCE

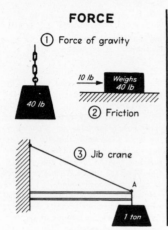

① Force of gravity

② Friction

③ Jib crane

WORK

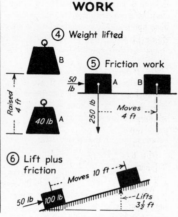

④ Weight lifted

⑤ Friction work

⑥ Lift plus friction

POWER

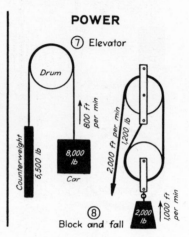

⑦ Elevator

⑧ Block and fall

▶ OPERATING ENGINEERS deal every day with force, work and power, but may not always be completely clear on the exact meaning of the terms. So here are some notes I hope will be helpful.

DEFINITIONS

Force is a push or pull. Engineers generally measure force in *pounds* or *tons*.

Work is done when a force acts on a moving body. The amount of work, in *foot pounds*, is the *product of the pounds* push or pull *by the number of feet the object moves in the direction of the force.*

Power is the rate of doing work, generally measured in foot pounds per minute, foot pounds per second, horsepower or kilowatts.

To get either of the first two measures of power divide the foot pounds of work by the minutes or seconds required to do the work.

One horsepower is the doing of work at the rate of 550 foot pounds per second, or 33,000 foot pounds per minute. One kilowatt is 1.33 horsepower.

Force Explained. In (1) the force of gravity acts straight down 40 lb on the weight and is balanced by the 40-lb

upward pull of the supporting cord. When a body is standing still, or moving in a straight line at constant speed, the forces on the body always balance.

In (2) the 10-lb force applied to push the block to the right is balanced by an equal friction force pushing on the block to the left. In this case the block is not lifted, so its 40 lb weight doesn't enter in.

Many forces act on the elements of a structure, such as the jib crane (3). At *A*, for example, the load pulls straight down one ton, the tie rod pulls more than one ton up left on a slant and the beam thrusts out horizontally to the right, also with a force over one ton. These forces balance at *A*.

Work Picture. A common case of work is lifting a weight straight up. When the 40-lb weight (4) is lifted four feet work done is 4 x 40 = 160 ft lb.

But note that work is not always *weight* times distance. In (5), for example, you don't consider the 250-lb weight, but only the 50-lb push and the 4-ft movement in the direction of the push. Work here is 50 x 4 = 200 ft lb.

Sliding a block up an inclined plane (6) combines lift work and friction work. Total work is 50 x 10 = 500

ft lb. The lifting fraction is 3½ x 100 = 350 ft lb.

Figuring Power. In (7) the net lift on the car is 8000 — 6500 = 1500 lb, so the power absorbed in lifting is 1500 x 800 = 1,200,000 ft lb per min or 1,200,000 ÷ 33,000 = 36.4 hp. Actually more motor power than this would be needed to allow for friction losses.

In (8) the rope pull would be 1000 lb if there were no friction. Here the actual work input is 1200 x 2000 = 2,400,000 ft lb per min and the *useful* work 2000 x 1000 = 2,000,000 ft lb per min.

We have figured power as force times distance divided by time. We can get the same result by changing the order of these operations to distance divided by time multiplied by force.

To put this another way, one horsepower is one pound acting at a speed of 550 ft per second or 33,000 ft per min.

Then we have these rules: *Horsepower* is pounds force times speed in feet per second divided by 550, or *horsepower* is pounds force times speed in feet per minute divided by 33,000.

So lifting a 400 lb weight 900 ft per minute requires 400 x 900 ÷ 33,000 = about 11 hp (plus power for friction).

Figuring Centrifugal Force

▶ You know about centrifugal force, but can you figure it? Let's try.

Car Rounding Curve. Fig. 1 shows three cases of a 2000-lb car rounding a curve. In case *A* the car goes 20 miles per hour around a 50-ft curve. Side thrust on the tires (another name for centrifugal force) is 1070 lb.

In case *B* the same car takes the same curve at 40 mph—twice as fast. But side thrust isn't just doubled, it's four times as great or 4280 lb.

Now what happens when this same car travels at same speed around a curve of twice the radius, Fig. 1*C*? Force drops to half or 2140 lb.

What's It Mean? In Fig. 1*B* we see that centrifugal force goes up as the square of the speed. It's four times for double speed, nine for triple.

Second, the force is inversely proportional to the radius of curvature. If you double the radius, you cut the force in half. This last rule compares two objects running at the *same linear speed* (not RPM).

Rule 1 is the formula. See if you can figure the side thrust on the cars in Fig. 1. Be sure to use the right *K*. My answers are rounded off.

In Terms of RPM. For most rotating machinery it's easier to figure centrifugal force from RPM than linear speed. Take a look at the 2-lb ball swinging around a shaft in Fig. 2. Compare these three cases with those in Fig. 1.

Going from Fig. 2*A* to *B*, we double the RPM and force goes up four times, just like doubling the carspeed in Fig. 1. Now compare Fig. 1*C* with Fig. 2*C*. In the first case doubling the radius halved the force. But in Fig. 2 doubling the radius doubled the force. Why? At constant RPM, increasing the radius directly increases linear rim speed.

Try the formula in Rule 2 for the case in Fig. 3. For even RPM the table is mighty handy. Use it to check my answers in Fig. 2.

If You're Talking Speed . . .

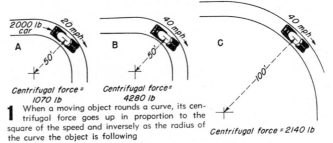

Centrifugal force = 1070 lb

Centrifugal force = 4280 lb

Centrifugal force = 2140 lb

1 When a moving object rounds a curve, its centrifugal force goes up in proportion to the square of the speed and inversely as the radius of the curve the object is following

RULE 1. If **S** = speed of moving object, **R** = radius of curve it follows and **F** = centrifugal force in pounds per pound of object weight, then . . .

$$\cdots \ \mathbf{F = K \times S \times S \div R}$$

When **S** is in miles per hours and **R** is in feet, **K** is 0.0671. When **S** is in feet per sec and **R** is in inches, **K** is 0.375. Total force is **F** times pounds object weighs

If You're Talking RPM . . .

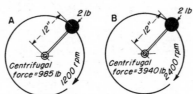

Centrifugal force = 985 lb

Centrifugal force = 3940 lb

Centrifugal force = 7880 lb

2 With rotating objects, centrifugal force goes up in proportion both to the radius and the square of the RPM. Increasing radius at constant RPM increases linear rim speed

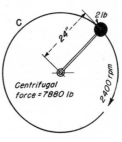

Centrifugal force = 87.2 lb

3 Here's what happens in an unbalanced rotating machine. Unbalance acts just like a weight set offcenter. Use the table (right) to get **F** for other speeds. For other weights, multiply by number of lb, also by inches of radius

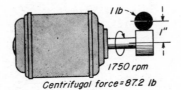

RPM	F, per lb, per inch radius
600	10.2 lb
900	23.0 "
1200	41.0 "
1800	92.1 "
2400	164 "
3000	256 "
3600	368 "

RULE 2. If **F** is centrifugal force in lb per lb of weight rotating at radius **R** . . .

$$\cdots \ \mathbf{F = RPM \times RPM \times R \div K}$$

When **R** is in ft, **K** is 2920. When **R** is in inches, **K** is 35100. Again multiply **F** by the number of pounds the rotating object weighs to get total centrifugal force

Tension, Compression and Shear

Tensile Stress. When a tensile test specimen, Fig. 1, is stretched to the breaking point the material tends to *neck down. Tensile strength* of the material in *pounds per square inch* (psi) is the total breaking tension divided by the *original* cross-section.

At any tension short of breaking, the *unit stress* is the actual total tension divided by this same original cross-section.

Up to a certain unit stress, called the *elastic limit,* the metal stretches in exact proportion to the applied load and the unit stress. Beyond that the metal starts to *give* a little and stretches more and more for a given increase in unit stress.

Compression. The *compressive strength* of a material is tested by crushing a short block. Total crushing load, divided by original cross-section, is the compressive strength in pounds per square inch.

Shear. In the same way, if the section is sheared across (as in a power shear) the *shearing strength* is the total force required divided by the cross-sectional area sheared.

Factor of Safety is ultimate strength divided by actual stress.

Bolts and Pins. Fig. 2 shows two applications. What is unit tensile stress at thread root of bolt at left? Cross-sectional area at the thread root of a 1-in. (U. S. Standard) bolt is 0.551 sq in. Then unit stress is $3,000 \div 0.551 = 5440$ psi. For Class B bolt steel of 60,000 psi tensile strength, factor of safety is $60,000 \div 5440 = 11.0$.

The ½-in. pin at right, Fig. 2, has a cross-sectional area $0.5 \times 0.5 \times 0.785 = 0.196$ sq in. Since pin is in *double shear*, total sheared area is $2 \times 0.196 = 0.392$, so shearing unit stress is $2500 \div 0.392 = 6,380$ psi.

Stress in Shell. In a seamless or welded tube or drum under internal pressure, the tendency to split lengthwise is double the tendency to split around. Thus, unless the circular seams are very weak, such shells or drums *always* fail by splitting lengthwise (longitudinally).

Figure the shell strength without seams, then correct for the efficiency of the longitudinal joint. Consider a half slice of shell one inch long, Fig. 3. Fluid pressure tends to split this into two halves, both of which are shown. If internal diameter (inches) is D and fluid pressure (psi) is P, the force on the shaded *piston area* is $P \times D$. This force is resisted and balanced by the pull of the two metal sections A. This resisting force is $2 \times t \times s$, where t is shell thickness and s is unit stress. From this relation (See notes in Fig. 3.) we derive the shell formula $P = 2 \times t \times s \div D$.

If diameter is 30 in., thickness 0.5 in. and safe unit stress 15,000 psi, safe pressure for a seamless shell is $2 \times 0.5 \times 15,000 \div 30 = 500$ psi. If the shell has a longitudinal riveted seam of 80% efficiency, the allowable pressure for 15,000 psi stress is $0.80 \times 500 = 400$ psi.

Heat Stresses. Almost any form of iron or steel (except cast iron) has a *modulus of elasticity* around 30,000,000. That means that a stress of 1 psi will stretch it 1/30,000,000 part of its own length. This relation holds right up to the elastic limit.

To put it another way, stretching iron or steel 1 in. per 100 ft. (1200 in.) produces a tensile stress of $\frac{1}{1200} \times 30,000,000 = 25,000$ psi. The same increase in length can be caused by a temperature rise of 120 F (say, from 80 to 200 F).

It follows that if a piece of steel rod or pipe of any length is heated to 200 F, then rigidly held at both ends, and cooled to 80 F the resulting tensile stress will be 25,000 psi, Fig. 4. This figures about 200 psi for each degree of change.

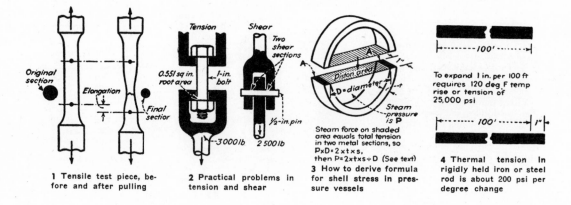

1 Tensile test piece, before and after pulling
— Original section — Elongation — Final section

2 Practical problems in tension and shear
— 0.551 sq in. root area — 1-in. bolt — Tension — Shear — Two shear sections — ½-in. pin — 3000 lb — 2500 lb

3 How to derive formula for shell stress in pressure vessels
— Piston area — D-diameter — Steam pressure is P — Steam force on shaded area equals total tension in two metal sections, so $P \times D = 2 \times t \times s$, then $P = 2 \times t \times s \div D$ (See text)

4 Thermal tension in rigidly held iron or steel rod is about 200 psi per degree change
— To expand 1 in. per 100 ft requires 120 deg F temp rise or tension of 25,000 psi

Short Cut to Square Roots

I offer to show how to get square roots without tables or slide rule—an easier method than the one taught in schools.

The following rule applies only to numbers between 1 and 100. Later I'll show how to bring greater or smaller numbers into this range.

THE RULE

1—By inspection set down the square root *to the nearest whole number only.* This will fall between 1 and 10.

2—Divide this into the original number and carry the answer (quotient) to *one decimal place only.*

3—Average the original whole-number square root and this quotient to *one decimal place only,* and call this the *approximate square root.*

4—Divide the original number by the approx square root and carry the quotient to *three decimal places.*

5—Average this latest quotient and the approx square root to three decimals and call this the "exact" square root.

Except for an occasional number between 1 and 3, the "exact" root thus obtained will never be in error by more than one in the third decimal place.

Problem: Find square root of 11.
Nearest whole-number root is 3
Then 11 ÷ 3 = 3.6 plus

Average of 3 and 3.6 is 3.3 = approx. root.
Then 11 ÷ 3.3 = 3.333
Average of 3.3 and 3.333 is 3.317 = "exact" square root of 11.

Problem: Find square root of 42.62. Nearest whole-number root is 6.
Then 42.62 ÷ 6 = 7.1 plus
Average of 6 and 7.1 is 6.5 = approx root
Then 42.62 ÷ 6.5 = 6.557
Average of 6.5 and 6.557 is 6.528 = "exact" square root.

NUMBERS 1 TO 3

For any numbers between 3 and 100 this rule will never give an error of more than one in the third decimal place. Between 1 and 3 error may sometimes be slightly greater. For example, the square root of 2 by this rule would be 1.416 instead of 1.414.

Where you want high accuracy in the square root of numbers between 1 and 3 insert the following step between step 3 and step 4 in the rule already given:

3A Divide the number by the approximate square root obtained in 3, and carry the quotient to *two decimal places only.* Average this quotient and the approx-square root of 3 *to two decimal places* and use the average as a *corrected approx square root* in step 4.

Problem: What is the square root of 2.304?
Nearest whole number is 1, say (or you can use 2 if you prefer).
Then 2.304 ÷ 1 = 2.3 plus
Average of 1 and 2.3 is 1.6
Then 2.304 ÷ 1.6 = 1.44
Average of 1.6 and 1.44 is 1.52
Then 2.304 ÷ 1.52 = 1.516
Average of 1.52 and 1.516 = 1.518 = "exact" square root.

NUMBERS BEYOND RANGE

For numbers greater than 100, or less than 1, move decimal point left or right two digits at a step until the number is in the range. Then take square root by rule already given.

Finally move the decimal point in the opposite direction a single digit for each two digits moved in the original number.

Problem: Find the square root of 23,000. Moving point *two double steps* to *left* gives 2.3. Square root of 2.3 is 1.517. Move point back *two single steps* to *right* to get 151.7 as the square root of 23,000.

Problem: Find the square root of .000032. Move point three double steps to right to get 32. Square root of 32 is 5.657. Then move point left three *single* steps to get 0.005657 as the square root of 0.000032.

SQUARE ROOTS

No.	1.0	1.1	1.2	1.3	1.4	1.5	1.6	1.7	1.8	1.9
Rt.	1.000	1.049	1.095	1.140	1.183	1.225	1.265	1.304	1.342	1.378

No.	2.0	2.1	2.2	2.3	2.4	2.5	2.6	2.7	2.8	2.9
Rt.	1.414	1.449	1.483	1.517	1.549	1.581	1.612	1.643	1.673	1.703

No.	3.0	3.1	3.2	3.3	3.4	3.5	3.6	3.7	3.8	3.9
Rt.	1.732	1.761	1.789	1.817	1.844	1.871	1.897	1.924	1.949	1.975

No.	4.0	4.1	4.2	4.3	4.4	4.5	4.6	4.7	4.8	4.9
Rt.	2.000	2.025	2.049	2.074	2.098	2.121	2.145	2.168	2.191	2.214

No.	5.0	5.1	5.2	5.3	5.4	5.5	5.6	5.7	5.8	5.9
Rt.	2.236	2.258	2.280	2.302	2.324	2.345	2.366	2.387	2.408	2.429

No.	6.0	6.1	6.2	6.3	6.4	6.5	6.6	6.7	6.8	6.9
Rt.	2.449	2.470	2.490	2.510	2.530	2.550	2.569	2.588	2.608	2.627

No.	7.0	7.1	7.2	7.3	7.4	7.5	7.6	7.7	7.8	7.9
Rt.	2.646	2.665	2.683	2.702	2.720	2.739	2.757	2.775	2.793	2.811

No.	8.0	8.1	8.2	8.3	8.4	8.5	8.6	8.7	8.8	8.9
Rt.	2.828	2.846	2.864	2.881	2.898	2.915	2.933	2.950	2.966	2.983

No.	9.0	9.1	9.2	9.3	9.4	9.5	9.6	9.7	9.8	9.9
Rt.	3.000	3.017	3.033	3.050	3.066	3.082	3.098	3.114	3.130	3.146

No.	10	11	12	13	14	15	16	17	18	19
Rt.	3.162	3.317	3.464	3.606	3.742	3.873	4.000	4.123	4.243	4.359

No.	20	21	22	23	24	25	26	27	28	29
Rt.	4.472	4.583	4.690	4.796	4.899	5.000	5.099	5.196	5.292	5.385

No.	30	31	32	33	34	35	36	37	38	39
Rt.	5.477	5.568	5.657	5.745	5.831	5.916	6.000	6.083	6.164	6.245

No.	40	41	42	43	44	45	46	47	48	49
Rt.	6.325	6.403	6.481	6.557	6.633	6.708	6.782	6.856	6.928	7.000

No.	50	51	52	53	54	55	56	57	58	59
Rt.	7.071	7.141	7.211	7.280	7.348	7.416	7.483	7.550	7.616	7.681

No.	60	61	62	63	64	65	66	67	68	69
Rt.	7.746	7.810	7.874	7.937	8.000	8.062	8.124	8.185	8.246	8.307

No.	70	71	72	73	74	75	76	77	78	79
Rt.	8.367	8.426	8.485	8.544	8.602	8.660	8.718	8.775	8.832	8.888

No.	80	81	82	83	84	85	86	87	88	89
Rt.	8.944	9.000	9.055	9.110	9.165	9.220	9.274	9.327	9.381	9.434

No.	90	91	92	93	94	95	96	97	98	99
Rt.	9.487	9.539	9.592	9.644	9.695	9.747	9.798	9.849	9.899	9.950

Useful Facts From Geometry: 1

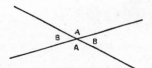

When two straight lines cross, the angles of the same letter are equal

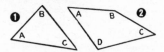

When a straight line crosses two parallel lines, the angles of the same letter are equal

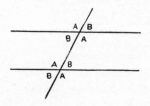

1 The interior angles, A, B, C of any triangle add up to 180 degrees. 2 The interior angles A, B, C, D of any quadrilateral (figure with four straight sides) add up to 360 degrees.

In any right triangle the two angles not right always add up to 90 degrees

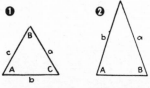

1 Any triangle with three equal angles, A, B, C, has three equal sides, a, b, c, and vice versa. Such triangles are called "equilateral." 2 Any triangle with two equal angles, A, B, has two equal sides, a, b, opposite the equal angles, and vice versa. Such triangles are called isosceles"

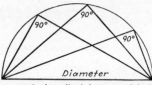

Any angle inscribed in a semicircle, as indicated here, is a right angle

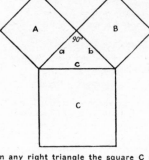

In any right triangle the square C on the hypotenuse equals A + B, sum of the squares on the legs. This holds either for the indicated areas or for the numerical squares of the sides, a, b, c. Thus, if a is 5 ft and b is 7 ft, c is $\sqrt{5 \times 5 + 7 \times 7} = \sqrt{25 + 49} = \sqrt{74} = 8.60$ feet

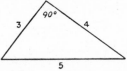

A triangle with sides proportional to 3, 4, 5 (say 3, 4, 5, or 6, 8, 10, or 9, 12, 15, or 30, 40, 50) will be a right triangle with the right angle between the 3 side and the 4 side. Proof: $3 \times 3 + 4 \times 4 = 5 \times 5$

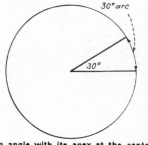

An angle with its apex at the center of a circle is measured by the intercepted arc. Thus, a 30-deg angle will intercept a 30-deg arc, and vice versa

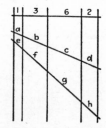

An angle with its apex anywhere on the circumference of a circle will be half the arc intercepted on the circumference, and vice versa. Here, each 30-degree angle intercepts a 60-degree arc

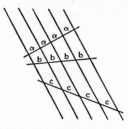

Equally spaced parallel lines will divide a line crossing them at any angle into equal parts. Thus if the a's are equal, the b's will be equal, and the c's will be equal.

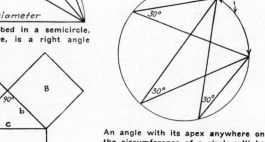

Unequally spaced parallel lines will divide a line crossing at any angle into segments proportional to the spaces between the parallel lines. Thus 1 is to 3 is to 6 is to 2 as a is to b is to c is to d or e is to f is to g is to h

Useful Facts From Geometry: 2

If a plane figure is enlarged without change of shape (as by photography, photostating or pantograph) the area increases in proportion to the square of any selected dimension. Try this for squares of 1, 2, 3 and 4 sides. Areas are 1, 4, 9 and 16, as in Fig. 1.

FIG. 1

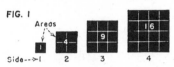

For a circle, the area is always less than the square of the diameter. Yet the area is still proportional to the square of the diameter. The 4-in. circle has exactly 16 times the area of the 1-in., and the 3-in. circle exactly 9 times, Fig. 2.

FIG. 2

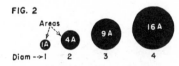

Areas of triangles pictured below in Fig. 3 (each half a square) are actually ½, 2, 4½ and 8, but still in the proportion 1, 4, 9, 16, as before.

FIG. 3

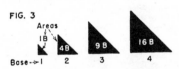

We begin to see that this relationship holds for any shape; for example, that shown below, in Fig. 4.

FIG. 4

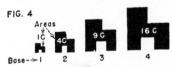

Now this same rule holds for the surfaces of solid bodies as well as for the shapes drawn on a flat surface. This is most obvious for a cube. It has six sides, each a square. So the surface of any cube is 6 times the square of one edge, **Fig. 5**.

FIG. 5

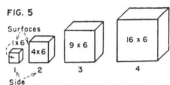

A sphere or ball evidently has somewhat less surface than a cube whose edge equals the sphere's diameter; yet the same proportion holds. Surface of a 4-in. ball is exactly 16 times that of a 1-in. ball, Fig. 6.

FIG. 6

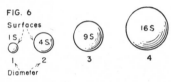

Now let's work a few problems illustrating effect of dimension changes on surface. First: What is effect on quantity of blueprint paper required if drawing scale is reduced from ½ in. per ft to ⅜ in. New dimension is ¾ of the first, so new area is ¾×¾ = 9/16, or 56%; saving is 44%.

Another problem: A shape cut from plywood weighs 12 lb. What would it weigh if a certain dimension were increased from 32 in. to 48 in. with no change in shape? Note that 48 ÷ 32 = 1.5. Then 1.5x1.5 = 2.25, and 2.25x12 = 27-lb weight.

How much lumber (area) and outside paint can be saved by cutting a building plan 12% in all dimensions? The final dimension is 88% of original; so the final surface is 0.88x0.88 = 0.7744, or 77.4%. Saving is 22.6%.

Problem: A certain part is a brass hex rod, ¾ in. across flats. How much is brass cost increased if rod is thickened to ⅞ in. across flats? Ratio of dimensions is 7/8 ÷ 6/8 = 7/6. Then 7/6x7/6 = 49/36 = 1.36, and cost increases 36%, assuming a fixed cost per lb for the hex rod.

If any solid shape is enlarged without changing the proportions, volume and weight go up in proportion to cube of any dimension. This holds for actual cubes. Volumes are 1, 8, 27 and 64 for edges of 1, 2, 3 and 4. Weights must be proportional to volumes. Fig. 7.

FIG. 7

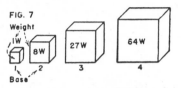

It is equally true for spheres or any other shapes, as the irregular pyramids shown below in Fig. 8.

FIG. 8

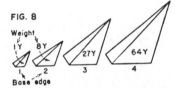

Problem: How much will weight of a steel cone be reduced by cutting off the point to a depth equal to ¼ length of cone? Both cutoff and original cones have same shape. Since little cone has ¼ the length, its relative weight will be ¼x¼x¼ = 1/64 = 0.0156. So weight loss is 1.56%. Fig. 9.

FIG. 9   Cut here

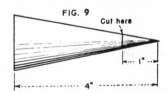

Problem: Conical coal pile is to be built up from fixed central spout until pile base is a circle with 80-ft dia. To find how much coal will be stored here, engineer makes model pile of the same coal with 10-ft dia. cone, by spilling it from a single spout and letting it find its own angle of repose (same as large pile). If coal in sample pile weighs 4560 lb, what is tonnage of big pile?

Ratio of dimensions is 8, so ratio of weights is 8x8x8 = 512. Pile holds 4560x512 ÷ 2000 = 1167 tons.

Useful Facts From Geometry: 3

10 - ELLIPSE

1 - SQUARE

2 - RECTANGLE

3 - PARALLELOGRAM

4 - TRAPEZOID

5 - RHOMBUS

6 - TRIANGLE

7 - REGULAR HEXAGON

8 - REGULAR OCTAGON

9 - CIRCLE

▶ HERE ARE handy rules for areas of common geometrical shapes:

1 Area of a *square* is the square of its side: Area $= A \times A$.

2 Area of a *rectangle* is the product of its two dimensions: Area $= A \times B$.

3 Area of a *parallelogram* (4-sided figure with opposite sides parallel and equal) is the product of any side by the perpendicular distance between that side and its parallel side: Area $= B \times H$.

4 Area of a *trapezoid* (4-sided figure with 2 sides parallel) equals product of distance between parallel sides by average length of parallel sides: Area $= H \times (B_1 + B_2) \div 2$.

5 Area of a *rhombus* (4-sided figure with all sides equal) is half the product of its diagonals: Area $= \frac{1}{2} \times d_1 \times d_2$.

6 Area of a *triangle* is half the product of base and height: Area $= \frac{1}{2} \times A \times H$, or $\frac{1}{2} \times B \times H$. Note: *Any* side may be taken as the base. Height is the perpendicular distance from apex to base (base-line extended where necessary).

7 Area of a *regular hexagon* is 86.6% of the square of the distance between flats:
Area $= 0.866 \times D \times D$.

8 Area of a *regular octagon* is 82.8% of square of distance between flats: Area $= 0.822 \times D \times D$.

9 Area of a *circle* is 78.54% of the square within which it fits. Area $= 0.7854 \times D \times D$. Note: For most engineering purposes 0.785 is close enough, and for many rough, offhand estimates 0.78 is okay.

10 An *ellipse* fills the same fraction (78.54%) of its inclosing rectangle as a circle fills of its inclosing square:
So area $= 0.7854 \times A \times B$.

11 Area of an *annulus* is difference between areas of inner and outer circles. Also, if average diameter (average of inner and outer diameter) is W, and width of band is W, area of the annulus is 3.1416 $\times$ average diameter $\times$ width: Area $= 3.1416 \times D \times W$. Use 3.14 for most practical problems.

12 *Area of a sector* is area of its circle multiplied by a fraction whose numerator is the angle of the sector in degrees, and whose denominator is 360 deg. This reduces to: Area $= 0.00873 \times$ angle of segment $\times$ side $\times$ side.

13-14 Many of the shapes in engineering may be broken up into the simple geometric forms. Thus *composite shape* of Fig. 13 can be broken up, Fig. 14, into triangles, rectangle and sector. Its area is:

area triangle AHC
plus area triangle CIE
plus area sector CHI
minus area rectangle BFGD

15 Where all other methods fail, an area can be measured by dividing it into small squares and counting them. First count full squares. Then estimate fractional squares. For precise results, use small squares.

16 Sometimes it is easier to subdivide the area into narrow strips of constant width. Add average lengths of all strips, and multiply this sum by width of strip to get the area. Lengths of strips are measured along their center lines. This is not theoretically exact, but generally gives a reasonably close answer. The narrower the strips, the less this error will be.

11 - ANNULUS

12 - SECTOR

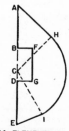

13 - COMPOSITE SHAPE

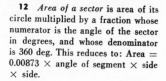

14 - ELEMENTS OF COMPOSITE SHAPE

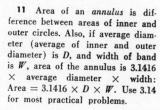

15 - IRREGULAR FIGURE

16 - IRREGULAR FIGURE

492

Dia In	Volume CuFt	Gal	Lb Water 60 F	Dia In	Volume CuFt	Gal	Lb Water 60 F	Dia In	Volume CuFt	Gal	Lb Water 60 F	Dia In	Volume CuFt	Gal	Lb Water 60 F	Dia In	Volume CuFt	Gal	Lb Water 60 F
10	.04545	.3399	2.835	26	.3072	2.298	19.16	42	.8017	5.997	50.00	58	1.529	11.44	95.36	78	2.765	20.69	172.5
11	.05499	.4114	3.430	27	.3313	2.478	20.66	43	.8404	6.286	52.42	59	1.582	11.83	98.67	81	2.982	22.31	186.0
12	.06545	.4896	4.082	28	.3563	2.666	22.22	44	.8798	6.582	54.88	60	1.636	12.24	102.0	84	3.207	23.99	200.0
13	.07683	.5741	4.792	29	.3822	2.859	23.84	45	.9204	6.885	57.41	61	1.691	12.65	105.5	87	3.440	25.73	214.6
14	.08908	.6666	5.556	30	.4091	3.060	25.52	46	.9617	7.194	59.98	62	1.747	13.07	109.0	90	3.682	27.54	229.6
15	.1022	.7650	6.374	31	.4367	3.267	27.24	47	1.004	7.511	62.62	63	1.804	13.49	112.5	93	3.931	29.41	245.2
16	.1163	.8700	7.254	32	.4654	3.482	29.03	48	1.047	7.833	65.30	64	1.862	13.93	116.2	96	4.189	31.33	261.3
17	.1313	.9824	8.189	33	.4950	3.702	30.87	49	1.091	8.163	68.05	65	1.920	14.36	119.8	99	4.455	33.32	277.9
18	.1472	1.102	9.181	34	.5254	3.930	32.77	50	1.136	8.500	70.85	66	1.980	14.81	123.5	102	4.729	35.37	294.9
19	.1641	1.227	10.23	35	.5567	4.165	34.72	51	1.182	8.843	73.72	67	2.040	15.26	127.2	105	5.011	37.48	312.5
20	.1818	1.360	11.34	36	.5891	4.406	36.74	52	1.229	9.193	76.65	68	2.102	15.72	131.1	108	5.301	39.66	330.6
21	.2004	1.499	12.50	37	.6222	4.655	38.81	53	1.277	9.550	79.65	69	2.164	16.19	135.0	111	5.600	41.89	349.3
22	.2200	1.646	13.72	38	.6563	4.910	40.93	54	1.325	9.913	82.64	70	2.227	16.66	138.9	114	5.906	44.18	368.4
23	.2404	1.798	14.99	39	.6913	5.171	43.12	55	1.375	10.28	85.76	71	2.291	17.14	142.9	117	6.221	46.54	388.0
24	.2618	1.958	16.33	40	.7272	5.440	45.36	56	1.425	10.66	88.88	72	2.356	17.63	146.9	120	6.545	48.96	408.2
25	.2841	2.125	17.72	41	.7640	5.715	47.65	57	1.477	11.05	92.12	75	2.557	19.12	159.5				

Handy Tables For Tank Capacity

▶ HERE ARE TABLES I worked up to ease the engineer's job of figuring how much oil, or other liquid, a given round tank will hold, either full or partly filled to a specified depth.

To get capacity of a flat-ended cylindrical tank, multiply length, in inches, by proper factor from Table 1. A convex head holds ⅔ as much as the same length of straight tank. Thus, for a tank with rounded ends, add to length of straight part ⅔ depth of each convex head (and deduct 2/3 depth of each concave head) to get equivalent length of flat-ended tank.

Full Tank—Problem: *What weight of water at 60 F will a cylindrical tank of 64 in. inside diameter hold if straight part is 120 in. long and two convex heads are each 6 in. deep?*

Each 6 in. head equals ⅔ × 6 = 4 in. of straight tank, so the equivalent straight length is 120 + 4 + 4 = 128 in. Capacity per inch of length (Table 1) is 116.2 lb of water. Total capacity is 116.2 × 128 = 14,870 lb water.

A horizontal cylindrical tank half full, or full, by depth is also exactly half full, or full (respectively) by volume. Factors for other fractions full by depth are listed in Table 2. Figures are decimals, not percentages.

Tank Party Full—Problem: *What weight of water will fill the preceding tank 25 in. deep?*

First, 25 ÷ 64 = 0.39, the fraction full by depth. Table 2 shows corresponding fullness fraction of 0.361. That is, a tank 39% full by depth is only 36.1% full by weight or volume.

Then 0.361 × 14,780 = 5,368 lb actual contents. A gage scale for any tank may be plotted from these two tables.

Table 2 shows exactly what happens when a horizontal cylindrical tank is being filled. Imagine a 10,000-gal tank with a diameter of 100 in. The only three places where depth measurements correspond directly with volume are empty, half full and full. That is, zero depth is zero gallons, 50-in. depth is 5,000 gal. and 100-in. depth is 10,000 gal.

Note that at quarter depth the tank is considerably less than quarter full—in fact, 19.55% full, or 1955 gallons.

By depth	By volume	By depth	By volume	By depth	By volume	By depth	By volume
.01	.0017	.26	.2066	.51	.5128	.76	.8155
.02	.0047	.27	.2179	.52	.5255	.77	.8263
.03	.0087	.28	.2292	.53	.5383	.78	.8369
.04	.0134	.29	.2407	.54	.5510	.79	.8474
.05	.0187	.30	.2523	.55	.5636	.80	.8576
.06	.0245	.31	.2640	.56	.5763	.81	.8677
.07	.0308	.32	.2759	.57	.5889	.82	.8776
.08	.0375	.33	.2878	.58	.6014	.83	.8873
.09	.0446	.34	.2998	.59	.6140	.84	.8967
.10	.0520	.35	.3119	.60	.6264	.85	.9059
.11	.0599	.36	.3241	.61	.6389	.86	.9149
.12	.0680	.37	.3364	.62	.6513	.87	.9236
.13	.0764	.38	.3487	.63	.6636	.88	.9320
.14	.0851	.39	.3611	.64	.6759	.89	.9401
.15	.0941	.40	.3736	.65	.6881	.90	.9480
.16	.1033	.41	.3860	.66	7002	.91	.9554
.17	.1127	.42	.3986	.67	.7122	.92	.9625
.18	.1224	.43	.4111	.68	.7241	.93	.9692
.19	.1323	.44	.4237	.69	.7360	.94	.9755
.20	.1424	.45	.4364	.70	7477	.95	.9813
.21	.1526	.46	.4490	.71	7593	.96	.9866
.22	.1631	.47	.4617	.72	.7708	.97	.9913
.23	.1737	.48	.4745	.73	.7821	.98	.9952
.24	.1845	.49	.4872	.74	.7934	.99	.9983
.25	.1955	.50	.5000	.75	.8045	1.00	1.000

At three-quarters depth, it's the other way around. The tank is then 80.45% full, and contains 8045 gal.

Table 2 shows several other interesting facts. A tank 5% full by depth is less than 2% full by volume. A tank 10% full by depth is 5.3% full by volume. A tank 30% full by depth is about one quarter full by volume—actually 25.23%.

How to Figure Water Heating

These quick and simple rules tell dollar-wise operators how much steam it takes to heat any given amount of water from any given temperature to any other in any kind of heating unit

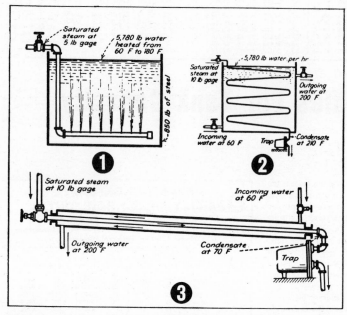

Fig. 1—Heating by mixing. The condensed steam cools to 180 F, so the heat supplied by 1 lb of steam is the total heat of the steam minus the heat in condensate at 180 F.

Fig. 2—Heating by coil in tank. In this case the heat given up per pound of steam is the heat in the steam minus heat in condensate at 210 F.

Fig. 3—Heating in counterflow exchanger. Here condensate is cooled to temperature of cold entering feedwater, so the heat given up per pound of steam is substantially higher than in Fig. 2, a gain of about 140 Btu per pound of steam.

▶ FROM ONE ANGLE OR ANOTHER, every operating engineer is interested in water heating. This article will show how much steam is needed to heat any given weight of water from any temperature to any other temperature. It's really simple if you don't have to split hairs.

How Much Tubing? A more difficult question—one that's often asked—is how many square feet of steel, copper or brass tubing are required to heat such and such a quantity of water per hour to such and such a temperature using steam at a specified temperature. There is generally no way to get a reliable answer to this question, except by

cut-and-try with actual tubing.

The reason is that conditions of water flow and arrangement of the coils make a whale of a lot of difference in the so-called "coefficient of heat transfer." This coefficient is merely the amount of heat that will pass through 1 sq ft of tube surface in one hour for each degree F of temperature difference between the steam inside and the water outside.

Professionals who are designing a certain kind of heat-transfer equipment day after day gradually learn what coefficients will give them fairly reliable results. But the operator generally lacks

this specialized design experience and the heat-transfer books won't help him much.

As I suggested in the beginning, the operator rigging steam coils to heat water may be forced to guess the amount of piping on the low side and then add piping if the original installation can't handle the load.

How Much Steam? Now let's get back to the original subject of this article. How much steam does it take to heat water? In round numbers, 1 Btu (British thermal unit) will raise the temperature of 1 lb of water one degree F. This applies closely regardless of the starting and final temperature.

For steam a rough-and-ready figure, plenty good enough for many rough estimates, is 1000 Btu given up for every pound condensed. This 1000 Btu is not too far off when the steam is supplied at pressures under 30 lb, as is usually the case when heating water.

Problem: How much steam will it take to heat 5000 gal of water per hour from 60 F to 120 F?

Allowing 8.3 lb per gal, the weight of water to be heated is 41,500 lb. The temperature rise from 60 F to 120 F required 60 Btu per pound, so the total heat to be supplied is $41,500 \times 60 =$ 2,490,000 Btu, say 2,500,000 Btu.

Since the condensing of 1 lb of steam gives up 1000 Btu, you divide 2,500,000 Btu by 1,000 to get the steam required — 2500 lb. per hour. If this steam costs 92 cents per thousand, the total steam cost for heating the water will be 92 cents × 2.5 = $2.30 per hour.

Actually, the condensation of 1 lb of steam rarely gives up exactly 1000 Btu. The amount given up varies with the steam pressure and with the final temperature to which the condensate is cooled.

Simple Steam Table. The abbreviated steam table here shown will handle almost any everyday problems of this sort. The table shows, for example, that saturated steam at 40 lb gage pressure has a temperature of 287 F. This is the temperature at which the

STEAM TABLE FOR WATER HEATING

Pressure psi, gage	Saturation temp, deg F	Heat (enthalpy), Btu per pound		
		Saturated water	Evaporation	Steam
0	212	180	970	1150
5	228	196	960	1156
10	240	208	952	1160
20	259	228	939	1167
30	274	243	929	1172
40	287	256	920	1176
50	298	267	912	1179
60	308	277	905	1182
70	316	286	898	1184
80	324	294	892	1186
90	331	302	886	1188
100	338	309	881	1190

water will boil into steam and at which it will condense from steam into water. The condensate from steam at 40 lb gage pressure can't be cooled below 287 F unless the water is first removed from immediate contact with the steam.

The table further shows that to heat water from 32 F up to this saturation temperature of 287 F requires 256 Btu. At this point, note that if you subtract 32 from 287 you get almost exactly this same figure of 256. For all practical purposes, with low and moderate steam pressures, the heat required to warm water up to the boiling pressure equals the boiling temperature minus 32.

After the water has been heated to the boiling temperature corresponding to 40-psi it takes an additional 920 Btu to evaporate it into steam. The sum of the 256 and 920 lb is 1176 Btu, the total heat (entirely) of the 40-psi steam.

This table works equally well for boiling and for condensing. When condensing, you start with a pound of saturated steam at 40-psi gage pressure, 287 F, with a total heat of 1176 Btu. If this steam condenses without cooling, it gives up 920 Btu. You then have a pound of water at 287 F containing 256 Btu. From then on, the steam gives up 1 Btu for every degree it drops in temperature.

There are two common ways to figure the heat given up by a pound of condensing steam. One is to take the heat in the initial steam and subtract from it the heat in the final condensate. The other is to set down the heat of evaporation (or condensation) and add to this the number of degrees the condensate is lowered below the saturation temperature.

Direct Steam Mixing. For a starter, consider the problem pictured in Fig. 1. The tank contains 5780 lb of water at 60 F. Steam is to be bubbled into this mass at 5 lb gage until the water is heated to 180 F. Temperature rise is 120 deg. Multiplying this by 5780 gives a total of about 694,000 Btu.

The final temperature of the condensed steam is 180 F, so its final heat is 180 − 32 = 148 Btu. According to the table, 5-lb steam contains 1156 Btu. The difference given up per pound is then 1156 − 148 = 1008 Btu. Dividing this into 694,000 Btu gives 689 lb of steam.

Coil Heater. Fig. 2 shows a slightly different case. The same weight of water (5780 lb) enters at 60 F, leaves at 200 F, after being heated by saturated steam, entering the coil at 10 lb gage. The condensate is discharged at 210 F.

In this case, the temperature rise of the water is 200 F − 60 F = 140, and the total heat required is 140 × 5,780 = 809,000 Btu. From the table, the initial heat in the steam is 1160 Btu per lb. The final heat in the condensate at 210 F is 210 − 32 = 178 Btu. Then the heat given up by each pound of steam before it leaves the coil is 1160 − 178 = 982 Mtu. Dividing 809,000 Btu by 982, we find that 824 lb of steam will be required.

Counterflow Heater. If the tube is well insulated against heat loss, the arrangement shown in Fig. 3, or some other in which the water to be heated flows countercurrent to the cooling condensate, gives the greatest efficiency because it cools the condensate down almost to the temperature of the entering cold water, let's say within 10 deg. In the case pictured, the condensate leaves at 70 F with only 70 − 32 = 38 Btu. The heat given off by the steam per pound is 1160 − 38 = 1122 Btu. Then the steam needed to heat 5,780 lb will be only 809,000 ÷ 1122 = 721 lb.

Table 1—Saturation—Temperatures

Temp	Abs press	Specific vol		Enthalpy ("Heat")		
F	Psi	Sat liquid	Sat vapor	Sat liquid	Evap	Sat vapor
32	0.08854	0.01602	3306	0.00	1075.8	1075.8
40	0.12170	0.01602	2444	8.05	1071.3	1079.3
50	0.17811	0.01603	1703.2	18.07	1065.6	1083.7
60	0.2563	0.01604	1206.7	28.06	1059.9	1088.0
70	0.3631	0.01606	867.9	38.04	1054.3	1092.3
80	0.5069	0.01608	633.1	48.02	1048.6	1096.6
90	0.6982	0.01610	468.0	57.99	1042.9	1100.9
100	0.9492	0.01613	350.4	67.97	1037.2	1105.2
110	1.2748	0.01617	265.4	77.94	1031.6	1109.5
120	1.6924	0.01620	203.27	87.92	1025.8	1113.7
130	2.2225	0.01625	157.34	97.90	1020.0	1117.9
140	2.8886	0.01629	123.01	107.9	1014.1	1122.0
150	3.718	0.01634	97.07	117.9	1008.2	1126.1
160	4.741	0.01639	77.29	127.9	1002.3	1130.2
170	5.992	0.01645	62.06	137.9	996.3	1134.2
180	7.510	0.01651	50.23	147.9	990.2	1138.1
190	9.339	0.01657	40.96	157.9	984.1	1142.0
200	11.526	0.01663	33.64	168.0	977.9	1145.9
212	14.696	0.01672	26.80	180.0	970.4	1150.4
220	17.186	0.01677	23.15	188.1	965.2	1153.4
240	24.969	0.01692	16.323	208.3	952.2	1160.5
280	49.203	0.01726	8.645	249.1	924.7	1173.8
300	67.013	0.01745	6.466	269.6	910.1	1179.7
340	118.01	0.01787	3.788	311.1	879.0	1190.1
380	195.77	0.01836	2.335	353.5	844.6	1198.1
400	247.31	0.01864	1.8633	375.0	826.0	1201.0

Table 2—Saturation—Pressures

Abs press	Temp	Specific vol		Enthalpy ("Heat")		
psi	F	Sat liquid	Sat vapor	Sat liquid	Evap	Sat vapor
0.50	79.58	0.01608	641.4	47.6	1048.8	1096.4
1.0	101.74	0.01614	333.6	69.7	1036.3	1106.0
5.0	162.24	0.01640	73.52	130.1	1001.0	1131.1
10	193.21	0.01659	38.42	161.2	982.1	1143.3
14.7	212.00	0.01672	26.80	180.0	970.4	1150.4
15	213.03	0.01672	26.29	181.1	969.7	1150.8
20	227.96	0.01683	20.089	196.2	960.1	1156.3
25	240.07	0.01692	16.303	208.5	952.1	1160.6
30	250.33	0.01701	13.746	218.8	945.3	1164.1
40	267.25	0.01715	10.498	236.0	933.7	1169.7
50	281.01	0.01727	8.515	250.1	924.0	1174.1
60	292.71	0.01738	7.175	262.1	915.5	1177.6
70	302.92	0.01748	6.206	272.6	907.9	1180.6
80	312.03	0.01757	5.472	282.0	901.1	1183.1
90	320.27	0.01766	4.896	290.6	894.7	1185.3
100	327.81	0.01774	4.432	298.4	888.8	1187.2
110	334.77	0.01782	4.049	305.7	883.2	1188.9
120	341.25	0.01789	3.728	312.4	877.9	1190.4
130	347.32	0.01796	3.455	318.8	872.9	1191.7
140	353.02	0.01802	3.220	324.8	868.2	1193.0
150	358.42	0.01809	3.015	330.5	863.6	1194.1
200	381.79	0.01839	2.288	355.4	843.0	1198.4
250	400.95	0.01865	1.8438	376.0	825.1	1201.1
300	417.33	0.01890	1.5433	393.8	809.0	1202.8
350	431.72	0.01913	1.3260	409.7	794.2	1203.9
400	444.59	0.0193	1.1613	424.0	780.5	1204.5

How to Use the Steam Tables—Part I

THIS IS THE FIRST of three simple, practical pages showing how an operating engineer can use steam tables to figure out problems involving heating water, making steam and superheating.

Values of the various properties of water and steam have been determined on the basis of experimental data. There are several sets of tables, but this one is from "Thermodynamic Properties of Steam," by J H Keenan and F G Keyes, published by John Wiley & Sons, Inc. (now out of print). However, the ASME and others have prepared similar tables, which may be found in standard engineering reference books.

Both Tables 1 and 2 give the properties of water and of saturated steam. The only difference is that you enter 1 with the boiling *temperature* and 2 with the boiler *pressure* (in psi absolute).

For example, Table 1 shows that, for water to boil at 100 F, the absolute pressure must be 0.95 psi abs.

Table 2 shows that at 40 psi abs water boils at 267 F. Note that we don't usually bother to put down all the digits given in the table. Most practical work doesn't require such hair-splitting accuracy. An operating engineer will rarely need to figure water temperature closer than the nearest degree, or heats or enthalpies closer than the nearest Btu.

After the first two, the columns are the same in both tables. "Sat liquid" means liquid water at the saturation or boiling temperature. "Sat vapor" means steam at the boiling temperature.

When water is boiling in a closed container, both the water and the steam over it are in a saturated condition.

The steam generated by a boiler with-out a superheater is saturated. For steam, saturated means steam that contains no liquid water, yet is not super-heated (still at boiling temperature).

Note that the absolute pressure is gage pressure plus about 15 lb.

Now, in Table 2, try reading across the line for 50 psi abs (35 psi gage).

Boiling temperature is 281 F. At this temperature one pound of water fills 0.0173 cu ft and one pound of saturated steam 8.51 cu ft. Specific vol is in cubic foot per pound of water or steam.

It takes 250 Btu to heat the pound of water from 32 F to the boiling point and another 924 Btu to evaporate it, making a total of 1174 Btu.

"Enthalpy" is what used to be called "heat" in the old steam tables. It's given in Btu per lb. They would have labeled the last three columns: "Heat of the liquid", "Heat of Vaporization" and "Total Heat".

What happens to a pound of water when changed to steam at various pressure levels

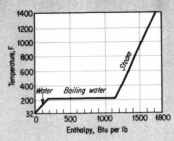

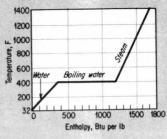

If we heat a pound of water in a cylinder with various pressures applied to the piston, we find that the same general sequence of events occurs at each pressure level. First, water temperature rises and its volume increases slightly. Then boiling begins, the volume increases greatly, and the tempera-ture stays constant. When all the water has changed to steam, the volume continues to increase on further heating and the temperature rises. This is called superheating.

The charts above demonstrate graphically that, as pressure increases, amount of heat required to bring water to the boil-

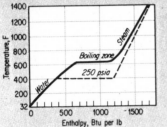

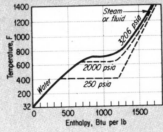

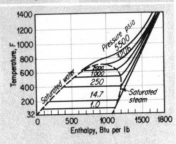

ing point goes up, while amount of heat necessary to vaporize it tends to get less. Eventually, at 3206.2 psia we reach a point where water turns to steam without boiling, and all the heat has gone into bringing the water up to this condition, called the *critical point*. The final graph summarizes the conditions shown on the preceding ones plus two other pressure levels.

If we heated water and its vapor in a sealed quartz tube, we would see, at any condition below the critical point, a clearly defined level, separating water and steam. Near the critical, water level becomes indistinct, then disappears.

How to Use the Steam Tables—Part 2

The most common type of practical problem involving the steam tables is figuring how much heat it takes to convert feedwater at any given temperature into saturated steam at any given pressure or temperature.

In all our sample problems we shall juggle the data to fit the few lines in the sample tables. The complete tables will give you any pressure or temperature you may need.

When heating water and making steam, the heat *supplied* is the "enthalpy" at the finish minus the enthalpy at the start. Or call it "heat," instead of enthalpy, if you find it simpler.

Take the case of a boiler generating saturated steam at 135-psi gage (150-psi abs). The enthalpy, or heat, of the final steam is 1194 Btu per pound.

The amount of heat required to produce this steam in an actual boiler will depend on the temperature of the feed-water.

Suppose the feedwater temperature is 180 F. Table 1 shows that the heat in the water is 147.9 Btu, say 148 Btu. Then the heat supplied to turn this water into steam is merely the difference, or 1194 − 148 = 1046 Btu.

It's an easy step from this to figuring the boiler efficiency. Let's say the boiler generates 10 lb steam per pound of coal burned and the coal contains 13,000 Btu per pound. Then, for every 13,000 Btu put in as fuel, there is delivered in steam, 10 × 1046 = 10,460 Btu.

The efficiency of any power unit is its output divided by its input, or (here)

10,460 ÷ 13,000 = 0.805 or 80.5%.

For most purposes you won't need Table 1 to get a close value of the heat of the liquid. Just subtract 32 from the water temperature.

For example, the enthalpy of water at 180 F is the heat required to raise it from 32 F to 180 F, or a difference of 148 deg. This takes about 148 Btu.

It won't work out quite so closely for very high temperatures. Take the case of water at 300 F. Table 1 gives 269.6 Btu, while our simple method gives 300 − 32 = 268 Btu, but still close enough for most purposes.

Below 212 F the rule is extremely ac-curate, never out by more than one-tenth of a Btu.

Condensing of steam in a condenser, heater, radiator or process is nothing but heating and evaporation worked backward.

TABLE 3—SUPERHEATED STEAM

Abs Pressure, psi (Sat Temp)		Sat Liquid	Sat Vapor	Temperature, degrees Fahrenheit							
				300	400	500	600	700	800	900	1000
15 (213.03)	v	0.016	26.29	29.91	33.97	37.99	41.99	45.98	49.97	53.95	57.93
	h	181.1	1150.8	1192.8	1239.9	1287.1	1334.8	1383.1	1432.3	1482.3	1533.1
20 (227.96)	v	0.016	20.09	22.36	25.43	28.46	31.47	34.47	37.46	40.45	43.44
	h	196.2	1156.3	1191.6	1239.2	1286.6	1334.4	1382.9	1432.1	1482.1	1533.0
40 (267.25)	v	0.017	10.498	11.040	12.628	14.168	15.688	17.198	18.702	20.20	21.70
	h	236.0	1169.7	1186.8	1236.5	1284.8	1333.1	1381.9	1431.3	1481.4	1532.4
60 (292.71)	v	0.017	7.175	7.259	8.357	9.403	10.427	11.441	12.449	13.452	14.454
	h	262.1	1177.6	1181.6	1233.6	1283.0	1331.8	1380.9	1430.5	1480.8	1531.9
80 (312.03)	v	0.018	5.472		6.220	7.020	7.797	8.562	9.322	10.077	10.830
	h	282.0	1183.1		1230.7	1281.1	1330.5	1379.9	1429.7	1480.1	1531.3
100 (327.81)	v	0.018	4.432		4.937	5.589	6.218	6.835	7.446	8.052	8.656
	h	298.4	1187.2		1227.6	1279.1	1329.1	1378.9	1428.9	1479.5	1530.8
150 (358.42)	v	0.018	3.015		3.223	3.681	4.113	4.532	4.944	5.352	5.758
	h	330.5	1194.1		1219.4	1274.1	1325.7	1376.3	1426.9	1477.8	1529.4
200 (381.79)	v	0.018	2.288		2.361	2.726	3.060	3.380	3.693	4.002	4.309
	h	355.4	1198.4		1210.3	1268.9	1322.1	1373.6	1424.8	1476.2	1528.0
300 (417.33)	v	0.0189	1.5433			1.7675	2.005	2.227	2.442	2.652	2.859
	h	393.8	1202.8			1257.6	1314.7	1368.3	1420.6	1472.8	1525.2
400 (444.59)	v	0.0193	1.1613			1.2851	1.4770	1.6508	1.8161	1.9767	2.134
	h	424.0	1204.5			1245.1	1306.9	1362.7	1416.4	1469.4	1522.4
500 (467.01)	v	0.0197	0.9278			0.9927	1.1591	1.3044	1.4405	1.5715	1.6996
	h	449.4	1204.4			1231.3	1298.6	1357.0	1412.1	1466.0	1519.6
600 (486.21)	v	0.0201	0.7698			0.7947	0.9463	1.0732	1.1899	1.3013	1.4096
	h	471.6	1203.2			1215.7	1289.9	1351.1	1407.7	1462.5	1516.7
800 (518.23)	v	0.0209	0.5687				0.6779	0.7833	0.8763	0.9633	1.0470
	h	509.7	1198.6				1270.7	1338.6	1398.6	1455.4	1511.0
1000 (544.61)	v	0.0216	0.4456				0.5140	0.6084	0.6878	0.7604	0.8294
	h	542.4	1191.8				1248.8	1325.3	1389.2	1448.2	1505.1
1200 (567.22)	v	0.0223	0.3619				0.4016	0.4909	0.5617	0.6250	0.6843
	h	571.7	1183.4				1223.5	1311.0	1379.3	1440.7	1499.2
1400 (587.10)	v	0.0231	0.3012				0.3174	0.4062	0.4714	0.5281	0.5805
	h	598.7	1173.4				1193.0	1295.5	1369.1	1433.1	1493.2

How to Use the Steam Table—Part 3

THIS THIRD EXAMPLE will finish up our lesson on the practical use of steam tables to help you generate steam efficiently. Table 3 also is reproduced by permission of the publisher of the preceding two tables; you'll find equivalent data in many engineering reference books.

First column gives the absolute pressure and (directly below it in parenthesis) the corresponding saturation temperature or boiling temperature.

In the next column *v* and *h* stand for volume of one pound and its heat content. For example, at 150 psi absolute the volume of one pound is 0.018 cu ft for liquid water and 3.015 cu ft

for the saturated steam. Corresponding heat contents of per pound are 330.5 Btu and 1194.1 Btu.

The temperature columns give the volume and heat content per pound for superheated steam at the indicated temperature. Take the case of steam at 150 psi and superheated to a total temperature of 600 F. Look in the 600 F column opposite 150 psi. The volume is 4.113, as against 3.015 for saturated steam at the same pressure. This is natural because steam expands like a gas when you superheat it.

Also the heat content is naturally higher—1325.7 Btu instead of 1194.1 Btu.

Note that this table gives the actual

temperature of the superheated steam rather than the degrees of superheat, which is a different thing. If the steam has been superheated from a saturation temperature of 358 F to 600 F, the superheat is 600 − 358 = 242 degrees.

These superheat tables are just as easy to use as the saturation tables described in the two previous installments.

Problem: How much heat does it take to convert one pound of feedwater at 205 F into superheated steam at 150 psi abs and 600 F? Heat in steam is 1325.7, Call it 1326. Heat in the water is 205 − 32 = 173 Btu. Then heat required to convert one pound to steam is 1326 − 173 = 1153 Btu.

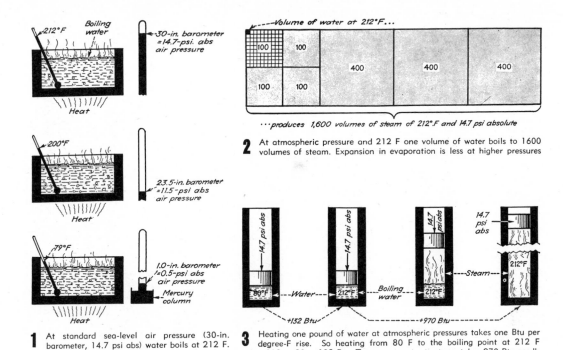

2 At atmospheric pressure and 212 F one volume of water boils to 1600 volumes of steam. Expansion in evaporation is less at higher pressures

1 At standard sea-level air pressure (30-in. barometer, 14.7 psi abs) water boils at 212 F. If pressure is cut to 0.5 psi abs, water boils at 79 F

3 Heating one pound of water at atmospheric pressures takes one Btu per degree-F rise. So heating from 80 F to the boiling point at 212 F takes 212 — 80 = 132 Btu. To convert water to steam takes 970 Btu per lb

Heating and Boiler Water

▶ THE STANDARD PRESSURE of the atmosphere at sea level is taken as 14.7 psi abs (14.7 pounds per square inch absolute). At this pressure the barometer reads 30 in. of mercury.

The temperature at which water boils rises with the pressure. At the standard atmosphere of 14.7 psi abs, water boils at 212 F, Fig. 1, top.

In airplanes, and on high mountains, the pressure is less, so water boils at a lower temperature. For example, Fig. 1, center, at 11.5 psi abs (23.5-in. barometer) water boils at 200 F.

If vacuum equipment brings the pressure down to 0.5 psi abs (that is 1-in. barometer, or 29-in. vacuum) water boils at 79 F. You can hold your hand in it comfortably, Fig. 1, bottom.

EXPANSION

When heated from 32 F to 212 F, liquid water expands about four per cent. Here are weights per cubic foot at various temperatures:

Degree F	Pounds per cubic foot
32	62.4
70	62.3
100	62.0
150	61.2
200	60.1
212	59.8

At 212 F one pound of water occupies only 0.0167 cu ft. When evaporated, this expands 1600 times to form 26.7 cu ft of steam at 212 F and 14.7 psi abs, Fig. 2.

Steam is compressible, about like air, so one pound weight of steam occupies less and less space as the pressure rises. At 100-psi gage pressure (114.7 psi abs) one pound of steam occupies 3.9 cubic feet. At 445 psi gage saturated steam weighs one pound per cubic foot.

For practical purposes, at temperatures under 400 F, one Btu (British thermal unit) is the heat required to warm one pound of liquid water one degree F. So, to heat 2000 lb of water from 80 F to 180 F (100 degrees rise) takes 100 × 2000 = 200,000 Btu. If this same water cools from 180 F to 80 F, it will give off 200,000 Btu.

In Fig. 3, to heat one pound of water from 80 F to 212 F takes 212 — 80 = 132 Btu. If further heat is supplied, with the fluid under atmospheric pressure through the floating piston, the water will start to turn into steam, raising the piston.

LATENT HEAT

When 970 Btu has been supplied, all the liquid will have been converted into saturated steam.

In this case the total heat required to convert one pound of water at 80 F into saturated steam at 212 F and 14.7 psi abs is 132 + 970 = 1102 Btu. The 970 Btu is the "latent heat of vaporization" for this pressure, temperature.

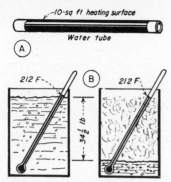

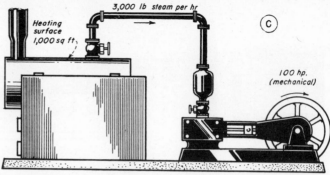

ONE RATED BOILER HORSEPOWER (A) is 10 sq ft of water-heating surface. One developed horsepower **(B)** is the evaporation of 34½ lb of 212F feedwater into steam at 212F and atmospheric pressure. Around 1870 a 1000-sq ft boiler (100 rated boiler hp) **(C)** could develop 3000 lb of steam per hr (100 developed boiler hp) to supply engine of 100 brake hp

Let's Talk "Boiler Horsepower"

▶ Many engineers, including me, have publicly denounced "boiler horsepower" as an old-fashioned and cumbersome measure that should be abolished. To my mind there is no doubt about the logic of this position, but I must admit that the old habit will be hard to break. Anyway, I'm going to tell what boiler horsepower is and how it got that way.

History. Back about 1870 an ASME Committee decided that a good non-condensing engine of that day would take about 30 lb of steam per hour to generate one horsepower—also that the average boiler of that period could generate about 3 lb of steam per square foot of water-heating surface.

Then 10 sq ft would generate 30 lb, so they concluded that the 10 sq ft should be rated as one boiler horsepower and that 30 lb of steam per hr, approximately, should be considered one developed boiler horsepower.

This last was more precisely stated as 34½ lb of steam per hr "from and at 212", meaning the evaporation of 34½ lb of water at 212 F into steam at the same temperature at atmospheric pressure.

Boilers don't usually run under these conditions, so the founding fathers of boiler horsepower invented a "factor of evaporation" to convert "actual evaporation" into "equivalent evaporation from and at 212".

Factor of Evaporation. For example, if the feedwater temperature is 184 F, and the steam is at 100 psi gage and 500 F, this factor is 1.16. This means that it takes 1.16 times as much heat to generate one pound of steam under these conditions as "from and at 212 F."

Now suppose a boiler operating under these conditions has 2000 sq ft of heating surface and generates 65,000 lb per hr. The rated boiler hp will be 2000 ÷ 10 = 200. The 65,000 lb per hr will be equivalent to 65,000 × 1.16 = 75,400 lb "from and at 212".

Dividing 75,400 by 34.5, we get 218 developed horsepower. Then 100 × 218 ÷ 200 = 109 "per cent rating".

Now I admit that there is nothing hard about this arithmetic except figuring the factor of evaporation. But I won't try to explain that here, for why should readers bother to learn an old fashioned method?

No Sense in This. Now let's apply a little common sense to what has been going on here. Ten square feet of water heating surface in a modern boiler will generate anywhere from 50 to 500 lb of steam per hour (the 500 lb in some big central-station boilers). Also, some large modern turbines use as little as 5 lb of steam per hp-hr instead of 30.

So we have a peculiar situation in which one boiler horsepower can serve anywhere up to 100 mechanical horsepower. Which doesn't make sense.

Under today's conditions rated boiler horsepower has become just a cumbersome way of stating the water-heating surface.

Developed boiler horsepower is a complicated way of reporting how much steam the boiler is generating.

The Right Way. Because of this ridiculous situation the makers and users of large boilers said goodbye to boiler horsepower long ago. Both the capacity and actual loads of such units is clearly stated as so many pounds of steam per hour—simple and sensible.

In the smaller plants boiler horsepower still hangs on from force of habit, and probably won't die out completely for years.

If you wonder why people keep on doing things the hard way just remember that English is easier to write than Chinese, but it's hard to sell the idea to the Chinese.

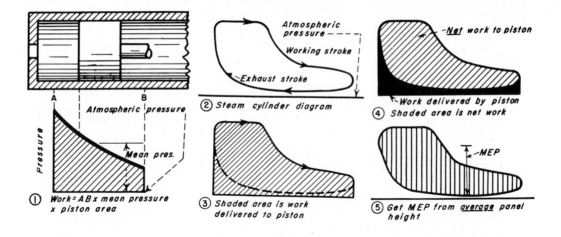

① Work = AB x mean pressure x piston area

② Steam cylinder diagram

③ Shaded area is work delivered to piston

④ Shaded area is net work

⑤ Get MEP from *average* panel height

Indicated Horsepower Explained

Most of us know that *work* is merely *force* times the distance moved. If you pull 50 lb to drag a box across the floor, and if you drag it 100 ft, the work is $50 \times 100 = 5000$ ft lb.

The *power* of a steam or internal combustion engine may refer to either the *brake horsepower* delivered at the shaft or the *indicated horsepower* as determined from the diagrams drawn by an engine indicator that shows the pressure at all points of the stroke.

The *indicated horsepower* of an engine is merely the net power delivered to the piston faces by the expanding steam or gas in the cylinder.

The amount of work so performed in one minute, divided by 33,000, will be the indicated horsepower.

The first and most important step is finding the net useful push on the piston. Remember that the work delivered to the piston during the working stroke is partly offset by the work the piston returns to the steam or gas during the exhaust stroke.

In Fig. 1 pressure falls as shown by the curve when the piston face moves from *A* to *B*. Find the average pressure from the average height of the curve between *A* and *B*. This average pressure, in pounds per square inch, multiplied by the area of the piston face in square inches, gives the average force or push on the piston. This force, multiplied by length *AB*, gives work done during the stroke.

Fig. 2 shows a steam-engine indicator diagram. The steam delivers work to the piston during the power stroke and takes back part of this work during the exhaust stroke.

In Fig. 3 the shaded area, all the way down to the atmospheric-pressure line, represents the work done on the piston during the working stroke.

In Fig. 4 the black area is the work returned during the exhaust stroke; so the shaded area in Fig. 4 is the net or useful work per working stroke.

Perhaps a simpler way to look at it is to consider that the average height of the diagram from the top line to the bottom line represents the net useful pressure. This is merely the average pressure on the working stroke minus the average pressure on the exhaust stroke and is called the mean effective pressure (MEP).

To get the MEP use a planimeter. If your plant does not have this very useful instrument, just figure the average height of a lot of narrow strips as shown in Fig. 5. Then the indicated horsepower for each piston face in the engine will be $P \times L \times A \times N \div 33{,}000$, where P is the *MEP*, L is the length of the stroke in *feet*, A is the net area of the piston face in square inches (subtract area of rod in case of crank end) and N is the number of working strokes per minute.

Problem: What is the indicated horsepower of a simple double-acting steam engine with the following data?

Piston diameter	10 in.
Rod diameter	2 in.
Stroke	12 in.
Speed	350 rpm
Head-end MEP	42 psi
Crank-end MEP	43 psi

Solution:

Head-end area	=	87.5 sq in.
Rod area	=	3.1 sq in.
Crank-end area	=	84.4 sq in.
Stroke	=	1 ft

Head-end indicated horsepower $= P \times L \times A \times N \div 33{,}000 = 42 \times 1 \times 87.5 \times 350 \div 33{,}000 = 38.9$ hp.

Crank-end indicated horsepower $= P \times L \times A \times N \div 33{,}000 = 43 \times 1 \times 84.4 \times 350 \div 33{,}000 = 38.5$ hp.

So the total indicated horsepower is $38.9 + 38.5 = 77.4$ hp.

How to Read a Line Chart

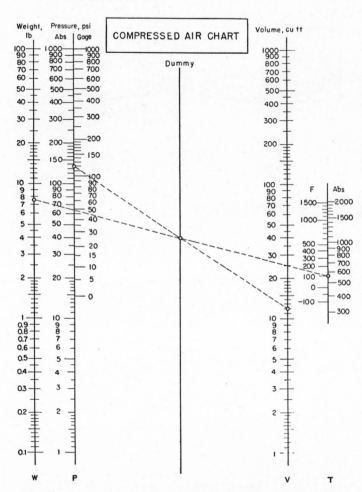

HERE we will kill two birds with one stone—get a handy chart for our files and learn how to use all charts of this very useful type.

For air, as for any other gas, there is a definite relation between weight, pressure, volume and temperature. If you know any three of these you can figure the fourth from a formula given in the thermodynamics books. If you don't want to sweat over an algebraic formula you can get the same answer from the chart on this page.

There isn't space here to explain how I plotted this chart, but here's how you use it:

Draw one dotted line from pressure on P to volume on V. Draw another from weight on W to temperature on T. In all cases the two lines must intersect at the dummy line.

If you know P and V connect them first.

If you know W and T connect them first.

This will locate the intersecting point on the dummy. Then draw a line from the remaining known quantity through the dummy intersection to the scale for the unknown. Where your line cuts that scale read the value of the unknown. If you proceed as shown by the dotted lines you can't miss.

Various engineering reference books contain hundreds of charts of this general type for solving all kinds of practical mathematicial problems. Usually dotted lines are drawn to show how to use the charts.

Charts of this kind, in which you draw a straight line across several separate scales, are known as *collinear diagrams*.

These are great time savers for busy engineers. You can take any scale as the unknown. If you complete the indicated construction, you always end up with a line crossing the unknown scale where you can then read the answer.

TO USE THIS CHART: Connect **P** to **V** and **W** to **T**. First connect two of the known quantities (**P** and **V** if you know both or **W** and **T** if you know them). Draw a second line from third known item through intersection of first line and dummy. Continue second line until it crosses column of unknown; then read your answer. Let's try a problem: **Problem:** A 12-cu ft tank holds air at 120 psi gage and 100 F. What is weight? **Solution** (shown by dotted lines): Connect 120 psi gage on **P** to 12 cu ft on **V**. From 100 F on **T** draw line through dummy intersection and extend until it cuts line **W**. Answer: 7.6 lb air in tank

HOW TO READ METERS CORRECTLY

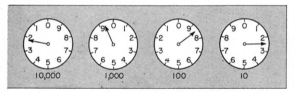

1982, NOT 2982, is correct reading here because 1000 dial pointer has not passed zero. So read 10,000 dial pointer as 1

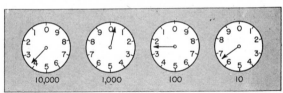

4026, NOT 3026, is correct reading here because 1000 dial pointer has passed zero. So 10,000 dial pointer should be past 4

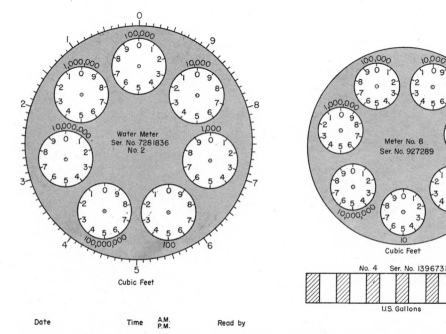

Date   Time   A.M. P.M.   Read by

PRINT SHEET of actual meter dials in plant without pointers but with all details as shown, depending on meters. Then operators taking reading can draw pointers in exact positions instead of writing down numbers. Clerk interprets readings

▶ IT SEEMS no two people read any meter the same. Quite often they're both wrong! Here are a few rules to remember:

(1) Begin at dial registering highest number of cu ft or gallons, and read from left to right. (2) Meter dials run clockwise on alternate dials and counter clockwise on dials in between. Always read smaller number if pointer is between two numbers. (3) Meters usually record cubic feet or gallons. One complete revolution of a dial means amount stamped under meter dial has passed through meter. (4) Al-ways read meter to exact point on lowest dial, using decimals and zeros if practical. (5) If there is doubt that pointer has passed over a number, check the next lower dial. If pointer has passed zero on this dial, then it has passed number on first dial. (6) Some-times wear and adjustments on meter cause pointers to pass a number before it should. Correct reading for first meter is 1982, not 2982. 2982 would be wrong as 1000-dial pointer has not passed zero. This means the 10,000-dial pointer must read 1 instead of 2. (7) Also, meter dials may not pass a number when they should. Correct reading for second meter is 4026, not 3026. Reason is 1000-dial pointer has passed zero; therefore the 10,000-dial pointer should have passed the 4.

Good System. One solution for meter-reading problems (common in some plants) is to have printed sheets show-ing dial of each meter in plant. Opera-tors can draw in dial pointers as well as record meter reading. This gives supervisory personnel a double check on meter readings without actually go-ing to each meter to check them. In long run, this will save headaches.

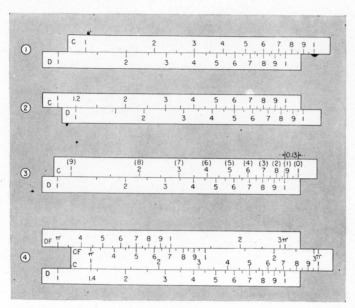

Some Slide-Rule Hints

▶ MOST OF THESE slide-rule tricks will be found in some instruction books, but many users tend to overlook them. So when they're needed, they aren't remembered. Here are some of the most useful:

1. When you have a long list of numbers all to be divided by the same number, *a*, it isn't necessary to reset the rule for each division. Simply perform the division 1/*a* by setting *a* on the C scale opposite the nearest index on the D scale, then read the answers on the D scale opposite the number being divided on the C scale.

Example: 1.5, 3, 4.5 and 6 are to be divided by 0.75. Find the answers. Set 7.5 on the C scale opposite the right index on the D scale, Fig. 1. (This gives a common multiplier of 1.333 on the D scale, but this need not be noted.) Opposite 1.5 on C find 2 on D, opposite 3 on C find 4 on D, opposite 4.5 on C find 6 on D, opposite 3 on C find 4 on D, opposite 4.5 on C find 6 on D, opposite 6 on C find 8 on D.

Example: 9, 6, 3 are to be divided by 1.2. Set 1.2 on the C scale opposite the left index on D, Fig. 2. On D opposite given numbers on C find answers, 7.5, 5, 2.5.

2. In some problems several numbers must each be divided by a quantity (1-*a*). Often the subtraction can be done mentally, but as a check the quantity *a* can be tallied off directly on the rule helping to avoid error in mental calculation. Look at the C scale, Fig.

3, and assume that it is numbered from 0 to 9, right to left, as shown by the numbers in parentheses. Find the number *a* on the renumbered C scale, and set it up against the right index of D. Then work on the numbers to be divided as in the first method.

Example: Divide 3, 4, 5, 6 and 7 by (1-0.13). On the C scale, Fig. 3, mentally renumber 9 to represent 0.1. Find 0.13 which corresponds to 8.7. Then opposite the given numbers find the answers, 3.45, 4.6, 5.75, 6.9, 8.05. (Note that the scale numbers and corresponding numbers in parentheses on the C scale each add up to 10.)

3. Those folded scales CF and DF starting and ending with π are of far more use than just for formulas involving π, Fig. 4. When properly used they effectively double the length of your rule. When you find the setting of your rule runs off the C or D scales for a given calculation, don't move the slide, but simply switch to the CF or DF scales for your answer.

Example: Multiply 8 by 1.4. Setting left C index to 1.4 on the D scale, for minimum slide movement, you'll find that 8 on the C scale is way off the calculating area on D. So now look for 8 on the CF scale, and opposite it on the DF scale find the answer 11.2.

In effect the folded scales are simply a convenient extension in both directions of the straight scales.

20
License examination
questions & answers

WHEN MANAGEMENT and operating engineer work out answers to problems together, running steam plant isn't tough

WATER TREATMENT sampling and analysis today is key to good-quality steam and efficient operation of high-pressure boilers

STEAM Q and A

. . . so you don't need licenses in your locale? You still need these answers to run plants safely . . .

Q 1—What attention should be given water level in boiler when lighting off cold boilers? Also when cutting out boiler that has been steaming?

A—Before lighting off cold boiler, have water level about one inch from bottom in gage glass. As boiler heats, water expands and level reaches half glass or over by time boiler is ready to cut in.

When fires are extinguished, before cutting out boiler steaming on line with other boilers, bring water level to at least half a glass. Heat stored in hot brick walls generates steam for some time. Steam stop valve on boiler must remain open while steam's generated with fires out, or safety valves will lift. During this time steam flows to main header and water drops in glass, so feed the boiler from time to time.

When water level in glass stops dropping, boiler is ready to cut out by closing main stop. If steam stop valve is not closed then, water will rise in glass because boiler stops generating steam and condenses steam from other boilers connected to it.

Q 2—What data should boiler's nameplate have?

A—Boiler nameplates should have: (1) Manufacturer's name, (2) Boiler's working pressure, (3) Hydrostatic pressure boiler's tested for, (4) Tensile strength of boiler metal, (5) Steam-drum safety-valve and superheater safety-valve lifting pressures, (6) Year boiler was built.

Q 3—In converting from coal to oil, what attention should be given the furnace for burning fuel oil efficiently?

A—The burners must be set so oil does not impinge on the furnace walls or on boiler surfaces. When oil strikes these surfaces, carbon forms because oil does not have a chance to mix thoroughly with air. Then furnace smokes as more carbon forms. Oil impinging on brick walls thins bricks quickly, causing costly maintenance. Furnace walls must be thick enough to prevent radiation and keep furnace hot for efficient combustion.

Furnace must be free from air leaks. Besides cooling furnace, air leaking in distorts fuel-oil spray, often causing impingement.

Keep furnace free from oil deposits. These are caused by dirty burners not allowing tips to atomize oil properly.

Examine fires often. If flame is "dirty", pull burner and change tip.

Q 4—Explain what precautions to take when lighting off boiler after doing extensive brickwork.

A—New brickwall must be dried slowly. That's because walls are wet and sudden heating to high temperature causes early failures from moisture turning to steam, causing tiny ruptures. Start small wood fire in furnace. A slow wood fire dries a new wall in about 24 hours.

If this is not practical, use smallest tip in center burner to keep fire away from walls. Raise steam as slowly as possible. Operating engineers often overlook importance of thoroughly drying new brickwork, then wonder why walls fail.

Q 5—What should the operating engineer know about installing and operating steam reducing valves?

A—If possible, locate steam reducing valves at point where line has driest steam. That is usually highest point in line. Connect drain line from reducing valve to trap. Place gate valves on both sides of reducer with bypass and valve around them. Then reducing valve can be bypassed when servicing by opening bypass and closing gate valves.

Gate valves on either side of reducers do not obstruct steam for normal operation, as would globe valves. Make sure reducing valve is installed with steam flow according to arrow on valve.

Before turning steam on new reducing valve the first time, drain line thoroughly. Crack steam valve to line on high pressure side to build up pressure gradually against reducer. Then adjust spring tension to desired pressure by watching pressure on reducer's low side.

Q 6—What are some causes of water bobbing up and down in boiler gage glass? How you would correct fault?

A—This condition is usually from foaming, especially in smaller boilers where contaminated condensate returns to boiler. Foaming can also be from excessive water hardness, high water density in boiler or from impurities forming scum on boiler water surface. In severe cases, water boils violently and carries over into steam line.

First step is to open surfaceblow valve to blow scum from waters surface. CAUTION! *Keep eyes on gage glass while blowing and hand on valve.* This habit keeps you from walking away from boiler and forgetting to shut off valve. You may have to raise water in glass and give additional blows to stop foaming.

If test shows boiler-water concentration is high, raise water to near top of glass and open bottomblow valve. Repeat feeding and blowing until water tests right and foaming stops. CAUTION! *Cut out fires while giving bottom blow to watertube boiler.* Make thorough check to find source of contamination. Grease extractors may be faulty or loaded with oil.

Q 7—What do you check when vacuum in steam condensing system drops?

A—When vacuum drops, first check the condenser-cooling water circulator. It may have slowed down. Then check water level in condenser's hotwell.

High water submerges condenser tubes, reducing their cooling surface.

Condensate pump's running too fast and keeping condenser's hotwell dry will also lower vacuum in systems where condensate pump discharges directly to open collecting tank.

Losing loop seal in intercondenser drain loop lowers vacuum. If this happens, close throttle valve in loop until seal returns to normal. If no sight glass in loop, liquid height can be located by feeling temperature of piping.

Air-ejector strainer may be clogged with dirt. Or there may be an air leak in condenser shell. Also low steam pressure on vacuum side of turbine sealing glands allows air leak into condenser.

Q 8—Horizontal steam engines are said to "run over" or to "run under". What does that mean?

A—These terms are used to indicate the circular motion of the crank. When an engine "runs over", crank is high when traveling away from cylinder. When engine "runs under", crank is low when traveling away from cylinder.

Q 9.—Name some causes that might make a duplex steam feed pump stop running.

A—One of the most common reasons for duplex pumps stopping is steam leaking into cylinder at both ends of the piston while piston is in middle of stroke. This is caused by ends of valve wearing round against a shoulder worn on steam valve chest. Valves must be free of shoulders and slightly longer than distance between outside of valve ports for this reason.

Another common reason for stopping is valve-stem threads wear off at valve

nuts. Then stem slides through nut without moving valve, and pump stops. Soft packing in liquid end is often installed without enough end and side clearances. Then hot water swells packing so piston can't move. At times steam leakage past piston and rings is so bad that pump stops. Then shoulders must be ground from pump and new piston rings and piston installed.

Q 10—Why is vacuum for reciprocating steam engines carried lower than for steam turbines?

A—Expanding steam to a very low pressure (high vacuum) in reciprocating engines is uneconomical for several reasons. High vacuum brings low-pressure end of engine down to very low temperature. Also steam expanded to very low pressure (high vacuum) occupies a lot more space; low pressure cylinder would have to be built extremely large. The cylinder condensation loss from steam washing the cooler cylinder surfaces would also be higher. The lower condensate temperature would need more preheating before entering boiler. Also more power would be used for turning main circulating pump faster.

A good practical vacuum for reciprocating engine is about 26 inches, which is a back pressure of about two pounds absolute.

Steam turbines can make good use of a higher vacuum because a turbine expands the steam down to the exhaust pressure, whereas pressure in steam engine cylinder is always well above back pressure when exhaust valve opens. A practical and economical turbine vacuum is about 28.5 inches of mercury.

Boiler Questions and Answers

These questions all refer to horizontal return tubular (HRT) boilers. They are similar to those asked in some of the better-arranged state examinations for licenses to operate steam boilers. Use them as an aid in interviewing and selecting power-house employees, and as a review of important features of boiler-room operating practice

Q—Why are diagonal stays not used instead of through-to-head stays below the tubes to brace the tube sheet in the same manner as above the tubes?

A—Because there is usually insufficient room for the proper number without placing them too close together, or using sizes larger than practical. Also, diagonal stays would tend to hold loose scale and sludge and prevent its free movement to blowdown.

Q—How much space would you want in front of an hrt boiler in planning its installation?

A—Enough room for tube replacement.

. Q—If you looked into the furnace and saw the bottom of the shell bulged, what would you do?

A—Shut the boiler down immediately and then have it inspected by an authorized inspector. Follow his recommendations before returning the boiler to service.

Q—Why are boilers over 72-in. diameter required to be supported by the outside suspension type of setting?

A—Because weight of the larger boilers may exceed safe load on brick-work. Crushing or buckling of walls might result from load of a large boiler full of water.

Q—What are two dangerous conditions offered by a weakened rear arch?

A—(a) If the arch collapses wholly or in part, upper part of the rear tube sheet may become overheated and damaged. (b) Anyone walking on top of the arch may cause it to collapse.

Q—What precaution is required with a flush-front-set hrt boiler?

A—The front arch protects the dry-sheet and front head seam from damage due to overheating. The arch should be kept in good condition.

Thousands of HRT boilers such as these are serving America today in industrial establishments, buildings and institutions everywhere

Q—Would loose brick lying in the firebox be of any interest to you? Why?

A—They might be from the closing-in line, or from some point where they are supposed to protect part of the boiler from overheating. Also, they might be from a point where their loss would weaken the walls.

Q—Name three points where brickwork should protect an hrt boiler from overheating.

A—Rear arch, closing-in line along sides of the shell, and front arch in the flush-front-set boiler.

Q—How and under what conditions may the segments of the heads above the tube dispense with diagonal or through stays?

A—In boilers not exceeding 36 in. diameter or 100 psi maximum allowable pressure, segments may be braced by stiffening with channel irons or angle irons (riveted back to back) riveted to the tube sheet. Specifications for structural form should conform to ASME Code rules.

Q—What dangers are there in settling of boiler foundation or supporting walls?

A—Severe stresses may be set up in piping, connections to the boiler, or in the boiler itself. Brickwork protecting parts from overheating may be dislodged. Sludge may settle to the front instead of to the rear and cause overheating. If serious, walls may become weakened and collapse.

Q—What would you look for at the holes in the brick wall where the blow-down pipe comes through?

A—To see that the pipe is not resting on the bottom of the opening (an indication that the boiler is probably settling) and to see that the pipe has sufficient clearance to allow free expansion and contraction.

Q—What is the lowest safe water level?

A—1 in. above the top row of tubes —at the lowest end.

Q—What is a strong-back?

A—It is a bar bolted to the tube sheet that prevents the sheet from buckling when stays are being installed before tubes are installed.

Q—If a boiler is pitched forward, slightly, what should be done?

A—Unless the installation was designed this way with the blowdown in front, the boiler should be reset with proper pitch towards the rear as soon as possible. In the meantime, the boiler should be shut down frequently and the manhole or handhole in the front head below the tubes removed. All sediment should be washed out. Frequency of these washouts depends upon rapidity with which sediment accumulates.

Q—What is the difference between a through brace and a through-to-head brace? Where is each used and why?

A—The through brace has washers and nuts on each end. The through-to-head brace has nuts and washers on one end; the other end is forged into an eye that is held by a pin or other construction to hold it clear of the rear head. Through stays may be used above the tubes in the rear; outside nuts are protected from burning off by the rear arch.

Q—If a fusible plug is used, where should it be located?

A—Near the centerline of the rear tube sheet, not less than 1 in. above the top row of tubes.

Q—In answer to last question, why near the centerline?

A—Because this is the highest part of the rear arch and this extra space is needed often to use a wrench.

Q—Where would you look to find the maximum pressure for which an hrt boiler was built?

A—On the standard stamping on the front tube sheet above the tubes.

Q—Why do flat segments of the tube sheets or heads require bracing?

A—Because internal pressure tends to bulge these areas outwards into spherical shape.

HOW SERIOUS IS AN ELECTRIC SHOCK? WELL, THAT DEPENDS ON . . .

. . .CURRENT FLOWING
- Moisture present
- Pressure between contact surfaces
- How much of body touches "hot" line

. . .PHYSICAL CONDITION OF THE VICTIM

. . .CURRENT PATH THROUGH BODY

Body resistance in ohms:	
Dry skin	100,000 to 600,000
Wet skin	11,000
Hand to foot (internally)	400 to 600
Ear to ear	100

. . .TOTAL TIME CURRENT FLOWS THROUGH BODY

HOW ELECTRIC SHOCK AFFECTS YOU

SAFE

1 ma (milli-ampere) or less	Can't be felt
1 to 8 ma	Shock not painful; can let go. Muscular control OK.

DANGEROUS

8 to 15 ma	Painful shock; can let go. Muscular control not lost.
15 to 20 ma	Painful shock; control of muscles lost. Cannot let go.
20 to 50 ma	Painful, severe muscular contractions, hard to breathe.
50 to 100 ma (possible) 100 to 200 ma (certain)	Ventricular fibrillation (A heart condition resulting in instant death—no known remedy).
200 ma and over	Severe burns, severe muscular contractions. Chest muscles clamp heart and stop it for duration of shock. (This prevents ventricular fibrillation).

Key questions and answers on boiler safety

Where pressure vessels are used, safe operation is as important as efficiency and economics. Code reflects current practice

The disastrous New York Telephone Co boiler explosion (POWER, January 1963, p 64) sparked many changes in ASME's Boiler and Pressure Vessel Code. As a result, more emphasis than ever before is placed on boiler safety. Here are key questions and answers about safety regulations from a book* that all boiler operators should know.

Q—Are state laws being passed to include furnace explosion protection in the existing Boiler Codes?

A—New York State has adopted a law for automatic heating boilers, including requirements on combustion safeguards.

EXAMPLE: Some of the following are now in the State Code:

1. Gas-fired boilers: (a) Pilot has to be proved, whether manual or automatic, before permitting the main gas valve to open, either manually or automatically, by completing an electrical circuit. (b) A timed trial for the ignition period is established on the basis of the input rating of the burner. For instance, for input rating of 400,-000 to 5,000,000 Btu/hr per combustion chamber, the trial for the ignition period for the pilot of automatically fired boilers cannot exceed 15 sec. And the main burner trial for ignition also cannot exceed 15 sec. (c) The burner flame-failure controls must shut off the fuel within a stipulated time, again depending upon the fuel input of the burner. For a burner rated with an input of 400,000 Btu/hr or higher, the electrical circuit to the main fuel valve must be deenergized automatically within 4 sec after flame failure. And the deenergized valve must close automatically within the next 5 sec.

2. Oil-fired boilers. Similar provisions have been adopted, with requirements on response time for controls to shut off the burner based on fuel input in gal/hr, instead of Btu/hr. The

*Elonka, S M and Kohan, A L "Standard boiler operators' questions and answers," McGraw-Hill Book Co, NY 1969.

flame must be supervised continuously by the controls.

No doubt, many states will incorporate requirements on safety combustion controls in the future as furnace explosions continue on their destructive path.

Q—Does the ASME Code require a low-water fuel cutoff on a cast-iron steam-heating boiler?

A—Yes. Each automatically fired steam or vapor system cast-iron boiler (or low-pressure-steam steel boiler) must be equipped with an automatic low-water fuel cutoff. It must be so located as to automatically cut off the fuel supply when the water level drops to the lowest safe waterline. The safe waterline cannot be lower than the lowest visible part of the water gage glass at the boiler.

Q—What two control parameters are usually considered in a flame failure system to prevent furnace explosions?

A—1. Control the input composition so that it cannot accumulate to an explosive batch or mixture, as graph in facing page describes. This is called input control.

2. Ignite in proper sequence all combustible combinations of fuel and air as they enter the furnace. This is called ignition control.

Input control depends on how much fuel can be put into a furnace before combustion must ensue, so that no dangerous mixture is present for a delayed light-off leading to an explosion. The time factor is called the grace period, which is the time needed to build up an explosive charge. It

varies with the design of the furnace: A small, fast-heating flash boiler might have a grace period of about 1 sec, while a large, commercial-type furnace might have 4 sec or more for ignition or input cutoff.

As a safety measure, ignition protection and shutdown of fuel and air into the furnace must be completed before the end of the grace period. As a rule of thumb, a fuel-air mixture greater in volume than 30% of the volume of the combustion chamber may be a damaging mixture.

Q—Describe a continuous, automatic check system for a flame-failure safety circuit.

A—Since it is possible to check the flame sensor by manually cutting off the flame, the same sort of check has been developed to do it automatically. Thus the flame safety circuit, including the sensor, is checked automatically and continuously by simulating flameout. As circuit diagram on opposite page shows, flame failure is simulated by means of a swinging shutter that interrupts the line of sight of an optical flame detector intermittently. Each time the sensor's view is blocked, a flameout is simulated long enough to prove out the sensor but not long enough to shut down the system. Should the sensor "see" flame while its view is blocked, the system immediately shuts down on malfunction. But if the sensor detects the intermittent flameout, its no-go signal of one-second duration is too brief to shut down the system, since the logic network has a time delay to prevent nuisance shutdowns. Thus, the system is constantly checking on its flame sensor, logic network and flame-failure safety circuit.

Q—Which inspectors make the legal inspections and reports (to a jurisdiction) that a boiler is safe or unsafe to operate, or that it requires repairs before it can be operated?

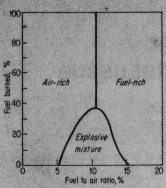

Methane-air mixture, showing flammability and explosive limits. Mixture is controlled to avoid explosive batch

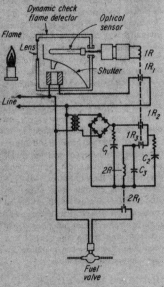

Automatic continuous check system of flame safety circuit uses shutter that simulates flameout by blocking sensor

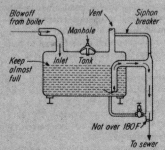

Tank for disposing of blowoff boiler water is necessary when no open space is available around boiler for blowoff

A—Three types of inspectors are:

1. State, province or city inspectors, who see that all provisions of the boiler and pressure vessel laws, and all the rules and regulations of the jurisdiction, are observed. Any orders these inspectors give must be complied with, unless the owner or operator petitions and is granted relief.

2. Insurance company inspectors qualified to make ASME Code inspections. If commissioned under the law of the jurisdiction where the unit is located, they can also make the required periodic reinspection. As commissioned inspectors, they require compliance with all the provisions of the law and rules and regulations of the authorities.

3. Owner-user inspectors, employed by a company to inspect unfired pressure vessels for direct use and not for resale by such a company. They must also be qualified under the rules of any state or municipality which has adopted the Code. Most states do not permit this group of inspectors to serve in lieu of state or insurance company inspectors.

Q—Why and when is a blowoff tank necessary for boilers?

A—Blowoff tank is necessary when there is no open space available into which blowoff from the boilers can discharge without danger of accident or damage to personnel or property. For example, discharging to a sewer would probably damage the sewer by blowing hot water under high pressure directly into it. As the schematic at bottom shows, a good blowoff tank installation is always full of water.

Q—Do not the integral chevron driers, separators and other hardware restrict the flow of steam outward from any opening in the drum of a steam generator, and thus restrict the opening to a safety valve?

A—The opening to the main steam line will usually be adequate to handle the design flow capacity. The ASME Boiler Code has definite rules on the opening to a safety valve. For example, internal collecting steam pipes, splash plates, or pans are permitted to be used near safety valve openings, provided "the total area for inlet of steam hereto is not less than twice the aggregate areas of the inlet connections of the attached safety valves. The holes in such collecting pipes must be at least ¼ in. in diameter."

In the case of steam scrubbers or driers, the ¼-in. diam opening does not apply, provided the "net steam inlet area of the scrubber or drier is at least 10 times the total area of the boiler outlets for the saftey valves." When inspecting drum internals, therefore, check the condition of the ¼-in. diam holes in the collecting pipes. They must be free and clear. The same applies to the openings on driers, because plugged driers on collecting pipes could lead to restrictions in the safety-valve openings. That would reduce relieving capacity flow, which could be dangerous.

Q—What type of safety valve should be installed on a boiler?

A—An ASME- or NB-approved and registered direct spring-loaded pop type, properly marked as to pressure and capacity and equipped with a testing lever. The pressure setting must match either the maximum allowable pressure for which the boiler is designed or, on older boilers, the maximum pressure allowed by state or city laws and codes.

Q—How can one recognize an ASME-approved safety valve, and how does one secure permission from ASME to make approved safety valves?

A—All ASME-approved safety valves are stamped with an ASME symbol, as shown below. Permission to use the symbol is granted by the ASME to any manufacturer complying with the provisions of the Boiler and Pressure Vessel Code. The manufacturer must agree, by filling out forms issued by the Society, that any safety valve to which the symbol is applied will be constructed in accordance with the Code and that it has the capacity stamped upon the valve under the stated conditions. The manufacturer must also agree that he will not misuse, or allow others to use, the stamp with which the symbol is applied. ●

Official symbol denoting ASME stamp of approval for safety valves built in accordance with Code requirements and limits

Why boilers explode

EXPLODED BOILER caused $228,000 in property damage and business interruption

We asked John Todd for the reasons, since he has been an authority on boiler explosions for many years. His experience points up one fundamental shortcoming—the assumption by owners that a boiler built to standards and furnished with automatic controls requires no expert attention.

Q With the many safety devices and automatic controls on present-day boilers, why are there so many boiler failures?

A The main reason, in my opinion, is the lowering or complete absence of operating standards. Too many owners have only one thought in mind when buying boilers provided with automatic controls—they hope to operate without the benefit of qualified operators. Please remember that until such time as all interested parties (these include governmental and insurance company inspectors, boiler manufacturers, code-making organizations and engineers) can convince boiler owners that they cannot shed all responsibilities toward safe operating practices by substituting automatic devices, boiler accidents will continue at the present alarming rate.

Q What is the most common type of boiler accident occurring today?

A The greatest accident producer today is low water which usually results in overheating and loosening of tubes, collapse of furnaces and, in some cases, complete destruction of the boiler. In certain classes of boilers a low-water condition can set the stage for a disastrous explosion that can cause serious loss.

Q Why is low water responsible for most boiler accidents in the past decade?

A Primarily it is due to a lack of boiler-operating and maintenance standards. This is true despite the fact that most boilers, especially heating ones, are provided with automatic feed devices, just one of the many automatic controls. But it is these devices that lull the owners into a false sense of security. Many of them feel that the boiler is fully automatic and completely protected from accidents. Not understanding fully how potentially dangerous a fired pressure vessel can be, the owner (or any one else) does not seem to take the slightest interest after installing one of these so-called automatic boilers.

The fact is that automatic feed devices, like all other automatic devices, will work perhaps a thousand times, perhaps many thousand times. But at some time they will probably fail, usually with disastrous results. That is why it's the duty of everyone concerned with boilers to realize that unless proper operating and maintenance standards are instituted, accidents are certain to occur.

Q You said that low-water accidents are caused primarily by failure of automatic feed devices. But isn't it true that, in practically all cases, low-water fuel cutouts are installed on all automatically-fired boilers to protect the boiler from overheating, in the event of failure of the feeding device? If such is the case, what causes the failure?

A That's correct. Practically all boilers today are provided with low-water fuel cutouts. What most people don't realize is that, as is true in most accidents regardless of the type of cutouts, a series of failures occurs. Thus, the basic fault may be failure of the automatic feed device to operate. Then we experience failure of the low-water fuel cutout. The net result is overheating and burning of the boiler metal. The failure of the automatic feed device and the subsequent failure of the low-water fuel cutout to operate stem from the same basic cause—lack of operating and maintenance standards practiced by the owner.

Q How should a low-water cutout be tested?

A The only positive method of testing a LWCO is by duplicating an actual low-water condition. This is done by draining the boiler slowly while under pressure. Draining is done through the boiler blowdown line. We find that many heating boilers are not provided with facilities for proper draining—an important consideration.

Many operators mistakenly feel that draining the float chamber of the cutout is the proper test. But this particular drain line is only provided for blowing out sediment that may collect in the float chamber. In most cases the float will drop when this drain is opened due to the sudden rush of water from the float chamber. Every boiler inspector can tell you of numerous experiences of draining the float chamber and having the cutout perform satisfactorily. But when proper testing was done by draining the boiler, the cutout failed to function.

Q What percentage of boiler losses are caused by low water?

A Approximately 75 per cent.

Q Various articles have been published on the subject of safety valves, particularly the low-pressure type with a setting of only 15 psi. If it is true that all boilers furnished today are provided with ASME-approved valves, why should anyone take exception to them and why shouldn't they work?

A In the first place, most people have a misconception of the term "ASME-approved." To set the record straight, the ASME itself does not approve a type of safety valve. By referring to the ASME Code, Section I—Power Boilers, and Section IV—Low Pressure Heating Boilers, you will see that both Codes contain limited-design criteria. The Codes also require a manufacturer to submit valves for testing. Such tests are solely for pressure-setting and relieving capacity.

In brief, the ASME symbol on a safety valve attests to the fact that the limited design criteria and the materials outlined in the code have been supplied by the manufacturer and that the relieving capacity and set pressure stamped on the valve have been proved.

Getting back to the first part of your specific questions, in some cases experience indicates that a particular type of safety valve has an inherent design weakness. After a short period of operation, the disk may be subject to sticking due to close clearances. This condition will render the valve useless, and the boiler will be without the benefit of over-pressure protection.

In regard to the second part of your question, the failure of safety valves to work is usually due to build-up of foreign deposits that result in "freezing." This is an indication that the valve has not been regularly tested or examined. One of the greatest causes of foreign deposit build-up is due to a "weeping" or leaking condition. The only way to be sure that a valve is in proper operating condition is to set up and adhere to a regular program of testing the valves by hand while the boiler is under pressure. Also, any weeping or leaking valve should be immediately replaced or repaired. This is important.

Q How often and in what manner should boiler safety valves be tested to make certain they are in proper working order?

A Low-pressure (15 psi) safety valves should be lifted at least once a month while the boiler is under steam pressure. The valve should be opened fully and the try lever released, so the valve will snap closed. For boilers operating between 16 and 225 psi, the safety valves should be tested weekly by lifting the valves by hand. On these higher-pressure boilers it is good practice to test the safety valves by raising the pressure on the boiler. This can usually be done when the boiler is being taken off the line. Then, if the valve feathers from improper seating, it can be corrected when the boiler is cold.

Q What can be done to prevent boiler failures?

A Boiler failures can at least be greatly reduced if boilers are placed under the custody of properly trained operators. This means that boiler owners must use sound judgment when employing boiler operators. Everyone with an interest in boilers must be encouraged in educating boiler owners and operators in proper operating procedures.

A very important step is establishing a regular program for testing of controls and safety devices, then faithfully following through.

Further, a program must be established for periodic maintenance of controls and safety devices. First thing to realize is that providing boilers with the most modern proved controls and safety devices is no guarantee that you will not have boiler failures. Any control or safety device is only as good as the testing and maintenance it receives. You cannot ever relax on these two.

Q What can a boiler owner do to assure that he has taken all possible steps to prevent boiler failure?

A First, purchase the best equipment available for a given service. Second, make certain that the boiler is properly installed and equipped with all the necessary appurtenances and safety devices. Third, before taking final acceptance, specify that the installation be inspected by a commissioned inspector in the employ of an insurance company or the state or municipality. By doing so, the owner will be assured that the equipment and the installation meet the legal requirements of the particular state or municipality. Fourth, the owner should provide the operator with a log book and a set of preventive maintenance and testing procedures. He should insist that such procedures be followed religiously and that the results of the tests and maintenance be recorded and be made a permanent part of the boiler room log.

Q 1—What is a resonance tube?

A—Photo shows resonance tube used to diagnose turbine "ailments". Turbines in normal operation have characteristic hum. One tip-off of trouble inside is unusual noise. Lengthening or shortening the tube "tunes it in" to noise frequency. Trouble sounds are charted and faults found by dividing the sound velocity by the frequency.

Q 2—Explain need for turbines strainers and their maintenance?

A—Strainer in photo is in main line ahead of turbine steam chest. Strainer catches particles carried over with steam such as valve parts, cotter pins or disks that could wreck turbine blades. This one is 50 per cent plugged with boiler compound that carried over. Inspect strainer at least once a year.

ping in pressure as each ring throttles it. Because steam must leak past a number of rings it drops in pressure each time, until reduced to a few pounds. On high-pressure units steam is bled off between rings and piped to low-pressure section of turbine to do work.

On the vacuum end, steam enters packing gland at about 8 psi and flows in two directions, toward the turbine, and out to the atmosphere. Steam tending to blow out of gland prevents air flowing in.

In labyrinth-type glands, the labyrinth rings clear the shaft by several thousandths of an inch. Carbon rings ride on shaft when turbine is stopped. Then when steam is turned on glands, the rings are raised off the shaft by steam blowing between shaft and ring. Clearance between carbon rings and shaft is then also a few thousandths of an inch.

Q 6—What precautions should be taken with a shut-down turbine?

A—The interior surfaces of shut-down turbines are subject to corrosion, especially if moisture is present or turbine is in a damp place.

After turbine is stopped, run air ejector for about an hour, keeping a moderate vacuum (about 12 inches). It's good practice to turn air ejector on for about fifteen minutes every day when turbine isn't running. This draws moisture out of casing.

While running air ejector, open the lines between turbine drains and ejector to remove moisture. Each day jack the turbine over a few revolutions and bring it to rest at a different position. Be sure to pump oil through bearings while jacking. Circulating oil prevents it guming and keeps water from collecting in bearings; a combination that corrodes shaft.

Q 7—What is a double-flow turbine?

A—The steam enters double-flow turbines at the center and divides, flowing in opposite directions. Each side is, in effect, a separate turbine, both mounted on a single shaft.

End thrust is divided about equally betwen the two halves, exerting little pressure on the thrust bearing. This naturally cuts down friction and reduces the size of thrust bearing needed.

Exhausts are connected together. This turbine is adapted for low-pressure work, because two sets of shorter blades can be used for passing larger steam volume.

Double-flow turbines are also used as the low-pressure unit of compound machines. The high-pressure unit is usually single-flow impulse-reaction.

TURBINE Q and A

Q 3—Give four important steps in warming up condensing steam turbine?

A—First, start the lubricating system and check oil flow to bearings. Also test low-oil alarm in system. Then engage and start the turbine's jacking gear.

Second, start condensate pump, cut in gland steam, and cut in second-stage air ejector to bring vacuum to about 20 in. Start main circulating pump.

Third, open all turbine and main steam-line drains. Next, bypass the main stops and warm up main lines slowly. Close drains when water is blown from lines.

Fourth, warm up the turbine slowly by cracking the warm-up valve. With jacking motor turning rotor and steam blowing through, turbines using moderate temp should be warmed in several hrs. Disengage jacking motor and open

throttle slowly. Check lube-oil system again.

Q 4—What are two main types of turbines and how do they differ?

A—Two main turbine types are impulse and reaction. In the impulse type, steam expands in nozzles leading from steam chest and is directed at the moving blades or buckets. Diaphragms separate the stage wheels. In the reaction turbine, steam expands in both the moving and stationary blades. In other words steam pressure drops in passing through both stationary and moving blades.

Q 5—How are turbine shafts sealed against loss of vacuum?

A—Two major types of sealing glands are used. One has soft-metal labyrinth rings, the other carbon rings.

On the high-pressure end steam leaks between gland rings and shaft, drop-

Q 8—Explain how turbine's speed is controlled?

A—Closing or opening valves controls amount of steam flowing through turbine. A governor regulates the turbine's speed in normal operation by opening or closing these valves ahead of nozzle groups.

An overspeed trip protects the turbine against run-away speeds by closing the main throttle or a clapper valve to shut off all steam to turbine. This is a small weight in shaft that's thrown against trip by centrifugal force.

Q 9—What is velocity compounding in an impulse turbine?

A—In velocity compounded stage there are two or more rows of blades secured to one rotor disk. Stationary blades are attached to the casing between the moving blades. These stationary blades reverse the steam flow and direct it against the next moving row. Velocity compounding is usually found in the first stage.

Advantage of velocity compounding is that shorter turbine is used because this stage absorbs much kinetic energy.

Q 10—What is a pressure stage and how does it differ from a velocity stage?

A—A pressure stage consists of one or more nozzles which speed up the steam before it passes through the single row of moving blades. All pressure drop takes place in the nozzles. Steam pressure remains constant while passing through moving blades.

In a velocity-compounded stage the nozzles speed up the steam faster than in a pressure stage. Steam pressure is constant after leaving nozzles.

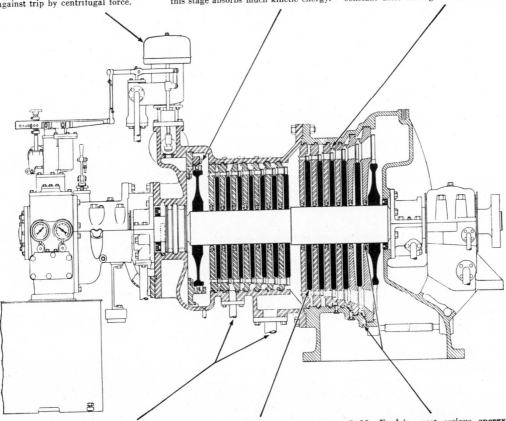

Q 11—What is an extraction turbine?

A—Steam passes through turbine's stages, expanding and doing work. When steam expands it decreases in pressure through each successive stage. Extraction ports bleed steam from one or more stages, depending on pressure or pressures needed.

Piping carries this bled steam for feedwater heating or for process. Regulating valves, either automatic or manual, adjust steam flow so turbine has enough steam for its load. Because bled steam has done work in turning turbine, its cost for process use is less than live steam. Also called a bleeder turbine.

Q 12—Describe how turbine disks are constructed for impulse turbines of the type shown here.

A—They are steel forgings, machined from a solid forging or built up from several small forgings.

Disks (wheels) are rough machined, annealed, heated and measured for trueness at 1000 F. Maximum distortion should be less than 0.001 in. from true center at any point. Then they are finish machined and tested for cracks.

The disks are shrunk or pressed on shaft of impulse turbines, with a spacer ring separating each wheel. Rotor is then bladed and dynamically balanced.

Q 13—Explain most serious energy losses in steam turbines.

A—In impulse turbines, steam leakage between diaphragms and shaft is important. In reaction turbines, leakage over blade ends through radial clearance passages averages about 5 per cent.

There is also frictional resistance between steam and nozzle sides, impact loss as steam enters moving blades, and eddying and fluid friction in blades. Friction caused by steam against wheels and blades is sometimes great, depending on steam density and wheel speed. Bearing friction is another loss.

WhyAmmonia Soaks Up Heat—

These questions and answers give a clear picture of what happens to liquid refrigerant and how it does its job while being pushed from one pressure to another by a compressor

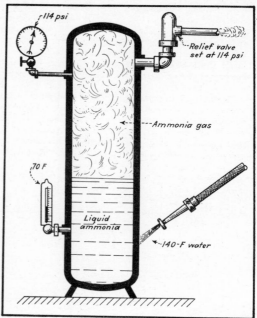

1 Under constant pressure heat boils ammonia without raising its temperature above that corresponding to pressure

▶ The what and how of refrigeration in a few simple explanations:

Q 1—Explain the term, one ton of refrigeration.

A—The American Society of Refrigerating engineers defines the standard commercial "ton" of refrigeration as the transfer of 200 Btu per min, or 12,000 Btu per hour. It is the basis of all refrigeration calculations, whether for cold storage, air conditioning, ice making or ice-cream manufacture. When a person says he has a 70-ton ice plant—that is something else. He means that he can make 70 tons of ice in 24 hours, and the rated refrigerating tonnage of his plant, as defined above, might be 150.

Q 2—What does superheat mean, and what causes ammonia to become superheated in the evaporator coils?

A—Pressure and temperature of saturated ammonia have a definite relationship. The temperature corresponding to a pressure of 20 psi is 5.5 F. This should be the gas temperature as it leaves the evaporator coils. However, a thermometer in the discharge line may show a temperature of 8 or 10 F. The difference is superheat.

If we could remove the gas as soon as it forms, giving it no chance to remain in the coils and pick up heat,

we would not have superheat at this point.

It is impossible to eliminate superheat entirely. A poorly-insulated suction line also helps superheat the gas.

Q 3—What happens to ammonia as it passes through an expansion valve? Is its boiling point changed?

A—When ammonia passes through an expansion valve, it expands because it passes from a higher to a lower pressure. As pressure drops, boiling temperature falls and some or all of the ammonia, depending on conditions, changes from liquid to gas.

Q 4—Why is it that a change in a substance's heat content does not always show up as a change in temperature?

A—Every substance has a definite boiling point, and when we say "boiling point" we mean the temperature at which the substance changes from liquid to gas. Boiling point changes with pressure.

A drum of ammonia at 70 F shows a pressure of 114 psi. If it has previously been stored in a cooler place, boiling takes place until it reaches 70 F and a pressure of 114 psi.

Suppose we put a relief valve, set at 114 psi, on this drum and spray hot water over it, Fig. 1. The ammonia

starts to boil and the gas thus formed escapes through the relief valve which holds a steady pressure.

As long as any liquid remains, its temperature holds at 70 F, all the supplied heat being absorbed in changing the liquid to a gas.

Q 5—What is the specific heat of ammonia? Also, explain how such heat affects the use of this liquid as a refrigerant.

A—Specific heat is actually a ratio, but we use the term to indicate the quantity of heat needed to raise one pound of a substance one F, or to lower one pound the same amount. Specific heat varies with temperature.

Generally speaking, when the ratio between latent heat and specific heat of the liquid is large, we have an ideal refrigerant.

Liquid ammonia is in this class, its specific heat is 1.129 at 70 F, and that of its gas at constant pressure corresponding to 86 F, is 0.4011.

Ammonia refrigerant serves to convey heat away from some substance that we wish to cool or freeze.

When liquid ammonia expands from a higher to a lower pressure, in an evaporator coil, we have to remove some of the heat content.

For example, if liquid reaches the

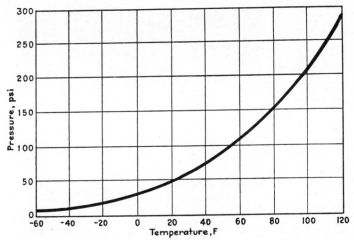

2 Ammonia vapor pressure-temperature characteristics. Its pressure increases slowly at low temperatures but more rapidly as the temperatures get higher

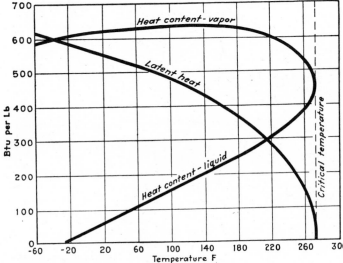

3 Latent heat of vaporization and heat content of saturated ammonia vapor and liquid. Heat contents of vapor and liquid are the same at critical temperature

expansion valve at 96 F and the evaporator coils are at 5 F, the liquid must be cooled to 5 F before it performs any useful work.

Temperature difference, 91 F here, multiplied by the specific heat, gives the quantity of heat to be removed. This can also be found in ammonia tables.

Q 6—Explain the term latent heat, when applied to ammonia.

A—Latent heat is the number of Btu that must be applied to and absorbed by ammonia to change it from liquid to gas. If a quantity of liquid am-

monia is just at its boiling point (−28 F at atmospheric pressure) each pound must absorb 589.3 Btu before it will change to a gas.

This is not a fixed quantity, but varies with pressure and temperature. For example, liquid ammonia at 70 F requires 508.6 Btu per lb to change it to a gas. The U. S. Bureau of Standards published Tables of Thermodynamic Properties of Ammonia; text books and manufacturer's catalogs contain tables of these characteristics.

Q 7—Explain the term, sensible heat,

when used in speaking of ammonia.

A—Sensible heat, applied to ammonia (or any other substance) shows up as temperature registered on the ordinary thermometer. The term, used without considering other factors, simply tells whether ammonia is hot or cold—as we think of these terms in everyday life. It is not a measure of total heat or Btu in ammonia.

Q 8—Give a simple explanation of the pressure-temperature relationship of ammonia in a closed cylinder?

A—Pressure of ammonia in a closed cylinder increases with its temperature—slowly at low temperatures, but rapidly at high temperature, Fig. 2.

Q 9—What is the critical temperature of ammonia and how does it affect the operation of a refrigerating plant?

A—Critical temperature is 280 F. At this temperature, latent heat becomes zero, and heat content of the liquid and the vapor is the same, Fig. 3. As the pressure at this temperature is well over 1600 psi, it is far above the normal operating range.

Q 10—What happens to ammonia in the evaporator coils?

A—Ammonia in the evaporator coil boils and does work of refrigeration. Actually what happens is this: Assume that condensing pressure—gage pressure of liquid going to the coils—is 185 psi, and that evaporator-coil pressure is 20 psi. Referring to ammonia tables we find: (1) heat content of the liquid at 185 psi (199.7 absolute) is 150.5 Btu per lb, (2) heat content of the liquid at 20 psi (34.7 absolute) is 48.9 Btu per lb.

Then heat removed to cool liquid to 5.5 F (20 psi) before doing work, is 150.5 − 48.9 = 101.6 Btu per lb. Since latent heat at 20 psi is 564.6 Btu per lb, and heat removed is 101.6 Btu per lb, the heat absorption-ability left to do work is 463.0 Btu per lb.

Q 11—How do you determine the amount of ammonia that must circulate per ton of refrigeration?

A—Find how much work or refrigerating effect we can expect from each pound of ammonia. This is the difference between the heat content of saturated liquid at condenser pressure and saturated vapor at suction pressure. It can also be expressed as the difference between the quantity of heat removed by the condenser water and the work done by the compressor.

Since one ton of refrigeration equals 200 Btu per minute, divide this number by the refrigerating effect of the ammonia in Btu per pound to get the ammonia need in lb per min per ton.

CLOSED SHELL AND TUBE ammonia condensers placed on roof. Oil separator traps oil in the gas; drains to the sump

WHAT'S WRONG HERE? Never place the stop valves under the safety valves in any system. This hookup violates every code

REFRIGERATION Q & A

27 questions often asked for unlimited refrigeration engineer's license

Q 1—What is a refrigerating system?

A—An apparatus or combination of mechanical equipment in which a refrigerant is circulated for cooling, or extracting heat from spaces or bodies.

Q 2—Name two types of refrigerating systems in general use?

A—Compression, absorption systems.

Q 3—In what type of systems are generators, analyzers, and weak-liquor coolers used?

A—In the absorption system.

Q 4—In what type of system is aqua ammonia used?

A—In the absorption system.

Q 5—What is a refrigerant?

A—A chemical compound or agent to produce refrigeration. Brine is not a refrigerant.

Q 6—What is a compressor as used in refrigerating work?

A—Compressor is a mechanical apparatus to draw evaporated gas from low-pressure side of system and to discharge it to high-pressure side, in a condensing

system. Unit may be reciprocating, rotary or centrifugal.

Q 7—A certain refrigerating system has a normal operating suction pressure of 150 psi gage and condensing pressure of about 1000 psi. What refrigerant is used? Ammonia, carbon dioxide or Freon?

A—Carbon dioxide (CO_2).

Q 8—Name several types of condensers in general use?

A—Shell and tube, double pipe, atmospheric and evaporator.

Q 9—Describe briefly an evaporating condenser.

A—It has a series of tubes or coils, usually the fin type, enclosed in a sheet-metal housing. Cooling surface is often divided into two sections. Hot discharge gases enter the top section, where part of the sensible heat from the hot gases is removed. Water is sprayed over lower section in sufficient quantities to keep the surfaces wet. A fan or blower drives the blast of air over the cooling surface, causing rapid evaporation. This condenser has advantage of large capacity in small area, eliminates cooling towers.

Q 10—What is meant by the term "direct refrigeration"?

A—A system where the refrigerant is circulated direct to the spaces or substances to be cooled. It is usually called a direct-expansion system.

Q 11—Is a direct-expansion system, with evaporating coils in the air-circulating system, permitted in public buildings?

A—No. But certain types of self-contained portable units with solid soldered connections are approved. See your local code.

Q 12—How are refrigerating systems usually classified as to capacity?

A—This varies in the different municipal codes. In New York City, *Class A* is a common system containing 1000 lb of refrigerant or more and having a capacity of 30 tons or over. *Class B* contains less than 1000 lb and more than 20 lb and has a capacity of less than 30 tons refrigeration. *Class C* contains not more than 20 lb refrigerant.

Q 13—What are hydrocarbon refrigerants?

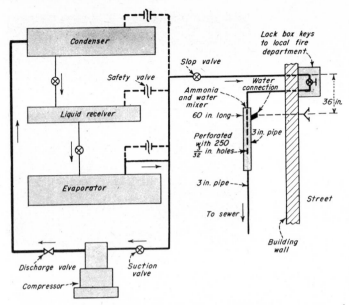

Lock box keys
to local fire
department

Slop valve

Safety valve

Water
connection

Ammonia
and water
mixer
60 in. long

Perforated
with 250
$\frac{3}{32}$ in. holes

Condenser

Liquid receiver

Evaporator

3 in. pipe

36 in.

3 in. pipe

Street

To sewer

Building
wall

Discharge valve

Suction
valve

Compressor

AMMONIA MIXER is important safety device for dumping large quantities of ammonia in emergency. Water absorbs the ammonia, washes it down into the sewer

A—They contain hydrogen and carbon, like ethane, propane and butane. These refrigerants, together with menthol chloride are flammable when mixed with air.

Q 14—Are hydrocarbon refrigerants permitted in Class A systems?

A—Many codes prohibit flammable refrigerants in large systems. See your local code.

Q 15—Describe method of detecting leaks in a system using Freon.

A—A halide torch is commonly used. This torch assembly has a copper plate enclosed in a metal shell, heated by an acetylene gas flame. The flame from this reaction plate changes to a greenish color when brought into contact with Freon gas. Leaks as small as .01% by volume may be detected.

Q 16—How are small ammonia leaks detected?

A—Gas leaks in open air are usually checked with sulfur sticks. These sticks consist of a heavy cotton cord coated with sulfur. When lighted and brought into contact with ammonia gas, the smoke given off turns white. Leaks in brine systems can be detected by a number of reagents.

Q 17—State a general rule for locating safety valves.

A—All pressure vessels, fired or unfired, that can be valved off as a sep-

arate unit, like brine coolers, receivers, accumulators and air tanks, must be protected by proper safety valves.

Q 18—What is maximum pressure to set safety valves on ammonia accumulators and shell-and-tube brine coolers?

A—150 psi gage.

Q 19—What pressure is allowed on ammonia system's high-pressure side?

A—250 to 300 psi.

Q 20—Where should safety valves discharge?

A—Discharge may be connected to low-pressure or suction side of the system or to atmosphere, depending on type and size of plant. Safety valves on brine coolers are usually discharged into the main suction line, just above the stop valve.

Q 21—State two other locations that require safety valves.

A—Ammonia compressors and generators must have pressure-relief valves connected below the main discharge valve, arranged to discharge into the main line above the stop valves. Many modern units have relief valves built into the header assembly.

Q 22—What other valves are required in the discharge line above the main stop valve?

A—A suitable check valve should be installed above the compressor, in the

discharge line. Consult your local code.

Q 23—Can you pack an ammonia stop-valve stem while under pressure?

A—Ammonia valves have a machine surface on top of valve disk and bottom of valve bonnet. When valves are wide open, these surfaces seat, cutting off pressure from the valve stem, permitting packing under pressure.

Q 24—In placing stop valve in compressor discharge lines, what precaution must be taken?

A—Valve must be so placed that discharge gases enter the valve under the valve disk. Reason: Should the valve disk become loose from the stem, it cannot drop down and close off the line.

Q 25—What is a pressure-limiting device?

A—Many codes require a high-pressure cutout, so arranged that when a predetermined pressure is reached, controllers or oil switches in the power lines to the motors will trip and cut off the power. Some codes require a remote pushbutton station from which the entire plant can be closed down in a fire or other emergency.

Q 26—What is an ammonia mixer and what is its use?

A—There are several types. One is manually operated, another automatic. Each consists of a system of valves so arranged with connections to the ammonia, water-supply and sewer systems, that the entire ammonia charge may be mixed with water and discharged into the sewer system in a fire. Some codes require that these manually operated mixers be mounted outside the plant, in suitable heavy-metal cabinet. Keys to cabinets are supplied to local fire stations. See local code for special requirements.

Q 27—State briefly how you would withdraw the ammonia charge from the plant, for reuse or transportation.

A—In pumping out a plant, refrigerant must be discharged into drums, approved by Interstate Commerce Commission. Place empty drums on an accurate scale, with flexible metallic hose, between liquid ammonia connection and drum for accurate weighing. Never fill drums to more than 90% of capacity by weight, which would allow for a slight inaccuracy in weighing and prevent overfilling, as gas space must be left for expansion. Drums should never be placed outside in the sun. Only such quantities of refrigerant are kept in storage in drums, as the local code permits. Most metropolitan areas restrict this quantity to less than 500 lb, some permitting only 300 lb,

REFRIGERATION Q & A

18 questions often asked for unlimited refrigeration engineer's license

AMMONIA COMPRESSOR has header with a discharge, suction, pumpout, bypass, safety valve, all cast in one compact unit

Q 1—What are the general requirements for ventilation in refrigerating machinery rooms?

A—This varies with amount of refrigerant in system, type of refrigerant, and whether ventilation is natural circulation or forced-draft. Thus, class-A system with over 1000 lb refrigerant needs (a) exhaust fan with capacity of at least 2000 cu ft per min or (b) window areas to outside air of 25 sq ft if on opposite walls or (c) 60 sq ft if windows open on one side of machinery room. See your local code.

Q 2—Can the ammonia condensers or discharge line be pumped out by the standard type of ammonia compressor in case of leaks?

A—Most compressors are fitted with bypass and pump-out valves, on machine header, besides the main suction and discharge valves. For normal operation, suction and discharge valves on header are open. Reversing their position permits pumping gas from discharge line, discharging it into the suction side of the system.

Q 3—Are open-flame lights permitted in the machinery rooms?

A—Arc lights or other open-flame lights are not permitted.

Q 4—How many helmets and gas masks are required in a refrigerating machinery room? Where should they be located?

A—Class-A systems, with over 1000 lb of refrigerant, should have two masks in the machinery room, near main exit, and one or two masks at convenient points, throughout the plant. All codes approve masks approved by U.S. Bureau of Mines for refrigerant in use.

Q 5—If you are testing a new system, using anhydrous ammonia as a refrigerant, what test pressures are required?

A—Most codes require air pressure of 300-psi gage on high-pressure side, 150-psi on low side. (IMPORTANT: Take great care in pumping air pressures on systems for testing.)

It is not advisable to use large ammonia compressors for pressures in excess of 150 psi. Above this point, use a small air compressor, designed for high-pressure work, and lubricated with air-compressor oil. Best procedure is: Open all valves on low-pressure side of system. This permits placing maximum allowable pressure on brine coolers and accumulators to equalize with that of high-pressure side. Entire system may then be tested at 150 psi.

Next step, close off low-pressure system and blow out air pressure, leaving a valve open at some point to atmosphere. This prevents possibility of placing a higher pressure than would be safe on low side of system, while testing high-pressure side. High pressure is then pumped up to the required pressure and tested for air leaks. Soapy water applied with a soft paint brush at all joints is a simple way to locate small leaks.

After checking over entire system by this method, blow out air pressure, close off atmospheric connections to low side and see that expansion valves are open.

Then pump a vacuum on the system, exhausting all air possible. Large compressors may be used for this purpose. System is now ready for charging. It is good policy to charge in only a small amount of ammonia, enough to raise gas pressure to 75 to 100 psi, and then again fairly test the system for small gas leaks. This may seem like a drawn-out process but, carefully carried out, it saves a lot of time and trouble.

Q 6—Are stop valves permitted between safety valves and receivers, accumulators and other pressure vessels to permit safety-valve repairs while system is under pressure?

A—No valves of any type are permitted whereby safety valves may be closed off, rendering them useless.

Q 7—If you were building a liquid ammonia receiver, say 16 in. in diameter welding in minus or plus heads, would you place the pressure on the concave or convex side?

A—On the concave side; 40% greater pressure is allowed on the concave side.

Q 8—Are brass and copper pipe fittings suitable for ammonia systems?

A—No. Ammonia fittings are made of cast steel or forgings. All piping should be steel or wrought iron.

Q 9—State size of safety valves required on 100-ton-capacity system using carbon dioxide.

A—One-half to ¾ inch.

Q 10—What type valves are required for liquid level gage glasses?

A—Gage glass valves must be fitted

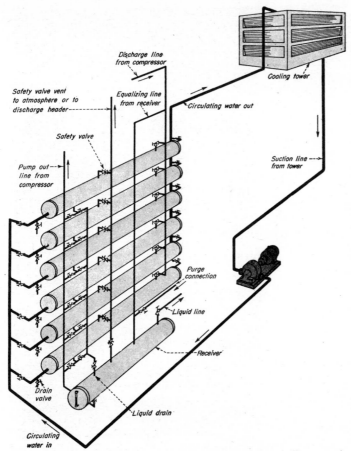

Discharge line from compressor

Cooling tower

Safety valve vent to atmosphere or to discharge header

Equalizing line from receiver

Circulating water out

Safety valve

Suction line from tower

Pump out line from compressor

Purge connection

Liquid line

Receiver

Drain valve

Liquid drain

Circulating water in

SHELL-AND-TUBE condensers have large capacity on small floor area. This is simple piping layout for easy operation and maintenance. Flanged heads remove easily

with a ball check, which closes automatically in case of glass failure. Metal screen or slotted metal casing must protect glasses.

Q 11—Are ammonia gas and air mixtures considered explosive?

A—Most authorities place explosive limits between 17 and 25% ammonia by volume. Most serious explosions from flame coming in contact with bad ammonia leaks have been caused by mixture of ammonia gas, air, and oil vapor. There is no danger from explosion in cutting into ammonia lines, receivers or accumulators with acetylene torches or making welded connections if system is open to atmosphere, ventilated.

Q 12—What is latent heat?

A—Latent heat is required to change the state of a body without change of temperature, as in changing melting

ice at 32 degrees to water of the same temperature.

Q 13—Is it possible in an atmospheric cooling tower to reduce temperature of the circulating water below that of surrounding atmosphere?

A—Yes. In a well-designed cooling tower, temperature of water off the tower often approaches wet-bulb temperature; 10 to 15 degrees below dry-bulb temperature is not uncommon, depending on relative humidity.

Q 14—Why are holes cast into the rims of flywheels and motor rotors on large units?

A—To help turn machine by hand when making repairs and adjustments. (NOTE: 60% of applicants state that these holes are for balancing purposes.)

Q 15—What would cause the discharge-gas temperature, leaving a com-

pressor, to suddenly become abnormally high without any noticeable change in suction or discharge pressures?

A—There are several causes: (a) broken suction or discharge valve (b) broken piston rings, causing heavy blowby or leaky bypass valves (c) passing hot discharge gas back into suction side of compressor.

Q 16—What would cause discharge temperatures to drop below normal, discharge line to become cold and compressor to frost over heavily?

A—Excessive amount of liquid refrigerant passing expansion valve, causing liquid to slop over from evaporating side to the compressor.

Q 17—If you were asked to take charge of a large refrigerating plant that had been closed down for some time, state briefly how you would proceed, what precautions you would take.

A—Assuming the plant had enough refrigerant in the system and was tight, first make and inspect all auxiliaries, such as water-supply system, pumps, brine coolers or expansion side of system and condensers. Check position of all valves on high- and low-pressure sides, see that they open and close easily.

If synchronous motors are used on high voltages, take Megger tests to determine condition of insulation on motors and feeders. Check contacts on oil switches and breakers, remove covers from vertical-enclosed machines, check condition of connecting-rod bearing bolts.

Check entire oiling system. Bar over by hand large reciprocating machines, one complete revolution, and check piston clearances. Check all rod adjustments and packing boxes. In starting compressors, open suction valves slowly, as liquid refrigerant may be lying in the suction lines. After starting, check all bearings to see that oil rings are turning, as they often fail to start when units are shut down for prolonged periods. See that your plant is supplied with adequate safety equipment—gas masks, proper fire extinguishers, first-aid kits.

Q 18—What is the purpose of a safety head in a vertical, single-acting ammonia compressor?

A—Safety head is a false head, resting on a shoulder on top of the compressor cylinder. Discharge valves are fitted into this head. Heavy springs, placed between the main cylinder head and the safety head hold the latter in position. In case of broken suction valves or heavy slug of liquid refrigerant, this head rises, preventing serious accident to the compressor.

THIS DIESEL will give trouble-free efficient service if the operating engineer knows equipment and keeps on his toes

DIESEL Q and A

FUEL-INJECTION system is the heart of any diesel. Operator must know it thoroughly before making the needed adjustments

Save diesels "timing" like this with your "time"

Diesel makes food do work in "one gulp" in cylinder

Q 1—If your four-cycle engine's camshaft is removed and the timing completely lost, how would you time the engine in a hurry?

A—A four-cycle engine's intake valve opens from 10 to 20 degrees before piston reaches top dead center (TDC) and exhaust valve closes from 10 to 20 degrees past top dead center.

For quick timing, jack piston of any cylinder to top dead center, with the camshaft gear out of mesh. Then turn camshaft until the inlet valve just starts to open and the exhaust valve just starts to close. Mesh gears while camshaft is in this position. Timing will usually be correct to one gear tooth.

Q 2—Give a short definition of an internal-combustion engine?

A—An internal combustion engine is any engine that changes the heat en-

ergy of fuel into mechanical energy within a cylinder. This includes gas turbines.

Q 3—If one cylinder of a diesel engine receives more fuel than the others, how is engine's operation affected and how would you prevent more fuel in one cylinder?

A—The cylinder receiving more fuel is affected in a number of ways, depending on conditions. That in turn affects the engine in various ways.

Excess fuel in one cylinder causes incomplete combustion in that cylinder because too much oil is injected for the air supplied. Then incomplete combustion causes a smoky exhaust. If excess fuel isn't stopped soon, carbon builds up around piston rings, sticking them in grooves. If rings stick, piston won't compress enough air (because

Pyrometer saves engine "blowing it's top" suddenly

of air leaking past rings) to keep temperature right in cylinder for proper combustion. That causes more smoke. Then stuck rings also cause combustion gas to blow past piston. Hot gas blows lube oil off cylinders, scoring them and even causing piston to seize.

Excess fuel may also cause combustion knock. Because of power loss in one cylinder from incomplete combustion, fuel is wasted. With remaining cylinders doing more than their share, engine starts overheating.

Best way to avoid excess fuel is to check cylinder exhaust temperatures with a pyrometer every hour. Then adjust fuel oil so all cylinder exhaust temperatures are within 10 F.

Q 4—What are the average compression pressures and temperatures in a full diesel engine?

A—Air is compressed from 400 to 500 psi in the cylinder, depending on engine design. At these pressures, it ranges from 800 to 1000 F. These temperatures are high enough to ignite fuel entering cylinder for combustion.

Q 5—How is diesel engine's power output increased without increasing the engines revolution or size?

A—The most popular method is by supercharging. This means delivering air above atmospheric pressure to the cylinder. Then same engine burns more fuel and increases mean effective pressure, without increasing cylinder size, or turning faster.

With supercharging, instead of engine drawing in atmospheric air, air is forced into cylinders by blower up to about 10 psi above atmospheric pressure. Then engines output is increased about 40 percent.

Supercharging is widely used at high altitudes. Because air is less dense, engine needs more air to burn same amount of fuel than at sea level.

Q 6—What is meant by injection lag and how is it caused?

A—Fuel is injected into the cylinder about ten degrees before piston reaches top dead center. But the fuel-pump cam is timed to force oil through the pump about 30 degrees before top dead center. This difference of 20 degrees is the injection lag.

Injection lag is caused by: (1) Expansion of fuel-oil discharge lines under high pressure, resulting in need for additional fuel volume to be displaced. (2) Compressibility of fuel. When fuel is near 3000 psi, it is compressed about 1 percent of its volume. (3) Leakage past the fuel-oil plunger.

So fuel-oil lag is actually a kind of lost motion and is compensated for by

pumping oil into the fuel-distributing system earlier than needed in the cylinder.

Q 7—Why are some diesel pistons oil cooled instead of water cooled?

A—Pistons are water cooled in the large older crosshead type marine engines. Modern engines use oil to prevent the crankcase lube oil being contaminated with water in case of leaks or cracks. When lube oil is contaminated with water, oil emulsifies and causes burned-out bearings.

Q 8—Name some causes of smoky exhaust.

A—A smoky exhaust is caused by one or more faults. Perhaps the most common is from fuel distributed unequally to cylinders. Or the engine's load might be excessive. If the cylinder or exhaust valve receives too much lube oil, engine will smoke because of burning lube oil. Also if fuel injection is late. Water in fuel or fuel not suitable for engine causes smoke. Fuel injectors not adjusted right are a sure cause. Weak or broken fuel-valve spring is another. Obstructed exhaust piping and low compression in cylinders are also common causes.

Q 9—Where should thermometers be installed on diesel plants?

A—There should be a thermometer at the following: Engine cooling-water inlet and outlet. Engine lube-oil inlet and outlet. Lube-oil sump tank. Air-compressor cooling water inlet and outlet. Cooling water outlet from each cylinder. Lube oil cooler's cooling-water inlet and outlet. In some installations thermometers are also placed on scavenging air.

Q 10—Why must fuel injectors be kept in perfect working condition?

A—A faulty injector is like a boiler's faulty oil-burner tip. Unless the oil sprayed into the cylinder is finely atomized, the oil cloud does not burn properly and some of the fuel is wasted in smoke.

Some diesels do have a coarse spray and still burn oil efficiently. But they depend on precombustion or turbulence chambers to mix the oil with air before entering cylinder.

The cloud of tiny oil drops sprayed directly into the cylinder ignites when heated to the right temperature. The smallest drops naturally heat up faster and therefore burn first. Because the larger particles heat slower, they ignite later and need more time to burn. If the particles are too large, they are exhausted before they burn completely. That wastes fuel and increases consumption. That's why oil injector maintenance is so important to efficiency.

Fuel "burns up" from punch-drunk air in cylinder

Supercharger puts extra "wallop" in diesel engines

Injection lag is fuel "lost motion" in the fuel system

Oil keeps hot piston from "freezing" in cylinder

Smoke can mean the operator is in a "fog" or groggy

Buying thermometers save "buying parts" for engine

Breaking up oil saves "breaking owner" in the end

RIGHT ANSWERS are important, whether diesel is large, or small one shown. Here's handy repair rack for small job

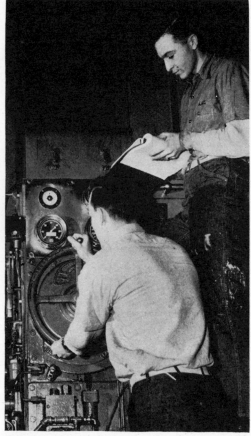

OPERATING DATA are jotted down by cadet midshipmen at Merchant Marine Academy diesel lab for study in classroom

DIESEL Q and A

Q 1—Why should the doors of a diesel-engine crankcase never, NEVER be opened after shutdown, until the engine has cooled?

A—In a totally enclosed diesel crankcase, air and oil vapor mixture is very rich. If the engine has a "hot spot," such as an overheated bearing or bushing, piston or any other part, with temperature high enough to form a seizure, there's danger of a crankcase explosion. Only reason there's no explosion before door is removed is because there's not enough air to mix with oil vapors.

But if crankcase door is opened while engine is still hot, fresh air rushing in through doors supplies needed oxygen for explosion. If operator removes door, he stands in direct line of blast and can be fatally hurt.

Q 2—Explain how to find cause if your diesel engine stops suddenly.

A—First check fuel-oil supply. Next check whether fuel system is airbound. Fuel filter might be plugged or fuel lines broken. Then check fuel system from oil-supply tank to strainer.

Check day tank for water in fuel. Do this by opening drain cock. If there is water, and drain valve is not large enough, remove valve bonnet from suction line to drain tank quickly, when engine is needed badly. But take precautions for fire hazard. Next check transfer pump. The valves, springs, gaskets, plunger, roller or guide may be at fault.

Diesel piston may have seized from poor lubrication. Check by trying to jack engine over. Remove piston and replace with new if it's badly scored. In severe cases, liner may have to be removed and piston jacked out. Be sure to change lube oil before starting engine with new piston and rings. Also clean oil pan, lines and filter.

If you find none of these conditions, engine may have stopped from bearing seizure. If bearing is not badly wiped, scrape enough to clean it up. If it is badly wiped, replace with new unit.

Or the engine's fuel-pump driving chain may be broken (if it has one). Then engine will have to be retimed.

On some engines the fuel pump's adjustable coupling can slip if not tightened properly. Check this.

Q 3—What is best way to start a diesel engine in extremely cold weather?

A—For quick starting, heat circulating water and lube oil to about 110 F. Then circulate through engine until it is warm. Some installations have permanent connection from hot-water line to cooling water for this purpose.

Or place resistor-type electric heater and fan in diesel's air intake. Heat air so it is warm for engine starting. Some engines have air pre-heater, "flame thrower" for cold-weather starting.

524

If compression on engine is poor from worn piston rings and liners, lift cylinder heads and smear heavy grease around top of pistons. This helps seal compression for those all-important revolutions before engine starts firing.

If engine burns heavy oil, run on kerosene for a few revolutions before shutting down, until heavy oil is cleaned out. After engine starts on kerosene next time, switch back to heavy oil.

Q 4—What is meant by compression ignition, and what pressures are needed for diesels?

A—Ignition in a diesel engine is caused by the heat of compression. This is done by the piston drawing outside air into the cylinder on the downstroke, and compressing it on the upstroke. Oil injected into the hot compressed air ignites without a spark, open flame or a hot bulb.

Pressures of 375 to 400 psi are enough for medium- and large-size cylinders with open combustion chambers. With engines using precombustion chambers, pressures around 450 to 560 psi are needed.

Reason for higher pressures on separate combustion-chamber-type engines is that the ratio of cooling surface to combustion space volume is greater. Then cooling effect of greater area must be overcome with higher pressures. Higher temperatures are also needed in small, high-speed engines where the fuel must be burned in a shorter time.

Q 5—How is fuel oil tested for carbon residue, and what effect does carbon have on the engine?

A—Fuel oil is tested for carbon residue by the Conradson test. This test is made by burning oil with a limited amount of air. Then carbon formed is measured.

Oil forming over 3% carbon causes rapid cylinder wear. A good distillate oil should run below 0.5% carbon content. Some residual oils form as much as 14% carbon.

Q 6—Many engines have valve seat inserts and valve stem guide inserts. Explain how inserts are renewed.

A—Inserts for valve seats are made of nickel, cast iron, stellite, copper-tungsten, or cast aluminum-bronze. These inserts are pressed or shrunk into valve cages or into cylinder heads.

Valve stem guide insert is either a bushing or sleeve made of close-grained cast iron. This is a drive fit into the valve stem guide of the valve cage. Because bushing is soft cast iron, it takes most of the wear caused by valve stem's up-and-down motion.

Valve-seat inserts are removed with a narrow-pointed cape chisel. First the insert is cut, then pried apart over its seat with the chisel. But take care not to harm the seat. Valve-stem inserts are removed by turning the valve cage bottom up. Then remove valve and stem, drive insert or bushing out with brass stock.

Q 7—What effect will a cocked connecting rod have on the engine, and when is a connecting rod said to be in alignment?

A—A cocked rod may make the piston operate in such a way that the piston ring faces are not parallel to the cylinder wall. Under such conditions, upper and lower portions of the ring face are worn away on opposite sides, but not the middle portion. This makes it impossible for one half of the ring to scrape oil on its downward stroke. Also, misalignment places a high load not only on the piston and cylinder wall, but also on the connecting rod bearing.

A connecting rod is in alignment when its piston's sliding surface is exactly at right angles or square with the connecting rod. This is checked by placing this rod vertically on a surface plate, so rod rests on machined bearing cap's surface. Then insert the wrist pin in the upper bearing and measure distance from surface plate to both ends of wrist pin with a surface gage, inside calipers or inside micrometer. If distance is same on both sides, rod is OK.

Q 8—What is a "jerk-pump" fuel-oil system for diesel engines?

A—Pump-injection system, sometimes called "jerk-pump" system, has an individual pump or cylinder pump serving each cylinder, and connected directly to the fuel nozzle.

The pump meters the oil charge and controls injection timing. Nozzle contains a valve, which opens when injection pressure is reached and closes when pump delivery ceases. Most of these systems contain a cam-operated pump, connected by a tube to the fuel nozzle.

Q 9—What would you look for if the backpressure on your engine was excessive?

A—First inspect the exhaust piping, mufflers and waste-heat boilers for flow restrictions, such as carbon deposits. Check for pulsations that might cause backflow of exhaust gas. Pulsations may be from faulty exhaust system.

Q 10—Give some general hints to follow before overhauling an engine.

A—Before overhauling an engine, it's good practice to observe it in operation. Watch cooling water and lube oil pressures, valve mechanism and governor operation.

If practical, indicate the engine and study cards for compression and combustion pressures, also for unusual signs. Listen for abnormal noises and note any condition that's not right.

When tearing down the engine, do it in an orderly manner by marking all mating parts not already marked. Place the large parts on boards to prevent marring them. Place smaller parts in boxes to prevent loss. First clean all parts with Stoddard solvent.

Try to work out repair schedules so only one cylinder is completely overhauled at a time. This has advantage of localizing to one cylinder running faults caused by overhaul. If entire engine is torn down, it's sometimes hard to find minor adjustment faults because everything worked on must be checked on all cylinders.

Q 11—Explain where heat generated in a diesel and in a gas engine goes.

A—*Diesel:* 34% for useful work. This is just another way of saying that 34% is the thermal efficiency. And 30% is jacket loss. It represents heat absorbed from pistons and cylinder walls by water in surrounding jackets. 27% is exhaust and radiation loss. This is amount of heat carried away in hot exhaust gas and heat radiated from engine and exhaust piping. 9% goes to friction loss and includes friction of crankshaft and connecting-rod, pistons, etc. This friction goes to heat and adds to other losses.

Gas engine: 25% is useful work; 30% is jacket loss; 38% exhaust and radiation loss; 7% friction loss.

14 Diesel Q&A

on today's maintenance and operating problems

Q 1—Some engineers renew connecting-rod bolts at set periods as a preventive measure. Is this necessary?

A—Experience shows it's good practice to renew these bolts whenever a piston seizes. During seizure, bolts are subjected to excessive stresses. Reason: They are not designed to stop the flywheel rotation suddenly and, at the same time, absorb forces being developed by combustion in other cylinders under such conditions.

If not under higher stress than normal, a modern well-designed alloy-steel connecting-rod bolt need not be replaced during engine's working life. Some manufacturers recommend renewing these bolts after so many hours of service. Time they suggest varies from 10,000 to 30,000 hr. But usually they don't take engine's speed into consideration. If engine runs twice as fast, bolts naturally do twice as much work. So total rpm and not engine hours should be deciding factor in the first place.

Changing bolts is a safety measure, usually for plants without trained diesel operators. Some engineers inspect root of connecting-rod bolt threads for hair cracks. These are caused by fatigue and should be watched.

Q 2—Are shims of various thicknesses as reliable in connecting-rod bearings as one solid shim?

A—One solid shim is more rigid. It does not have the deflections of many thin shims. It's best to use only one thick shim here because this bearing really takes a beating if there's any give in the shims. Use one solid shim of the laminated type. Then you can peel off a thin sheet to take care of wear when overhauling bearings.

Another bad feature in using many loose, thin shims is that they get dirty, burred and bent. They may seem tight when bolted together, but such shims, by causing bearings to loosen under constant pounding, can wreck engine.

Q 3—What are some causes of dry-cylinder liners that lead to piston seizure and costly engine damages?

A—Operators often overlook the fact that after an engine has been idle for some time lubricating oil drains away. A little of the oil may get trapped between piston rings and cylinder liner. But seizure often starts if cylinder liner is dry when starting. This takes place before oil is supplied by lubricators or before it splashes up from crankcase.

Answer is to turn engine over slowly a few revolutions so all parts are thoroughly lubricated. If engine has a hand pump, always charge lube system with it before starting engine.

Of course, a very common cause of piston seizure is stuck piston rings and combustion gas blowing past them, burning lube oil off cylinder walls.

Q 4—What are some causes of lost power in a diesel engine?

A—Engine may gradually lose power from cam wearing, timing chains or gears wearing, injection timing being retarded, or faulty fuel-oil spray, etc. Because engine wears gradually with use, some operators give it the original acceptance tests every few years.

You can't always tell work that each of the cylinders does by comparing their exhaust temperatures. Leaky exhaust valve increases exhaust temperature also. Some of the other reasons for unbalanced exhaust temperatures are: defective spray, filter choking off air supply, retarded injection.

Best way to find cause of high exhaust temperature is to use an indicator. With a little experience, an alert operator can tell many engine faults by studying indicator cards.

Q 5—What are some engine governor faults to check from time to time?

A—Governors may tend to hunt. Cause is usually from lost motion due to wear. Hunting can cause rapid governor wear and may even fracture the spring attachment. Wear on the weights' fulcrum pins may be excessive from bearing on same spot, or because these parts don't get oil.

Balls in ball bearings of governor tend to bed into the race at normal working positions. This is known as false Brinelling. This condition has caused governors to fail when load is suddenly removed. Check ball bearings on your governors and renew them if they show signs of this fault.

Q 6—How does change in fuel-injection timing affect firing pressures?

A—Early combustion before piston reaches end of compression stroke is from too-early fuel injection. Or, if combustion is greatly delayed, pressure rise in cylinder is abnormally high.

Then shock is hard on bearings, cylinders and piston rings. Fuel timing must be right for engine condition, grade fuel used, altitude, water-jacket temperatures, etc. Otherwise, engine will knock.

Q 7—When do early and delayed combustion knocks occur?

A—Both take place while diesel piston is at top end of compression stroke, or before piston starts down on power stroke. In either case, entire engine-cylinder assembly is abnormally strained.

Spark-ignited engine causes a knock, usually referred to as detonation. This takes place when piston is in lower portion of cylinder. Detonation is usually caused by final explosion of accumulated, unburned fuel because of incorrect air-to-fuel ratio or ignition qualities of fuel.

Q 8—Diesel may knock when piston is at top end of stroke from high pressure rise per degree of crank angle. What causes this condition?

A—There are a number of causes: (1) Ignition lag is too long, resulting in delayed combustion from low compression. Low compression is usually from leakage past piston rings. (2) Fuel that is poorly atomized also causes delayed combustion. (3) Jacket-water temperature is too low. (4) Air-to-fuel ratio may be incorrect. (5) Fuel-injection timing may be incorrect because of overloading, causing too early combustion. (6) Engine may be out of balance, causing too early combustion in some cylinders. (7) Another cause of incorrect injection timing may be from fuel timing set incorrectly for grade of fuel used, giving too early or too late combustion.

Q 9—What are some reasons for high lubricating-oil consumption in a diesel engine?

A—Common cause is too much lube oil reaching cylinder walls, where it's spread into combustion area and burned. This often starts from excessive bearing clearances, which lower lube-oil pressure because oil leaks out of such bearings. Then operator increases oil volume at pump to bring up pressure. This may hold bearing temperatures right, but it floods cylinder walls.

Another reason for high lube-oil consumption is ineffective oil-ring control. Under this heading look for: excessive piston clearance, cylinder or ring distortion, worn cylinder or rings, insufficient ring gap, stuck piston rings, scraper edge of oil ring worn round, or clogged oil ring.

Third reason: New oil in engine may be too light or lube oil may be diluted with fuel oil. High oil temperature, oil leaks, wrist-pin plates not oil tight, excessive suction on crankcase breather or light-load operating conditions—all boost lube-oil consumption of your diesel engine.

Q 10—What are advantages or disadvantages of the positive-displacement-type scavenging pump over blower (centrifugal) type?

A—Positive-displacement type has some advantages over centrifugal type. For one thing, displacement volume is not affected by humidity, variations in barometric pressure or temperature. Also, positive-displacement blower responds immediately to changes in engine speed when engine driven. These blowers are usually engine driven, and power for driving decreases as engine speed decreases.

Q 11—Explain how to measure the compression pressure in diesel cylinders.

A—Cut off fuel from one cylinder at a time and indicate that cylinder with an indicator, while running remaining cylinders on fuel or starting air.

Q 12—How do you adjust engine compression pressure?

A—Decreasing linear clearance between piston and cylinder head lowers compression pressure, while increasing clearance raises compression pressure.

Changing the compression pressure depends on the engine. Easiest way to do this on large engines is to remove or add shims under the connecting-rod foot. That lowers or raises the piston. Compression pressures rarely have to be changed. If they do, it's just a case of decreasing space between cylinder head and piston so air is crowded into less space—raising its pressure.

Q 13—Do diesel cylinders corrode and pit, or is all the lost metal due to wear?

A—Corrosion in internal-combustion cylinders is often indicated by a pitting of the metal surfaces. At times this pitting is visible to the unaided eye, but frequently only under magnification. Often corrosion is indicated only by higher-than-normal wear rates of cylinders and piston rings. Corrosion wear may be heaviest at top of compression-ring travel or at lower part of cylinder.

Here's reason why corrosion in cylinder is often taken for wear. When a metal surface is attacked by corrosion, a film of corrosion gets thicker and protects underlying metal. This slows rate of corrosion attack. But in engine cylinder, rubbing action of piston rings on cylinder walls continuously removes corrosion products as they form. So new metal is exposed for further attack. With this condition, operators often attribute high rate of lost metal to wear only, overlooking corrosion.

Most corrosion trouble in engine cylinder is from combustion process. Examine gaseous materials present during and resulting from combustion if corrosion is bad—or wear rate is high.

Q 14—Does a 2-cycle diesel need same size flywheel as a 4-cycle?

A—No. Two-cycle engines have smaller flywheels than 4-cycle engines.

Usually the more cylinders, the smaller the wheel. There are exceptions to this general rule.

Diesel has relatively high ratio of maximum pressure to mean pressure in cylinder. While inertia smooths this out, turning moment impressd on shaft is irregular and these engines need fairly heavy flywheels to smooth out jerky turning moment.

Slow-speed engine naturally needs heavier flywheel than high-speed engine. Another consideration is the ac generator unit that must run in parallel. In this case, flywheel size is also based on electrical factors.

18
DIESEL
Q and A

MODERN supercharged diesel gives efficient power to City of Menasha, Wisconsin. Operating know-how will keep this unit running for many years of very hard service

1 Should bearing clearances be same in 4- and 2-cycle engines?

Some manufacturers have found that reducing 4-cycle engine bearing clearances by almost 50% improves engine operation. A 2-cycle engine has no reversal on main or crankpin bearings. But a 4-cycle engine has inertia forces at higher speed on the exhaust stroke that reverse the load. That causes an impact condition that breaks down the oil film locally. Proper engine alignment and closer bearing clearances in modern engines reduce bearing troubles on 4-cycle engines.

2 How do the new, vertical, radial diesel engines work on gas?

For burning gas, these engines operate on spark ignition at about 260 psi. Each cylinder has a gas-inlet valve, above the scavenging ports. These valves are actuated by rocker arms and push rods from a cam on crankshaft's upper portion.

Valves introduce low-pressure gas into cylinders in path of incoming scavenging air to assure thorough mixing. Two spark plugs are in a vertical plane. They fire at different crank angles, before top dead center. Amount of gas delivered to cylinders varies (depending on load on engine) controlled by valve in gas header that's connected to governor. Volume of scavenging air is also proportioned to load by throttling blower's suction.

3 Does fuel oil in a diesel cylinder stop burning when fuel injector closes and stops spraying oil?

No. There may be considerable after-burning after the injector closes, although theoretically there is none.

After-burning depends on ignition temperature inside cylinder, how well fuel is atomized, other factors.

4 Why don't diesel engines have eccentrics instead of cams to open and close the valves?

Eccentric operation is too slow for a diesel engine. The cam, on the other hand, gives quick or slow opening and slow or quick closing, if desired, depending on shape of cam.

5 What is meant by term "torque at firing speed"?

Torque at firing speed is torque demanded by engine at speed that engine starts to run under its own power. This torque is largely used to overcome friction, but also includes torque needed to drive engine auxiliaries, such as fans, supercharger, fuel, water and lube-oil pumps.

Another thing that increases torque at firing speed is effect of compression and expansion. While a large part of energy put into compression is regained on expansion, recovery is not 100%. So additional torque must be supplied to make up these losses.

6 What is the dewpoint temperature of diesel exhaust gases?

Dewpoint of internal-combustion engine exhaust gas is usually from 110 to 130 F at atmospheric pressure. At 500 psi it is about 300 F. Dewpoint temperature may be much higher, depending on atmospheric relative humidity. Temperature also depends on exact composition of combustion products.

7 Where do knocks occur in diesels, and what are some causes?

Knocks often develop in large bearings, valve gear or inside diesel cylinders. Bearing knocks come from too much clearance. This may be from gradual wear or overheated bearings, after bearing metal fails. Valve gear may knock because of excessive rocker-arm roller clearance. Also from worn or broken rollers or from sticking valve stems.

Knock inside the cylinder is from too-early injection. Also from excessive spray-valve lift at low speeds, spray air pressure too high, or from stuck piston rings. Another knock might be from water in cylinder if liner or head is cracked.

8 Detergent oils are widely used in internal combustion engines today. Do they improve life of all engines?

Some designs, under certain conditions, work best with highly detergent oils.

Other diesel engines work best with medium detergent-disperant activity, while still others of conservative design and rating work about same as with good, straight mineral oils. So it all depends on type of engine.

9 What are best safeguards against fuel-nozzle troubles?

Most nozzle troubles are from dirty fuel and from nozzle overheating, not from breakage. First step is to use clean fuel. That means centerfuging or filtering. Next use a 2- or 3-stage filtering system between fuel-supply pump and injector pump. While engine is running, drain filters and strainers of water at set periods. When overhauling injection pumps and nozzles, keep foreign particles out of equipment.

SEAL-TYPE piston rings, *upper grooves,* have kept the hot gases from blowing by

10 Older slow-speed engines were operated with cooling-water discharge and lube-oil discharge temperatures of 100 to 120 F. Today, engines often run with water and oil temperatures 180 to 195 F. Is this good practice?

Low-water temperatures of 120 F are below dewpoint. That causes lube-oil sludging problems. High cooling-water temperatures give better fuel economy. But there are often other problems because of too high temperatures.

Raising lube oil above 150 F often increases oxidation rate of oil. With additive oils, inlet-oil temperature of 140 F and outlet of 160 F is good practice. Keep water temperature above lube-oil temperature if engine has air-cooled pistons. Then cylinder will expand more than piston, thus holding clearance between piston and cylinder right. Run with water inlet of at least 160 F and outlet of 170 F. But keep these two temperatures within 15 degrees, if possible.

11 What are some common faults to look for in defective fuel nozzles?

Look for incorrect spring tension. Perhaps next most common fault is fuel injector valve not seating properly. Check for stuck nozzle valve-stems and for erosion or grooving along stems. Next inspect orifices for defects. Often nozzle tips are clogged or carbonized. Easiest to find are broken springs or broken valves.

12 When diesel bearing fails, is it better to replace one pair or should you replace complete set?

Some manufacturers recommend replacing bearings in sets. Remove a failed bearing at once, or it may cause other parts to fail. Reason is that bearings are inexpensive compared to crankshafts and other large parts. Best answer to replacing bearings in sets is knowing your engine. Past experience might indicate that only failed bearing need be changed.

13 Do piston-ring combinations have to be changed as an engine ages?

Some firms say yes. Changing to different ring combination as an engine ages might be an advantage, especially on larger engines.

While first set of rings is used, liner wear tapers gradually. There is also groove wear. Study engine performance. If oil consumption is high or engine output low, correct by using right rings for problem when replacing. There are many ring combinations to solve various problems.

14 What type oil filters do diesels have and do they remove all dirt from oil that enters or is produced by the engine during normal operation?

Filters are usually either full-flow or shunt type. One passes all oil through, while other bypasses part of oil. Full flow must be large to handle the volume, while shunt flow can be much smaller.

No filter can remove dirt unless it reaches filter. For example, some dirt settles in sump and remains there. Usually there is little settling if engine is operated normally. Only way an engine can be clean is for filter to remove all dirt. Fuel-flow filtering is popular today as it removes dirt sooner.

15 What is main cause of noisy combustion in a diesel engine?

Chief reason is high rate of pressure rise in cylinder during early combustion stages. During these early stages, increase rate should be about 30 psi per degree of crank angle. Experience shows higher rates usually cause a "rough-running" engine. Best way to know rate of pressure rise is to take "jerk" or "pull" cards. Do this by pulling cord of indicator drum by hand during compression-firing revolution of crank. This cannot be done on high-speed engine.

Such a card shows compression and firing pressure as a peak at top dead center of card. Line starts at zero degree crank angle at bottom center and rises very gradually to about 60 deg. Then rate of rise is more rapid until peak is reached at top dead center. On quiet engine, look for smooth combustion line with about same slope or pressure rise per degree of crank angle as the compression curve. Compression curve is lower part of this curve just before fuel is injected.

16 What determines the lubricating-oil consumption of a diesel engine?

Mechanical condition of engine is important. If piston rings are in poor shape, too much oil is pushed up the cylinder walls and so is lost. If rings fit poorly or are stuck, or if liner is tapered badly, hot combustion gas blows past rings and burns lubricating oil off the liner and piston. These are only some of the more common causes.

17 What are some of main causes of corrosion wear in engine cylinder?

Corrosion wear in diesel cylinders is always present. Under some conditions it is much greater than abrasive wear. Carbonic, sulfurous and sulfuric acids form when fuel oil is burned. These are most serious causes.

Best way to fight them is to use corrosion-resisting materials for cylinders and piston rings—and pistons. Then keep jacket cooling water high as possible and use lube oils with additives to fight your condition. High sulfur diesel-fuels make corrosion a constant problem in any type engine.

18 If your 4-cycle engine's camshaft is removed and the timing completely lost, how do you time the engine in a hurry?

A 4-cycle engine's intake valve opens from 10 to 20 degrees before piston reaches top dead center (TDC) and exhaust valve closes from 10 to 20 degrees past TDC.

For quick timing, jack piston of any cylinder to TDC, with the camshaft gear out of mesh. Then turn camshaft until the inlet valve just starts to open and the exhaust valve just starts to close. Mesh gears while camshaft is in this position. Timing is usually correct to one gear tooth, which is close.

Stop contamination . . .

in hydraulic systems. It's the largest *single* cause of trouble. Next is operating at incorrect temperatures. Here are ten ways to reduce pump repairs and downtime in these systems: (1) stop leaks (2) check oil daily for contaminants, water, foam (3) listen for unusual noises (4) operate between 90 and 120 F (5) check oil level, keep reservoir full (6) clean or change filters and strainers at least every six months (7) use only oil recommended by pump firm (8) add only clean, new oil through a strainer (9) use only clean, well-marked containers for adding oil, and (10) test oil periodically.

5 timely Q&A on diesel engines to help you do a better operating job

1. What is meant by cetane number of fuel oil?

A. Cetane scale is applied to fuel oils, gasoline, diesel fuel, etc, to show oil's ignition quality. This quality means oil's ability to ignite in diesel cylinder —or under similar conditions. Fuel with good ignition quality will auto-ignite at low temperatures and is, therefore, preferred. This quality affects engine starting, smoking and knocking.

Hydrocarbon family ranges from simple combinations, such as methane gas, through various solids, liquids and gases. Some have complicated chemical structures. All have various ignition characteristics. Formula for cetane is $C_{16}H_{34}$.

Cetane is an excellent diesel fuel in its pure form, but isn't easy to separate completely from other hydrocarbons. Cetane content for various crudes differs greatly. $C_{10}H_7CH_3$ is another hydrocarbon, but it has poor ignition qualities. Cetane number is based on proportions of these two hydrocarbons.

Cetane number of 100 means ignition quality equal to pure cetane, while zero means one equal to that of alpha-methyl-naphthalene ($C_{10}H_7CH_3$). Ignition number of 60 means mixture equivalent to one composed of 60% cetane and 40% $C_{10}H_7CH_3$.

2. How is cetane number determined?

A. Ignition-delay method on single cylinder test engine is used to determine cetane number. Fuel to be tested is used in engine, and compression ratio is raised until ignition-delay reaches a standard length. Poor-quality fuels' ignition quality needs a high compression to shorten the delay period, while fuels of good quality take lower compression for the standard delay.

Once compression ratio for test fuel is found, various mixtures of cetane and alpha-methyl - napthalene are burned. From this, compression ratio needed for each fuel, to bring delay period to standard figure, is found.

Cetane number of test fuel equals percentage of cetane in mixture that requires same compression ratio as test fuel, to produce standard delay period.

3. What is meant by right- and left-hand engines. Buyers ordering two engines for same plant often specify one each. Is this good practice?

A. Engine is classified as right or left when flywheel is on right or left sides, viewing engine from operating side. Reason one engine of each is specified is because owners believe such an arrangement is easier to attend if controls are on same aisle between engines.

Diesel Engine Manufacturers Association says there may have been reason for such an arrangement years ago before the days of remote governor control, but not with today's common use of such control and the little attention modern engines need.

Such specifications require one engine to be nonstandard. Disadvantage to owner is loss of interchangeability, which is so important to avoid carrying large stocks of spare parts. Then, again, any advantage is lost if a third diesel is installed.

4. If engine fails to start, what would you look for?

A. Engine rarely fails to start, but such difficulty can be divided into two groups: (1) Starting mechanism fails to turn engine through an operating cycle. (2) Engine turns through operating cycle but fails to fire. If starting mechanism doesn't work, check starting air-pressure or look for rundown or weak battery—depending on starting system.

If air-starting pressure is high enough, air-starting valves may not open fully or at right time. Or by sticking, leaky exhaust or inlet valves may prevent engine from quickly raising compression high enough to fire. Leakage from stuck rings or badly grooved cylinders may have same effect. In cold weather, heat of compression may be too low unless cooling water and (in extreme cases) even lube oil are heated.

At times, bearings are too tight (after overhaul) or there is no lubrication on cylinder walls, etc. All this puts extra load on starting mechanism.

5. Should operators spend time listening to diesel engines to become familiar with normal noises so they can tell when unusual sounds develop?

A. Yes. It's good practice to spend a few minutes each day listening to natural beat of engine. Listen at each cylinder to detect knock in crank, wrist pin, etc. Also, listen to gear train or to drive chain. Backlash in gears and sprockets can be detected, also loose chains.

Listen for vibration at supercharger. Vibration is caused by dirt on impeller, too much bearing clearance or increased end thrust. To avoid supercharger trouble, run a spin test every six months. Just remove supercharger silencer and spin air rotor by hand. If there's noticeable drag, dismantle supercharger. Usually all that it needs is just a good cleaning.

Change in air manifold pressure at given load shows supercharger needs attention. Trouble may be air leaks in intake connections. Watch air-manifold pressure, or engine will starve for air— causing many headaches for you.

METRIC SYSTEM

The metric system was legalized by the United States government in 1866, but progress in adopting it has been slow. Today, with imports of foreign machinery skyrocketing, the plant operator often finds himself repairing and replacing machine components made to the metric system. While any machinery handbook should have complete conversion tables, here we present a few that should be useful in an emergency.

Remember these three principal units: *meter*, the unit of length; *liter*, the unit of capacity; *gram*, the unit of weight.

Q What is ISO and DIN?

A ISO is the International Standard Organization composed of representatives of the various national standard institutes. Its purpose is to coordinate engineering standards, specifications and recommended procedures to ensure interchangeability of parts and to improve technical communication. As the United States makes the transition to the metric system, it is logical and therefore probable that the future engineering standards will closely follow ISO recommendations.

DIN (Deutsche Industrie Normen) are the West German Industrial Standard Specifications. Since they were the most complete and detailed of all the metric standard systems, other countries have been largely following them. ISO standards are, therefore, closely patterned after DIN standards. Since World War II, the DIN standards have been gradually modified, following ISO recommendations. Although differences between the systems still exist (like nominal diameters of some small threads), they are insignificant for all practical engineering purposes.

CONVERSION

FRACTIONAL INCHES — DECIMAL INCHES — MILLIMETERS

Fractional Inches	Decimal Inches	mm.	Fractional Inches	Decimal Inches	mm.
1/64	.015625	0.397	33/64	.515625	13.097
1/32	.03125	0.794	17/32	.53125	13.494
3/64	.046875	1.191	35/64	.546875	13.891
1/16	.0625	1.588	9/16	.5625	14.288
5/64	.078125	1.984	37/64	.578125	14.684
3/32	.09375	2.381	19/32	.59375	15.081
7/64	.109375	2.778	39/64	.609375	15.478
1/8	.125	3.175	5/8	.625	15.875
9/64	.140625	3.572	41/64	.640625	16.272
5/32	.15625	3.969	21/32	.65625	16.669
11/64	.171875	4.366	43/64	.671875	17.066
3/16	.1875	4.762	11/16	.6875	17.462
13/64	.203125	5.159	45/64	.703125	17.859
7/32	.21875	5.556	23/32	.71875	18.256
15/64	.234375	5.953	47/64	.734375	18.653
1/4	.25	6.350	3/4	.75	19.050
17/64	.265625	6.747	49/64	.765625	19.447
9/32	.28125	7.144	25/32	.78125	19.844
19/64	.296875	7.541	51/64	.796875	20.241
5/16	.3125	7.938	13/16	.8125	20.638
21/64	.328125	8.334	53/64	.828125	21.034
11/32	.34375	8.731	27/32	.84375	21.431
23/64	.359375	9.128	55/64	.859375	21.828
3/8	.375	9.525	7/8	.875	22.225
25/64	.390625	9.922	57/64	.890625	22.622
13/32	.40625	10.319	29/32	.90625	23.019
27/64	.421875	10.716	59/64	.921875	23.416
7/16	.4375	11.112	15/16	.9375	23.812
29/64	.453125	11.509	61/64	.953125	24.209
15/32	.46875	11.906	31/32	.96875	24.606
31/64	.484375	12.303	63/64	.984375	25.003
1/2	.5	12.700	1	1	25.400

DECIMAL EQUIVALENTS OF MILLIMETERS

mm.	Inches	mm.	Inches	mm.	Inches
..01	.00039	.20	.00787	10	.39370
.02	.00079	.30	.01181	20	.78740
.03	.00118	.40	.01575	30	1.18110
.04	.00157	.50	.01969	40	1.57480
.05	.00197	.60	.02362	50	1.96850
.06	.00236	.70	.02756	60	2.36220
.07	.00276	.80	.03150	70	2.75590
.08	.00315	.90	.03543	80	3.14960
.09	.00354	1.00	.03937	90	3.54330
.10	.00394			100	3.93700

$$1 \text{ km} \quad (\text{kilometer}) \quad = \quad 1000 \quad \text{m} \quad (\text{meter}) \quad = \quad 0.621 \text{ miles}$$
$$1 \text{ m} \quad (\text{meter}) \quad = \quad 10 \quad \text{dcm} \quad (\text{decimeter}) \quad = \quad 1.094 \text{ yards}$$
$$1 \text{ dcm} \quad (\text{decimeter}) \quad = \quad 10 \quad \text{cm} \quad (\text{centimeter}) \quad = \quad 3.937 \text{ inches}$$
$$1 \text{ cm} \quad (\text{centimeter}) \quad = \quad 10 \quad \text{mm} \quad (\text{millimeter}) \quad = \quad .394 \text{ inches}$$
$$1 \text{ mm} \quad (\text{millimeter}) \quad = \quad .039 \text{ inches}$$

MOST COMMON DIN NUMBERS

DIN	DESCRIPTION	DIN	DESCRIPTION	DIN	DESCRIPTION
1	Taper Pin	916	Socket Set Screw Cup Point	1084	Spanner Nut (Lock Nut)
7	Dowel Pin (Unground)	931	Hex Head Bolt Part Thread (Coarse Thread)	6325	Precision Ground Dowel Pin
84	Cheese Head Machine Screw	933	Hex Head Screw Full Thread (Coarse Thread)	6912	Low Head Socket Head Cap Screw (Socket with guide)
85	Pan Head Machine Screw	934	Hex Nut		
125	Flat Washer	935	Castle Hex Nut	7980	Lock Washer (Smaller o.d. for Socket Head cap screw)
127	Lock Washer	936	Hexagon Jam Nut		
137	Spring Washer	939	Stud	7984	Low Head Socket Head Cap Screw
551	Slotted Set Screw	960	Hex Head Cap Screw Part Thread (Fine Thread)	7985	Phillips Head Machine Screw Fillister Head Machine Screw
912	Socket Head Cap Screw	961	Hex Head Bolt Full Thread (Fine Thread)	7986	Phillips Head Pan Head Machine Screw
913	Socket Set Screw Flat Point	963	Flat Head Machine Screw (Slotted)	7987	Phillips Head Flat Head Machine Screw
914	Socket Set Screw Cone Point	985	Stop Nuts with Nylon Lock Ring	7991	Flat Head Socket Cap Screw
915	Socket Set Screw Dog Point	1481	Roll Pins (Spring Pin)		

ISO METRIC - U.S. UNIFIED SCREW THREAD COMPARISON

Unified Scr Thd	Decimal Equiv.	Nominal Diameters ISO & Unified Converted mm.	Unified Scr Thd	Decimal Equiv.	Nominal Diameters ISO & Unified Converted mm.
—	.0394	1.	7/16	.4375	11.113
—	.0043	1.1	—	.4724	12.
—	.0047	1.2	1/2	.5	12.7
—	.0055	1.4	—	.5512	14.
#0	.060	1.524	9/16	.5625	14.288
—	.063	1.6	5/8	.625	15.875
—	.0709	1.8	—	.6299	16.
#1	.073	1.854	11/16	.6875	17.463
—	.078	2.	—	.7087	18.
#2	.086	2.184	3/4	.75	19.05
—	.0866	2.2	—	.7874	20.
—	.0984	2.5	13/16	.8125	20.638
#3	.099	2.515	—	.8662	22.
#4	.112	2.845	7/8	.875	22.225
—	.1181	3.	15/16	.9375	23.813
#5	.125	3.175	—	.9449	24.
—	.1378	3.5	1.	1.	25.4
#6	.138	3.505	1-1/16	1.0625	26.988
—	.1575	4.	—	1.063	27.
#8	.164	4.166	1-1/8	1.125	28.575
—	.177	4.5	—	1.811	30.
#10	.190	4.826	1-3/16	1.875	30.163
—	.1969	5.	1-1/4	1.25	31.75
#12	.216	5.486	—	1.2992	33.
—	.2362	6.	1-5/16	1.3125	33.338
1/4	.250	6.35	1-3/8	1.375	34.925
—	.2756	7.	—	1.4173	36.
5/16	.3125	7.938	1-7/16	1.4375	36.513
—	.315	8.	1-1/2	1.5	38.1
3/8	.375	9.525	—	1.5354	39.
—	.3937	10.			

ISO - METRIC COARSE THREADS DRILL SIZES

Nominal Diameters mm 1st Choice	2nd Choice	Pitch mm/Thread	Recommended Drill Size mm
1		0.25	0.75
	1.1	0.25	0.85
1.2		0.25	0.95
	1.4	0.3	1.1
1.6		0.35	1.25
	1.8	0.35	1.45
2		0.4	1.6
	2.2	0.45	1.75
2.5		0.45	2.05
3		0.5	2.5
	3.5	0.6	2.9
4		0.7	3.3
	4.5	0.75	3.75
5		0.8	4.2
6		1	5
8		1.25	6.75
10		1.5	8.5
12		1.75	10.25
	14	2	12
16		2	14
	18	2.5	15.5
20		2.5	17.5
	22	2.5	19.5
24		3	21
	27	3	24 or 1-5/16"
30		3.5	26.5 or 1-3/64"
	33	3.5	29.5 or 1-5/32"
36		4	32 or 1-1/4"
	39	4	35 or 1-3/8"
42		4.5	37.5 or 1-31/32"
	45	4.5	40.5 or 1-19/32"
48		5	43 or 1-11/16"

STRENGTH PROPERTIES (Commercial Steel)

CLASS		New Designation	4.6	4.8	5.6	5.8	6.6	6.8	6.9	8.8	10.9	12.9	14.9
		Formerly	4 D	4 S	5 D	5 S	6 D	6 S	6 G	8 G	10 K	12 K	—
Brinell		min.	110		140		170			225	280	330	390
		max.	170		215		245			300	365	425	—
Rockwell	HRB	min.	63		78		88			—	—	—	—
		max.	88		97		102			—	—	—	—
	HRC	min.	—		—		—			18	27	34	40
		max.								31	38	44	49
Yield Point psi			45,000		56,000		76,000		76,000	91,000	128,000	153,000	180,000
Tensile Strength psi		min.	56,000		70,000		85,000		85,338	113,784	142,230	170,000	200,000
		max.	78,000		100,000		113,000		99,561	128,000	170,676	200,000	230,000

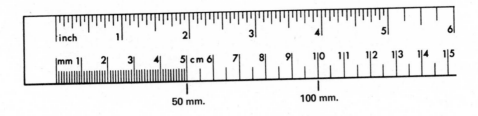

50 mm.   100 mm.

Index